Geological Society Special Publications
Series Editor A.J. FLEET

GEOLOGICAL SOCIETY SPECIAL PUBLICATION NO. XX

Volcanism Associated with Extension at Consuming Plate Margins

EDITED BY

J.L. SMELLIE
British Antarctic Survey
Cambridge, UK

1994
Published by
The Geological Society
London

THE GEOLOGICAL SOCIETY

The Society was founded in 1807 as the Geological Society of London and is the oldest geological society in the world. It received its Royal Charter in 1825 for the purpose of 'investigating the mineral structure of the Earth'. The Society is Britain's national society for geology with a Membership of 7500 (1993). It has countrywide coverage and approximately 1000 members reside overseas. The Society is responsible for all aspects of the geological sciences including professional matters. The Society has its own publishing house which produces the Society's international journals, books and maps, and which acts as the European distributor for publications of the American Association of Petroleum Geologists and the Geological Society of America.

Fellowship is open to those holding a recognized honours degree in geology or cognate subject and who have at least two years relevant postgraduate experience, or who have not less than six years relevant experience in geology or a cognate subject. A Fellow who has not less than five years relevant postgraduate experience in the practice of geology may apply for validation and subject to approval, may be able to use the designatory letters C. Geol (Chartered Geologist).

Further information about the Society is available from the Membership Manager, The Geological Society, Burlington House, Piccadilly, London W1V 0JU, UK.

Published by The Geological Society from:
The Geological Society Publishing House
Unit 7
Brassmill Enterprise Centre
Brassmill Lane
Bath BA1 3JN
UK
(*Orders*: Tel. 0225 445046
Fax 0225 442836)

First published 1994

The publisher makes no representation, express or implied, with regard to the accuracy of the information contained in this book and cannot accept any legal responsibility for any errors or omission that may be made.

British Library Cataloguing in Publication Data

A catalogue record for this book is available from the British Library

ISBN 1–897799–17–9

Typeset by Type Study, Scarborough, UK

Printed by Alden Press Ltd,
Oxford, UK

Distributors

USA
AAPG Bookstore
PO Box 979
Tulsa
OK 74101–0979
USA
(*Orders*: Tel. (918)584–2555
Fax (918)584–0469)

Australia
Australian Mineral Foundation
63 Conyngham Street
Glenside
South Australia 5065
Australia
(*Orders*: Tel. (08)379–0444
Fax (08)379–4634)

India
Affiliated East-West Press PVT Ltd
G-1/16 Ansari Road
New Delhi 110 002
India
(*Orders*: Tel. (11)327–9113
Fax (11)326-0538)

Japan
Kanda Book Trading Company
Tanikawa Building
3–2 Kanda Surugadai
Chiyoda-Ku
Tokyo 101
Japan
(*Orders*: Tel. (03) 3255–3497
Fax (03) 3255–3495)

Contents

Volcanism associated with extension at consuming plate margins

J.L. SMELLIE

British Antarctic Survey, NERC, High Cross, Madingley Road, Cambridge CB3 0ET, UK

The title of this volume is self explanatory. But for some, an association of arc-related volcanism with extension will still come as a surprise. The concept of subduction evolved as a logical consequence of seafloor spreading (to keep the surface area of the earth constant). It is a contracting phenomenon in the sense that the distance between two points on opposing plates shortens. Nothing in the plate tectonic paradigm in its simplest form (rigid plates, all deformation concentrated at narrow plate boundaries) leads the observer intuitively to expect extension to be the dominant tectonic regime at consuming margins. Indeed, in the 1970s, conventional views stated that subhorizontal compressive stresses were transmitted throughout the over-riding plate because of plate convergence. The principal mode of deformation was regarded as shortening in the overriding plate parallel to the convergence direction. Contemporaneous pub-lications proliferated with genetically sug-gestive, sometimes ill-conceived terms such as '*compressive* arcs', 'the *thrusting* of one plate below another', 'ridge *push*', etc. Much tectonic speculation and geophysical modelling of sub-duction has been built on these false assump-tions, and they are still propagated in many current textbooks and research papers.

Geoscientists need to familiarize themselves with the characteristics of actual plate motions. A popular current thesis relates the state of stress to the degree of coupling of the two convergent plates. Recent two-dimensional, finite element modelling has demonstrated how the stress in the overriding plate is critically dependent on whether the subduction zone is locked (i.e. high degree of coupling) or unlocked. Thus, a compressional stress regime occurs in locked systems, and extensional stress in unlocked systems; a gradient from compression in the fore-arc to extension in the back-arc is said to characterize partially locked systems. However, studies of the strain field in real cases (modern arc regions) clearly indicate the prevalence of subhorizontal extension orientated nearly perpendicular to the arc. Implicit in these real observations is the sug-gestion that stresses resulting from plate coupling are either *not* transmitted to the volcanic arc, or else locked subduction zones are uncommon and extension may be the *normal* mode of behaviour.

This thematic volume presents the rationale and evidence for extension and coeval volcanism at consuming plate margins. Using mainly Cenozoic case histories for oceanic and con-tinental margin arcs, structural evidence is presented, which demonstrates that volcanism in most arc systems is contemporaneous with normal faulting and subsidence *as a general case*. Representative, modern geochemical data sets are provided and integrated in a petrological discussion for each case history. The western Pacific, now widely regarded as a type region for the study of convergent plate margin processes, is particularly well represented. Also included is evidence from less commonly investigated re-gions, where subduction has ceased or is no longer obvious but an association with an arc is unequivocal. By these means, it is hoped that this volume contains a representative cross-section of current research on extension-related arc volcanism in the broadest sense, and that it will stimulate further discussion and research into the complex tectonics and petrology of convergent plate margins.

I am grateful to the many people and organizations who contributed to the production of this volume. These include the Director and staff of the British Antarctic Survey, the staff at the Geological Society Publishing House, the contributors (most of whom tried to keep to deadlines), and the numerous referees (listed below) for their careful reviews. Stefan Keymer and Roger Missing are gratefully acknowledged for solving problems of incompatible computer disk formats. Finally, special thanks go to Gill McDonnell and Lesley Ward for their patient secretarial support.

Referees: P.F. Barker, S.H. Bloomer, J.W. Cole, A.J. Crawford, J.F. Dewey, S.M. Eggins, J.G. Fitton, M.F.J. Flower, P.A. Floyd, C.J. Hawkesworth, M.J. Hole, R.J. Korsch, P.R. Kyle, P.T. Leat, M. McCulloch, P. Morris, M.A. Morrison, E. Nakamura, D.W. Peate, G. Rogers, N.W. Rogers, A.D. Saunders, J. Scarrow, M. Storey, D.R. Tappin, J. Tarney, S.D. Weaver, D.G. Woodhall, G. Worner

From Smellie, J.L. (ed.), 1995, *Volcanism Associated with Extension at Consuming Plate Margins*, Geological Society Special Publication No. 81, 1.

Subduction systems and magmatism

WARREN B. HAMILTON

Branch of Geophysics, US Geological Survey, Denver, Colorado 80225, USA

Abstract: Most published subduction modelling and much palaeotectonic speculation incorporate the false assumption that subducting oceanic plates slide down fixed slots. In fact, hinges roll back into oceanic plates and slabs sink more steeply than the inclinations of the Benioff zones which define transient positions of the slabs. The lower parts of overlying mantle wedges sink with the slabs, pulling away from partial-melt zones higher in the wedges. The complex behaviour of arc systems can be comprehended in terms of this mechanism of subduction. The common regime in overriding plates is extensional, and leading edges are crumpled only in collisions. Shear coupling between subducting slabs and overriding plates is limited to shallow depths and varies widely, with corresponding variations in tectonic erosion, accretion, and regurgitation of high-P subducted materials. Arcs can advance, lengthen, change curvature, festoon around obstacles, rotate while deforming, and fold and pinch shut. Two arcs can collide as an intervening oceanic plate is subducted simultaneously beneath both, or they can migrate apart as new lithosphere is formed between them. Subduction cannot occur simultaneously beneath opposite sides of a rigid plate because impossible retrograde subduction would be required beneath one of them. Histories, including inception ages, collisions, polarity reversals and stage of petrological evolution, vary greatly along continuous arc systems. Long-continuing steady-state systems are uncommon.

Magmatic arcs are properly viewed as features migrating with sinking lower plates, not as fixed features of upper plates. Hot inclined zones within mantle wedges, midway between sinking slabs and overriding crust, are avenues for replenishment of mantle pulled away with subducting plates and also are sites of generation of arc protomelts as volatiles rise into them from dehydrating slabs. Back-arc basins form by spreading behind migrating arcs; strips of arcs may be abandoned in the spreading systems. An arc can migrate so rapidly that it plates out oceanic lithosphere rather than producing a welt.

Exposed sections of the upper mantle and basal crust of arcs show that the Mohorovičić discontinuity is a self-perpetuating density filter and that the already-evolved basaltic and melabasaltic melt that leaves the mantle forms great basal-crust sheets of norite, gabbro and granulite. All more-evolved rock types in these sections are generated in the crust by fractionation, secondary melting and contamination (and this falsifies much petrological modelling).

Behaviour of arc systems

All lithosphere plates move relative to all others. All boundaries between plates also move and most of those boundaries change shape and length as they move. The nature of these interactions invalidates many of the tectonic and magmatic models which clutter the literature. Few of the geologists and petrologists who work with the structures, magmatic rocks and 'allochthonous terranes' produced by convergent-plate interactions, and few of the geophysicists who model subduction, have familiarized themselves with the characteristics of actual plate systems.

Causes of plate motions

Plate motions represent the Earth's primary current mode of heat loss, but the immediate plate drives are mechanical and gravitational.

Correlations between relative plate rotations and velocities, bounding structures and other parameters indicate that gravitational subduction and subduction-generated mantle flow move both subducting and overriding plates, drawing them together, and that velocities tend to increase with age, hence thickness and density, of subducting slabs (Carlson 1983; Spence 1987; Jurdy & Stefanick 1991; Tao & O'Connell 1993). Slabs sink because they are denser than the mantle they displace, and because, being colder, they undergo pressure-phase transformations to denser rocks at depths shallower than do the displaced rocks (Anderson 1987). Subducting and overriding plates are commonly coupled seismically only to depths of about 30–50 km (Tichelaar & Ruff 1993), below which slabs may be effectively strengthless, deforming internally by extension down to depths of about 300 km and by shortening below

From Smellie, J.L. (ed.), 1995, *Volcanism Associated with Extension at Consuming Plate Margins*, Geological Society Special Publication No. 81, 3–28.

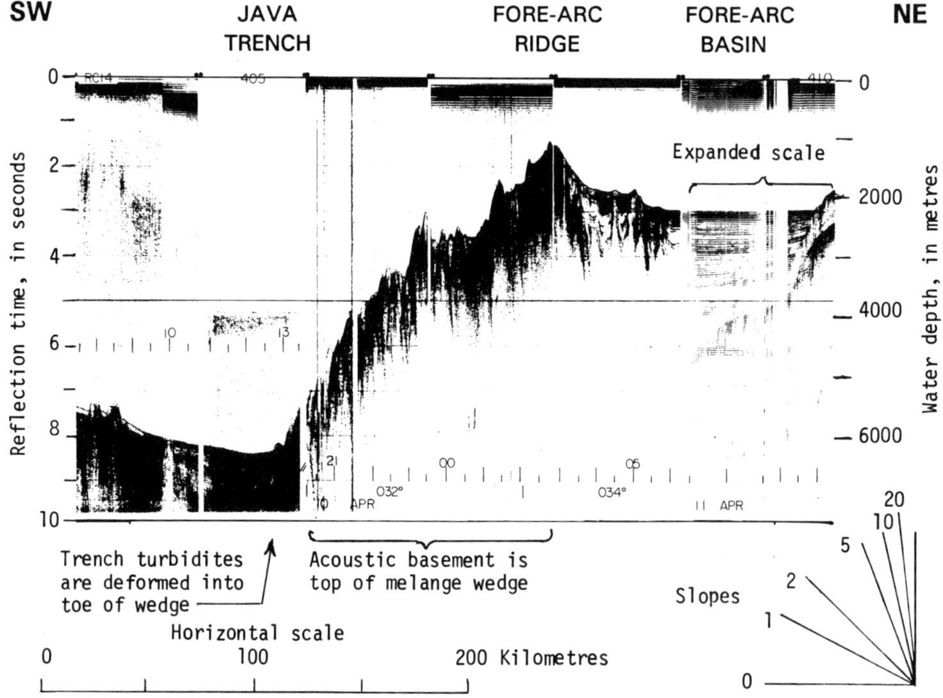

Fig. 1. Single-channel seismic-reflection profile across the active trench and fore-arc ridge and basin near the south end of Sumatra. Note undeformed character of landward part of fore-arc basin. What appear to be tight folds in the outer part of the fore-arc basin are broad shale-cored anticlines (cf. Fig. 7). Vertical exaggeration *c.*35× (note slope scale); cf. Figs 2 & 3. Profile provided by Lamont-Doherty Earth Observatory of Columbia University.

that (Tao & O'Connell 1993); the shortening may record transformational faulting accompanying the pressure-phase change from olivine to spinel (Kirby *et al.* 1991). Ridge slide (commonly misnamed 'ridge push') is of less importance in moving plates; it is due much more to the 80 km of relief at the base of an oceanic plate, against less-dense asthenosphere, than to the few km of bathymetric relief at the top of the plate. Spreading ridges form where plates move apart and hot mantle wells into the gap and ridges migrate and change shape and length at widely varying rates. Motion is retarded by the presence of thick continents on plates. The return flow that compensates for lithosphere motion may occur mostly in the asthenosphere.

Most Euler poles for large plates are located at present high latitudes, so much of the relative motion of modern plates represents differences in their rotational velocities relative to the spin axis. 'The rotational motion of the earth could be the ultimate cause of tectonic plate motion' (Heirtzler 1991).

Mechanism of subduction

Most trenches are gentle-sided features with slopes of only a few degrees (Figs 1 & 2). Marine geological papers are often illustrated with reflection profiles with extreme vertical exaggerations, which convey the illusions of steep slopes and tight deformation (Fig. 1). Subducting plates inflect to broad curves seaward of trenches and dip gently beneath thin accretionary wedges of imbricated debris in front of overriding plates. The base of the overriding plate, at the back of the exposed part of the accretionary wedge, is only 10 km or so beneath the sea floor, and the subducting plate continues with gentle dip beneath the front of the overriding plate (Fig. 3). The bathymetric trench is the dihedral angle between the subducting plate and the accretionary wedge and seldom is bedrock in contact between opposed plates. The gentle dip of the subducting plate, and its outer rise seaward of the trench, are elastic responses to the gravitational loading by the accretionary wedge and the leading part of the overriding

Fig. 2. Multichannel seismic-reflection profile across the Java Trench and accretionary wedge south of Bali. The top of the subducting Indian Ocean plate produces high-amplitude low-frequency reflections beneath the thinly tapered accretionary wedge and lies only 11 km beneath the crest of the wedge at the fore-arc ridge. Profile is depth converted (horizontal and vertical scales are equal and depths are true) but not migrated; hints of north-dipping imbrication can be seen in the accretionary wedge but many diffractions are present. Profile provided by Shell Internationale Petroleum Maatschapij N. V.

plate. The fundamental subduction hinge, or slab-bend zone, occurs beneath the overriding plate and is a broad flexure mostly between depths of about 20 and 40 to 100 km (Spence 1987; Taber *et al.* 1991). Most slabs dip between 30° and 70° at greater depth, with extreme values ranging between about 10° and vertical (Spence 1987).

Many geologists and geophysicists tend to think, wrongly, of plate boundaries as fixed in position within two-dimensional cross sections. Much published tectonic speculation and most geophysical modelling of subduction has been built on the false assumption that a subducting plate rolls over a stationary hinge and slides down a slot that is fixed in the mantle, and that commonly overriding plates are shortened compressively across their magmatic arcs and fore-arcs (Fig. 4).

These assumptions are disproved both by the behaviour and characteristics of modern convergent-plate systems, in which the subducting plate is of normal oceanic lithosphere and the Benioff seismic zone has a moderate to steep inclination, and by analyses of 'absolute' plate motions. Hinges commonly retreat – roll back – into subducting oceanic plates as overriding plates advance, even though at least most subducting plates are also advancing in 'absolute' motion. It cannot be overemphasized that subducting slabs sink more steeply than the inclinations of Benioff seismic zones, which mark transient positions, not trajectories, of slabs. Perhaps the most obvious evidence for hinge rollback is that the Pacific Ocean is becoming smaller as flanking continents and marginal-sea plates advance trenchward over ocean-floor plates and as the Atlantic and Arctic oceans become larger. Collisions between facing island arcs (as are now underway in the Molucca Sea and the western Solomon Sea: Hamilton 1979, McCaffrey 1982, Silver *et al.* 1991) and reversals of subduction have explanations only in terms of rollback. Diverse evidence for rollback has been presented by, among others, Carlson & Melia (1984), Chase (1978), Dewey (1980), Garfunkel *et al.* (1986), Hamilton (1979), Hawkins *et al.* (1984), Kincaid & Olson (1987), Malinverno & Ryan (1986), Molnar & Atwater (1978), Spence (1987), Stern & Bloomer (1992), Tao & O'Connell (1993) and Uyeda & Kanamori (1979). As most of these authors have variously emphasized, overriding plates are pulled toward retreating hinges and the typical regime in overriding plates is one of extension, not shortening.

Retrograde motion could occur only if a dense slab could push light mantle forward and upward

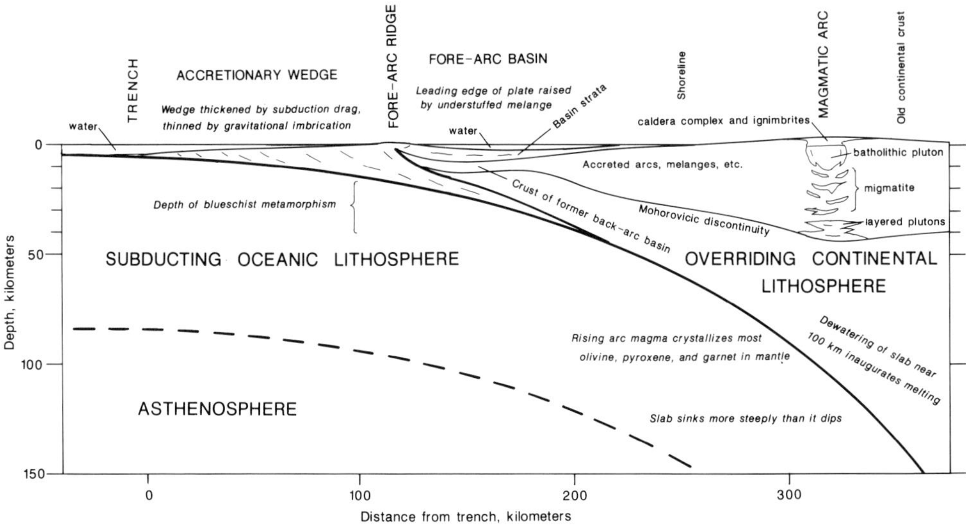

Fig. 3. Section across a continental-margin subduction system, scaled for modern Sumatra and for Cretaceous California. Constraints discussed by Hamilton (1979, 1988, 1989*a*, 1989*b*).

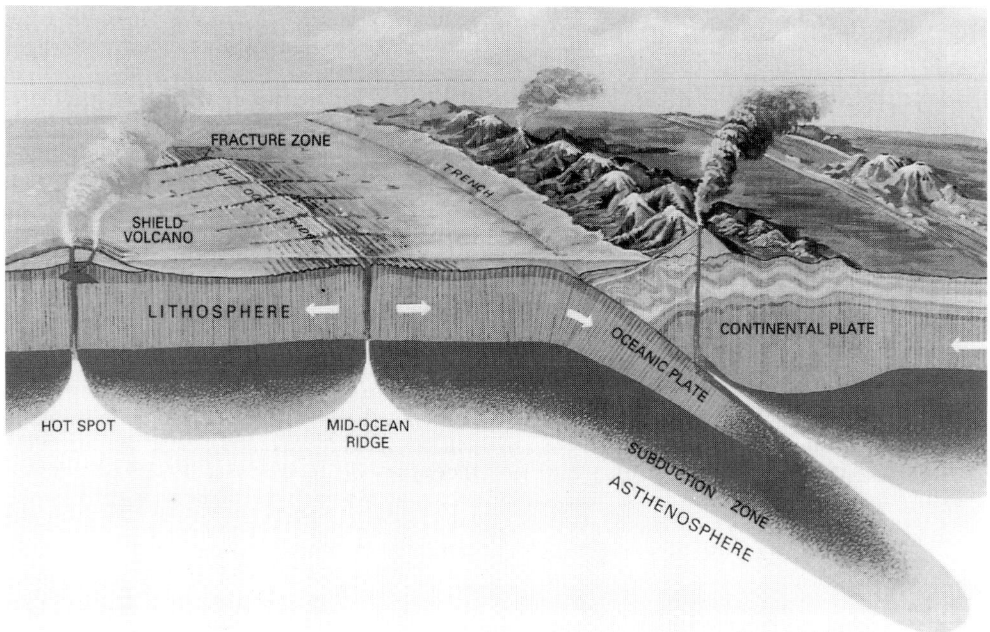

Fig. 4. Typical unscaled popular depiction of subduction. Oceanic lithosphere is depicted as moving away from a fixed ridge, inflecting at a fixed hinge at a trench and sliding down a slot fixed in the mantle, as an advancing overriding plate is crumpled against the subducting plate. All of these concepts are false. From Simkin *et al.* (1989); similar misconceptions typify most textbooks and many research papers.

out of its way – an impossibility in a gravity-dominated system – or if, implausibly, the mantle beneath both overriding and subducting plates were moving with the latter. Exceptions that can be argued to counter these interpretations are of doubtful validity. Palinspastic cartoons by many geologists notwithstanding, directly-inward subduction cannot occur simultaneously beneath opposite sides of an internally rigid plate. Subduction now occurs inward beneath both sides of the Caribbean region – Antilles on the east, Central America on the west, but poorly understood plate boundaries intervene. Subduction may now occur at both east and west sides of southern Mindanao, but the trajectories, and in part even boundaries, of the various small plates in that region are so poorly constrained that this cannot yet be evaluated properly.

Mantle-wedge mechanics and magmatism

As subducting slabs sink, the frontal parts of overriding plates tend to move forward to keep pace with the rolling-back hinges; but what happens in the mantle wedges above the sinking slabs? Critical new constraints on thermal structure, and hence kinematics and magmatism, are provided by the seismic tomography by Zhao *et al.* (1992, in press) and Zhao & Hasegawa (1993). Figure 5 shows their tomography of *P*-wave velocity anomalies, hence presumably of thermal structure, for northern Honshu. The cold, high-velocity slab, about 80–90 km thick, is clearly delineated; one zone of seismicity occurs near its upper surface, a second parallel zone within it. In the overlying mantle wedge, velocity is normal near the slab. Higher in the wedge, low velocities define an inclined irregular zone 50 km or so thick, subparallel to the slab and with the largest negative velocity anomalies about 50 km above the slab. The low velocity zone is imaged westward downdip to a depth of about 175–200 km, where the top of the subducting slab is about 225–250 km deep. (There are no data further west.) Only shallower than 50–75 km does the low-velocity zone steepen upward to define hot material beneath and near the volcanic zones. *S*-wave velocity anomalies also show a low-velocity zone inclined in the mantle wedge in the same direction as the slab.

The negative velocity anomaly is 3–6% in the centre of both the inclined and subvolcanic parts of the low-velocity zone, which probably is approximately at its solidus temperature (cf. Sato *et al.* 1989). The low-velocity zone must have very low strength relative to the rest of the wedge and so presumably is the primary sector in which the wedge is being pulled apart by the subducting slab (Fig. 6). It thus appears that at shallow depth the slab slides past the base of the wedge while simultaneously the coupled slab and basal wedge sink away from the low-velocity

zone. At greater depth there may be little shear between slab and mantle entrained with it (cf. Tao & O'Connell 1993). The sinking of a subducting slab requires that mantle displaced downward beneath it somehow be balanced by mantle replaced above the slab. The inclined low-velocity zone may represent such compensating flow of rising hot mantle which reaches its solidus temperature as it is depressurized. The nature of mantle circulation relative to the slab is otherwise undefined.

Arc volcanoes occur typically about 100 km above the tops of sinking slabs (Figs 5 & 6). As many scholars have long assumed (e.g. Peacock 1990), water released by dehydration of slab rocks may produce partial melts in overlying wedge rocks, such melts rising and evolving to the volcanic belt; but the relevant dehydration reactions occur through a broad interval of temperature and pressure, not at a constant depth (Pawley & Holloway 1993). The seismic tomography precludes a continuous low-velocity column from slab to volcanoes, so the lower part of the wedge beneath the volcanoes is not the site of the requisite melting. The mantle melts which evolve into arc magmas may be generated primarily where water rising from the dehydrating slab (and carrying incompatible elements from subducted sediments: Plank & Langmuir 1993) reaches the replenishing fertile, hot, obliquely-rising mantle in the low-velocity part of the wedge. It is an obvious inference that partial melts generated by volatiles in the inclined low-velocity zone migrate up the zone to enter the crust primarily in the generally narrow magmatic arc (Fig. 6). The arc appears in these terms to be localized by the inclined low-velocity zone and hence to have an explanation more in kinematics than in a narrow depth window of slab dehydration.

Among many problems with the injection-down-the-slot model of subduction is that it cannot account for steady-state arc magmatism. Dehydration of a sinking slab can release water continuously for flux but a constant mantle wedge would soon be cooled below temperatures at which melt could form and also would be depleted of fusible material. The rise of new hot mantle material into the low-velocity zone of the wedge provides a continuing resupply of both heat and fertile material. The continuing rise of water from the dehydrating slab should help keep the migrating low-velocity material above its solidus temperature.

Deformation of overriding plates

The dominant regime above subducting slabs is

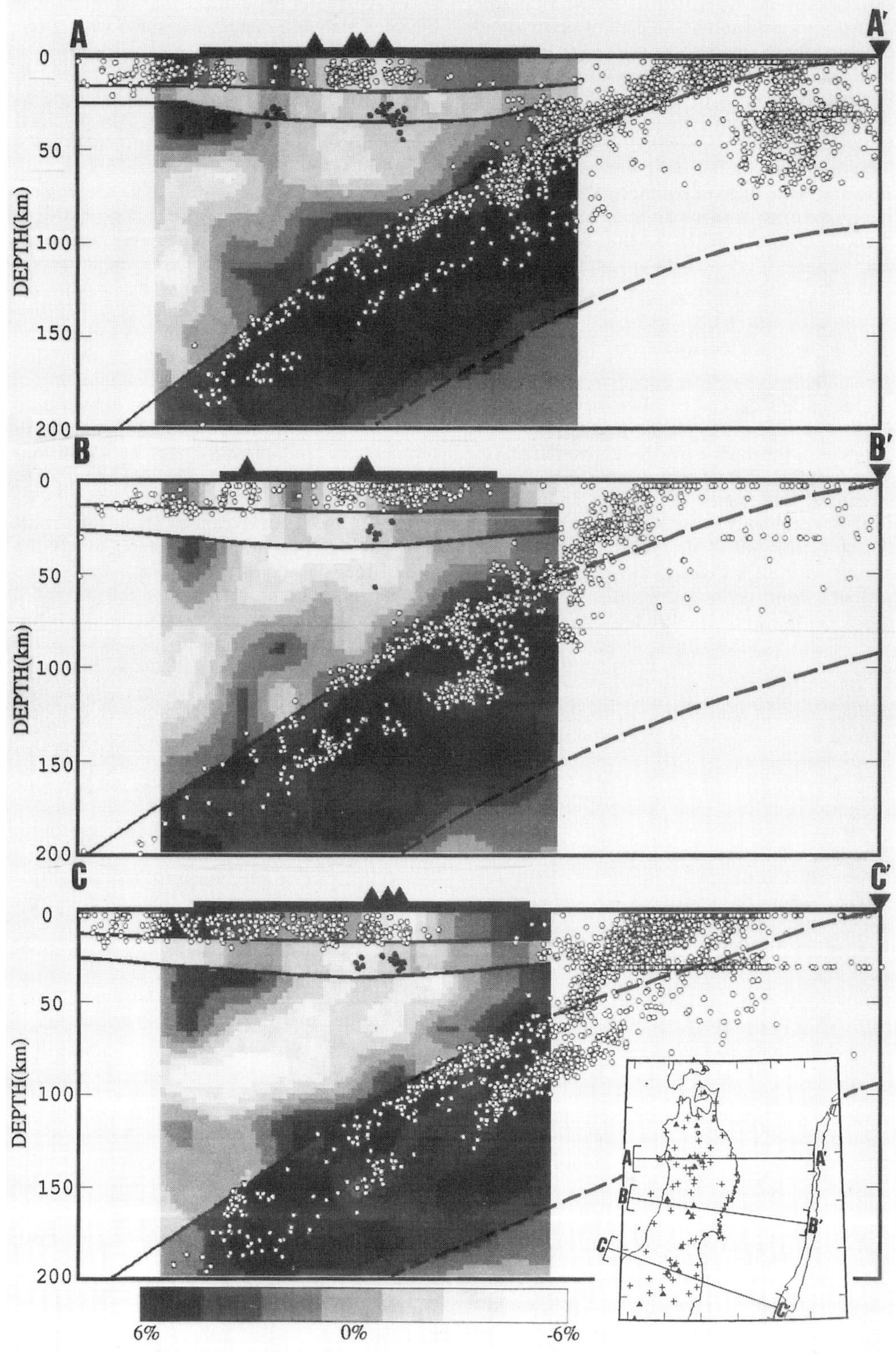

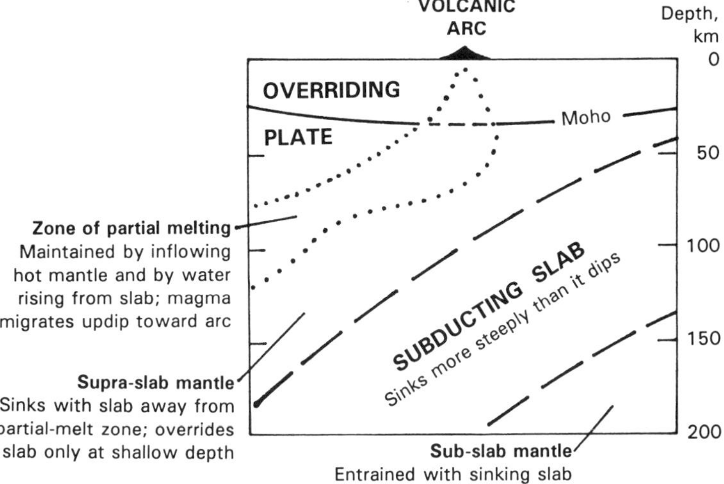

Fig. 6. Kinematic and magmatic components of a subduction system, deduced from seismic tomography (e.g. Fig. 5) and considerations discussed in text. The subducting slab shears past only the shallow, updip part of the overlying mantle wedge. Otherwise, the lower part of the mantle wedge sinks with the slab and pulls away from the inland and upper part of the wedge across a zone of partial melting.

extensional and not, as in popular fantasy, compressional. Cartoons incorporating the false concept that the fronts of overriding plates are crumpled and shredded against abruptly-downturned incoming plates nevertheless are presented in every current physical-geology textbook the author has seen and also appear in many advanced textbooks and research papers. These unscaled cartoons bear no quantitative relationship to any actual subduction system. Compare Figs 3 & 4.

Fore-arc basins

That even the fragile leading edges of overriding plates indeed are not crumpled is shown by the character of the fore-arc basins developed atop them. Fore-arc basins (synonym, outer-arc basins; also some trench-slope basins) are formed on the leading edges of most overriding continental plates and of many island-arc plates. Many examples were listed and illustrated by Coulburn & Moberly (1977), Dickinson & Seely (1979), Hamilton (1979) and Seely (1979).

Fore-arc basins are defined by their basin structure. Their surfaces, varying with rates of deformation and sedimentation, may be bathymetric basins, continental shelves or coastal plains, or continental slopes. Their strata lap landward onto continental or island-arc basement, and seaward onto basement which rises to a structural high, the fore-arc, or outer-arc, ridge or basement high. This basement can be seen in some outcrop examples to be ophiolite, arguably formed in back-arc basin settings before island arcs now landward of them collided with the continents to which they now are accreted (Hamilton 1979, 1988, 1989*b*, and references therein). Many observers infer that fore-arc basin strata are deposited upon old, stabilized parts of accretionary wedges, but the author knows of no example in which this can be demonstrated either in outcrop or in marine geological or geophysical data. For example, Nasu *et al.* (1980 figs 2, 3, 6) drew imbricated accretionary-wedge basement beneath the fore-arc basin off northern Honshu, but their poor-quality reflection profiles actually show only

---

Fig. 5. Seismic tomography of Honshu, northeast Japan, showing deviations of P-wave velocities from those normal for depth, after Zhao *et al.* (1992; their version of this figure is in colour). Land area indicated along top by thick line; active volcanoes marked by triangles; Sea of Japan to left, Pacific Ocean to right, axis of Japan Trench at right edge. Horizontal and vertical scales equal. Small circles are hypocentres of earthquakes, which are much better located beneath the land than beneath the sea. The dense, cold, subducting slab has a high-velocity anomaly. A low-velocity zone lies subparallel to the slab in the mantle wedge above it and is inferred to mark hot and relatively fertile mantle flowing into the disrupting part of the wedge as the slab sinks below it. Index map shows active volcanoes (▲) and seismic stations (+). Figure provided by Dapeng Zhao.

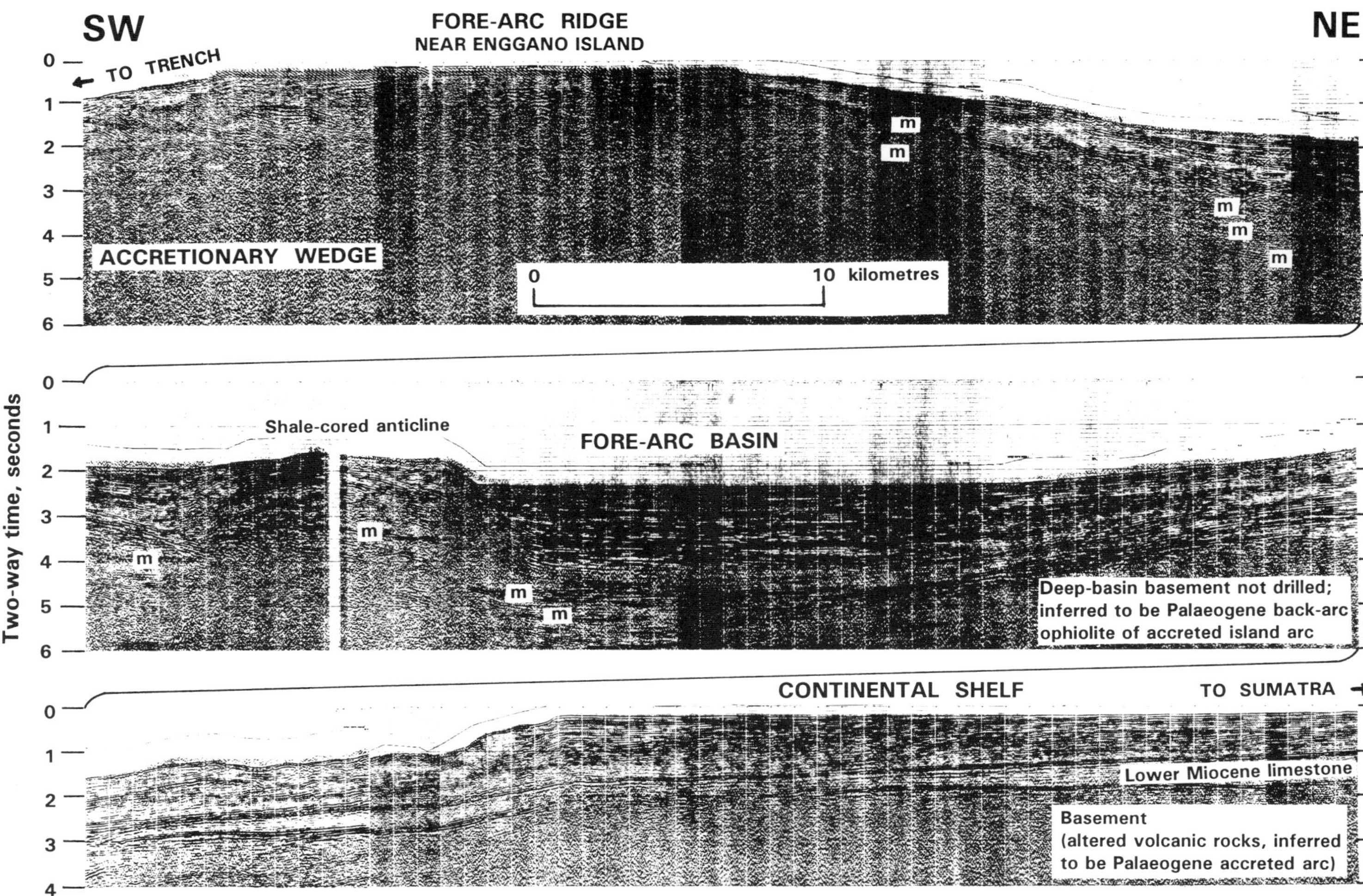
SW
NE
FORE-ARC RIDGE
NEAR ENGGANO ISLAND
TO TRENCH
ACCRETIONARY WEDGE
0   10 kilometres
Two-way time, seconds
Shale-cored anticline
FORE-ARC BASIN
Deep-basin basement not drilled;
inferred to be Palaeogene back-arc
ophiolite of accreted island arc
CONTINENTAL SHELF
TO SUMATRA
Lower Miocene limestone
Basement
(altered volcanic rocks, inferred
to be Palaeogene accreted arc)

diffractions, multiples and processing artifacts in the time intervals at issue.

Figure 7, a reflection profile across the Sumatra fore-arc, illustrates the lack of shortening across a typical fore-arc. (On the basis of single-channel profiles with large vertical exaggeration, Diament *et al.* (1992) speculated that this Sumatra fore-arc is sliced by great NW-trending strike-slip faults: in the many profiles in the author's files, there is no evidence supporting this kinematically implausible conjecture.) The subducting Indian Ocean plate lies close beneath this almost undeformed fore-arc basin.

The subducting oceanic Nasca plate in part has an uncommonly gentle dip beneath western South America, so this is an excellent place to test hypotheses of shortening of overriding continental plates against subducting plates. Here it is obvious that at least the thin leading edge of South America is not being shortened, and further that little young shortening affects the western half of the Andes at the surface. There is no shortening of the fore-arc sedimentary basin that lies atop most of the leading edge of the overriding plate along the narrow continental shelf (Hayes 1974; Coulbourn & Moberly 1977; Moberly *et al.* 1982), only a few tens of kilometres above the subducting plate (Cahill & Isacks 1992). Basal basin strata in the south are as old as Late Cretaceous (Hayes 1974). Obviously this preserved part of the thin leading edge of the overriding plate has not been crumpled during Late Cretaceous and Cenozoic time, and any grinding away of the front of the overriding plate against the subducting plate has been limited to zones further seaward or at the base of the plate. A little-deformed fore-arc basin is present also along southern Chile, where the Antarctic plate is being subducted beneath South America (Cande & Leslie 1986). The well-developed and uncrumpled fore-arc basins of South America (Coulburn & Moberly 1977), Hispaniola (Biju-Duval *et al.* 1982), the Aleutian Islands (Harbert *et al.* 1986) and west Luzon (Lewis & Hayes 1984) are among many that show landward migration with time of depocentres. This migration and other features indicate that the basins are formed primarily by rise of the fore-arc ridges, with concurrent elastic depression of the basins, as accretionary-wedge materials accumulate beneath the thin leading edges of overriding plates (Fig. 3).

Overriding plates

The lack of crumpling of the overriding South American plate, where again popular conjecture postulates severe shortening, is made obvious by the geology of Peru, where most of the young deformation from the coast across the Western Cordillera and the Altiplano to the crest of the Eastern Cordillera is extensional (Moberly *et al.* 1982; Schwartz 1988; Sébrier *et al.* 1988). Miocene strata are almost undeformed across the Altiplano between the Western and Eastern Cordillera (Kono *et al.* 1989). The eastern slope of the Andes (i.e. of the Eastern Cordillera), by contrast, records severe eastward thrusting, whereby a pre-existing stratal wedge is being thrust eastward onto the craton over a developing foreland basin. At least much of this east-flank thrusting must be a gravitational-spreading response to the thickening of the crust represented by the high Andes; but what is the cause of that thickening? My own preference is to emphasize the thickening effect of arc magmas rising from the mantle, and, as an end-member model, to rationalize that there may be no net shortening across the range (except for the basement thrusting in the sub-Andean ranges still farther east). Geological structure apparently precludes shortening distributed across the entire Andes in upper-crustal rocks. Perhaps thickening as a response to drag of the continental plate on the subducting plate is limited to the deep crust across most of the range and is transmitted to the surface only in the east. Perhaps the crust beneath the Andes is thickened by tectonic underplating by crust dragged back from beneath the front of the overriding plate. There is in any case little

---

Fig. 7. Seismic-reflection profile across the fore-arc ridge, fore-arc basin and continental shelf of south Sumatra. The top of the subducting Indian Ocean plate lies about 12–15 km beneath the SW end of the profile and 40 km beneath the NE end. The fore-arc ridge is the top of the accretionary wedge (cf. Figs 2 & 3). The lack of deformation of shelf and fore-arc strata shows that the thin leading edge of the overriding plate has not been crumpled during the last 18 m.y. of continuous subduction. Basin is inferred to have developed on back-arc-basin ophiolite formed behind Cretaceous and Palaeogene island arc that collided with Sumatra in middle Tertiary time (cf. Fig. 3; Hamilton 1989*b*). The old island arc advanced in the direction now northeast and produced a suture in medial Sumatra; the present southwest-facing subduction system broke through the lithosphere that had been behind that arc. No vertical exaggeration for sediments with Vp = 3 km s^{-1}; exaggeration 2× for water depth (1 s reflection time = 750 m). Some obvious multiples are marked (m). Profile provided by White Shield Oil Co.

apparent support for speculation such as that by Sheffels (1990) that enormous subduction-linked shortening has affected the upper crust across the entire Andes.

Active extensional faulting similarly affects northern coastal Chile on structures mostly subparallel to the trench nearby to the west (Armijo & Thiele 1990).

Accretionary wedges

Pervasive imbrication and shear of course affect the weak accretionary wedges formed in front of overriding plates from sediments and other materials scraped from subducting oceanic lithosphere. Trench fill can be seen on reflection profiles to be scraped off at the toes of wedges; deeper materials are underplated progressively further back under them (e.g. many reflection profiles in Hamilton 1979). A wedge is a thin, dynamic debris pile (Fig. 2), thickened by underplating and by dragging back of its base, and thinned by gravitational spreading. Drilling and outcrop studies show mélange and broken formation to typify wedge materials, but proportions and types of these and other materials vary greatly with such factors as supply of sea-floor and trench sediments and convergence rates.

Earthquakes within accretionary wedges commonly represent thrust faulting with slip vectors normal to trenches or between the direction of plate convergence and those normals (DeMets *et al*. 1990; DeMets *et al*. 1992; McCaffrey 1992). Although complex kinematic explanations are commonly given for such divergences between directions of plate convergence and of coseismic slip, my preferred general explanation is that wedges are thickened by slow, aseismic plate drag, in the plate-convergence directions, and are thinned by coseismic spreading in the gravitational direction, toward the trenches. Wallace *et al*. (1993) presented field and theoretical evidence that extension occurs in active accretionary wedges.

Blueschists and other high-P/low-T metamorphic rocks can form only under the fronts of overriding plates, for the 8–15 kbar pressures they require cannot be reached in the thin accretionary wedges in front of the plates (Fig. 3). Where clasts of high-P rocks occur, as often they do, in low-P, low-T mélange of accretionary wedges, or where sheets of high-P rocks occur beneath thin leading edges of overriding plates, a return flow is required as part of the subduction process. It is not yet clear how such circulation is produced in the subducted part of the accretionary wedge and under what conditions of subduction it develops.

Subduction erosion

Where plate convergence is rapid and where young, buoyant lithosphere is being subducted, the front and underside of the overriding plate can be ground or sliced away against the subducting slab. Elsewhere (Hamilton 1989*a*) the example of latest Cretaceous and early Palaeogene southwestern United States is discussed, wherein frontal and lower parts of the overriding plate were removed by subduction and accretionary-wedge materials were plated directly against and beneath what had previously been middle- and lower-crustal magmatic-arc complexes. Obviously this process requires more coupling between subducting and overriding plates than does the far more common case of decoupling with which this paper is primarily concerned.

Arc and back-arc extension

Earthquake first-motion study shows the dominant stresses in volcanic island arcs to be extensional, perpendicular to the arcs (Apperson 1991). Young extensional faulting in an active volcanic arc is illustrated by Fig. 8. Active extension is demonstrated by geological and geophysical data in many arcs (e.g. Izu-Bonin: Taylor *et al*. 1991; Aleutian: Geist *et al*. 1988; New Zealand: Darby & Williams 1991; Hamilton 1988 and references therein).

Karig (e.g. 1972, 1975) demonstrated that the Mariana island arc has migrated Pacificward as new back-arc-basin oceanic crust formed behind it. Karig and many others since (e.g. Taylor & Karner 1983) have found that island arcs generally migrate in such fashion. Some back-arc spreading is accomplished by regular or irregular seafloor spreading behind the entire arc (Caress 1991). New lithosphere can form by the rapid migration of a magmatic arc which plates out a variable-thickness sheet of arc crust rather than forming a full island-arc welt of thick crust (Hawkins *et al*. 1984; Shervais & Kimbrough 1985; Stern & Bloomer 1992). Thus the magmatic welt can move forward with the advancing part of the overriding plate, can be abandoned as a remnant arc on the relatively retreating part, or can be split longitudinally between them as the forward half migrates away from the rear half, and these processes can vary markedly with time in a single arc system. The relative importances of these contrasted processes remain to be established but they can be visualized in terms of

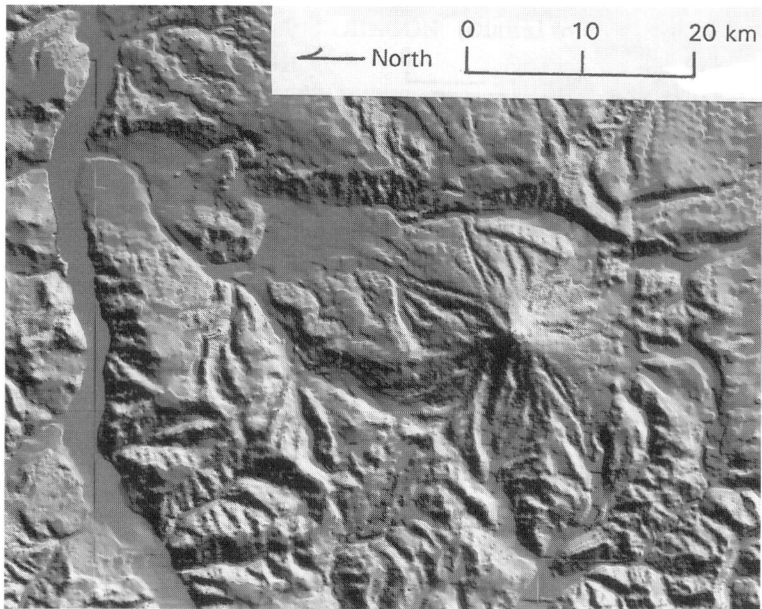

Fig. 8. Extensional faulting in an active volcanic arc. Shaded-relief rendition of digital topography of late Quaternary stratovolcano of Mt Hood (right of centre), Cascade Range, Oregon, which rises from a wedge-shaped late Quaternary graben (bounding faults converge toward upper left). Columbia River on left. Ilumination from southeast (upper right). Provided by US Geological Survey Flagstaff Image Processing Facility.

strength variations in mantle wedges above subducting slabs, as discussed previously. Oceanic island arcs do not bound rigid plates of old lithosphere, but instead mark the fronts of plates of young lithosphere that are widening in the extensional regimes above sinking slabs. An island arc should be viewed as a product of a subducting slab rather than as a fixture of an overriding plate. A belt of arc-magmatic rocks forms above that part of a subducting slab whose top is 100 km or so deep, and migrates to track that contour as the slab falls away.

Oceanic arcs commonly are not inaugurated by the breaking of subduction through old oceanic crust, but rather break through near boundaries between thin and thick crust and migrate over the plates of thin crust (Hamilton 1979, 1989b; Karig 1982).

Arc histories

Arcs are not steady-state tectonic systems, but instead evolve and change complexly and rapidly. Different parts of a single, continuous arc can have grossly different histories and characteristics. Complex sequences of collision, aggregation, reversal, rifting, and internal deformation are the rule, and aggregates of collided bits can be assembled far from their final resting places. Petrological and crustal features evolve as activity continues in a given sector. Oceanic sectors of arcs migrate and lengthen with time, and one sector of a continuous arc can have been inaugurated tens of millions of years later than another sector. Major plate-convergence complexes record subduction at rates of the order of 10 cm/a, or $100 \text{ km}/10^6\text{a}$. Large motions and great complexity are the common case. Such characteristics can be illustrated from many modern arc systems, examples of which have been discussed at length (Hamilton 1979, 1988, 1989b).

Collisions and reversals progress along strike with time, and strike-slip and oroclinal deformation are common. Collisions do not occur between neatly matched shapes; rather, irregular masses are jostled together with highly variable deformation. Large plates commonly continue to converge after a collision, and a common result is the breaking of a new subduction system through oceanic crust at the back of one or the other of the collided light crustal masses; often this represents a reversal of polarity of subduction as well as a jump in position. Arcs are commonly inaugurated by such subduction reversals consequent on collisions between other arcs and light crustal masses and collision histories vary greatly along trend. Subduction of oceanic lithosphere beneath a continental plate commonly begins as a consequence of a plate collision. Convergence between megaplates continues but the light crust on the subducting plate is too low in density to be subducted, so a new subduction system breaks through outboard of the continental plate as enlarged by the subduction. Such post-collision reversals are now underway in the Timor and Molucca regions; dozens of others are recorded in Circum-Pacific onshore and offshore geology. The Solomon–Admiralty arc complex displays two reversals, one of which presently is progressing along strike as the composite arc slides past a trench-trench-transform triple junction.

Terrane accretion

Light crustal masses become sutured together when all intervening oceanic lithosphere is subducted beneath one or both of the converging masses. The resulting complex juxtapositions of disparate crustal materials can be in turn overprinted by the magmatic products of new subduction systems, or can themselves be rifted or re-sutured in still more complex arrays.

Plate tectonics provides a genetic, relational framework for many major geological features as accretionary wedges, ophiolites, fore-arc basins, and oceanic, transitional and continental magmatic arcs. Unfortunately, many of the writers now designating 'allochthonous terranes' in orogenic belts are unaware of such frameworks and hence do not test their too-often *ad hoc* postulates of miniplate motions against the relational predictions implicit in those postulates (Hamilton 1990; Sengör & Dewey 1990).

Magmatism

Modern magmatic arcs mostly form about 100 km above subducting slabs, regardless of the type of crust represented by the upper plate. The relationship of magmagenesis to subduction was discussed previously. Arc lavas erupted at the surface vary from olivine basalt to rhyolite. How is such contrast generated? Much modern petrology is an attempt to deduce answers from the compositions of the final volcanic products. Emphasized here instead is the critical evidence provided by actual vertical variations in exposed sections through magmatic-arc crust (see later).

Models for magmagenesis

Approaches to determining the evolution of the melts that produce the volcanic rocks and upper-crustal plutons define a spectrum between deductive and empirical end members. Much popular deductive modelling incorporates assumptions that are dubious or false.

Petrological modelling. Most petrological modelling represents non-unique deduction. The petrology, chemistry and isotopes of, say, subaerial volcanic rocks are studied and sequences that could have produced the final magmas from assumed parent magmas by assumed processes are deduced by logic and algebra. Common assumptions among deductive petrologists are that the variations in arc-volcanic rocks record primarily melting of varied subducted-slab and mantle-wedge materials to generate diverse basaltic melts which rise, with little modification within the mantle, into the crust and either erupt directly or fractionate to yield more aluminous or felsic melts. Recent papers advocating such unrealistic models include Brophy (1989), Crawford *et al.* (1987), Defant & Drummond (1990), Hawkesworth *et al.* (1991), Leeman *et al.* (1990) and Miller *et al.* (1992). Many of these models are incompatible with experimental petrology (Wyllie 1984), and others with observed crustal geology. O'Hara & Mathews (1981) argued that conclusions from such modelling are quite non-unique even for the simplest of systems, that of mid-ocean-ridge basalts, and that even these represent evolved, not primitive, melts. Langmuir (1989) held out hope that more sophisticated modelling can discriminate between the various processes.

Much petrological and isotopic modelling is based on the assumption that a uniform and known mantle reservoir has evolved by progressive fractionation since cold accretion of the Earth. That this assumption is false is argued briefly in the subsequent section on the early Earth. 'Model ages' and 'mantle growth curves' are artifacts of this invalid assumption.

Empirical method: look under the volcanoes. A contrary empirical approach, much less used, is to study sections exposed through continental and island-arc crust to see what materials are actually present and to characterize the processes that relate them. In the following sections, some of the evidence is presented showing that primitive gabbro, itself a fractionate rather than a primary mantle melt, thickly underplates magmatic-arc crust and is the primary conveyor of heat into the crust, and that complex processes of fractional crystallization, assimilation, secondary melting, and magma mixing proceed from there. Underplating also is a major process in crustal rifting. Among papers emphasizing the evidence for, and importance of, such underplating are DeBari (1992), DeBari & Coleman (1989), DeBari & Sleep (1991), Fountain (1989), Furlong & Fountain (1986), Hamilton (1988, 1989*a*), and Voshage *et al.* (1990). Increasingly, other petrologists are invoking models that incorporate intrusion into the crust of only basalt, followed by complex combinations of fractional crystallization, assimilation, secondary melting and magma mixing. Recent papers from this group, minimally constrained by actual crustal cross sections, include Asmerom *et al.* (1991), Baker *et al.* (1991), Chen & Tilton (1991), DePaolo *et al.* (1992), Hyndman & Foster (1988), Johnston & Wyllie (1988), McBirney *et al.* (1987), Reid & Hamilton (1987), Skjerlie & Johnston (1992) and Ussler & Glazner (1989).

Volcanic arc rocks

Volcanic rocks are erupted commonly from centres standing 100 km or so above subducting plates of oceanic lithosphere. The composition of volcanic rocks in modern arc systems varies with the thickness and petrological maturity of the crust through which they erupt (e.g. Hamilton 1979, 1988; Ewart & Le Maitre 1980; Hildreth & Moorbath 1988; Wörner *et al.* 1992). Volcanic rocks in young intra-oceanic island arcs of small crustal volume are dominantly low-Al olivine-tholeiitic basalts which differ from spreading-ridge basalts primarily in their generally lower contents of such high-field-strength elements as Ti, Zr and Hf. Rocks erupted in mature oceanic island arcs are typically calc-alkalic basalt, andesite and dacite, within which the proportion of intermediate rocks commonly increases with crustal volume of the arc on which they form. Plagioclase-phyric two-pyroxene high-Al basalt and andesite are abundant. Volcanic arc melts erupted through continental crust, or through thick terrigenous sedimentary rocks proxying for such crust, are commonly much more silicic in bulk composition; basalt is uncommon; and bulk and isotopic compositions of the rocks reflect thickness and age of the pre-existing crust. Rocks of both young and mature oceanic arcs commonly are markedly more primitive isotopically than are those coming through continental crust.

These broadly systematic variations in lavas and the crust through which they are erupted occur along continuous magmatic arcs which cross from oceanic to continental crust: Kermadec–New Zealand; Banda–Java–Sumatra; Mariana–Japan; Kuril–Kamchatka; Aleutian–Alaska; Central America–Mexico. They occur also where the age of inception of continuous oceanic arcs becomes older along their lengths, as from Banda to Java. Continuous oceanic lithosphere is subducted so whatever dehydration or other processes affect the subducting slab presumably vary little along strike (although subducted sediment will vary with the source of its clastic component), and thus the melts that arrive at the surface have been greatly modified by interactions with the crust and possibly with the mantle wedge above the slab.

It is an obvious inference from these considerations alone that arc magmas that enter the crust are at least as primitive as olivine tholeiite and that aluminous and silicic arc rocks owe their character to processes operating within the crust. That this indeed is the case, and that the melts reaching the Mohorovičić discontinuity are more primitive than tholeiite, is shown by observed variations with depth in continental and oceanic crust.

Magmatic underplating

The known character of basal crust is too often overlooked in syntheses of the origin of upper-crustal plutonic and volcanic rocks. Continental basal crust and uppermost mantle are exposed *in situ* in at least three crustal sections: the Ivrea Zone of the western Italian Alps, the Kohistan arc of northern Pakistan and the Talkeetna arc of southern Alaska. The three examples are of Phanerozoic age, expose sections from the upper crust to the upper mantle, and were ramped onto subduction complexes during Cretaceous and Palaeogene collisions. The character of these lowest crustal and uppermost mantle rocks and their relationships to overlying crustal rocks constrain the origin of silicic igneous rocks and the petrological evolution of continental crust. In all three cases, the basal crust is dominated by stratiform two-pyroxene gabbro and mafic granulite. The Mohorovičić discontinuity as it would be defined geophysically lies within, not beneath, these underplated complexes. The dominance of arc magmatism in the evolution of these crustal sections is indicated to the author by the apparent arc-magmatic character of the upper-crustal plutonic and volcanic rocks in each and by the character of the deep mafic rocks themselves. Kohistan and Talkeetna apparently represent island-arc crust whereas Ivrea represents evolved continental crust. The author's

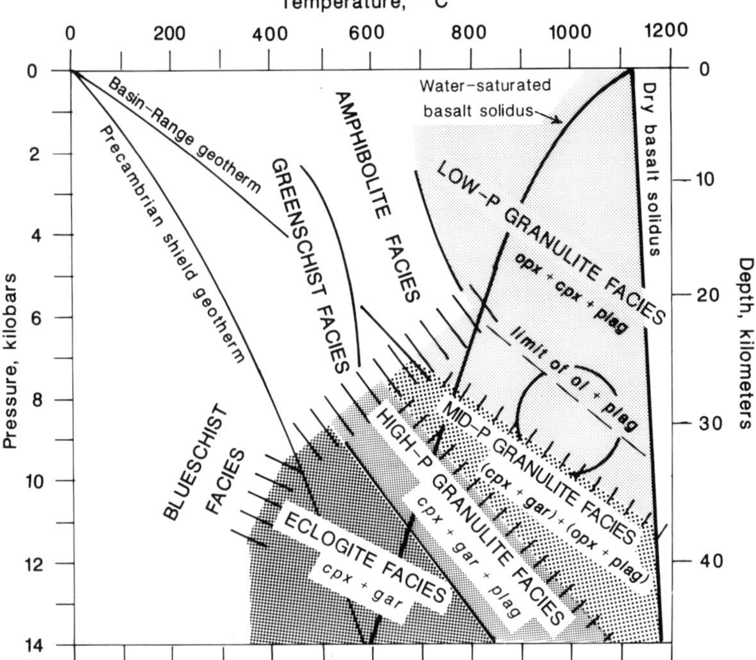

Fig. 9. P/T diagram of mineral assemblages relevant to the lower continental crust. Boundaries approximate those for mafic and intermediate rocks but vary with bulk composition; coexisting minerals vary in composition across each facies. Amphibolite–granulite reaction represents hydration–dehydration, hence availability of water; a garnet amphibolite facies, not shown, often intervenes at $P > 6$ kbar. Reactions between granulite assemblages represent decreasing molar volume (primarily decreasing stability of plagioclase and increasing stability of garnet and clinopyroxene) with increasing P/T ratios. Eclogite–granulite boundary is very poorly constrained at low T. Abbreviations: cpx, clinopyroxene; gar, garnet; ol, olivine; opx, orthopyroxene; plag, plagioclase. Circle marks a field often represented by exposed deep-crustal plutonic rocks. Adapted from many published papers including Hansen (1981), Johnson *et al.* (1983), Newton & Perkins (1982), and papers referred to by each.

syntheses of Kohistan and Ivrea are biased by the application of crustal-zoning and facies concepts and incorporate observations made during field trips led by M. Q. Jan and R. D. Lawrence in the Kohistan section and by A. C. Boriani, Luigi Burlini, V. J. Dietrich, D. M. Fountain, E. H. Rutter and Silvano Sinigoi in the Ivrea section.

Some observers believe the dominant cause of the magmatic underplating seen in the Kohistan, Ivrea, and Talkeetua examples discussed here to be in magmatism related to crustal rifting rather than in arc magmatism. The arguments for rifting are primarily statements of philosophy: gabbroic underplating probably accompanies severe rifting of continental crust (the author concurs), so demonstrated underplating likely indicates rifting (*non sequitur*; see discussions of xenoliths in this report as well as of these crustal sections). As the associated middle and upper crustal magmatism in the cases at issue appears to be of arc rather than rift type to those who advocate a rift origin of the deep rocks, the arguments for deep-crustal rift-related magmatism seem forced. Some of the rift-origin arguments (e.g. Quick *et al.* 1992) for Ivrea are based on structural evidence for moderate extension within the deep-crustal rocks, so, again, it is emphasized that the general setting of arc magmatism is extensional and is not accompanied by shortening as many geologists assume.

Kohistan island-arc crustal section. The Kohistan crustal section is bounded on both north and south sides by north-dipping subduction sutures (Coward *et al.* 1987). Reconnaissance dating shows the rocks to be at least mostly of Late Cretaceous and early Palaeogene age; the older rocks are petrologically primitive and formed in an oceanic island arc, whereas the younger are more evolved and may have formed after accretion of the arc to the composite northern (Eurasian) plate but mostly before collision with the southern (Indian) plate (Petterson & Windley 1985; Treloar *et al.* 1989). The structurally deepest rocks are ramped up to moderate north dips at the south edge of the complex whereas the shallowest rocks are in the north part of the complex, but major very large, tight folds are present within the intervening region. The synthesis here integrates the estimates of initial depths, as deduced from mineral assemblages (Fig. 9) described in papers by others and also as noted by me in the field by the author, with inferences of megafolds as deduced from the broad symmetry of rock assemblages, depth indicators, and attitudes seen in the field. The conclusions are in general, but not detailed, agreement with the structure of the complex deduced from field data by Petterson & Windley (1985), whereas they are in sharp disagreement with the conflicting interpretations by Bard (1983), Coward *et al.* (1987) and Jan (1988).

Initial thickness of the Kohistan crustal section was about 40 km, of which approximately the lower 30 km are preserved. About 5 km of what appear to be *in-situ* mantle rocks are preserved beneath the crustal section and consist of serpentinized spinel peridotite (mostly harzburgite?) cut by voluminous veins and dikes of spinel-bearing diopsidite; other ultramafic rock types are less abundant (Jan & Howie 1981; author's observations). Above this is a great stratiform sheet of gabbroic rocks which had an initial thickness of perhaps 10 km. The entire gabbro is preserved in the north-dipping southern section, which includes the Jijal complex and the southern third of the Kamila amphibolite in the terminology of Jan (1988). All but the basal part of the gabbro is exposed farther north in a great isoclinal anticline, which includes the Chilas complex of Jan (1988) and the granulitic layered lopolith of Bard (1983). The basal 0.5 km of the section consists of garnet-pyroxene rock which Loucks *et al.* (1990) regarded as a primary igneous cumulate although others have considered it to be high-pressure granulite metamorphozed from mafic gabbro. The next several km of the gabbroic sheet are dominated by garnet-pyroxene-plagioclase rock but include both garnet-free gabbro and plagioclase-free garnet-clinopyroxene rock (Gansser 1979; Jan & Howie 1981; Loucks *et al.* 1990; author's observations). The rest of the gabbro sheet consists almost entirely of flow-layered two-pyroxene, olivine-free gabbro, with or without hornblende, of which the

lower part is partly metamorphozed at high- to middle-pressure granulite and garnet-amphibolite facies and the upper part is metamorphozed, less pervasively, at garnet-amphibolite facies (Jan & Howie 1981; Bard 1983; Jan 1988; Khan *et al.* 1989). The igneous clinopyroxene has arc, not ophiolitic, Al/Ti ratios (Loucks 1990). The plagioclase-free basal rocks have a mineralogy appropriate for mantle geophysical properties so the Mohorovičić discontinuity is about 0.5 km higher in the section than is the base of the magmatic section that otherwise represents the lower crust. Pyroxenite and other young magmatic rocks also are present with presumably older rocks in the deeper part of the mantle complex. High in the gabbro are small complexes of layered cumulates of two-pyroxene gabbro, olivine gabbro, dunite, troctolite, and anorthosite (Khan *et al.* 1989). Olivine and plagioclase crystallized out together in these small complexes but are everywhere separated by thick reaction rims of pyroxenes (Jan *et al.* 1984; author's observation), so crystallization was at a depth probably near 30 km (cf. Fig. 9). The lack of olivine with plagioclase in mafic gabbro deeper in the mafic complex is indicative of depths there greater than about 30 km, and the garnet amphibolite above the upper part of the complex indicates crystallization deeper than about 20 km. Vertical compositional variations presumably occur in the main gabbro and metagabbro but are not defined by published data.

Above the great stratiform gabbro and metagabbro is an upward-varying section, perhaps initially also 10 km thick, of amphibolite and migmatite (the northern two thirds of the Kamila amphibolite of Jan (1988), and also the Northern Amphibolitic Series of Bard (1983)). The lower part of this section is dominated by strongly layered amphibolite, variably garnetiferous and migmatitic, whereas the upper part is a migmatitic complex of amphibolite and diorite and voluminous sheets of tonalite, trondhjemite, granodiorite, and pegmatite (Bard 1983; author's observations). No granitic rocks occur within the underlying stratiform gabbro, so the granitic rocks in the migmatitic amphibolite complex presumably represent partial melts of pre-existing rocks, the restites from which are the amphibolites low in the supra-gabbro section. The Ivrea crustal section, described subsequently, presents an example of a similar process operating upon metapelitic rocks.

The next 10 km or so of the Kohistan crustal section (the top 10 km are missing) consists of basalt, andesite, dacite, rhyodacite and sedimentary rocks, metamorphozed at greenschist and lower amphibolite facies and intruded by cross cutting plutons of tonalite, granodiorite, and quartz monzonite (Bard 1983; Jan & Asif 1983; Petterson & Windley 1985; Shah & Shervais 1991). This assemblage is to be expected in an island arc.

Talkeetna island-arc crustal section. The southern Alaskan crustal section of the Lower and Middle Jurassic Talkeetna island arc, ramped in Cretaceous time in the direction now southward upon accretionary complexes, was recognized and described by DeBari & Coleman (1989) and DeBari & Sleep (1991). (For further description, without emphasis on depth and crustal-section significance, see Burns 1985 and Plafker *et al.* 1989.) The lowest preserved part of the section studied by DeBari consists of 2 km of tectonized residual-mantle harzburgite and dunite. Above this is 2 km of cumulate ultramafic rocks, mostly pyroxenite but including dunite and peridotite (wehrlite). Next higher is 200 m or so of garnet-pyroxene granulite and garnet gabbro, which grades upward into about 2 km of two-pyroxene gabbro with upward-increasing hornblende; olivine is lacking in this deep section so crystallization was at a depth probably greater than 30 km. Above this (with intervening crustal section faulted out?) is about 4.5 km of two-pyroxene gabbro, with either hornblende or olivine (hence formed shallower than 30 km), and also tonalite. Above this in turn, with an intervening middle-crust section faulted out, are lithic graywackes and basaltic, andesitic and silicic volcanic rocks, intruded by tonalites.

The geophysical Mohorovičić discontinuity lies within the underplated Jurassic magmatic section, between cumulate ultramafic and garnet-pyroxene rocks beneath and gabbros above. The parental melts of the underplated section were high in Mg and low in Al; the deep-crustal gabbroic magmas were enriched in Al by the removal of Mg and other components in the crystallization of the underlying pyroxenites and ultramafic rocks (DeBari & Sleep 1991). This may be an important process in generating the high-Al basalts common in evolving island arcs.

Ivrea continental magmatic-arc crustal section. A crustal section, dominated by late Palaeozoic igneous rocks which the author regards as of arc-magmatic origin, dips subvertically to gently southeastward above high-pressure subduction-complex rocks west of Lake Maggiore in the western Italian Alps. The lower part of this section is commonly termed the Ivrea Zone, or Ivrea-Verbano Zone, whereas various terms, among them Strona-Ceneri Zone and Laghi Series, are applied to the upper parts. Most reliable radiometric and stratigraphical dates indicate the young igneous rocks that dominate the evolution of the crust to be of Late Carboniferous(?), Permian and Triassic(?) ages and to have been intruded into pre-existing continental crust with a complex early Palaeozoic history (Voshage *et al.* 1988, 1990; papers by various authors in Quick & Sinigoi 1992; Rutter *et al.* 1993). Both top and bottom of the late Palaeozoic crustal section are preserved: extrusive andesite to rhyolite at the top, Mohorovčić discontinuity at the bottom. Extensional deformation probably is recorded within the complex (Hodges & Fountain 1984; Quick *et al.* 1992; Rutter *et al.* 1993) but even if the extension is synmagmatic this could have occurred within an extending arc rather than in a rifting continental margin as assumed by Fountain (1989), Handy & Zingg (1991) and various authors in Quick & Sinigoi (1992).

Ultramafic mantle rocks occur in lenses mostly along and near the exposed base of the complex (Schmid 1967; Rutter *et al.* 1993). The best-studied of these, the Balmuccia peridotite (Shervais 1979*a*, *b*; Garuti *et al.* 1980; Voshage *et al.* 1988), apparently exposes the sheared Mohorovičić discontinuity on the east side of a lens whose west side is faulted against basal-crust mafic rocks. The lens consists of spinel peridotite (locally garnetiferous; mostly contains two magnesian pyroxenes; presumably represents residual mantle of undefined age) injected by dykes and veins of late Palaeozoic chrome diopsidite and more aluminous spinel-two pyroxene pyroxenite. In another lens, the uppermost ultramafic rocks are layered cumulate peridotites, part of the young layered igneous complex.

Geopetally above these mantle lenses is a stratiform mass, 6–10 km thick, of vertically-varying gabbro and its granulitic equivalents (Schmid 1967; Garuti *et al.* 1980; Pin & Sills 1986; Rutter *et al.* 1993). Magmatic fabrics and mineralogy are well preserved in some sectors but in others have been obliterated by retrogression, likely isobaric, to high-pressure granulite (garnet + clinopyroxene ± plagioclase) in deep sections, and to middle- or low-pressure granulite in shallow sectors. The basal 1 km of the great gabbro sheet consists of interlayered cumulate gabbro, pyroxenite, and subordinate peridotite, representing a crust-mantle transition within the layered complex; and above this is another 1 km of layered cumulate gabbro. Next higher is 5 km or so of obscurely layered two-pyroxene gabbro, locally anorthositic, which becomes more feldspathic upward and grades into a zone, about 2 km thick, of leucogabbro, diorite, and hybrid monzonite and quartz-bearing rocks. Isotopes and trace elements show increasing crustal contamination upward in this sequence (Pin & Sills 1986; Voshage *et al.* 1990). Much of this upward variation records contamination by felsic melts generated from overlying crustal rocks (Voshage *et al.* 1990) rather than extreme fractionation of mantle melts or crystallization of discrete melts from diverse sources (as was suggested by Pin & Sills 1986).

This great stratiform gabbroic sheet lies beneath metapelites and other pre-magmatic rocks, and lenses of such rocks also are enclosed within the sheet. Metapelites within and close to the sheet are restites that are rich in garnet (up to 30%) and often sillimanite but are mica-free. These restites commonly contain only two of the three components quartz, K-feldspar and plagioclase, and have lost about half of their initial material to hydrous granitic melts that have migrated elsewhere (Schmid 1979), as is to be anticipated from experimental studies (Vielzeuf & Holloway 1988). The process of partial melting can be observed in metapelites more distant from the sheet, where similar restites are the melasome component of migmatites extremely rich in leucosomes of pegmatite and leucogranite. Still more distant metapelites are micaceous schists and gneisses that have not been conspicuously degranitized.

The preserved upper crust of the Ivrea section contains cross-cutting granites with contact-metamorphic aureoles, and extrusive volcanic rocks. The middle crust has been partly cut out tectonically but presumably included sheets of two-mica granite such as are common in other sections of the middle continental crust.

Other examples. The basal 10 km or so of crust in the examples cited is of gabbro, mafic gabbro, and its granulitic equivalents. Most granulite terranes around the world expose only the higher non-gabbroic part of the lower crust (Bohlen & Mezger 1989), but a number of other crustal sections do apparently expose at least the upper parts of gabbroic lower-crustal complexes like those of Kohistan and Ivrea.

Fiordland. Very deep crust (with *in-situ* mantle beneath it?) of an isotopically primitive Early Cretaceous island arcs are exposed in Fiordland, southwest New Zealand (references in Hamilton 1988; Bradshaw 1989, 1990). On the basis of inferences from petrologic assemblages of the depths of formation, it is suggested that the rocks could be fitted into a crustal section, similar to that of Kohistan, recording thick underplating by gabbro. The crustal section has been transpressively ramped up westward along the active Alpine fault and has been eroded obliquely.

Two-pyroxene gabbro, diorite and tonalite dominate the deep, western part of the section, within which lenses of ultramafic rocks increase downward in abundance; leucogabbro, calcic anorthosite and granodiorite are subordinate. Magmatic crystallization was in the middle-pressure granulite facies (two pyroxenes; plagioclase stable with orthopyroxene but not with olivine; no garnet). At the deepest crustal levels exposed, these rocks were widely retrograded to variably garnetiferous granulite-facies gneisses, and locally to eclogite facies; somewhat shallower rocks widely preserve igneous fabrics or were retrograded at garnet-amphibolite facies. Facies relationships permit the inference that magmatism and retrogression were essentially isobaric and that the deepest rocks exposed formed at a depth of about 35 km. Ultramafic rocks occur at the structural base of the complex north of Milford Sound, beneath retrograded mylonitic garnet-clinopyroxene-hornblende granulite, and are faulted against high-pressure metasedimentary rocks (Bradshaw 1990); the uppermost-mantle part of the terrane may be exposed here, thrust over subducted sedimentary rocks. In shallow parts of the complex elsewhere in Fiordland, olivine and plagioclase crystallized together in mafic plutonic rocks, metavolcanic and calc-silicate gneisses are present, and retrogression occurred at amphibolite and garnet-amphibolite facies; isobaric magmatism and retrogression at depths of about 20–25 km can both be inferred. Both massive and layered plutonic rocks are present at both lower- and mid-crustal levels.

Vancouver Island. All but the base is present of a crustal section of a Jurassic island arc on Vancouver Island, British Columbia (DeBari 1992), and it is much like Kohistan. Mafic non-cumulate deep-crustal gabbros are of island-arc-tholeiite composition (16–17% Al_2O_3, 7–9% MgO). Partial melting of pre-existing amphibolite produced tonalitic melts, which in part hybridized with gabbroic melt to produce diorites. Dioritic, tonalitic and granodioritic melts rose into the upper crust, and from them andesitic to rhyolitic volcanic rocks were erupted. The rocks at all levels follow calc-alkaline trends. 'The paucity of mafic compositions in the upper levels of the arc suggest that the lower crust was an effective "filter" where mantle derived magmas stalled, fractionated, and mixed with migmatization products before moving to higher levels' (De-Bari 1992).

Anorthosite-norite complexes. The stratiform anorthosites above norites that are the deepest rocks exposed in a number of Proterozoic granulite terranes may be the upper parts of underplated basal-crustal complexes. Examples occur in western Norway and in the Adirondack Mountains of New York State.

Fast-migrating arcs. Most large ophiolite masses emplaced tectonically on continents and island arcs may represent the crust and upper mantle of fast-migrating arcs, not of mid-ocean spreading ridges (Evans *et al.* 1991; Hamilton 1988 and references therein; Stern & Bloomer 1992). These arc ophiolites much resemble the underplated complexes of thicker arc sections except that they are capped only by thin sections of extrusive and hypabyssal rocks of primitive-arc types. The geophysical Mohorovičić discontinuity is within, not beneath, these complexes, which unlike mid-ocean ridge basaltic rocks typically carry orthopyroxene as well as clinopyroxene. The arc-ophiolite crustal sections are erratically thicker than those typically formed at mid-ocean ridges, though thinner than that of island arcs whose volcanoes stand above the sea.

The Eocene Acoje ophiolite of western Mindanao was described, as a 'nascent island arc', by Hawkins and Evans (1983). The moderately dipping Acoje section exposes the entire crust, about 9 km thick, and about 10 km of underlying mantle. All but the top 1 km of the mantle section consists of serpentinized and tectonized residual harzburgite and subordinate dunite and chromite containing late clinopyroxene-rich pods. The upper 1 km or so of geophysical mantle is the basal 1 km of the arc-magmatic section and consists of undeformed cumulates of olivine and clinopyroxene. These are intercalated, over a thickness of several hundred metres, with the basal part of the gabbroic rocks that make up the lower 7 km or so of the overlying crust. Most of this gabbroic section consists of layered-cumulate two-pyroxene gabbro, which grades upward into massive gabbro and norite, about 1 km thick, high in which are abundant small plutons and dykes of hornblende tonalite and leucotonalite. The top 1 or 2 km of the crustal section consists of dykes, sills, and pillow flows of basalt like that of modern primitive island arcs rather than like spreading-ridge basalt.

Nonexistent 'Conrad discontinuity'. The tops of the deep-crustal gabbroic sections do not correspond to the much-shallower mid-crustal 'Conrad discontinuity' of some seismologists, nor does that purported 'discontinuity' correspond to any general abrupt change in exposed crustal sections. The 'discontinuity' may commonly be an artifact of the first-arrival methodology of refraction seismology as applied to a crust with a graded downward increase in seismic velocity.

Xenoliths. Kimberlites and alkaline volcanic rocks often contain xenoliths of lower crustal and mantle rocks. The xenoliths commonly indicate that basal crust of all ages is dominated by two-pyroxene gabbro and mafic granulite, and hence records underplated gabbroic magmatic rocks, in both magmatic-arc and rift settings (Bohlen 1987; Griffin & O'Reilly 1987*b*; Bohlen & Mezger 1989; Rudnick 1992). The uppermost mantle consists of ultramafic or garnet-clinopyroxene rocks formed as part of the igneous complex, so, as in the exposed crustal sections discussed previously, the geophysical Mohorovičić discontinuity commonly lies within these underplated complexes, not beneath them (Griffin & O'Reilly 1987*a*). Proportions of crust-contaminated gabbros and of rocks restitic after removal of partial melts vary widely.

An example worth special mention is the xenolith-sampled deep crust beneath the central part of the vast felsic Cretaceous magmatic-arc batholith of the Sierra Nevada. This deep crust consists of gabbroic granulites crystallized in the olivine-free part of the granulite facies, hence deeper than about 30 km, and also cumulate pyroxenite, peridotite and garnet pyroxenite that equilibrated at depths near 40 km (Dodge *et al.* 1986).

Seismic properties of the crust-mantle transition. The basal crust and uppermost mantle, commonly, are seismically reflective, whereas the rest of the upper mantle is acoustically more or less transparent (Hale & Thompson 1982; Mooney & Brocher 1987; Hauser & Lundy 1989; Holbrook 1990). These characteristics accord with the outcrop and xenolith evidence for the general presence of a basal crust and uppermost mantle of underplated layered ultramafic, garnet-granulitic and gabbroic rocks and for the presence of the geophysical Mohorovičić discontinuity within, not beneath, the underplated section.

Delamination. The lower part of the underplated crustal and uppermost mantle sections is of rocks so dense that delamination and sinking into the mantle of such rocks is likely to occur (DeBari & Sleep 1991). Such delamination may

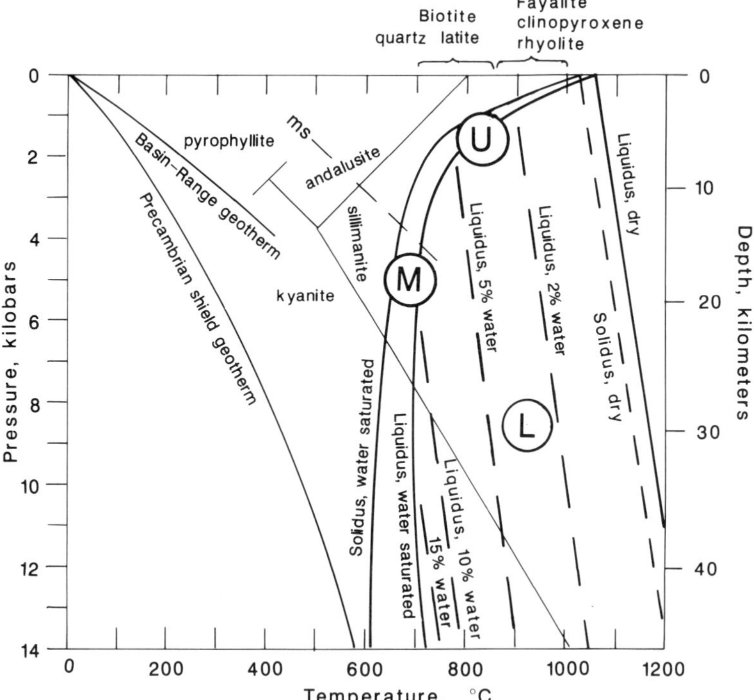

Fig. 10. Crystallization relationships for leucogranite, showing the melting interval for water-saturated granite and contours on the liquidus surface for undersaturated melt, after Huang & Wyllie (1981). Fields of Al_2SiO_5 polymorphs from Holdaway (1971). Primary muscovite can crystallize from granitic magma only deeper than the stability limit of muscovite (ms; after Tracy 1978; precise position varies with composition of melt and volatiles). Typical fields are shown by circles for late-stage granitic magmas of the upper (U), middle (M), and lower (L) crust. Both wallrocks and magmas are relatively anhydrous in the lower part of the lower crust. The middle crust, rich in hydrous minerals, is raised toward granite-solidus temperature by transiting magma. Magmas that equilibrate with, or that are produced by partial melting of, micaceous wallrocks, become richly hydrous. Hydrous magmas once formed cannot generally rise to shallow depths, for they cross the water-saturated solidus curve as they are depressurized. Low-water magmas that reach the upper crust without such middle-crust equilibration do not become saturated until crystallization is advanced, at which time either pegmatites or tuffs are expelled.

be a common process of recycling of ultramafic and mafic-granulitic fractionates back into the mantle. Evidence is increasing that recycling of crustal materials back into the mantle by combinations of subduction and delamination is a relatively rapid process (Class *et al.* 1993).

Origin of arc magmas The basal 10 km or so of the crust of continents and mature island arcs typically is dominated by gabbro and its granulitic equivalents. Observed vertical variations in and above the more widely exposed upper part of the lower crust can be discussed in these terms. Heat is introduced into the crust in gabbroic magmas and is transferred to whatever rocks preceded them. Magmas evolve by exothermically crystallizing refractory minerals and by endothermically assimilating, melting and breaking down fusibile ones, and the reactions involved vary with the changing compositions of melts and solids, with the availability of water in wallrock minerals and in melt, and with pressure and temperature.

Arc proto-melts generated in the mantle by processes related to subduction (or to rifting) are much more mafic than basalt. The proto-melts must undergo modification as they rise through the upper mantle (Kelemen 1986). The melts that reach the top of the mantle are much evolved; the melts that enter the crust as 'primitive' arc basalts, basal-crust gabbros, and mid-ocean-ridge basalts are further evolved. The Mohorovičić discontinuity is a self-perpetuating density filter, the shallow limit of crystallization of voluminous rocks of ultramafic composition or plagioclase-free mineralogy;

melts cannot rise past it until they have crystallized much of their olivine, pyroxene, and garnet components within the mantle and have evolved to gabbroic compositions. Hot, dry residual gabbroic magmas rise into the dry lower crust. The density of these gabbroic melts inhibits rise through low-density crust, although they readily reach the surface through the thin, dense crust of immature oceanic island arcs. In mature arcs, either oceanic or continental, the mafic melts crystallize in the deep crust to produce great stratiform gabbroic complexes variably complicated by fractionation and assimilation and by post-magmatic recrystallization at granulite and garnet-amphibolite facies.

Felsic melts are derived by widely varying combinations of fractionation, of secondary partial melting of pre-existing crustal rocks (which can be older arc-magmatic rocks including basal-crust gabbro) by the mafic melts that reach the basal crust and of hybridization and magma mixing. The secondary melts are dominantly tonalitic and mafic-granodioritic where formed in petrologically primitive arc crust (and trondhjemitic where such crust contains abundant hornblende); charnockitic and other dry granitic melts where formed in previously dehydrated continental crust; and hydrous granodioritic to granitic melts where formed in crust containing abundant mica (Fig. 10). These secondary melts are variably hybridized with the melts of mantle origin, and the composite melts, which can contain abundant crystals from earlier and deeper crystallization, tend to rise and evolve further. Although little mantle material need be present in melts that reach high crustal levels, mantle heat carried by evolving magma is the primary source of warming of the crust above equilibrium geotherms. Progressive partial equilibrations – crystallization of refractory minerals, melting and assimilation of fusible components – produces progressively more evolved and hydrous melts. Granites and migmatites in the lower crust typically form from hotter and drier melts than do those of the middle crust, a contrast that reflects primarily the availability of water from the dominant wallrocks. The content of combined water in wallrocks tends to increase upward in the crust, and particularly in the middle crust wallrock water is abundantly available yet also pressure is high enough to permit the existence of richly hydrous, low-temperature melts (Fig. 10). Magmas that equilibrate in the middle crust with micaceous wall rocks, or that are derived from them by secondary melting, are hydrated and cooled, and then are quenched by expulsion of volatiles as they rise to the levels, or crystallize to

the points, at which water pressure exceeds load pressure. The resulting granites often are peraluminous and typically are in complexes of migmatites and sheets of pegmatites and hydrous-magma granites.

Felsic melts that cross the middle crust without equilibrating there can continue to rise if they remain relatively hot and dry, and this probably occurs primarily after the middle crust has been heated and partly dehydrated by earlier magmas stopping there. Voluminous magmas that reach the upper crust spread as shallow batholiths above the deeper migmatites and concordant granites, and erupt as ash-flow sheets and as far-travelling volcanic ash when their own rise and crystallization produce water saturation. Most magmas of upper crustal batholiths probably contain less than 1.5% water and reach water saturation only at shallow depth after considerable crystallization, with resultant expulsion of pegmatites and volcanic materials (Maaloe & Wyllie 1975), whereas granites in the middle crust mostly crystallize from magmas with 3–5% water, cross the water-saturated solidus at greater depth, and solidify there rather than rising to the upper crust.

Contacts between and within granitic and metamorphic rocks typically are steep in shallow complexes and undulating in deep ones. The tendency toward gentle dips in the middle and deep crust is accentuated by the variably pervasive flattening that accompanies retrograde metamorphism in many regions. Much such metamorphism and deformation probably records extension or shortening and probably much also records gravitational flow of heated rocks that are displaced outward and downward by rising magmas and then flow beneath shallow-spreading batholiths. Depression to deep crustal levels of supracrustal rocks may be due primarily to the repeated injection and eruption of less-dense magma above them. The common contrast between 'syntectonic' granites of the middle crust and 'post-tectonic' granites of the upper crust is a manifestation of the gravity-driven deformation of the early hydrous middle-crust granites beneath the later shallow-spreading, less-hydrous granites of the upper crust.

The early Earth

The widely accepted geochemical and petrological assumption that a cold-accreted Earth of chondritic composition gradually fractionated to produce crust throughout geological time biases much petrological modelling of arc magmatism and yet is incompatible both with cosmological and planetological evidence and with Archaean geology. This is documented elsewhere (Hamilton 1993) and the argument is only summarized here.

Current quantitative theories agree that the Earth was largely or entirely melted, perhaps superheated and devolatilized, by heat of accreting planetesimals, impacts including the Mars-size impact which splashed out the Moon, separation of the core and radiogenic heat. This history invalidates cold-start models and, considering also the Earth's position in the compositionally graded solar system, requires the bulk Earth to be more refractory than asteroid-belt chondrite. Retained water and CO_2 probably were added by impactors after the Moon formed; the mantle is not a source of primordial volatiles but rather is a sink that has depleted the hydrosphere. As scaling and velocity considerations require the Earth's impact history to be more intense than that of the Moon, the outer shell of the Earth must have been repeatedly recycled by impacts before 3.8 Ga.

Dominant models of Archaean tectonics and magmatism invoke plate-tectonic mechanisms, yet Archaean terranes display no viable analogues for ophiolites, magmatic arcs, subduction mélange, or rifted and sutured plates. Archaean magmatism was concurrent over vast tracts, not limited to arc-like belts. Voluminous liquidus ultramafic lava (komatiite) indicates upper mantle to have contained much melt beneath large provinces. (Contrary arguments for Archaean geotherms much like modern ones are based on 'model ages' of xenoliths in kimberlite erupted through Archaean crust, but these 'ages' are merely rationalizations of the assumptions at issue, constrained mantle-equilibration ages being no older than the much younger kimberlites themselves.) Although no basal-crust sections are known in outcrop, the basal Archaean crust as sampled by xenoliths is apparently dominated by noritic assemblages variably metamorphozed to high- and middle-pressure granulite facies, so basal Archaean crust probably records magmatic underplating like that of younger age. Only komatiitic and basaltic magma entered the crust from the mantle. Intermediate and felsic melts were produced in the crust by variably hydrous contamination, fractionation and secondary and primary partial melting. Granite-and-greenstone terranes consist of coalescing volcanoes and regionally semi-concordant ultramafic, mafic and felsic igneous and sedimentary accumulations, domed by batholiths generated by province-wide partial melting of deep crust by introduced and radiogenic heat.

Not until Proterozoic time did plate-tectonic mechanisms become prevalent. The Archaean Earth lost heat primarily by voluminous magmatism; heat loss through windows produced by separating plates became progressively more important subsequently. Important recycling of crust into mantle in Archaean time may have been accomplished by delamination of dense lower crustal and upper mantle rocks.

Concluding remarks

The tectonic and magmatic processes which have shaped the post-Archaean continents are dominated by convergent plate tectonics. It is now more than 25 years since the conceptual revolution of plate tectonics began to give us the framework within which to comprehend genetic relationships among these processes. Most geoscientists nevertheless are still unfamiliar with the characteristics and behaviour of actual plate systems (and there is still no good textbook on the subject), and the literature is cluttered with speculations based on false assumptions.

Much petrological modelling similarly is negated by the powerful constraints provided by exposed partial and complete sections eroded obliquely through the Earth's crust. The actual products of depth-varying processes can be studied in these sections and many popular hypotheses (such as those proposing generation of intermediate magmas in the mantle) can thereby be falsified. Much petrological and geochemical conjecture would never have been published had its authors tested their implicit predictions against observable variations with depth.

I am indebted to Dapeng Zhao for permission to incorporate an illustration of his exciting new seismic tomography. The manuscript was much improved as a result of thoughtful reviews by J. Quick, J. Smellie, W. Spence and J. Tarney.

References

ANDERSON, D.L. 1987. Thermally induced phase changes, lateral heterogeneity of the mantle, continental roots, and deep slab anomalies. *Journal of Geophysical Research*, **92**, 13968–13980.

APPERSON, K.D. 1991. Stress fields of the overriding plate at convergent margins and beneath active volcanic arcs. *Science*, **254**, 670–678.

ARMIJO, R. & THIELE, R. 1990. Active faulting in northern Chile–ramp stacking and lateral decoupling along a subduction plate boundary? *Earth and Planetary Science Letters*, **98**, 40–61.

ASMERON, Y., PATCHETT, P.J. & DAMON, P.F. 1991. Crust-mantle interaction in continental arcs–

inferences from the Mesozoic arc in the south-western United States. *Contributions to Mineralogy and Petrology*, **107**, 124–134.

BAKER, M.B., GROVE, T.L., KINZLER, R.J., DONNELLY-NOLAN, J.M. & WANDLESS, G.A. 1991. Origin of compositional zonation (high-alumina basalt to basaltic andesite) in the Giant Crater lava field, Medicine Lake volcano, northern California. *Journal of Geophysical Research*, **96**, 21819-21842.

BARD, J.-P. 1983. Metamorphic evolution of an obducted island arc – Example of the Kohistan sequence (Pakistan) in the Himalayan collided range. *University of Peshawar Geological Bulletin*, **16**, 105–184.

BIJU-DUVAL, B., BIZON, G., MASCLE, A. & MULLER, C. 1982. Active margin processes – Field observations in southern Hispaniola. *American Association of Petroleum Geologists Memoir*, 34, 325–344.

BOHLEN, S.R. 1987. Pressure-temperature-time paths and a tectonic model for the evolution of granulites. *Journal of Geology*, **95**, 617–632.

—— & MEZGER, K. 1989. Origin of granulite terranes and the formation of the lowermost continental crust. *Science*, **244**, 326–329.

BRADSHAW, J.Y. 1989. Origin and metamorphic history of an Early Cretaceous polybaric granulite terrain, Fiordland, southwest New Zealand. *Contributions to Mineralogy and Petrology*, **103**, 346–360.

—— 1990. Geology of crystalline rocks of northern Fiordland: details of the granulite facies Western Fiordland Orthogneiss and associated rock units. *New Zealand Journal of Geology and Geophysics*, **33**, 465–484.

BROPHY, J.G. 1989. Basalt convection and plagioclase retention – A model for the generation of high-alumina arc basalt. *Journal of Geology*, **97**, 319–329.

BURNS, L.E. 1985. The Border Ranges ultramafic and mafic complex, south-central Alaska: cumulate fractionates of island-arc volcanics. *Canadian Journal of Earth Sciences*, **22**, 1020–1038.

CAHILL, T. & ISACKS, B.L. 1992. Seismicity and shape of the subducted Nazca plate. *Journal of Geophysical Research*, **97**, 17503–17529.

CANDE, S.C. & LESLIE, R.B. 1986. Late Cenozoic tectonics of the southern Chile Trench. *Journal of Geophysical Research*, **91**, 471–496.

CARESS, D.W. 1991. Structural trends and back-arc extension in the Havre Trough. *Geophysical Research Letters*, **18**, 853–856.

CARLSON, R.L. 1983. Plate motions, boundary forces, and horizontal temperature gradients – Implications for the driving mechanism. *Tectonophysics*, **99**, 149–168.

—— & MELIA, P.J. 1984. Subduction hinge migration. *Tectonophysics*, **102**, 399–411.

CHASE, C.G. 1978. Extension behind island arcs and motions relative to hot spots. *Journal of Geophysical Research*, **83**, 5385–5387.

CHEN, J.H. & TILTON, G.R. 1991. Applications of lead and strontium isotopic relationships to the petrogenesis of granitoid rocks, central Sierra Nevada batholith, California. *Geological Society of America Bulletin*, **103**, 439–447.

CLASS, C., GOLDSTEIN, S.L., GALER, S.J.G. & WEIS, D. 1993. Young formation age of a mantle plume source. *Nature*, **362**, 715–721.

COULBOURN, W.T. & MOBERLY, R. 1977. Structural evidence of the evolution of fore-arc basins off South America. *Canadian Journal of Earth Sciences*, **14**, 102–116.

COWARD, M.P., BUTLER, R.W.H., KHAN, M.A. & KNIPE, R.J. 1987. The tectonic history of Kohistan and its implications for Himalayan structure. *Journal of the Geological Society, London*, **144**, 377–391.

CRAWFORD, A.J., FALLOON, T.J. & EGGINS, S. 1987. The origin of island arc high-alumina basalts. *Contributions to Mineralogy and Petrology*, **97**, 417–430.

DARBY, D.J. & WILLIAMS, R.O. 1991. A new geodetic estimate of deformation in the Central Volcanic Region of New Zealand. *New Zealand Journal of Geology and Geophysics*, **34**, 127–136.

DEBARI, S.M. 1992. Petrology and field relations of a lower- to upper-crustal section from a Jurassic island arc, Vancouver Island, Canada (Abstract). *EOS*, **73** (no. 43 suppl.), 642.

—— & COLEMAN, R.G. 1989. Examination of the deep levels of an island arc – Evidence from the Tonsina ultramafic assemblage, Tonsina, Alaska. *Journal of Geophysical Research*, **94**, 4373–4391.

—— & SLEEP, N.H. 1991. High-Mg, low-Al bulk composition of the Talkeetna island arc, Alaska – Implications for primary magmas and the nature of arc crust. *Geological Society of America Bulletin*, **103**, 37–47.

DEFANT, M.J. & DRUMMOND, M.S. 1990. Derivation of some modern arc magmas by melting of young subducted lithosphere. *Nature*, **347**, 662–665.

DEMETS, C. 1992. Oblique convergence and deformation along the Kuril and Japan trenches. *Journal of Geophysical Research*, **97**, 17,615–17,625.

——, GORDON, R.G., ARGUS, D.F. & STEIN, S. 1990. Current plate motions. *Geophysical Journal International*, **101**, 425–478.

DEPAOLO, D.J., PERRY, F.V. & BALDRIDGE, W.S. 1992. Crustal versus mantle sources of granitic magmas – a two-parameter model based on Nd isotopic studies. *Royal Society of Edinburgh Transactions: Earth Sciences*, **83**, 439–446.

DEWEY, J.F. 1980. Episodicity, sequence and style at convergent plate boundaries. *Geological Association of Canada Special Paper*, 20, 553–573.

DIAMENT, M. *et al.* 1992. Mentawai fault zone off Sumatra – A new key to the geodynamics of western Sumatra. *Geology*, **20**, 259–262.

DICKINSON, W.R. & SEELY, D.R. 1979. Structure and stratigraphy of forearc regions. *American Association of Petroleum Geologists Bulletin*, **63**, 2–31.

DODGE, F.C., CALK, L.C. & KISTLER, R.W. 1986. Lower crustal xenoliths, Chinese Peak lava flow, central Sierra Nevada. *Journal of Petrology*, **27**, 1277–1294.

EVANS, C.A., CASTENADA, G. & FRANCO, H. 1991. Geochemical complexities preserved in the volcanic rocks of the Zambales Ophiolite, Philippines. *Journal of Geophysical Research*, **96**, 16,251–16,262.

EWART, A. & LE MAITRE, R.W. 1980. Some regional compositional differences within Tertiary-Recent orogenic magmas. *Chemical Geology*, **30**, 257–283.

FOUNTAIN, D.M. 1989. Growth and modification of lower continental crust in extended terrains – The role of extension and magmatic underplating. *American Geophysical Union Monograph*, **51**, 287–299.

FURLONG, K.P. & FOUNTAIN, D.M. 1986. Continental crustal underplating – Thermal considerations and seismic-petrologic consequences. *Journal of Geophysical Research*, **91**, 8285–8294.

GANSSER, A. 1979. The division between Himalaya and Karakorum. *University of Peshawar Geological Bulletin*, **13**, 9–21.

GARFUNKEL, Z., ANDERSON, C.A. & SCHUBERT, G. 1986. Mantle circulation and the lateral migration of subducted slabs. *Journal of Geophysical Research*, **91**, 7205–7223.

GARUTI, G., RIVALENTI, G., ROSSI, G., SIENA, F. & SINIGOI, S. 1980. The Ivrea-Verbano mafic-ultramafic complex of the Italian western Alps – Discussion of some petrologic problems and a summary. *Rendiconti Società Italiana di Mineralogia e Petrologia*, **36**, 717–749.

GEIST, E.L., CHILDS, J.R. & SCHOLL, D. 1988. The origin of summit basins of the Aleutian Ridge Implications for block rotation of an arc massif. *Tectonics*, **7**, 327–341.

GRIFFIN, W.L. & O'REILLY, S.Y. 1987a. Is the continental Moho the crust-mantle boundary? *Geology*, **15**, 241–244.

—— & —— 1987b. The composition of the lower crust and the nature of the continental Moho–xenolith evidence. *In*: NIXON, P.H. (ed.) *Mantle xenoliths*. John Wiley & Sons, Chichester, 413–431.

HALE, L.D. & THOMPSON, G.A. 1982. The seismic reflection character of the continental Mohorovičić discontinuity. *Journal of Geophysical Research*, **87**, 4625–4635.

HAMILTON, W.B. 1979. Tectonics of the Indonesian region. *U.S. Geological Survey Professional Paper*, **1078**, 345 p.

—— 1988. Plate tectonics and island arcs. *Geological Society of America Bulletin*, **100**, 1503–1527.

—— 1989a. Crustal geologic processes of the United States. *Geological Society of America Memoir*, **172**, 743–781.

—— 1989b. Convergent-plate tectonics viewed from the Indonesian region. *In*: SENGÖR, A.M.C. (ed.) *Tectonic evolution of the Tethyan region*. Kluwer Academic Publishers, 655–698.

—— 1990. On terrane analysis. *Philosophical Transactions of the Royal Society, London*, **A–331**, 511–522.

—— 1993. Evolution of Archean mantle and crust. *In*: REED, J.C., JR. *et al.* (eds) *The geology of North America, v. C–2, Precambrian of the conterminous United States*. Geological Society of America, 597–614, 630–636.

HANDY, M.R. & ZINGG, A. 1991. The tectonic and rheological evolution of an attenuated cross section of the continental crust – Ivrea crustal section, southern Alps, northwestern Italy and southern Switzerland. *Geological Society of America Bulletin*, **103**, 236–253.

HANSEN, B. 1981. The transition from pyroxene granulite facies to garnet clinopyroxene granulite facies – Experiments in the system CaO-MgO-Al$_2$O$_3$-SiO$_2$. *Contributions to Mineralogy and Petrology*, **76**, 234–242.

HARBERT, W., SCHOLL, D.W., VALLIER, T.L., STEVENSON, A.J. & MANN, D.M. 1986. Major evolutionary phases of a forearc basin of the Aleutian terrace – Relation to North Pacific tectonic events and the formation of the Aleutian subduction complex. *Geology*, **14**, 757–761.

HAUSER, E.C. & LUNDY, J. 1989. COCORP deep reflections – Moho at 50 km (16 s) beneath the Colorado Plateau. *Journal of Geophysical Research*, **94**, 4071–7081.

HAWKESWORTH, C.J., HERGT, J.M., McDERMOTT, F. & ELLAM, R.M. 1991. Destructive margin magmatism and the contributions from the mantle wedge and subducted crust. *Australian Journal of Earth Sciences*, **38**, 577–594.

HAWKINS, J.W. & EVANS, C.A. 1983. Geology of the Zambales Range, Luzon, Philippine Islands – Ophiolite derived from an island arc-back arc basin pair. *American Geophysical Union Geophysical Monograph*, **27**, 95-123.

HAWKINS, J.W., BLOOMER, S.H., EVANS, C.A. & MELCHIOR, J.T. 1984. Evolution of intra-oceanic arc-trench systems. *Tectonophysics*, **102**, 174–205.

HAYES, D.E. 1974. Continental margin of western South America. *In*: BURK, C.A. & DRAKE, C.L. (eds) *The geology of continental margins*. New York, Springer-Verlag, 581–598.

HEIRTZLER, J.R. 1991. Magnetic anomalies and early plate tectonics at Lamont-Doherty Geological Observatory. *Tectonophysics*, **187**, 27–36.

HILDRETH, W. & MOORBATH, S. 1988. Crustal contributions to arc magmatism in the Andes of central Chile. *Contributions to Mineralogy and Petrology*, **98**, 455–489.

HODGES, K.V. & FOUNTAIN, D.M. 1984. Pogallo Line, South Alps, northern Italy – an intermediate crustal level, low-angle normal fault? *Geology*, **12**, 151–155.

HOLBROOK, W.S. 1990. The crustal structure of the northwestern Basin and Range province, Nevada, from wide-angle seismic data. *Journal of Geophysical Research*, **95** 21843–21869.

HOLDAWAY, M.J. 1971. Stability of andalusite and the aluminum silicate phase diagram. *American Journal of Science*, **271**, 97–131.

HUANG, W.L. & WYLLIE, P.J. 1981. Phase relationships of S-type granite with H$_2$O to 35 kbar – muscovite granite from Harney Peak, South Dakota. *Journal of Geophysical Research*, **86**, 10515–10529.

HYNDMAN, D.W. & FOSTER, D.A. 1988. The role of tonalites and mafic dikes in the generation of the Idaho batholith. *Journal of Geology*, **96**, 31–46.

JAN, M.Q. 1988. Geochemistry of amphibolites from

the southern part of the Kohistan arc, N. Pakistan. *Mineralogical Magazine*, **52**, 147–159.

—— & Asif, M. 1983. Geochemistry of tonalites and (quartz) diorites of the Kohistan-Ladakh (Transhimalayan) granitic belt in Swat, NW Pakistan. *In*: Sham, F.A. (ed.) *Granites of Himalayas, Karakorum and Hindukush*. Punjab University, 355–376.

—— & Howie, R.A. 1981. The mineralogy and geochemistry of the metamorphosed basic and ultrabasic rocks of the Jijal complex, Kohistan, NW Pakistan. *Journal of Petrology*, **22**, 85–126.

——, Parvez, M.K. & Khattak, M.U.K. 1984. Coronites from the Chilas and Jijal-Patan complexes of Kohistan. *University of Peshawar Geological Bulletin*, **17**, 75–85.

Johnson, C.A., Bohlen, S.R. & Essene, E.J. 1983. An evaluation of garnet-clinopyroxene geothermometry in granulites. *Contributions to Mineralogy and Petrology*, **84**, 191–198.

Johnston, A.D. & Wyllie, P.J. 1988. Interaction of granitic and basic magmas – experimental observations on contamination processes at 10 kbar with H$_2$O. *Contributions to Mineralogy and Petrology*, **98**, 352–362.

Jurdy, D.M. & Stefanick, M. 1991. The forces driving the plates – constraints from kinematics and stress observations. *Philosophical Transactions of the Royal Society, London*, **A–337**, 127–139.

Karig, D.E. 1972. Remnant arcs. *Geological Society of America Bulletin*, **83**, 1057–1068.

—— 1975. Basin genesis in the Philippine Sea. *Initial Reports of the Deep Sea Drilling Project*, **31**, 857–879.

—— 1982. Initiation of subduction zones – Implications for arc evolution and ophiolite development. *Geological Society, London, Special Publication*, **10**, 563–576.

Kelemen, P.B. 1986. Assimilation of ultramafic rock in subduction-related magmatic arcs. *Journal of Geology*, **94**, 829–843.

Khan, M.A., Jan, M.Q., Windley, B.F., Tarney, J. & Thirlwall, M.F. 1989. The Chilas mafic-ultramafic igneous complex; The root of the Kohistan island arc in the Himalaya of northern Pakistan. *Geological Society of America Special Paper*, **232**, 75–94.

Kincaid, C. & Olson, P. 1987. An experimental study of subduction and slab migration. *Journal of Geophysical Research*, **92**, 13,832–13,840.

Kirby, S.H., Durham, W.B. & Stern, L.A. 1991. Mantle phase changes and deep-earthquake faulting in subducted lithosphere. *Science*, **252**, 216-225.

Kono, M., Fukao, Y. & Yamamoto, A. 1989. Mountain building in the central Andes. *Journal of Geophysical Research*, **94**, 3891–3905.

Langmuir, C.H. 1989. Geochemical consequences of *in-situ* crystallization. *Nature*, **340**, 199–205.

Leeman, W.P., Smith, D.R., Hildreth, W., Palacz, Z. & Rogers, N. 1990. Compositional diversity of late Cenozoic basalts in a transect across the southern Washington Cascades – Implications for

subduction zone magmatism. *Journal of Geophysical Research*, **95**, 19561–19582.

Lewis, S.D. & Hayes, D.E. 1984. A geophysical study of the Manila Trench, Luzon, Philippines. 2. Fore arc basin structural and stratigraphic evolution. *Journal of Geophysical Research*, **89**, 9196–9214.

Loucks, R.R. 1990. Discrimination of ophiolitic from nonophiolitic ultramafic-mafic allochthons in orogenic belts by the Al/Ti ratio in clinopyroxene. *Geology*, **18**, 346–349.

——, Ashraf, M., Awan, A.M. & Khan, S. 1990. The Jijal Complex – Layered mafic-ultramafic arc cumulates from the crust-mantle boundary, Pakistani Himalayas. *EOS*, **71**, 664.

McBirney, A.R., Taylor, H.P. & Armstrong, R.L. 1987. Paricutin re-examined – a classic example of crustal assimilation in calc-alkaline magma. *Contributions to Mineralogy and Petrology*, **95**, 4–20.

McCaffrey, R. 1982. Lithospheric deformation within the Molucca Sea arc-arc collision – evidence from shallow and intermediate earthquake activity. *Journal of Geophysical Research*, **87**, 3663–3678.

—— 1992. Oblique plate convergence, slip vectors, and forearc deformation. *Journal of Geophysical Research*, **97**, 8905–8915.

Maaloe, S. & Wyllie, P.J. 1975. Water content of a granite magma deduced from the sequence of crystallization determined experimentally with water-undersaturated conditions. *Contributions to Mineralogy and Petrology*, **52**, 175–191.

Malinverno, A. & Ryan, W.B.F. 1986. Extension in the Tyrrhenian Sea and shortening in the Apennines as result of arc migration driven by sinking of the lithosphere. *Tectonics*, **5**, 227–245.

Miller, D.M., Langmuir, C.H., Goldstein, S. & Franks, A.L. 1992. The importance of parental magma composition to calc-alkaline and tholeiitic evolution – Evidence from Umnak Island in the Aleutians. *Journal of Geophysical Research*, **97**, 321–343.

Moberley, R., Shepherd, G.L. & Coulbourn, W.T. 1982. Forearc and other basins, continental margin of northern and southern Peru and adjacent Ecuador and Chile. *Geological Society, London, Special Publication*, **10**, 171–189.

Molnar, P. & Atwater, T. 1978. Interarc spreading and Cordilleran tectonics as alternates related to the age of subducted oceanic lithosphere. *Earth and Planetary Science Letters*, **41**, 330–340.

Mooney, W.D. & Brocher, T.M. 1987. Coincident seismic reflection/refraction studies of the continental lithosphere – A global review. *Reviews of Geophysics*, **25**, 723–742.

Nasu, N. & 5 others. 1980. Interpretation of multichannel seismic reflection data, legs 56 and 57, Japan Trench Transect, Deep Sea Drilling Project. *Initial Reports of the Deep Sea Drilling Project*, **56–57**, 489–503.

Newton, R.C. & Perkins, D., III. 1982. Thermodynamic calibration of geobarometers based on the assemblages garnet–plagioclase–orthopyroxene (clinopyroxene)–quartz. *American Mineralogist*, **67**, 203–222.

O'Hara, M.J. & Mathews, R.E. 1981. Geochemical

evolution in an advancing, periodically replenished, periodically tapped, continuously fractionated magma chamber. *Journal of the Geological Society, London*, **138**, 237–277.

PAWLEY, A.R. & HOLLOWAY, J.R. 1993. Water sources for subduction zone volcanism – New experimental constraints. *Science*, **260**, 664–667.

PEACOCK, S.M. 1990. Numerical simulation of metamorphic pressure-temperature-time paths and fluid production in subducting slabs. *Tectonics*, **9**, 1197–1211.

PETTERSON, M.G. & WINDLEY, B.F. 1985. Rb-Sr dating of the Kohistan arc-batholith in the Trans-Himalaya of north Pakistan, and tectonic implications. *Earth and Planetary Science Letters*, **74**, 45–57.

PIN, C. & SILLS, J.D. 1986. Petrogenesis of layered gabbros and ultramafic rocks from Val Sesia, the Ivrea Zone, NW Italy – Trace element and isotope geochemistry. *Geological Society, London, Special Publication*, **24**, 231–249.

PLAFKER, G., NOKLEBERG, W.J. & LULL, J.S. 1989. Bedrock geology and tectonic evolution of the Wrangellia, Peninsular, and Chugach Terranes along the Trans-Alaska Crustal Transect in the Chugach Mountains and southern Copper River Basin, Alaska. *Journal of Geophysical Research*, **94**, 4255–4295.

PLANK, T. & LANGMUIR, C.H. 1993. Tracing trace elements from sediment input to volcanic output at subduction zones. *Nature*, **362**, 739–743.

QUICK, J.E. & SINIGOI, S. (eds) 1992. *Ivrea-Verbano Zone Workshop, 1992*. U.S. Geological Survey Circular 1089, 30 pp.

——, ——, NEGRINI, L., DEMARCHI, G. & MAYER, A. 1992. Synmagmatic deformation of the underplated igneous complex of the Ivrea-Verbano zone. *Geology*, **20**, 613–616.

REID, J.B., JR. & HAMILTON, M.A. 1987. Origin of Sierra Nevadan granite – evidence from small scale composite dikes. *Contributions to Mineralogy and Petrology*, **96**, 441–454.

RUDNICK, R.L. 1992. Xenoliths: Samples of the lower continental crust. *In*: FOUNTAIN, D.M., ARCULUS, R.J. & KAY, R.W. (eds) *The continental lower crust*. Elsevier, Amsterdam, 269–316.

RUTTER, E.H., BRODIE, K.H. & EVANS, P.J. 1993. Structural geometry, lower crustal magmatic underplating and lithospheric stretching in the Ivrea-Verbano zone, northern Italy. *Journal of Structural Geology*, **15**, 647–662.

SATO, H., SACKS, I.S. & MURASE, T. 1989. The use of laboratory velocity data for estimating temperature and partial melt fraction in the low-velocity zone – Comparison with heat flow and electrical conductivity studies. *Journal of Geophysical Research*, **94**, 5689–5704.

SCHMID, R. 1967. Zur Petrographie und Struktur der Zone Ivrea-Verbano zwischen Valle d'Ossola und Grande (prov. Novara, Italien). *Schweizerische mineralogische und petrographische Mitteilungen*, **47**, 935-1117.

—— 1979. Are the metapelites of the Ivrea-Verbano zone restites? *Università di Padova Memorie di Scienze Geologiche*, **33**, 67–69.

SCHWARTZ, D.P. 1988. Paleoseismicity and neotectonics of the Cordillera Blanca fault zone, northern Peruvian Andes. *Journal of Geophysical Research*, **93**, 4712–4730.

SÉBIER, M. *et al.* 1988. The state of stress in an overriding plate situated above a flat slab – The Andes of central Peru. *Tectonics*, **7**, 895–928.

SEELY, D.R. 1979. The evolution of structural highs bordering major forearc basins. *American Association of Petroleum Geologists Memoirs*, **29**, 245–260.

SENGÖR, A.M.C. & DEWEY, J.F. 1990. Terranology – vice or virtue? *Philosophical Transactions of the Royal Society, London*, **A–331**, 456–477.

SHAH, M.T. & SHERVAIS, J.W. 1991. Petro-chemical evolution of the Dir metavolcanic sequence, Kohistan island arc terrane, northern Pakistan. *Geological Society of America Abstracts with Programs*, **23**, A–391.

SHEFFELS, B.M. 1990. Lower bound on the amount of crustal shortening in the central Bolivian Andes. *Geology*, **18**, 812–815.

SHERVAIS, J. 1979*a*. Ultramafic and mafic layers in the Alpine-type lherzolite massif at Balmuccia, N.W. Italy. *Università di Padova Memorie di Scienze Geologiche*, **33**, 135–145.

—— 1979*b*. Thermal emplacement model for the Alpine lherzolite massif at Balmuccia, Italy. *Journal of Petrology*, **20**, 795–820.

—— & KIMBROUGH, D.L. 1985. Geochemical evidence for the tectonic setting of the Coast Range ophiolite – A composite island arc-oceanic crust terrane in western California. *Geology*, **13**, 35–38.

SILVER, E.A. *et al.* 1991. Collision propagation in Papua New Guinea and the Solomon Sea. *Tectonics*, **10**, 863–874.

SIMKIN, T., TILLING, R.I., TAGGART, J.N., JONES, W.J. & SPALL, H. 1989. *This dynamic planet – World map of volcanoes, earthquakes, and plate tectonics*. U.S. Geological Survey, Denver.

SKJERLIE, K.P. & JOHNSTON, A.D. 1992. Vapor-absent melting at 10 kbar of a biotite- and amphibole-bearing tonalitic gneiss – Implications for the generation of A-type granites. *Geology*, **20**, 263–266.

SPENCE, W. 1987. Slab pull and the seismotectonics of subducting lithosphere. *Reviews of Geophysics*, **25**, 55–69.

STERN, R.J. & BLOOMER, S.H. 1992. Subduction-zone infancy – Examples from the Eocene Izu–Bonin–Mariana and Jurassic California arcs. *Geological Society of America Bulletin*, **104**, 1621–1636.

TABER, J.J., BILLINGTON, S. & ENGDAHL, E.R. 1991. Seismicity of the Aleutian Arc. *The Geology of North America, Decade Map Volume 1*, Geological Society of America, 29–46.

TAO, W.C. & O'CONNELL, R.J. 1993. Deformation of a weak subducted slab and variation of seismicity with depth. *Nature*, **361**, 626–628.

TAYLOR, B. & KARNER, G.D. 1983. On the evolution of marginal basins. *Reviews of Geophysics and Space Physics*, **21**, 1727–1741.

—— *et al.* 1991. Structural development of

Sumisu Rift, Izu-Bonin Arc. *Journal of Geophysical Research*, **96**, 16113–16129.

TICHELAAR, B.W. & RUFF, L.J. 1993. Depth of seismic coupling along subduction zones. *Journal of Geophysical Research*, **98**, 2017–2037.

TRACY, R.J. 1978. High grade metamorphic reactions and partial melting in pelitic schist, west-central Massachusetts. *American Journal of Science,* **278**, 150–178.

TRELOAR, P.J. *et al*. 1989. K-Ar and Ar-Ar geochronology of the Himalayan collision in NW Pakistan – Constraints on the timing of suturing, deformation, metamorphism and uplift. *Tectonics*, **8**, 881–909.

USSLER, W., III & GLAZNER, A.F. 1989. Phase equilibria along a basalt-rhyolite mixing line – implications for the origin of calc-alkaline intermediate magmas. *Contributions to Mineralogy and Petrology*, **101**, 232–244.

UYEDA, S. & KANAMORI, H. 1979. Back-arc opening and the mode of subduction. *Journal of Geophysical Research*, **84**, 1049–1061.

VIELZEUF, D. & HOLLOWAY, J.R. 1988. Experimental determination of the fluid-absent melting relations in the pelitic system. *Contributions to Mineralogy and Petrology*, **98**, 257–276.

VOSHAGE, H. *et al*. 1988. Isotopic constraints on the origin of ultramafic and mafic dikes in the Balmuccia peridotite (Ivrea Zone). *Contributions to Mineralogy and Petrology*, **100**, 261–267.

—— *et al*. 1990. Isotopic evidence from the Ivrea Zone for a hybrid lower crust formed by magmatic underplating. *Nature*, **347**, 731–736.

WALLACE, S.R., PLATT, J.P. & KNOTT, S.D. 1993. Recognition of syn-convergence extension in accretionary wedges with examples from the Calabrian Arc and the Eastern Alps. *American Journal of Science*, **293**, 463–495.

WÖRNER, G., MOORBATH, S. & HARMON, R.S. 1992. Andean Cenozoic volcanic centers reflect basement isotopic domains. *Geology*, **20**, 1103–1106.

WYLLIE, P.J. 1984. Constraints imposed by experimental petrology on possible and impossible magma sources and products. *Philosophical Transactions of the Royal Society, London*, **A–310**, 439–456.

ZHAO, D. & HASEGAWA, A. 1993. P wave tomographic imaging of the crust and upper mantle beneath the Japan islands. *Journal of Geophysical Research*, **98**, 4333–4353.

——, CHRISTENSEN, D. & PULPAN, H. In press. Tomographic imaging of the Alaska subduction zone. *Journal of Geophysical Research*, **98**.

——, HASEGAWA, A. & HORIUCHI, S. 1992. Tomographic imaging of P and S wave velocity structure beneath northeastern Japan. *Journal of Geophysical Research*, **97**, 19909–19928.

Volcanism and sedimentation in a rifting island-arc terrain: an example from Tonga, SW Pacific

PETER D. CLIFT[1,2] & ODP Leg 135 Scientific Party*

[1] *Department of Geology and Geophysics, Grant Institute, the University of Edinburgh, West Mains Road, Edinburgh, EH9 3JW, Scotland.*

[2] *Present Address, Ocean Drilling Program, Texas A&M University, College Station, Texas 77845–9547, USA.*

Abstract: Scientific drilling of narrow sub-basins within the Lau back-arc basin system of the SW Pacific has recovered uppermost Miocene to Recent volcaniclastic sediment and pelagic nannofossil oozes. Pliocene sediment gravity flow and turbidite sands from the western Lau Basin indicate a local source for the sediment, probably intrabasinal seamount volcanoes active during the initial stages of arc rifting. Derivation of abundant material from either the remnant volcanic arc (Lau Ridge) or the new Tofua arc is ruled out by the rugged topography of the basin and the proximal nature of the facies. Sediments from the Tonga platform, adjacent to the present day Tofua arc, indicate a peak in volcanic activity prior to and during the generation of the first back-arc basin crust at 5.25 Ma. A 2.0 Ma hiatus in arc volcanism on the trench side of the basin after rifting was brought to an end by the foundation of the Tofua arc at 3.0 Ma (Late Pliocene). On the basis of the sedimentary, geochemical and seismic data it is suggested that basin rifting involved an initial stage of extension of the original island arc, accompanied by volcanism in the form of major seamount volcanoes within the basin. These produced volcanic ash by submarine eruption, which was then reworked into adjacent sub-basins by slumping, gravity flow or turbidity current. Basin opening proceeded with a trenchward migration of extension and volcanism with time. This system was disrupted by the southward propagation of the Eastern Lau Spreading Centre into the southern Lau Basin at 1.5–1.0 Ma. This resulted in extension and volcanism being concentrated along the median valley of the spreading centre and a cessation in explosive volcanism of wide compositional range. Sedimentation in the Lau Basin since that time has been principally pelagic with minor ash layers mostly derived from the Tofua arc.

Changes in the nature of arc volcanism during the rifting of oceanic volcanic arcs has been a matter of some controversy for several years (e.g. Karig 1970; Hawkins *et al.* 1984). In particular, a determination of the volume, chemistry and location of active volcanism during the rifting of arcs to form back-arc basins is fundamental to an understanding of the tectonic evolution of these complicated plate boundary regions. In this paper the authors present a variety of geological data to show that Late Miocene rifting of the Tonga arc in the southwest Pacific was characterized by a two-stage history of arc extension and intrabasinal seamount volcanism, followed by the initiation of island arc volcanism shortly before the arrival of a southward-propagating oceanic spreading ridge into the southern Lau Basin during the Early Pleistocene. The sediments which form the basis of this study were recovered from a transect of six sites in the Lau Basin and one on the Tonga platform (Fig. 1) drilled during Leg 135 of the Ocean Drilling Program between December 1990 and February 1991 (Parson *et al.* 1992).

The Lau Basin lies 500 km north of New Zealand and has for some twenty years been recognized as a classic marginal or back-arc basin and oceanic island arc, the Tofua arc (e.g. Karig 1970). Present subduction of the Pacific plate is towards the WNW, perpendicular to the Tonga trench, which forms the plate boundary between the Pacific and Indian–Australian plates. Subduction along this margin of the

* Ocean Drilling Program, Leg 135 Scientific Party: L.M. Parson, J.W. Hawkins, J.F. Allan, N. Abra-hamsen, U. Bednarz, G. Blanc, S.H. Bloomer, R. Bøel, T.R. Bruns, W.B. Bryan, G.C.H. Chaproniere, P.D. Clift, A. Ewart, M.G. Fowler, J.M. Hergt, R.A. Hodkinson, D. Lavoie, J.K. Ledbetter, C.J. MacLeod, K. Nilsson, H. Nishi, C.E. Pratt, P.J. Quinterno, R.R. Reynolds, R.G. Rothwell, W.W. Sager, D. Schops, S. Soakai & M. Styzen.

From Smellie, J.L. (ed.), 1995, *Volcanism Associated with extension at Consuming Plate Margins*, Geological Society Special Publication No. 81, 29–51.

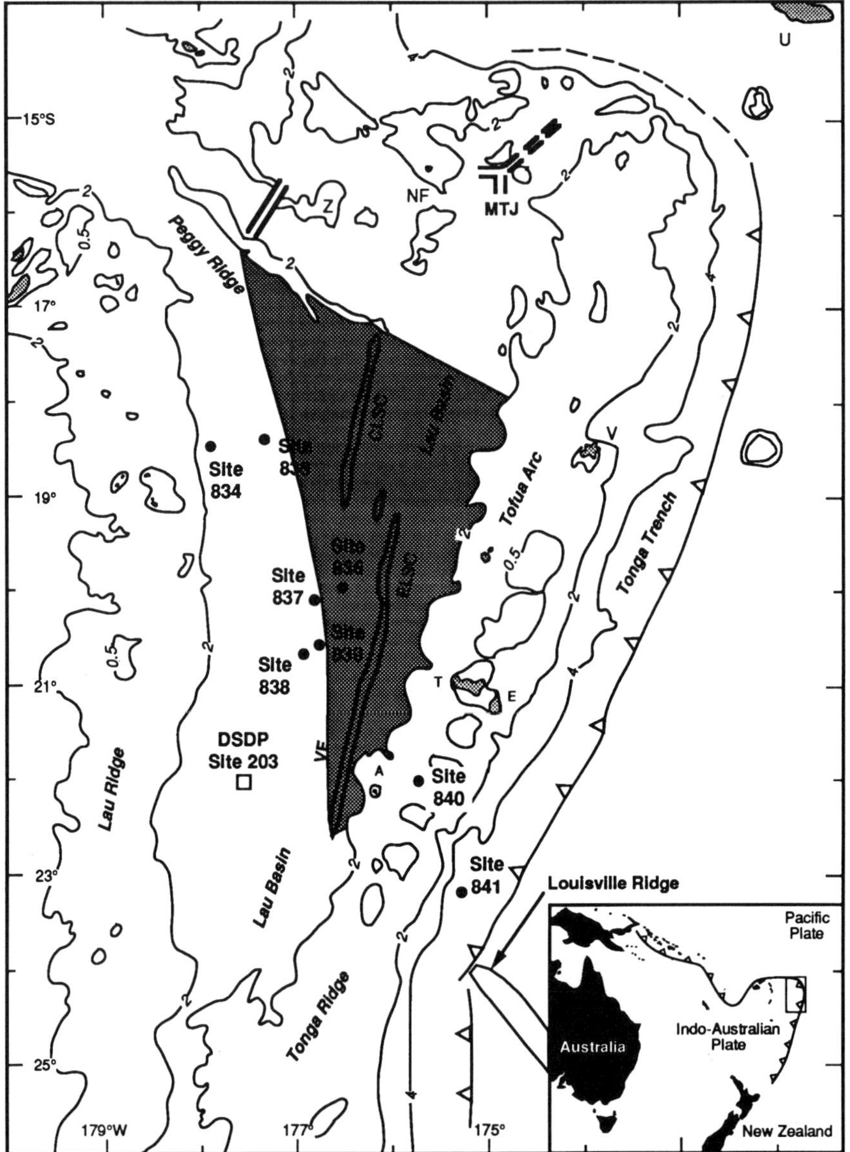

Fig. 1. Location map showing the Lau Basin in the SW Pacific and the location of drill sites 834–840 in the basin and adjoining Tonga platform. Bathymetry in kilometres. Line with triangles denotes plate boundary, with triangles on the overriding plate. Z, Zephyr Shoal; MTJ, Mangatolu Triple Junction; NF, Niua Fo'ou; CLSC, Central Lau Spreading Centre; ELSC, Eastern Lau Spreading Centre; VF, Valu Fa ridge; U, Upola; V, Vava'u; T, Tongatapu; E, 'Eua; A, Ata. Dark shaded area represents crust formed by the Central and Eastern Lau Spreading Centres as they propagated south.

Indian–Australian plate has continued since at least the middle Eocene (*c.* 46.0 Ma; e.g. Packham 1978), with the Lau Basin forming the most recent of two back-arc basins created during this time period, as a result of arc rifting. The earlier basin is the Oligocene-aged South Fiji Basin (e.g. Weissel & Watts 1975). The history of the uppermost Miocene-Recent Lau Basin, as recorded by its sedimentary fill, provides a classic example of the tectonic and magmatic response of an island arc to processes of extension at a plate boundary in an intra-oceanic setting.

Tectonic setting

The Lau Basin

Tectonically the Lau Basin may be subdivided into two regions on the basis of topographical and geophysical data (Figs 1 & 2; Parson *et al.* 1992). The western, older part has a disorganized topography caused by extensional faulting and localized magmatism both within basins and on seamount volcanoes. In contrast, the eastern part of the basin has a smoother topography and magnetic and gravity responses typical of crust produced at an oceanic spreading ridge (Parson *et al.* 1992). The Lau Basin is believed to have formed by an initial period of regional extension of pre-existing arc crust. As extension proceeded, an oceanic spreading centre (Eastern Lau Spreading Centre) formed at the southern termination of the Peggy Ridge, a leaky transform fault (Fig. 1), and began to propagate south. Later a second centre, the Central Lau Spreading Centre, formed at the Peggy Ridge and also began to propagate south. The southward propagation of the Central Lau Spreading Centre is accommodated by the retreat of the northern termination of the Eastern Lau Spreading Centre (Parson *et al.* 1990). The Lau Basin is thus composed of two distinct types of crust: that generated at one of the two backarc spreading centres, and that formed by extension of the original Miocene arc crust. The drill sites, which all occupy small sub-basins, thus fall within one of these two major tectonic groupings (Fig. 2). The type of crust upon which a basin is founded can be seen to have a profound effect on the nature of the sedimentary fill at each site. Sites 834, 838 and 839 fall clearly within the western part of the basin formed of extended arc crust, while Sites 836 and 837 lie on crust generated at the Eastern Lau Spreading Centre. Site 835 lies near the transition between the two crustal types, but seems to have greatest similarity with the extended arc crust.

In all cases the sub-basins drilled have a number of important characteristics in common. They are all of relatively small size (2–5 km diameter), and generally do not exceed 10–15 km in length before being offset by transverse faults. On either side of each basin, topographical highs rise up to 1.0 km above the basin

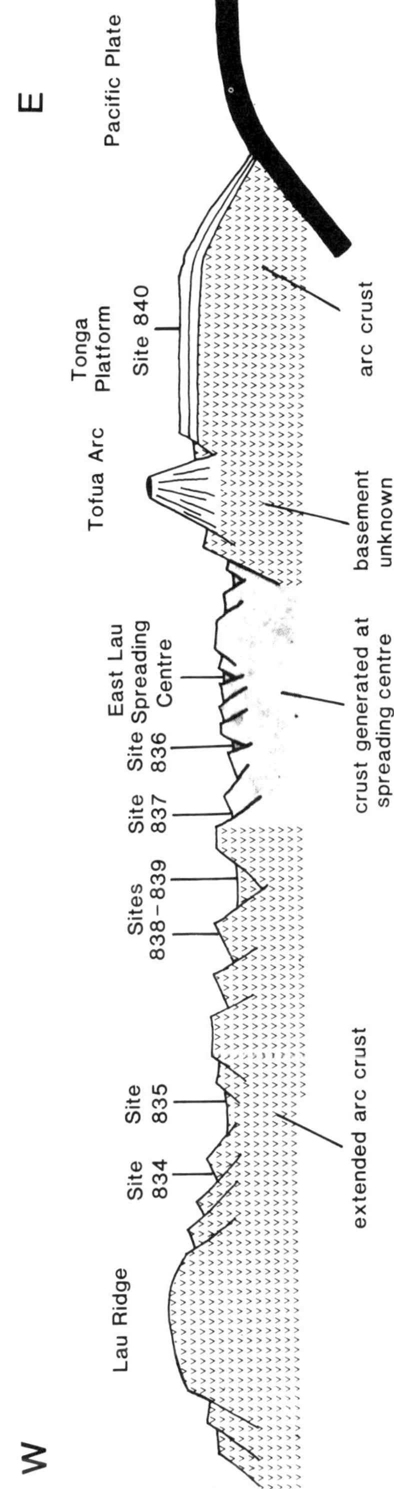

Fig. 2. Schematic cross section of the Lau Basin and Tonga arc showing the location of the ODP drill sites (Sites 834–840). Note the distinction between crust generated along the Eastern Lau Spreading Centre and that formed by extension of pre-existing arc crust. Not to scale.

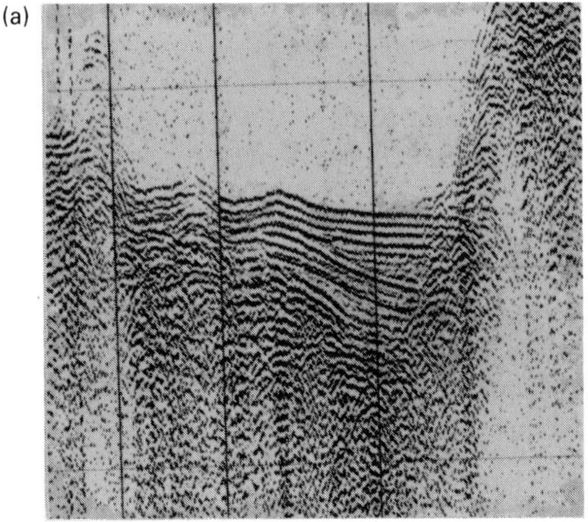

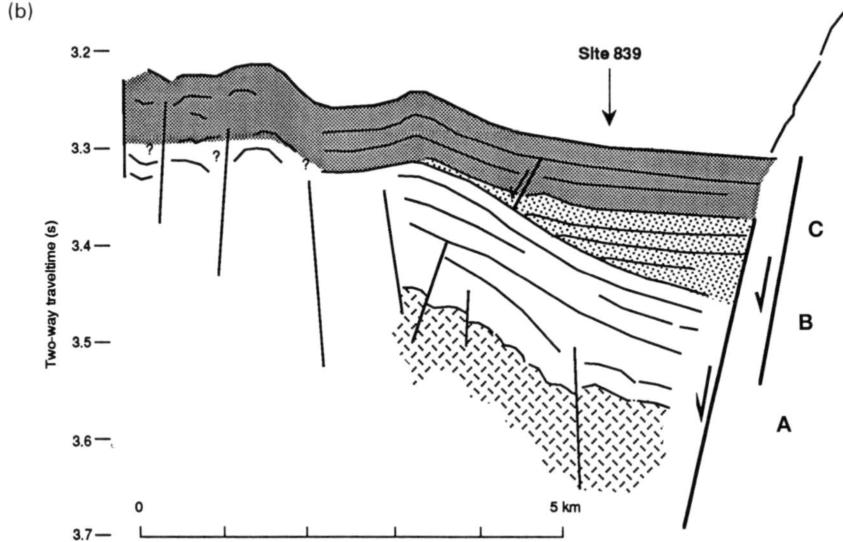

Fig. 3. (a) Seismic section of Site 839 showing the partial sediment fill, surrounding basement highs and strong tectonic control. (b) Interpretation of (a), showing the large normal fault bounding the right (east) side of the basin. Intrabasinal reflectors show the basement/cover contact (A) and two intrabasinal unconformities caused by rotation of the basin fill by faulting (B and C).

floor. These are commonly normal fault-bounded structures, composed of rifted Miocene arc basement, although post-rift volcanic edifices are also recognized (Hawkins 1989; Parson *et al.* 1990). Seismic profiles shot across the sub-basins usually show a passive infilling of sediment, suggestive of a rapid rifting event. However in one case, Site 839 (Fig. 3), an intrabasinal unconformity was recognized. Tilting and erosion of the earlier sediments deposited in the basin was interpreted from the seismic data (Parson *et al.* 1992) and testify to the continuation of extension and seismic activity in rifted parts of the Lau Basin after the initial stretching event. Dating of sediment above and below the unconformity at Site 839 shows the same age, i.e. uppermost Pliocene (nannofossil zone CN 13b). This indicates not only rapid sedimentation, but also that the tilting and erosion of the lower sedimentary sequence took place in a short time span. This demonstrates that rates of extension and sedimentation at Site 839 were very high during the initial basin formation. The concentration of seismic activity and extension on the modern Eastern and Central Lau Spreading Centres highlights the change in tectonic style from the regime in which Site 839 was created. Both spreading centres are believed to operate in a similar way to spreading centres in large ocean basins, with asthenospheric upwelling below the ridge crest resulting in crustal accretion by eruption from and crystallization within axial magma chambers (Collier & Sinha 1990). $\beta = \infty$ within the 5–10 km wide rift valley at the crest of the spreading ridge, with half spreading rates being calculated at 100 mma^{-1} (Parson & Hawkins 1994). Furthermore, the Central Lau Spreading Centre is propagating south at an average rate of 110–120 mma^{-1} (Parson & Hawkins 1994), making it one of the fastest-propagating ridges known from modern oceans. In effect, as the Eastern Lau Spreading Centre has migrated south through the Lau Basin, there has been a change-over from arc rifting and basinwide extension to more narrowly focused oceanic spreading. This observation has important implications for the interpretation of redeposited sediments described below.

The Tonga Platform

At Site 840, lying on the Tonga Platform close to the active Tofua arc, sediment has accumulated in a relatively simple tectonic setting since at least the Late Miocene. The Tonga Platform is probably founded on rifted arc crust and has experienced continuous extensional deformation since its formation in the Middle Eocene (Packham 1978). This has resulted in normal faulting and subsidence. Previous workers have suggested that the Tonga Platform has experienced several periods of uplift coinciding with the rifting of the Lau Basin (e.g. Herzer & Exon 1985) and the subduction of the Louisville seamount chain (Tappin et al. 1993). Ledbetter & Haggerty (1994) cite palaeo-water depth indicators which suggest as much as 2500 m of uplift at Site 840 prior to Lau Basin rifting, based on the corroded state of preservation of the microfauna and changes in the size of the ichnofauna. However, Clift (1994) came to a different conclusion when he used seismic and palaeontological data from the Tonga Platform

(Austin et al. 1989) to limit the amount of pre-rift uplift to <300 m. In effect, corrosion of some microfauna in the Miocene section at Site 840 is believed to indicate deposition within the lysocline by Ledbetter & Haggerty (1994), but to be the result of diagenetic dissolution by Clift (1994).

In spite of the controversy concerning the uplift record, Site 840 can still be used as a useful reference point to monitor the volcanic history of the modern Tofua arc. Biostratigraphical dating of the core from Site 840 indicates no long period hiatus in sedimentation since 7.0 Ma. In addition, the position of the Tonga platform adjacent to the modern Tofua arc places it in an ideal position to collect volcanic debris being shed from the volcanic arc at any time since its formation. In a similar fashion, prior to rifting, the Tonga platform must have lain adjacent to the Eocene-Miocene Tonga arc. Since the Tonga platform is in a remote intra-oceanic setting it is difficult to see how any other source except the volcanic arc can have contributed significantly to the volcaniclastic sedimentation there. This simplifies the arguments regarding sediment provenance and implies that the sedimentary section might be able to shed some light on the age of the Tofua arc. While previous authors (e.g. Hawkins et al. 1984) have agreed that there was a time gap between rifting of the Lau Basin and the start of volcanism along the Tofua arc, there is no consensus just how long this period was. Hawkins et al. (1984) considered the present Tofua arc to have been active for only the last 1.0 Ma, compared to an age of basin rifting at 5.25 Ma (Parson et al. 1992). Periods without volcanism in other western Pacific arc/back-arc basin systems vary considerably. A gap of 3 Ma is recognised in the Bonin arc during rifting of the Shikoku Basin (Taylor 1990). In the Mariana arc, controversy exists as to the length of the volcanic gap. Scott et al. (1981) proposed a gap of 2–3 Ma during spreading of the Mariana Trough, and up to 9 Ma during rifting of the earlier Parece Vela Basin. In contrast Hussong & Uyeda (1982) suggested a simple subsidence of the arc and no actual cessation of activity during formation of the Mariana Trough.

Sedimentology

The Lau Basin: Sites 834–839

The sediments that infill the sub-basins of the Lau Basin have a number of characteristics in common (Fig. 4). The sections recovered at Sites 834 to 839 all have an overall fining-upward trend, with the basal sediments comprising

                                    P.D. CLIFT *ET AL.*

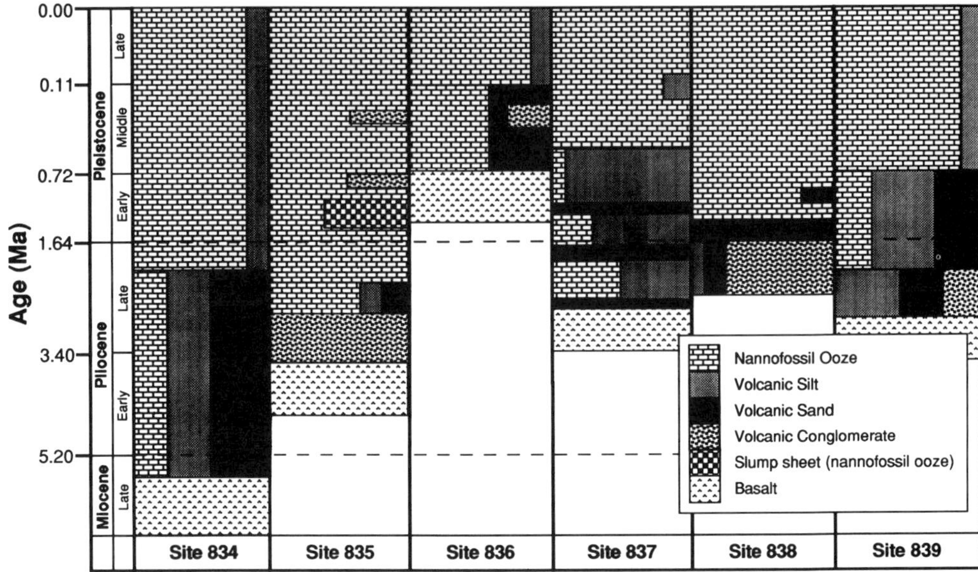

Fig. 4. Summary of dominant sedimentary facies deposition with time across the Lau Basin. Columns are divided vertically in proportion to the sediment types contained over a given interval.

variable proportions of massive volcaniclastic conglomerates and interbedded coarse sands. These fine up into a series of massive and graded, thick-bedded sands and silts. The top of each sedimentary sequence is topped by dark brown nannofossil oozes, with minor volumes of thin, graded, airfall ash deposits. This trend may be seen in the published detailed sedimentary logs (Parson *et al.* 1992), and in logs interpreted from down-hole resistivity images taken with the formation microscanner (FMS; Clift 1994). Figure 4 shows a simplified version of these detailed sedimentary columns, with a schematic representation of each section subdivided vertically to show the relative proportions of the principal lithologies over a given time interval.

Despite this simplification, it is possible to see that the time at which volcaniclastic sedimentation was replaced by pelagic-dominated sedimentation within each sub-basin is not synchronous. The dates used for this observation are based on a combination of bio- and chronostratigraphical data (Parson *et al.* 1992). The diachronous nature of the sedimentation is important in supporting the suggestion from the topographical data that the sediment for each basin is locally derived. If a single source, such as the Lau Ridge, was responsible for producing all the volcanic debris redeposited in the Lau Basin then the end of volcanism and thus sediment production would presumably result in the end

of volcaniclastic sedimentation over a short period of time basin-wide, a fact not observed. Even if the rough topography was responsible for a delay in the cessation of volcaniclastic sedimentation at individual sites, it might be expected that sites far from the Lau Ridge would show an earlier decline in volcaniclastic sedimentation compared to those positioned closer to the source. Figure 4 clearly shows the opposite trend, with volcaniclastic sedimentation finishing earliest in the oldest parts of the basin, adjacent to the Lau Ridge, and most recently in the centre of the basin.

Conglomerate and gravel. Volcaniclastic conglomerates generally form only a small part of each section by volume and are found towards the base of the section at each site, above a basal unconformity with the basaltic basement. At some sites (e.g. Site 838; Fig. 5b), however, larger proportions of the sedimentary cover are composed of gravels or gravelly sands. The conglomerates are normally graded or massive and poorly sorted. Thicknesses of recovered conglomerates and gravels range from 50 cm up to 18.0 m within the Lau Basin, although the amount of recovery of these sediments is often rather poor. Individual clasts are angular to subrounded in texture and range up to 5 cm across. Compositionally they comprise a range of volcanic rock types, from rhyolitic to basaltic andesite (Parson *et al.* 1992). Dacite is the

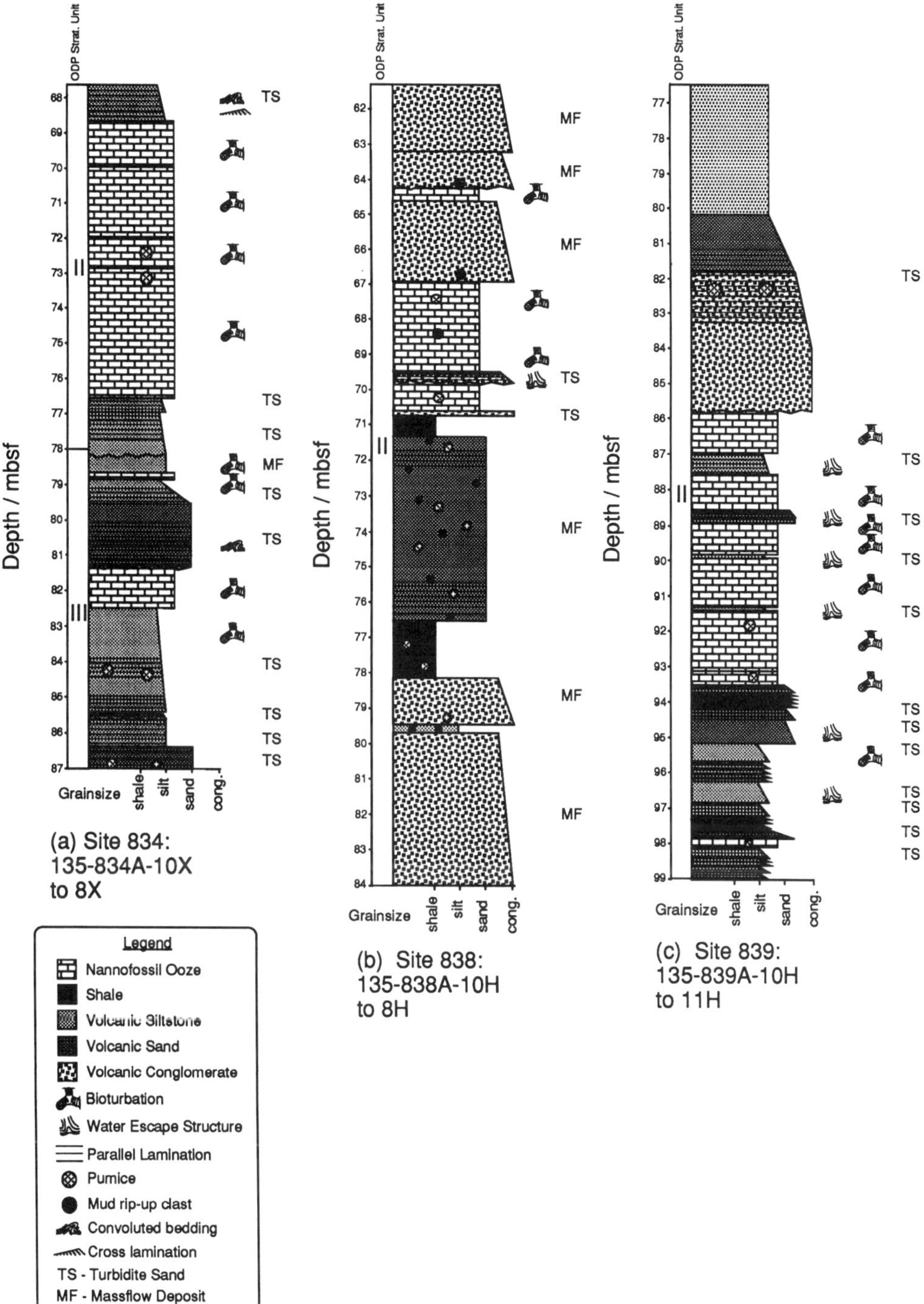

Fig. 5. Sedimentary logs of volcaniclastic sediments from Sites 834, 838 and 839, showing the range of structures and thicknesses which characterize the volcaniclastic facies. Depths in metres below seafloor (mbsf).

dominant rock type with lesser amounts of basaltic material being recognized. The contribution of andesitic volcanic rocks is variable across the basin. Volcaniclastic material dominates almost completely, although reworked clasts of nannofossil chalks are found in minor amounts throughout. Larger volumes of reworked nannofossil chalks were only identified at Site 835, where they were found in beds of mud-clast conglomerate up to 5.3 m thick. There is no contribution to the sediments from quartzose or high-grade metamorphic material. Graded gravels and conglomerates often pass up into volcanic sand, and are seen at the base of thick turbidite beds. Figure 5c shows a sedimentary log of part of the sequence recovered at Site 839, showing the relationship of the conglomerates to the surrounding sandstones in this part of the Upper Pliocene section. Drilling disturbance may be responsible for the destruction of sedimentary structures in unlithified gravels and for the occasional development of inverse grading. However, it is often impossible to say in these cases whether this might be a primary structure.

An unusual volcanic gravel breccia was recovered at Site 836. It is dominated by sharp, fresh shards of basaltic material, compared to the dominant dacitic character found elsewhere. The texture of the shards and the proximity to the basaltic basement suggests that these sediments are primary hyaloclastites produced by the shattering of freshly erupted basalt. The presence of a conglomeratic facies and the frequent occurrence of delicate, well preserved hyaloclastite and pumice shards, as well as volcanic lapilli, within them indicate a proximal setting for deposition at Site 836 and all of the other backarc sites.

Sand and sandstone. Volcanic sands make up one of the most important volcanic sediments at all of the back-arc sites (e.g. Fig. 5a & c). They consist of two facies types. The first is characterized by massive beds of medium- to coarse-grained sand (Fig. 6a). Sorting is moderate to good, although occasional larger, matrix-supported clasts of pumice are also present. Non-volcanic components in the sand include planktonic foraminifers, radiolarians and coccoliths. Both the tops and bases of individual beds are sharp, and internally there is little or no structure visible. These sands are interpreted as the products of sediment gravity flows, derived a relatively short distance from the site of deposition (<20 km). The second principal sand facies is a turbidite facies. Whereas complete Bouma cycles are rarely seen, the sediments are clearly turbidites as they are characterized by sharp erosive bases and an upward fining of grain size. Compositional grading is also observed. The coarse bases of many turbidites show a preferential concentration of denser basaltic andesite clasts rather than the lighter dacitic glass fragments, which dominate in the upper parts of each bed. Pyroxene and plagioclase feldspar crystals are also concentrated along the base of individual turbidites. This transition is manifested in an upward colour change from a dark base, 3–10 cm thick, which gradually becomes lighter above (Fig. 6b). Parallel lamination is common (Fig. 6c) but other sedimentary structures are not abundant (e.g. rare cross lamination). Water escape structures and soft sedimentary slump folding (Fig. 6c) testify to the water-saturated, and thus rapidly deposited, nature of the sediment on deposition. The presence of slump folding in particular indicates sedimentation on a slope. Seismic activity within the basin, as evidenced by pulsed extension recorded at Site 839, may be an important trigger for the slumping and redeposition seen in the volcaniclastic sediments and also for some nannofossil oozes, which include large-scale slump sheets and associated redeposited conglomerates (e.g. Site 835).

Volcanic silts and siltstone. Volcanic silts were recovered from each of the sites and are typically found confined to the upper parts of volcanic turbidites or as thinner, graded turbidite layers towards the top of the volcaniclastic-dominated part of each section. They display many of the sedimentary features of the sands in being graded and commonly parallel laminated.

Nannofossil oozes, chalks and associated ash layers. The upper part of the sedimentary section at each site, generally of Mid–Late Pleistocene age, is dominated by nannofossil oozes (Fig. 4). These oozes are typically structureless and stained dark brown by contamination with manganiferous oxides and pelagic clays. Throughout the section at each site, the tops of turbidite beds are characterized by nannofossil oozes grading up from silts. A mottling effect, the result of bioturbation, is often observed around the transition zone. Bioturbation may explain the lack of lamination within the majority of the pelagic sediments. In addition to the pelagic biogenic sediments, minor amounts of volcanic material (<1.0% by volume) are found, both as disseminated particles and as continuous, thin, graded ash layers. Discrete airfall ash layers are characterized by

Fig. 6. (a) Core photograph of structureless, mass-flow sand from 135–837A–8H–1; (b) A graded volcanic turbidite from 135–837A–8H–4, showing compositional grading and sharp erosive basal contact with underlying nannofossil ooze; (c) Parallel laminated and slump-folded sands in a turbidite at 135–834A–9H–6. Scales graduated in centimetres.

sharp bases (Fig. 7a) and grade up into nanno-fossil ooze, often through a transitional biotur-bated zone, similar to that seen in the volcaniclastic turbidites. Extensive bioturbation is a likely cause of the dissemination of the minor amounts of volcanic material found in nanno-fossil ooze throughout the section. The thickness of the ash layers rarely exceeds 10 cm and is more typically less than 5 cm thick. Com-positionally, the airfall ash layers show little lithological variation compared to the reworked volcaniclastic sediments, and they contain either dacitic or basaltic andesite clasts. This feature can be used to distinguish thin, fine-grained, volcaniclastic turbidites from airfall ash layers.

In addition to the airfall ash layers and rare, fine-grained turbidites, the Pliocene–Pleisto-cene nannofossil oozes are characterized by the common occurrence of rhyolitic pumice clasts and, more rarely, blocks of basaltic andesite, generally 2–5 cm across (Fig. 7b). While the basaltic andesite clasts are generally unaltered,

the rhyolitic pumice is often chemically de-graded to clays, presumably as a result of the low chemical stability of rhyolitic glass and the extremely high surface area to volume ratio, which would promote rapid hydration. These volcanic clasts are dispersed randomly through-out the section.

In summary, sedimentation in the back-arc basin is characterized by an early phase of proximal gravity flow deposits and associated turbidites derived from a local source close to each individual sub-basin. The later sedimentary history at each site is dominated by pelagic nannofossil oozes with minor quantities of volcanic ash and pumice. The switch from proximal volcaniclastic to pelagic sedimentation is seen to occur at younger times in the easterly sub-basins.

Tonga Platform: Site 840

The sediments, which have accumulated on the

(a)　　　　　　　　　　　　　　　　(b).

Fig. 7. (a) Thin, graded airfall ash layer within Pleistocene nannofossil ooze from 135–834A–2H–4.; (b) Basaltic clasts within nannofossil ooze from 135–834A–12X-CC. Scales graduated in centimetres.

Tonga platform and were sampled at Site 840, were divided by Parson *et al.* (1992) into three stratigraphical sub-units (Fig. 8). These three sub-units correspond to two fining-upward cycles of Late Miocene (sub-unit III) and uppermost Miocene to Recent age (sub-units I and II). Sediments cored at Site 840, have a number of features in common with those found at Sites 834–839, despite an obvious change in the colour of the nannofossil oozes and chalks from dark brown to white. This change is attributed to a lack of contamination by hydrothermal manganiferous oxides on the platform.

Water depths on the Tonga Platform are much shallower than in the Lau Basin: 750 m at Site 840 compared to 2500–3000 m in the Lau Basin. Despite this, a broadly similar set of facies were recorded, i.e. massive and graded volcaniclastic silts, sands and gravels, interbedded with pelagic oozes and chalks. The coarse-grained volcanic sediments are interpreted to be the result of sediment gravity flow, either in the form of debris flows for the massive, poorly-sorted conglomerates, or as proximal, high-concentration turbidites in the case of many of the

sandstones and graded conglomerates. Sedimentary structures are abundant, with both thick (>0.5 mm) and thin (<0.5 mm) planar laminations being found in both sandstones and siltstones, indicating their current-deposited nature. While the siltstones are commonly observed to be laminated throughout (Bouma T_d), planar laminations are preferentially found only in the lower parts of sandstone beds (Bouma T_b). These planar laminated sandstones are in turn overlain by trough cross laminated and wavy laminated sandstones (Bouma T_c). Sedimentation rates during these periods were often high, as evidenced by the appearance of climbing ripple laminations in siltstones, as well as common soft sediment deformation in rapidly deposited sandstones. Convoluted, slumped, laminated sandstones up to 85 cm thick testify to the water-saturated nature of the sediment on deposition, as do abundant water escape structures (e.g. disrupted laminae) and minor sediment diapirs.

The sedimentary section at Site 840 can thus be interpreted as a series of sediment gravity flow deposits and turbidites, deposited against a

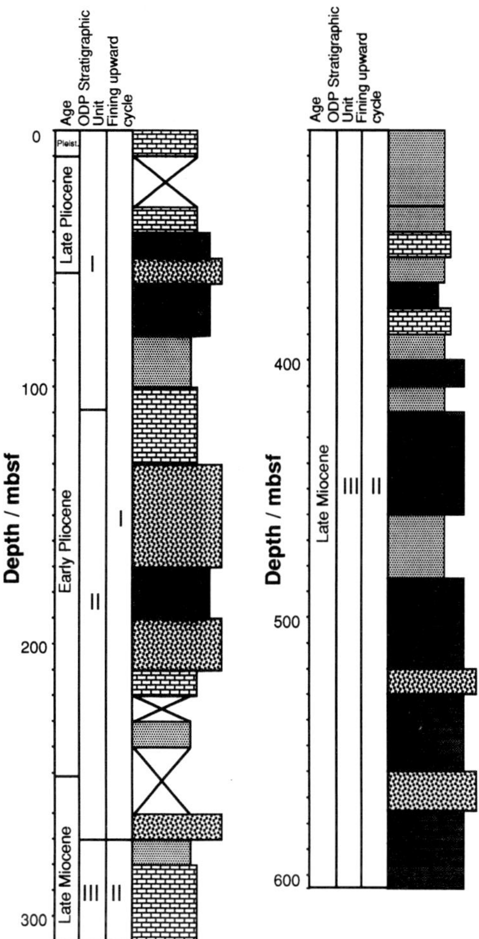

Fig. 8. Simplified sedimentary log of the sequence deposited at Site 840, showing the two fining-upwards cycles described in the text. Note the domination of pelagic sediments and the condensed nature of the sequence deposited since the Early Pliocene. Symbols used are the same as in Fig. 4.

background sedimentation of pelagic chalk. Since the Tonga platform lies adjacent to the modern Tofua arc, and is presumed to have had a similar relationship with the Miocene Tonga arc, it is reasonable to infer that the volcaniclastic debris was derived from erosion of the volcanic arc. The volcaniclastic sediments were deposited as a sediment apron, in the form of a submarine fan system, in front of the volcanic arc. The two fining-upward cycles are thus an important record of the volcanic and tectonic evolution of part of the arc system. The older fining-upward cycle (sub-unit III) is the product of mass wasting from an active volcanic arc

redepositing into a rapidly subsiding forearc basin (Clift 1994). Following a period of pre-rift uplift (e.g. Tappin *et al.* in press), a second fining-upwards cycle was generated by tectonism and rapid subsidence in the forearc during rifting of the Lau Basin. The coarse volcaniclastic sediments at the base of sub-unit II may also partially reflect increased syn-rift volcanism (Tappin *et al.* 1994) comparable to that recorded in the Sumisu Rift of the Izu-Bonin arc (e.g. Taylor 1990). The fining-upward transition from sub-unit II to I is then taken to reflect separation of the Tonga platform from the Lau Ridge and the cessation of volcanism on the east side of the basin immediately after rifting.

Sedimentation by gravity flow and high density turbidity currents occurred in the lower half of sub-unit III (Upper Miocene) and in sub-unit II (uppermost Miocene–Early Pliocene). Subsequent sedimentation during the Pliocene–Pleistocene of sub-unit I changed in character to bioturbated pelagic chalks with thin (*c.* 10–15 cm-thick), fine-grained, planar laminated volcanic sand and silt turbidite interbeds. Coeval sections on the islands of Tongatapu and 'Eua are of purely carbonate composition throughout this time interval (e.g. Tappin & Ballance 1991). Within the Late Pliocene–Pleistocene section at Site 840 a number of thin, graded, primary, airfall ash layers are identified. In common with the primary airfall ash beds recorded in the Pliocene–Pleistocene of the Lau Basin, the layers found at Site 840 show a strong compositional grading, with crystal-rich lapilli being found at the base of each bed, passing up into volcanic glass shards and then into a mixed nannofossil ooze/ash sediment.

Sediment textures

The nature of the sediment grains comprising the volcanic sands and silts within the Lau Basin has been studied through use of a Cambridge S90B scanning electron microscope at Edinburgh University. Figure 9a is a photograph of a SEM image showing an example of the delicate dacitic glass shards typically present. The abundance of elongate shards, as well as fragile clasts of highly vesicular dacitic pumice, suggests that much of the sediment in the back-arc basin is derived directly from the reworking of submarine volcanic eruptions. The material is clearly very fresh. Such delicate shard morphologies would not easily be preserved during long distance, water-born transport in a turbidite or sediment gravity flow. This again suggests local redeposition of material produced by a subaerial or submarine eruption plume.

(a)

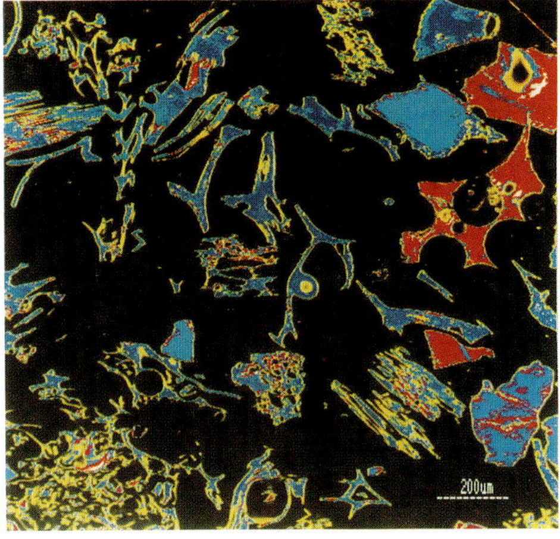

(b)

Fig. 9. (**a**) SEM image of elongate dacitic glass shards and fragile pumice clasts from 135–834A–10H–2, 35 cm, suggestive of local derivation. (**b**) False colour electron backscattered image of volcanic sand from 135–835A–15–6, 83 cm. Dark blue colour indicates dacitic composition, red indicates basaltic andesite clasts; minor quantities of foraminifers, clinopyroxene and plagioclase are also present.

Importantly, the morphologies argue against a reworking of older arc material, but are in favour of active volcanoes during the Early Pliocene close to the site of deposition, i.e. within the Lau Basin. While the presence of fresh shards alone does not necessitate contemporaneous volcanism, the rough topography of the Lau Basin does not support the possibility of volcanic debris stored in a shelf area and redeposited later.

Volcaniclastic geochemistry

As suggested by the light colour of most of the volcanic sands, the principal type of sediment grain found in Lau Basin sediments is dacitic glass. The composition of the sediments was determined during ODP Leg 135 using refractive index reference oils according to the method of Church & Johnson (1980). Following this procedure, electron microprobe analyses were

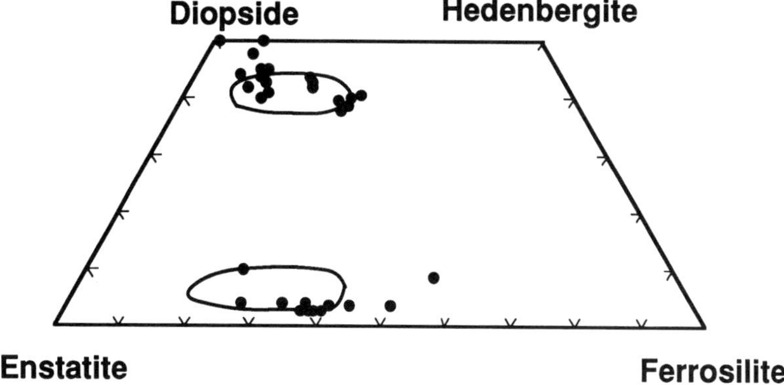

Fig. 10. Pyroxene compositional trapezoid showing the bimodal nature of Lau Basin detrital pyroxene grains. Fields indicate range of known compositions from the modern Tofua arc (data from Ewart *et al.* 1973).

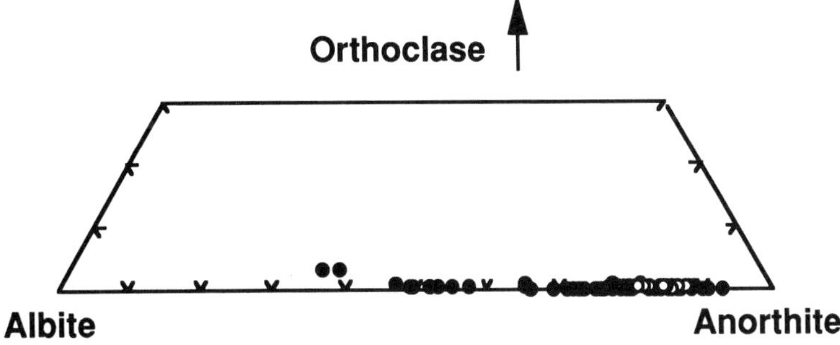

Fig. 11. Feldspar compositional trapezoid for Sites 834–839 (open circles) showing the concentration of values at 70–90% anorthite. Black circles indicate range of known compositions from the modern Tofua arc (data from Ewart *et al.* 1973).

undertaken by Clift & Dixon (1994) and Bednarz & Schmincke (1994). The results of the analyses are represented by typical examples shown in Tables 1 (volcanic glass), 2 (pyroxene) and 3 (plagioclase). Figure 9b shows a false colour electron backscattered image produced on the Edinburgh University *Camebax Microbeam* electron microprobe. The image is from a representative turbiditic sand and shows the dominance of dacitic glass (deep blue colour). The highly vesicular and angular nature of the glass fragments is again apparent, as is the uniform image of the glass, reflecting both the general lack of phenocrysts and of strong alteration. Vesicular basaltic andesite glass fragments (coloured red) form a minor population of clasts.

Electron microprobe analysis shows that clinopyroxene crystals from the Lau Basin occupy a range of compositions from augite to diopside (Fig. 10). Orthopyroxene crystals are also present in small numbers and are typically of hypersthene composition. Probe analyses of feldspar have only identified high-Ca plagioclase

compositions (Fig. 11), with anorthite contents ranging from 70–90%, compared with 80–90% for the Recent Tofua arc (Ewart *et al.* 1973). The bimodal pyroxene population and calcic plagioclase compositions are comparable to the sands analysed by Cawood (1991) from the Tonga forearc and are compatible with a low-K tholeiitic volcanic source for the sediment (Clift & Dixon 1994; Bednarz & Schmincke 1994). Although the tholeiitic nature of the source does not change over the period of arc rifting, other changes are noted and permit information about the sedimentary provenance to be deduced.

Volcanic glass silica contents

Total silica contents for glass grains analysed from all over the Lau Basin are shown in Fig. 12. The total composition of individual grains, as measured by the electron microprobe, rarely sums to 100%. The lower totals encountered may be explained by three factors, other than analytical error. Alteration of the glass by

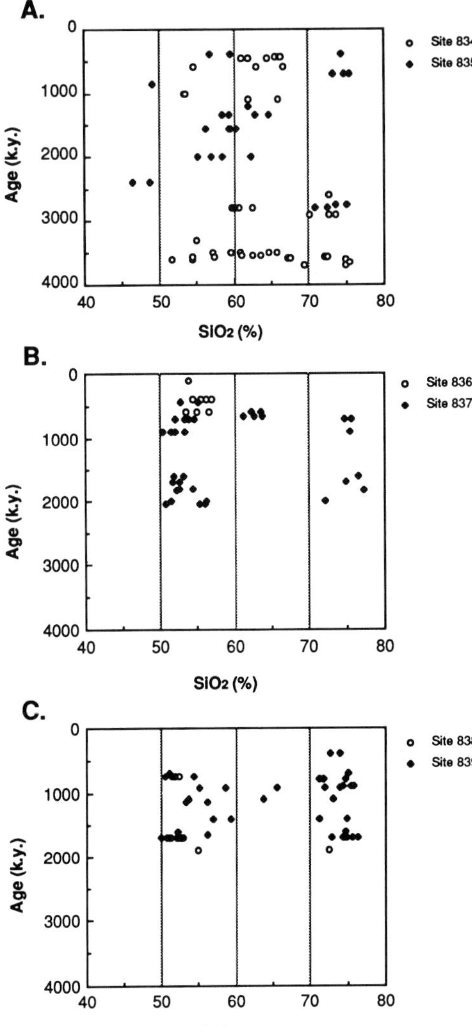

Fig. 12. Diagrams showing the variation in total silica values for the back-arc basin Sites 834–839 (after Clift & Dixon 1993). (**a**) Sites 834 and 835; (**b**) Sites 836–837; (**c**) Sites 838–839.

hydration, particularly towards the base of the section at the older sites, may cause low totals. However, as mentioned above, optical examination and electron backscattered imaging suggests that this is not often a major factor. Instead, original magmatic water may be present within the glass. Burnham & Jahns (1962) have shown experimentally that basaltic glass can contain up to 6% H_2O, while andesitic glass may contain up to 8.5% H_2O. This correlates well

with the generally lower totals found in the higher-silica glass analyses (Table 1). A third factor affecting analytical totals is that of alkali loss during analysis, as a result of volatilization by the 10 nA beam. In alkali-rich glasses (often also silica-rich) such loss may be as high as 3%, causing a corresponding increase in apparent silica totals of 2–3%. All of these factors combine to affect the higher-silica glasses more strongly than low-silica glasses. However, rather than make rather inexact corrections for each of these factors, the uncorrected compositions are used in Fig. 12 because the increase in silica due to alkali loss is approximately counterbalanced by the presence of water in the glass matrix.

Figure 12 shows a pattern to the silica contents of the glasses from the back-arc basin. There is no trend in the silica values from top to the bottom of each hole, although there is a clear distinction across the basin between Sites 834 and 835 and Sites 837 to 839. Whereas Sites 834 and 835 show a broad range of silica contents from 48–80%, Sites 837–839 are bimodal, with one group at 52–57% SiO_2 ('basaltic andesite') and another group at 70–75% SiO_2 (rhyolitic). This implies separate sources for the sediments derived at Sites 834–835 compared to those at Sites 837–839. The same sediment sources seem to have continued to provide sediment to each group of sites throughout much of the basin history without any mixing across the basin, as might be expected from the sedimentary evidence cited above. Site 836 contains only basaltic clasts and is presumed to represent derivation from a third sediment source.

For comparison, Fig. 13 shows the spread of silica contents analysed at Site 840 (Tonga platform) and whole rock analyses from the modern arc (data from Ewart *et al.* 1973). These glasses subdivide neatly into two groupings: analyses of glasses deposited prior to *c.* 5.0 Ma (then in the forearc of the Miocene Tonga arc) and those deposited since that time (now in the forearc of the Tofua arc). Prior to 5.0 Ma, a broad spectrum of silica contents is recorded, whereas since that time the limited number of analyses show a bimodal spread. This pattern indicates a change in the major element chemistry of the island arc at the time of basin rifting. It is noteworthy that the glass grains at Sites 834 and 835 are chemically comparable to the pre–5.0 Ma Miocene glasses at Site 840, while those at Sites 837–839 are more similar to the Pliocene–Pleistocene of Site 840. Site 836 does not correlate well with any known period of volcaniclastic sedimentation at Site 840.

Table 1. *Representative analyses of volcanic glass grains from Sites 834–840*

Hole / Core, section / interval (cm)	834A-2H-1, 37	834A-1H-2, 37	834A-1H-3, 82	834A-2H-1, 37	834A-3H-1, 72	834A-5H-2, 20	834A-7H-4, 78	834A-11X-1, 40	835A-2H-1, 79	835A-4H-5, 57	835A-7H-3, 140	835A-7H-3, 140	835A-8H-2, 140	835A-8H-2, 140	835A-10H-1, 26	836A-2H-6, 34	836A-3H-6, 134	837A-1H-3, 86	837A-2H-4, 103	837A-2H-7, 50	837A-4H-6, 67	837A-6-4, 132	837A-9H-1, 53	837A-9H-3, 93	838A-7H-2, 89	839A-3H-4, 22
Age (ka)	450	450	470	600	1000	2600	3300	3650	400	860	1350	1350	1550	1550	2400	400	600	450	600	650	900	1600	1825	2050	1700	750
Depth (mbsf)	1.87	1.87	3.82	8.79	17.82	39.30	60.38	93.60	10.29	35.07	61.40	61.40	69.40	69.40	85.67	9.08	19.43	3.86	13.53	17.50	35.17	51.82	66.99	78.43	53.39	18.72
SiO_2	54.61	66.34	61.85	63.09	53.28	72.68	54.93	75.52	56.76	49.19	58.46	64.59	60.32	59.39	46.30	56.95	56.66	52.79	63.65	61.10	51.98	76.52	77.45	56.14	52.91	50.56
TiO_2	0.69	0.66	0.48	0.95	0.72	0.44	0.94	0.28	0.80	0.67	1.23	1.08	1.24	1.23	0.93	0.94	1.08	0.65	0.76	0.83	0.77	0.42	0.32	1.14	1.08	1.20
Al_2O_3	14.38	12.99	12.58	13.60	13.54	10.88	14.63	11.22	17.59	15.02	13.91	13.15	13.38	13.68	18.06	13.66	14.10	14.78	13.46	12.23	14.09	11.28	10.99	14.14	13.94	14.10
Fe_2O_3	10.11	7.52	9.87	7.29	12.11	5.19	6.90	3.39	6.55	8.83	8.91	6.47	8.55	8.45	6.55	10.48	9.82	10.12	8.23	11.95	8.59	2.33	1.65	10.38	10.44	10.62
MnO	0.26	0.17	0.14	0.18	0.25	0.16	0.22	0.14	0.13	0.33	0.18	0.21	0.21	0.21	0.29	0.17	0.22	0.31	0.19	0.23	0.17	0.09	0.05	0.27	0.20	0.22
MgO	5.17	1.58	2.46	2.05	5.05	0.69	5.26	0.17	1.18	7.43	2.45	1.32	1.93	2.15	6.81	3.40	2.74	5.58	1.68	3.51	6.71	0.42	0.36	3.04	4.59	5.50
CaO	9.81	5.57	7.12	5.78	10.47	3.84	8.48	1.86	7.75	13.14	6.54	4.56	5.86	6.23	14.17	8.00	6.93	10.21	6.10	8.10	11.72	2.38	1.47	1.28	9.19	10.04
Na_2O	1.89	1.56	1.96	2.29	1.28	1.10	3.00	1.12	2.85	1.48	1.64	1.72	1.68	1.62	2.20	1.97	1.81	1.36	1.35	1.42	1.56	0.99	0.86	0.46	1.96	2.18
K_2O	0.52	1.25	0.73	0.58	0.41	0.78	2.36	0.38	0.80	0.33	0.92	1.11	1.04	0.94	0.09	0.72	0.38	0.58	0.47	0.36	0.32	0.73	0.98	7.90	0.50	0.28
P_2O_5	0.21	0.31	0.19	0.27	0.21	0.19	0.41	0.07	0.31	0.20	0.53	0.35	0.41	0.38	0.14	0.28	0.32	0.20	0.21	0.32	0.33	0.06	0.09	0.19	0.36	0.28
Total	97.65	97.97	97.39	96.08	97.32	95.78	96.91	94.14	94.72	96.62	94.78	94.55	94.61	94.29	95.51	96.56	94.06	96.57	96.09	100.04	96.24	95.22	94.22	94.94	95.16	94.97

Hole / Core, section / interval (cm)	839A-4H-3, 35	839A-5H-2, 15	839A-6H-1, 117	839A-4H-4, 38	839A-5H-2, 15	839A-6H-1, 117	839A-7H-2, 101	839A-11H-6, 49	840A-1H-1, 102	840A-1H-1, 102	840B-1X-CC, 33	840A-1H-3, 45	840A-1H-3, 45	840C-1H-2, 114	840C-2H-1, 37	840B-12X-5, 50	840B-12-5, 50	840B-17X-CC, 2	840C-10H-1, 3	840C-13H-5, 79	840B-33X-1, 21	840B-34X-1, 86	840B-43X-3, 133	840B-50X-1, 134	840B-51X-5, 10	840B-55X-4, 94	840B-56X-4, 108	840B-61X-3, 131
Age (ka)	920	1100	1400	920	1100	1400	1700	1700	350	350	400	450	450	1725	1800	5000	5000	5125	5300	5350	5650	5800	6130	6475	6500	6700	6750	6950
Depth (mbsf)	28.35	34.65	43.67	28.35	34.65	43.67	54.51	99.49	1.02	1.02	1.02	3.50	3.50	40.64	47.87	1100.50	110.48	153.82	171.50	253.80	308.40	318.70	398.33	463.44	477.80	515.42	525.67	572.61
SiO_2	65.50	63.70	57.07	65.50	53.71	59.34	74.37	76.35	63.47	63.47	53.97	54.10	56.50	55.28	71.50	52.81	54.48	67.01	52.95	55.66	54.23	50.39	66.37	53.87	71.16	63.93	53.83	50.82
TiO_2	0.82	1.04	1.28	0.82	1.00	1.08	0.43	0.37	0.71	0.71	1.28	0.96	1.27	1.31	0.69	1.08	1.32	0.86	0.95	1.04	1.27	0.92	0.87	1.24	1.47	0.11	1.33	1.29
Al_2O_3	14.57	11.67	13.15	14.57	13.13	14.67	11.53	11.11	14.21	14.21	12.40	14.73	11.89	12.75	11.58	12.65	12.72	15.98	14.21	14.84	14.23	13.94	14.83	14.56	10.45	20.06	13.91	16.52
Fe_2O_3	5.79	9.52	10.40	5.79	12.49	8.53	2.61	3.36	8.16	8.16	13.64	12.24	4.38	4.05	0.76	4.48	0.95	[illegible]	11.81	10.25	11.46	10.66	0.33	10.56	0.49	0.41	11.67	2.78
MnO	0.14	0.15	0.18	0.14	0.26	0.14	0.08	0.10	0.17	0.17	0.28	0.27	0.34	0.36	0.23	0.28	0.37	0.37	0.27	0.24	0.23	0.22	0.40	0.21	0.54	0.07	0.19	0.21
MgO	0.65	1.65	2.60	0.65	4.16	2.15	0.47	0.31	1.34	1.34	3.76	4.50	14.06	14.23	4.88	5.68	13.23	5.95	5.55	3.47	2.76	6.57	5.04	3.49	6.96	1.57	3.56	11.33
CaO	0.61	0.68	0.97	6.35	9.13	6.70	0.80	0.59	5.52	1.60	8.66	9.21	9.25	8.89	3.73	9.16	5.57	4.43	9.41	8.31	7.24	11.78	0.98	7.86	1.72	8.23	8.27	10.44
Na_2O	6.35	5.84	6.92	2.39	1.77	0.97	2.67	2.85	2.11	5.52	1.62	1.99	1.13	1.92	1.94	1.62	2.09	3.32	1.71	1.78	1.94	1.63	4.78	2.09	2.96	2.68	1.56	4.69
K_2O	2.39	1.90	1.45	0.61	0.28	0.72	0.99	1.12	1.60	2.11	0.70	0.85	1.92	1.01	1.71	1.55	3.32	1.09	0.08	0.33	0.47	0.29	3.06	0.50	1.07	3.20	0.33	1.71
P_2O_5	0.25	0.24	0.38	0.25	0.26	0.39	0.13	0.15	0.23	0.23	0.33	0.30	0.27	0.28	0.14	0.39	0.26	0.14	0.21	0.26	0.29	0.31	0.15	0.39	0.16	0.07	0.26	0.21
Total	97.05	96.40	94.40	97.05	96.19	94.69	94.10	96.33	97.51	97.51	96.63	99.14	99.78	97.33	94.78	97.03	96.08	99.85	97.16	96.15	94.11	96.70	96.80	94.76	96.97	100.32	94.91	95.27

Table 2. *Representative analyses of pyroxene grains from Sites 834–840*

Hole Core, section interval (cm)	834A- 1H-2, 37	834A- 3H-1, 72	834A- 3H-4, 40	834A- 6H-2, 35	834A- 7H-4, 78	835A- 4H-5, 57	835A- 4H-5, 57	835A- 4H-5, 109	836A- 1H-1, 84	837A- 2H-4, 103	837A- 4H-6, 67	837A- 6H-4, 132	837A- 8H-2 49	838A- 11X-4, 30
Age (ka)	600	1200	1250	2800	3300	900	900	950	100	550	900	1500	1900	1900
Depth (mbsf)	1.87	17.82	22.00	47.45	60.38	35.07	35.07	35.59	0.84	13.53	35.17	51.82	66.99	94.00
SiO_2	51.48	52.13	51.22	49.19	52.54	51.11	49.76	47.76	52.36	51.11	51.98	52.51	51.28	50.02
TiO_2	0.21	0.12	0.18	0.19	0.18	0.37	0.33	1.05	0.13	0.30	0.33	0.18	0.30	0.23
Al_2O_3	1.07	1.32	1.28	0.47	1.21	3.31	4.21	6.49	1.55	1.22	1.26	0.51	1.21	0.54
Fe_2O_3	22.38	15.02	11.06	30.92	3.15	4.09	5.45	6.62	16.77	15.31	10.29	23.49	15.29	28.46
MnO	0.74	0.44	0.70	1.41	0.10	0.11	0.14	0.18	0.37	0.59	0.47	1.09	0.56	1.56
MgO	21.22	23.89	13.91	12.61	17.12	16.92	15.80	15.21	24.89	12.30	13.48	19.97	11.98	15.83
CaO	2.00	5.07	20.46	4.14	23.87	21.12	21.32	19.74	2.14	18.46	21.01	1.35	19.32	1.66
Na_2O	0.01	0.04	0.21	0.03	0.18	0.13	0.17	0.24	0.14	0.16	0.25	0.02	0.14	0.04
K_2O	0.01	0.01	0.01	0.01	0.00	0.01	0.00	0.01	0.02	0.01	0.00	0.00	0.01	0.00
P_2O_5	0.04	0.09	0.32	0.07	0.40	0.26	0.48	0.29	0.03	0.24	0.36	0.03	0.30	0.04
Total	99.35	98.24	99.53	99.21	98.93	98.52	98.25	98.63	98.51	100.00	99.76	99.32	100.66	98.61

Hole Core, section interval (cm)	839A- 3H-3, 144	839A- 6H-5, 136	839A- 6H-7, 59	839A- 10H-3, 63	839A- 11X-6, 82	840A- 1H-1, 102	840A- 1H-3, 45	840C- 1H-2, 29	840B- 2X-6, 144	840C- 4H-4, 20	840B- 12X-5, 50	840C- 10H-1, 3	840B- 33X-1, 88	840B- 42X-3, 3	840B- 55X-4, 94	840B- 62X-5, 42
Age (ka)	800	1450	1500	1700	1750	350	450	1700	1700	2900	5000	5300	5700	6100	6700	7000
Depth (mbsf)	18.44	49.86	52.09	84.13	98.32	1.02	3.50	39.79	18.44	71.20	110.50	171.50	309.10	388.80	515.40	584.40
SiO_2	49.73	51.72	50.90	52.30	52.41	50.45	53.92	50.14	49.73	50.64	52.01	51.47	52.75	49.99	52.52	50.14
TiO_2	0.62	0.19	0.38	0.13	0.26	0.16	0.10	0.61	0.62	0.36	0.30	0.35	0.36	0.68	0.34	0.45
Al_2O_3	4.08	0.81	1.32	0.78	1.06	0.46	0.95	3.13	4.08	1.25	0.88	2.11	1.32	2.87	1.24	2.93
Fe_2O_3	7.46	21.81	10.73	20.14	4.47	28.45	2.76	11.28	7.46	13.97	20.60	10.10	19.75	15.16	18.99	11.33
MnO	0.18	0.97	0.57	0.62	0.24	1.03	0.08	0.34	0.18	0.59	1.11	0.31	0.51	0.53	0.59	0.32
MgO	15.88	21.49	13.45	22.27	14.97	16.53	17.59	14.34	15.88	12.04	21.80	14.41	22.29	18.30	22.51	15.94
CaO	19.44	1.44	20.38	1.90	24.83	1.82	23.42	18.48	19.44	19.27	1.62	20.53	2.55	10.71	2.26	16.75
Na_2O	0.26	0.06	0.25	0.04	0.12	0.03	0.12	0.20	0.26	0.28	0.03	0.20	0.04	0.17	0.03	0.19
K_2O	0.01	0.01	0.00	0.01	0.00	0.01	0.00	0.01	0.01	0.00	0.00	0.00	0.00	0.02	0.00	0.01
P_2O_5	0.35	0.03	0.36	0.04	0.41	0.00	0.38	0.29	0.35	0.00	0.06	0.31	0.07	0.18	0.00	0.25
Total	98.64	98.71	98.72	98.33	99.04	99.09	99.42	99.44	98.09	98.76	98.69	100.14	100.00	99.27	98.82	98.76

Table 3. *Representative analyses of plagioclase grains from Sites 834–840*

Hole Core, section interval (cm)	834A- 1H-2, 37	834A- 3H-4, 40	834A- 6H-4, 58	834A- 7H-4, 78	834A- 12X-2, 88	835A- 4H-2, 54	835A- 4H-5, 109	835A- 6H-2, 45	835A- 14X-6, 84	835A- 15X-6, 83	837A- 1H-3, 86	837A- 2H-7, 50	837A- 3H-2, 124	837A- 4H-6, 52	837A- 8H-2, 49	837A- 9H-1, 53
Age (ka)	450	1100	2900	3300	3700	900	1000	1200	2600	2700	450	650	710	900	1825	200
Depth (mbsf)	1.87	22.00	50.68	60.38	105.28	31.95	35.59	50.95	131.84	141.13	3.86	17.50	20.24	35.02	66.99	75.03
SiO_2	46.93	46.68	44.41	47.32	45.66	46.11	50.32	45.92	58.59	57.35	44.53	47.21	57.76	48.24	47.30	56.36
TiO_2	0.02	0.03	0.04	0.02	0.01	0.05	0.05	0.03	0.02	0.04	0.03	0.02	0.04	0.03	0.03	0.03
Al_2O_3	31.86	31.84	33.61	31.73	34.20	32.44	29.01	32.61	25.14	25.41	33.39	32.44	26.69	31.20	33.23	27.30
Fe_2O_3	0.90	0.87	0.55	0.72	0.83	0.83	0.73	0.82	0.40	0.31	0.83	0.86	0.41	0.52	0.75	0.36
MnO	0.01	0.00	0.01	0.01	0.05	0.05	0.01	0.00	0.02	0.01	0.03	0.01	0.03	0.03	0.02	0.04
MgO	0.08	0.09	0.02	0.07	0.08	0.30	0.28	0.16	0.02	0.02	0.16	0.07	0.07	0.06	0.07	0.06
CaO	16.86	16.77	18.39	16.17	18.29	17.63	14.16	17.23	7.43	7.93	18.52	17.39	9.55	15.55	17.50	10.11
Na_2O	1.93	1.93	1.26	2.37	1.29	1.62	3.99	1.77	7.10	6.84	1.12	1.91	5.87	3.05	1.61	5.67
K_2O	0.04	0.03	0.05	0.03	0.04	0.02	0.02	0.05	0.60	0.56	0.04	0.02	0.12	0.01	0.02	0.06
Total	98.63	98.24	98.34	98.44	100.44	99.02	98.56	98.58	99.33	98.47	98.65	99.93	100.54	98.70	100.54	100.00

	838A- 7H-2, 69	838A- 7H-2, 69	838A- 11X-4, 30	839A- 1H-3, 44	839A- 4H-3, 126	839A- 6H-1, 117	839A- 7H-2, 101	839A- 12X-1, 1	839A- 18X-CC, 6	840B- 1X-CC 33	840C- 1H-2, 29	840C- 4H-6, 34	840C- 10H-1, 3	840B- 38X-1, 17	840B- 51X-5, 10	840B- 62X-5, 42
Age (ka)	1700	1700	2000	300	1000	1500	1650	1700	1750	400	425	3000	5300	5950	6500	7000
Depth (mbsf)	53.39	53.39	94.00	3.44	29.26	43.67	54.51	99.51	167.26	1.68	2.25	74.34	171.50	347.30	477.80	584.40
SiO_2	54.23	55.03	54.95	48.93	49.23	48.15	49.53	46.03	53.07	53.03	48.45	45.59	46.52	46.68	56.01	55.21
TiO_2	0.04	0.06	0.03	0.02	0.03	0.01	0.04	0.04	0.05	0.05	0.05	0.03	0.04	0.04	0.05	0.06
Al_2O_3	27.32	27.23	27.39	30.67	31.19	31.51	30.80	32.96	28.81	28.63	32.26	33.70	32.93	33.32	27.46	27.13
Fe_2O_3	0.54	0.50	0.56	0.70	0.52	0.69	0.76	0.65	0.55	1.40	0.77	0.86	1.25	0.69	0.55	0.63
MnO	0.02	0.01	0.02	0.00	0.02	0.00	0.02	0.00	0.00	0.03	0.02	0.02	0.03	0.01	0.00	0.01
MgO	0.06	0.07	0.06	0.07	0.04	0.08	0.07	0.03	0.04	0.22	0.06	0.18	0.11	0.19	0.08	0.07
CaO	10.97	10.41	10.85	15.14	15.08	15.58	14.78	17.23	12.10	13.52	16.25	18.61	17.70	17.89	10.67	10.70
Na_2O	5.37	5.90	5.54	3.08	3.24	2.72	3.27	1.91	4.91	3.93	2.38	1.06	1.54	1.62	5.83	5.54
K_2O	0.09	0.08	0.05	0.02	0.04	0.03	0.03	0.01	0.05	0.11	0.02	0.02	0.04	0.03	0.09	0.08
Total	98.64	99.29	99.43	98.63	99.38	98.78	99.29	98.86	99.57	100.92	100.26	100.06	100.14	100.47	100.73	99.43

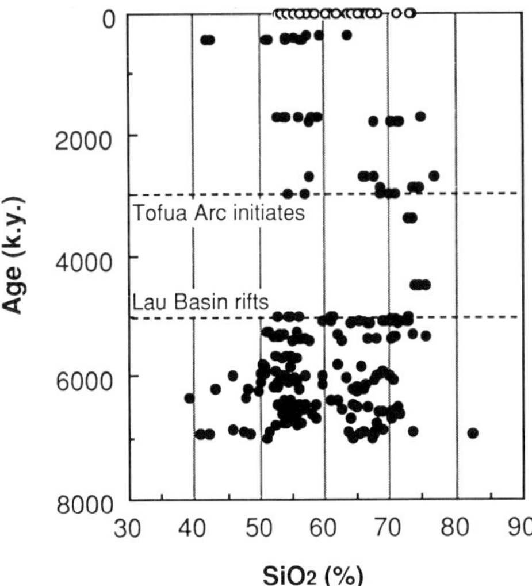

Fig. 13. Diagram showing the apparent change from basalt–rhyolite to bimodal volcanism during basin rifting, as recorded at Site 840, Tonga Platform. Close circles indicate analyses from Site 840, open circles indicate modern Tofua arc (data from Ewart *et al.* 1973)

Interpretation and discussion

Arc volcanism

A number of important constraints can be derived from these data regarding the sedimentary and volcanic evolution of the Tonga/Tofua arc system during rifting of the Lau Basin. As far as the evolution of arc volcanism during this time period is concerned, rapid sedimentation at Site 840 during the Late Miocene (7.0–5.0 Ma) confirms the presence of an active arc adjacent to the Tonga Platform at this time. However, during the rifting of the Lau Basin the Miocene Tonga arc became separated from the Tonga Platform and is now preserved as the Lau Ridge on the west side of the Lau Basin. Therefore, there must have been a period between the separation of the Miocene Tonga arc and Tonga Platform and the start of activity of the Tofua arc, when there was no arc volcanism on the east side of the Lau Basin. The marked reduction in the volume of volcanic sediment deposited at Site 840 during the Early Pliocene is suggestive of a lack of arc volcanism in this position at that time. If this is so, the presence of any volcaniclastic sediment at that time requires explanation. It is difficult to see how submarine volcanic eruptions within the Lau Basin could have made any contribution to sedimentation on the Tonga Platform. It is therefore most probable that Early Pliocene volcaniclastic material

at Site 840 was derived as airfall ash from the Lau Ridge, which was then actively erupting the Korobasaga Volcanic Group (Gill 1976). The appearance of coarser, thicker ash layers at 3.0 Ma (Late Pliocene) is taken to indicate the start of activity on the Tofua arc. The ashes at Site 840 were probably derived from a nearby volcano, such as Ata, located 40 km away. Alternative sources include other volcanoes along the length of the Tofua arc. The Taupo Volcanic Province of New Zealand may also have contributed some material to the Tonga Platform. Pleistocene acidic tuffs from Taupo are known to be over 5 cm thick up to 300 km from their source (Walker 1980) and are recognized up to 1000 km from source in the Pacific Ocean (Shane & Froggatt 1991). However, the prevailing winds tend to disperse these ashes towards the east, restricting their potential for deposition on the Tonga Platform.

Back-arc volcanism

Further information about rift-related volcanism can be derived from the volcaniclastic sediments of the Lau Basin. The contrast in the silica contents at different back-arc basin sites reinforces the sedimentary and topographical information suggesting strong sedimentary subdivision of the basin. Furthermore, the fresh nature of the sediment grains indicates that the sources of this sediment were active volcanoes. Several lines of evidence point to these volcanoes being sited within the Lau Basin itself. Since the activity on the Tofua arc dates from 3.0 Ma this is too young to have produced the sediments found at the base of Sites 834–835 and 837–839. Even today, when the Tofua arc has grown to its maximum size, its contribution to sedimentation in the Lau Basin is limited to thin airfall ashes and it is clearly incapable of providing the thick gravity flow sands and turbidites found at the base of each back-arc site. The Lau Ridge was active throughout the early rifting history and it must have provided airfall material throughout the back-arc basin. However, its contribution was volumetrically small since coarse volcaniclastic sedimentation adjacent to the Lau Ridge ceased prior to that in the centre of the basin. In addition, the geochemical differences between sediments in the basin centre (Site 837–839) and those adjacent to the Lau Ridge (Sites 834–835) confirms their different provenances and suggests that the Lau Ridge, was not a major sediment source for the whole basin. The evidence points strongly in favour of sediment being produced by seamount volcanoes active within the Lau Basin during its early rifting. These would have commenced activity close to the Lau Ridge and subsequently migrated east as the basin opened in order to explain the eastward younging in the locus of coarse clastic sedimentation.

The sedimentary and geochemical differences between individual sub-basins within the Lau Basin system provide important information about their source volcanoes. The presence of locally derived basaltic and basaltic andesite hyaloclastite breccia at Site 836 is in accord with this site's position within basaltic crust produced at a back-arc spreading centre. In contrast, Sites 838 and 839 lie within a region produced by extension of the original arc crust. The texturally fresh nature of the grains suggests derivation of the sediment from a seamount volcano close to Sites 838 and 839, which was active during and shortly after the rifting of these two basins. Erosion of the older arc basement can be ruled out both on textural and chemical grounds, as the older (i.e. Miocene) rifted arc basement might be expected to show the complete spread of silica values shown by the Miocene arc detritus at Site 840. Instead the bimodal silica chemistry of the grains found at these sites indicates that the source volcano was not of this type.

Site 837, while it is apparently situated on crust produced at the Eastern Lau Spreading Centre, has sedimentation strongly influenced by the same or a similar intra-basinal source as Sites 838–839, presumably because it lies very close to Sites 838 and 839, as well as the transition between rifted and true oceanic crust. No suitable source for the dacite-dominated, bimodal sediment found at Site 837 is believed to exist within the basaltic spreading centre-type crust on which the site is apparently located. Thus, the sediment deposited at Site 837 is inferred to have been derived from seamount volcanoes found within the stretched arc crust close-by. Seamount volcanoes within the rifted arc crust would have been active up to the point at which the propagating Eastern Lau Spreading Centre arrived in the area. It is thus reasonable to expect some of their tephra to be transported the short distance on to the new spreading centre-type crust. It is the proximity of Site 837 in distance and time to the seamount volcanoes of the western Lau Basin that accounts for the difference in sedimentary facies between Sites 836 and 837, even though they are sited on the same type of back-arc basin crust.

Sites 834 and 835 lying further to the north and closer to the Lau Ridge clearly constitute a separate grouping. In addition to the sedimentary evidence for local sediment sourcing at these sites, the andesitic nature of some of the sediment grains, and thus the source volcanoes, is seen to be distinctive from Sites 837–839. The volcanoes close to Sites 834 and 835 have major element compositions comparable to the pre–5.0 Ma Tonga arc. The sediments within the Lau Basin show that the oldest sites (834–835) share the spread in silica values seen in the Miocene Tonga arc, while the younger sites (837–839) have bimodal silica compositions, similar to the Pliocene–Pleistocene Tofua arc. The change in volcanic chemistry from andesitic to bimodal must have occurred between the creation of Site 835 (3.25 Ma) and the oldest site of the new bimodal trend, Site 838 (2.7 Ma). It thus appears that the coarse volcaniclastic sediment deposited immediately after the creation of each sub-basin reflects both the tectonic setting in which the crust was generated and the nature of the rift-related magmatism.

While the coarse volcaniclastic sediment at each site provides information about local volcanism, fine-grained airfall ashes and pumice within nannofossil sediment higher in each section (generally the Pleistocene) record the nature of volcanism over a wider area. Their origin is thus more difficult to constrain. Ash layers 5–10 cm thick are probably mostly derived from a relatively proximal source (i.e. the Lau Ridge or Tofua arc). However, the Taupo Volcanic Province, as a centre of extremely violent eruptions, could be the source of some of the thinner acidic ash layers, although as noted above, the distance and regional wind direction will tend to minimize this contribution. The Lau Ridge and Tofua arc are the most likely sources for the acidic, and particularly the basic, airfall ash layers, as eruptions of basic magma are generally less explosive and less widely distributed. The limited electron microprobe data support this interpretation, since the uniform low-K tholeiite glasses of the Lau Basin cannot be reconciled with the acidic calc-alkaline glasses of the Taupo province (Shane & Froggatt 1991) or even with the alkalic volcanism of the Pleistocene Mago Volcanic Group of the Lau Ridge (Gill 1976).

The source of pumice fragments within the Lau Basin is even more difficult to constrain. The ease with which pumice may be rafted considerable distances following an eruption on the modern Tofua arc or within the basin means that a single source is unlikely for all of the pumice. The subaerial and submarine Tofua arc, Lau Ridge, and seamounts within the Lau Basin, as well as the Taupo Volcanic Province are all possible contributors. As with the airfall ash layers, prevailing trade winds towards the present east do not favour the Tofua arc or

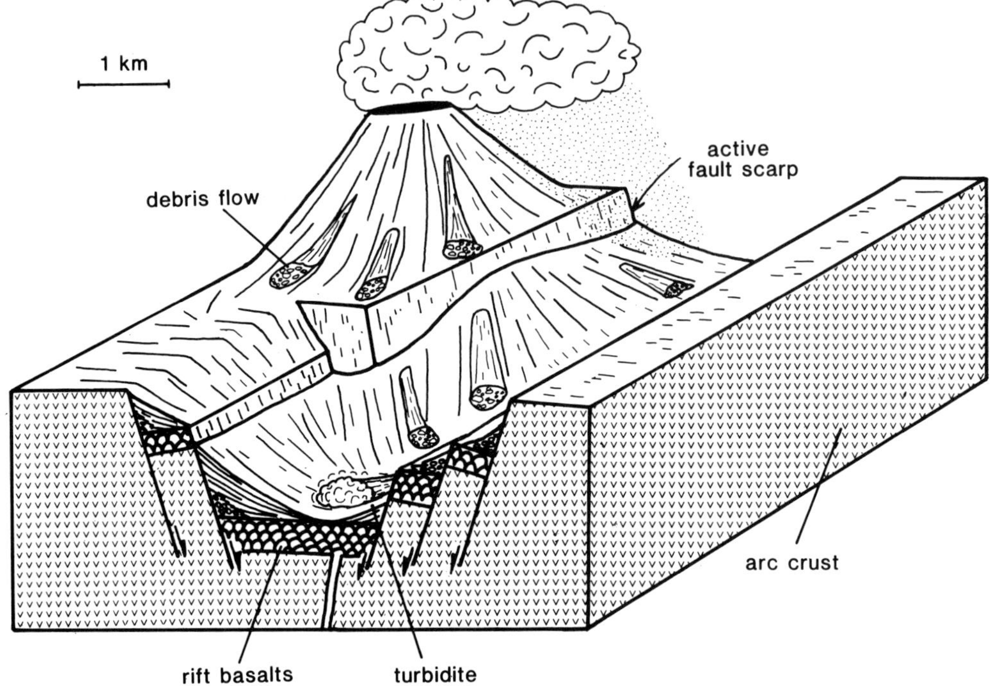

Fig. 14. Schematic representation of the early history of a sub-basin within the Lau back-arc basin, showing the generation of volcaniclastic sediment by explosive eruption at nearby seamount volcanoes. The sediment is subsequently reworked as debris flows and turbidites and transported into the central part of the graben.

Taupo province as major contributors. The Pleistocene Mago Volcanic Group of the Lau Ridge is typified by Strombolian cones and flows, with a smaller number of tuff cones (Cole *et al.* 1985) and is thus an unlikely source of voluminous pumice. The limited evidence thus favours seamount volcanoes within the Lau Basin as the major producers of pumice in the Lau Basin sediments. However, no geochemical constraints exist from this material to verify this hypothesis or to constrain the sources further.

Conclusions

The sediments of the Lau Basin and Tonga platform provide important new constraints on the nature of arc volcanism during the rifting of marginal basins. Sediments from Site 840 indicate that arc volcanism prior to rifting was characterized by a wide range of silica contents and probably culminated in a voluminous burst of activity immediately prior to the separation of the Tonga Platform from the Miocene Tonga arc (Lau Ridge). The early rift history is characterized by continued volcanism on the remnant arc (Lau Ridge) and extension of the Miocene arc

crust. Extension within the basin was accompanied by eruptions of basalt in graben centres and a wider range of volcanic compositions from seamount volcanoes on the flanks of individual sub-basins (Fig. 14). Localized slumping, gravity flow and turbidity current activity redeposited the dacite-dominated ash from these volcanoes into adjacent sub-basins, where it ponded. As extension proceeded, the locus of extension and volcanism migrated eastwards. After a gap of approximately 2.0 Ma, arc volcanism recommenced on the trench side of the basin at 3.0 Ma, although its products are only identifiable on the Tonga Platform. While the deposition of Tofua arc ash must have occurred in the Lau Basin, the chemical similarity (compositionally bimodal tholeiites) with the intrabasinal volcanoes makes its recognition there impossible with the present data. Shortly after the establishment of the Tofua arc, a southward propagating back-arc spreading ridge (Eastern Lau Spreading Centre), which had nucleated on the Peggy Ridge in the northern Lau Basin, reached the vicinity of Sites 838 and 839. The arrival of the Eastern Lau Spreading Centre focused extension and volcanism along

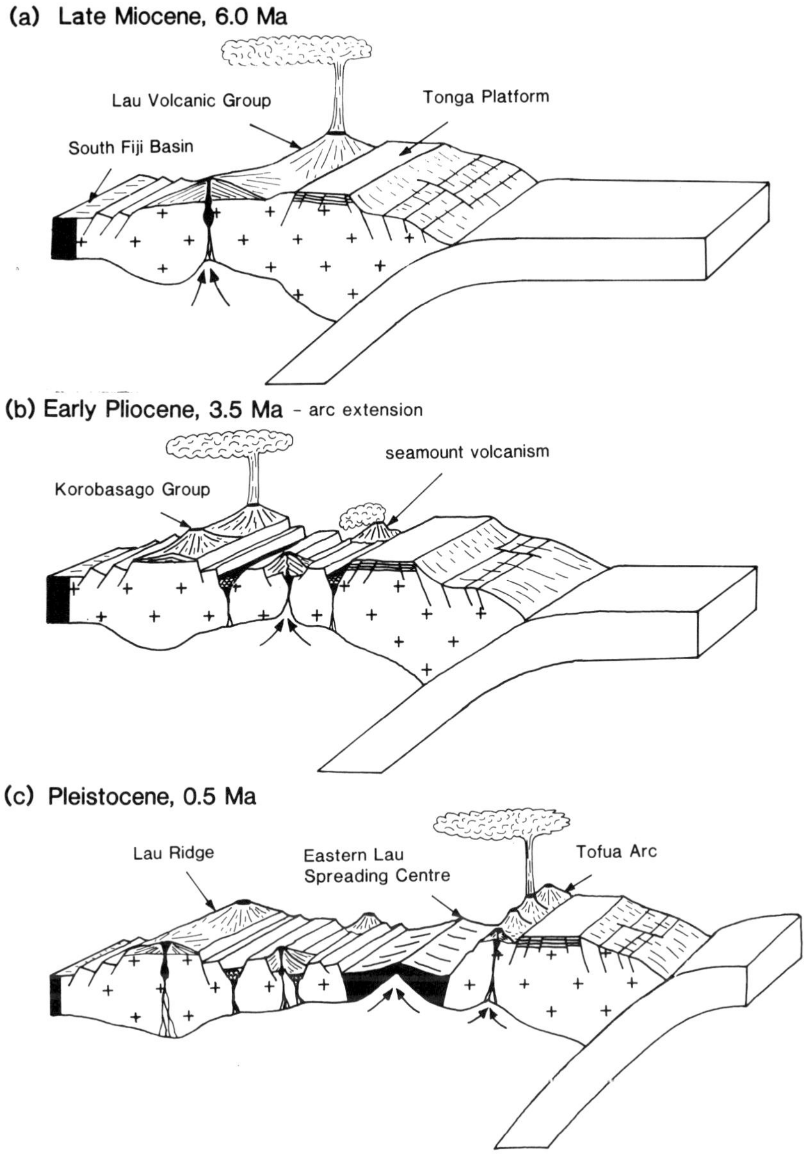

Fig. 15. Proposed model of basin evolution, with the immediate post-rift period being marked by intrabasinal seamount volcanism (**b**), prior to foundation of a new island arc chain (**c**).

its median valley, similar to a mid-ocean ridge in a major ocean basin. The subsequent demise of large intra-basinal seamount volcanoes resulted in a change in the nature of sedimentation within the Lau Basin. New sub-basins (e.g. Site 836) were no longer filled by dacite-dominated gravity flow deposits and turbidites, but by thin sequences of basaltic hyaloclastite. Throughout the basin nannofossil oozes accumulated, with minor volumes of primary airfall ash and

pumice, principally derived from the Tofua arc and intra-basinal seamounts north of the Peggy Ridge. Thus, Lau Basin rifting divides into two phases (Fig. 15). The early rifting is characterized by extension of arc crust, cessation of arc volcanism on the trench side of the basin after a volcanic climax and tholeiitic rhyolitic to basaltic magmatism within the new basin. The later history features a stabilization of the system with extension and volcanism concentrated along back-arc spreading centres and re-establishment of arc volcanism on the trench side of the basin. While it is possible that advance of the propagating spreading centre caused the foundation of a new arc, a time delay of approximately 1.5 Ma between the start of activity on the Tofua arc (3.0 Ma) and the arrival of the Eastern Lau Spreading Centre in the southern Lau Basin (1.5 Ma) makes such a relationship difficult to demonstrate.

P.C. would like to thank D. Tappin and D. Woodhall for their informative reviews which did much to improve the quality of this paper.

References

AUSTIN, J., TAYLOR, F.W. & CAGLE, C.D. 1989. Seismic stratigraphy of the central Tonga Ridge. *Marine and Petroleum Geology*, **6**, 71–92.

BEDNARZ, U. & SCHMINCKE, H.-U. 1994. Composition and provenance of volcaniclastic sediments in the Lau Basin (SW Pacific), ODP Leg 135 (Sites 834 to 839). *In:* HAWKINS, J.W., PARSON, L.M., ALLAN, J.F. et al. (eds) *Proceedings of the Ocean Drilling Program, Scientific Results*, **135**, 51–74.

BURHAM, C.W. & JAHNS, R.H. 1962. A method of determining the solubility of water in silicate melts. *American Journal of Science*, **260**, 721–745.

CAWOOD, P.A. 1991. Nature and record of igneous activity in the Tonga arc, SW Pacific, deduced from the phase chemistry of derived detrital grains. *In:* MORTON, A.C., TODD, S.P. & HAUGHTON, P.D.W. (eds) *Developments in Sedimentary Provenance Studies*. Geological Society, London, Special Publications, **57**, 305–323.

CHURCH, B.N. & JOHNSON, W.M. 1980. Calculation of the refractive index of silicate glasses from chemical composition. *Geological Society of America Bulletin*, **91**, 619–625.

CLIFT, P.D. 1994. Controls on the sedimentary and subsidence history of an active plate margin; An example from the Tonga Arc, SW Pacific. *In:* HAWKINS, J.W., PARSON, L.M., ALLAN, J.F. et al. (eds) *Proceedings of the Ocean Drilling Program, Scientific Results*, **135**, 173–190.

—— & DIXON, J.E. 1994. Variations in arc volcanism and sedimentation related to rifting of the Lau Basin, SW Pacific. *In:* HAWKINS, J.W., PARSON, L.M., ALLAN, J.F. et al. (eds) *Proceedings of the Ocean Drilling Program, Scientific Results*, **135**, 23–50.

COLE, J.W., GILL, J.B. & WOODHALL, D. 1985. Petrological history of the Lau Ridge, Fiji. *In:* SCHOLL, D.W. & VALLIER, T.L. (eds) *Geology and offshore resources of the Pacific island arcs – Tonga region*. Circum-Pacific Council for Energy and Mineral Resources Earth Science Series, **2**, 379–391.

COLLIER, J. & SINHA, M. 1990. Seismic images of a magma chamber beneath the Lau Basin back-arc spreading centre. *Nature*, **346**, 646–648.

EWART, A., BRYAN, W.B. & GILL, J.B. 1973. Mineralogy and geochemistry of the younger volcanic islands of Tonga, SW Pacific. *Journal of Petrology*, **14**, 429–465.

GILL, J.B. 1976. Composition and age of Lau Basin and Ridge volcanic rocks: Implications for evolution of an inter-arc basin and remnant arc. *Geological Society of America Bulletin*, **87**, 1384–1395.

HAWKINS, J.W. 1989. *Cruise report – ROUNDABOUT Expedition, Legs 14, 15, R/V Thomas Washington*. Scripps Institution of Oceanography reference series, La Jolla, California, **89–13**.

—— BLOOMER, S.H., EVANS, C.A. & MELCHIOR, J.T. 1984. Evolution of intra-oceanic arc-trench systems. *Tectonophysics*, **102**, 175–205.

HERZER, R.H. & EXON, N.F. 1985. Structure and basin analysis of the southern Tonga forearc. *In:* SCHOLL, D.W. & VALLIER, T.L. (eds) *Geology and offshore resources of the Pacific island arcs – Tonga region*. Circum-Pacific Council for Energy and Mineral Resources Earth Science Series, **2**, 55–74.

HUSSONG, D.M. & UYEDA, S. 1982. Tectonic processes and the history of the Mariana Arc: A synthesis of the results of Deep Sea Drilling Project Leg 60. *In:* HUSSONG, D.M., UYEDA, S. et al. (eds) *Initial Reports of the Deep Sea Drilling Project*, **60**, 909–929.

KARIG, D.E. 1970. Ridges and Basins of the Tonga Kermadec island arc system. *Journal of Geophysical Research*, **75**, 239–254.

LEDBETTER, J.K. & HAGGERTY, J. 1994. Sedimentation history of the Tonga forearc at Ocean Drilling Program Site 840. *In:* HAWKINS, J.W., PARSON, L.M., ALLAN, J.F. et al. (eds) *Proceedings of the Ocean Drilling Program, Scientific Results*, **135**.

PACKMAN, G.H. 1978. Evolution of a simple island arc: the Lau Tonga Ridge. *Bulletin of the Australian Society of Exploration Geophysicists*, **9**, 133–140.

PARSON, L.M. & HAWKINS, J.W. 1994. Two stage ridge propagation and geological history of the Lau back-arc basin. *In:* HAWKINS, J.W., PARSON, L.M., ALLAN, J.F. et al. (eds) *Proceedings of the Ocean Drilling Program, Scientific Results*, **135**, 819–828.

——, ——, ALLAN, J.F. et al. 1992. *Proceedings of the Ocean Drilling Program, Initial Results*, **135**, College Station, Texas.

——, —— & HUNTER, P.M. 1992. Morphotectonics of the Lau Basin Seafloor – Implications for the opening history of back-arc basins. *In:* PARSONS, L.M., HAWKINS, J.W., ALLAN, J.F. et al. (eds) *Proceedings of the Ocean Drilling Program, Initial Reports*, **135**, 81–82.

——, PEARCE, J.A, MURTON, B.A. et al. 1990. Role of ridge jumps and ridge propagation in the tectonic evolution of the Lau back-arc basin, southwest Pacific. *Geology*, **18**, 470–473.

SCOTT, R.B., KROENKE, L., ZAKARIADZE, G. & SHARASKIN, A. 1981. Evolution of the South Philippine Sea: Deep Sea Drilling Project Leg 59 results. *In:* KROENKE, L., SCOTT, R.B., et al. (eds) *Initial Reports of the Deep Sea Drilling Project*, **59**, 803–817.

SHANE, P.A.R. & FROGGATT, P.C. 1991. Glass chemistry, paleomagnetism and correlation of middle Pleistocene tuffs in southern North Island, New Zealand and western Pacific. *New Zealand Journal of Geology and Geophysics*, **34**, 203–211.

TAPPIN, D.R. & BALLANCE, P.F. 1991. Contributions to the sedimentary geology of 'Eua Island – Kingdom of Tonga: reworking in an oceanic forearc. *In*: STEVENSON, A., HERZER, R.H. & BALLANCE, P. (EDS) *Contributions to the marine and onland geology and resource assessment of the Tonga-Lau-Fiji region*. SOPAC technical bulletin **8**, Springer Verlag, Berlin, 1–20.

——, HERZER, R.H. & STEVENSON, A.J. 1993. Structure and history of an oceanic forearc – The Tonga Ridge 22° to 26° south. *In*: STEVENSON, A., HERZER, R.H. & BALLANCE, P. (eds) *Contributions to the marine and onland geology and resource assessment of the Tonga-Lau-Fiji region.* SOPAC TECHNICAL BULLETIN **8**, Springer Verlag, Berlin, 20–42.

——, BRUNS, T., GEIST, E.L. & LAVOIE, D. 1994. Correlation of regional seismic stratigraphy with the sedimentary sequence at Site 840. *In*: HAWKINS, J.W., PARSON, L.M., ALLAN, J.F. *et al.* (eds) *Proceedings of the Ocean Drilling Program, Scientific Results*, **135**, 367–372.

TAYLOR, B. 1990. Rifting and the volcanic-tectonic evolution of the Izu-Bonin-Mariana Arc. *In*: TAYLOR, B., FUJIOKA, K., JANECEK, T.R. *et al.* (eds) *Proceedings of the Ocean Drilling Program, Scientific Results,* **126**, 627–652.

WEISSEL, J.K. & WATTS, A.B. 1975. Tectonic complexities in the south Fiji marginal basin. *Earth and Planetary Science Letters*, **28**, 121–126.

WALKER, G.P.L. 1980. The Taupo Pumice: product of the most powerful known (ultraplinian) eruption. *Journal of Volcanology and Geothermal Research*, **8**, 69–94.

Geochemistry of Lau Basin volcanic rocks: influence of ridge segmentation and arc proximity

JULIAN A. PEARCE[1], MICHELLE ERNEWEIN[2], SHERMAN H. BLOOMER[3], LINDSAY M. PARSON[4], BRAMLEY J. MURTON[4] & LYNN E. JOHNSON[5]

[1] *Department of Geological Sciences, University of Durham, Durham DH1 3LE, UK*

[2] *Deptartment of Geology, University of Geneva, 1211 Geneva 4, Switzerland*

[3] *Department of Geology, Boston University, Boston, Mass. 02215, USA*

[4] *Institute of Oceanographic Sciences, Wormley, Godalming GU8 5UB, Surrey, UK*

[5] *Naval Research Laboratory, MC 5110, Washington DC 20375–5000, USA*

Abstract: Side-scan sonar imaging of the Central Lau Basin (SW Pacific) has revealed a Central Lau Spreading Centre (CLSC) propagating southwards at the expense of an Eastern Lau Spreading Centre (ELSC) with a small Intermediate Lau Spreading Centre (ILSC) forming a 'relay' between the two. Volcanic rocks sampled along these spreading centres, and from two adjacent seamounts, are glassy to fine-grained pillow lavas and sheet flows of basalts, ferrobasalts and andesites. The evolved rocks are mostly confined to the propagating tip of the CLSC and can be explained by a high rate of cooling relative to magma supply, as invoked for magma genesis at propagating ridges elsewhere. Compared with equivalent rocks from the eastern Pacific, the most evolved members of the CLSC suite require similarly high degrees (>90%) of fractional crystallization from their basaltic parents. Fractional crystallisation cannot, however, account for the compositional differences between CLSC and ELSC lavas. Whereas the composition of the CLSC lavas lies just within the compositional spectrum of typical N-MORB, the ELSC lavas are distinctly enriched in alkali and alkaline earth elements, reach oxide and apatite saturation at lower Fe, Ti and P concentrations and generally show greater vesicularity despite slightly greater depths of eruption, all indicative of a water-rich subduction component. They also have lower contents of Ni, Sc, Na and Fe and higher contents of Si at a given MgO concentration that indicate a more depleted and more hydrous mantle beneath the ELSC compared with the CLSC. These results provide further evidence that, beneath the Central Lau Basin, the source composition changes progressively from MORB-type to island arc tholeiite type as the subduction zone is approached, both eastwards from the CLSC to ELSC and southwards along the ELSC to the Valu Fa Ridge. They also indicate that the composition of the subduction component may vary systematically away from the arc, with Th, LREE, Ba, Rb and H (as H_2O) all present close to the arc, only Ba, Rb and H_2O present at intermediate distances and just H_2O perceptible at the furthest distances.

The Lau Basin is a triangular-shaped, actively spreading back-arc basin separating the Lau Ridge remnant arc from the active Tofua (Tonga) arc (Karig 1970, 1971; Sclater *et al.* 1972; Hawkins 1974) (Fig. 1a). It is characterized by shallow water depths (2000–3000 m), a block-faulted, rough topography, high heat flow, poorly defined magnetic anomalies, a thin sediment cover and active basaltic volcanism (Sclater *et al.* 1972; Hawkins 1974, 1976; Weissel 1977). Various models such as diffuse spreading (Lawver & Hawkins 1978) and ridge jumps (Weissel 1977) have been proposed to explain the mode of crustal accretion in the basin. Well-defined spreading segments had, until recently, been identified only locally (Weissel 1977; Morton & Sleep 1985; Foucher 1986; von

Stackelberg *et al.* 1988). However, a GLORIA side-scan survey of the central part of the basin (Parson *et al.* 1990) enabled the active spreading axes in the Central Lau Basin to be recognized as three well-defined 'neovolcanic' zones (Fig. 1b): the Central Lau Spreading Centre (CLSC), which runs down the central part of the basin and is propagating southwards; the Eastern Lau Spreading Centre (ELSC), which is located in the eastern part of the basin, some 75 km closer to the active arc than the CLSC, and is retreating southwards; and the Intermediate Lau Spreading Center (ILSC), which is interpreted as a small 'relay' spreading segment located between the CLSC and ELSC. The ELSC forms the northern extension of the previously-identified Valu Fa Ridge (Scholl & Vallier 1985; Foucher

From Smellie, J.L. (ed.), 1995, *Volcanism Associated with Extension at Consuming Plate Margins*, Geological Society Special Publication No. 81, 53–75.

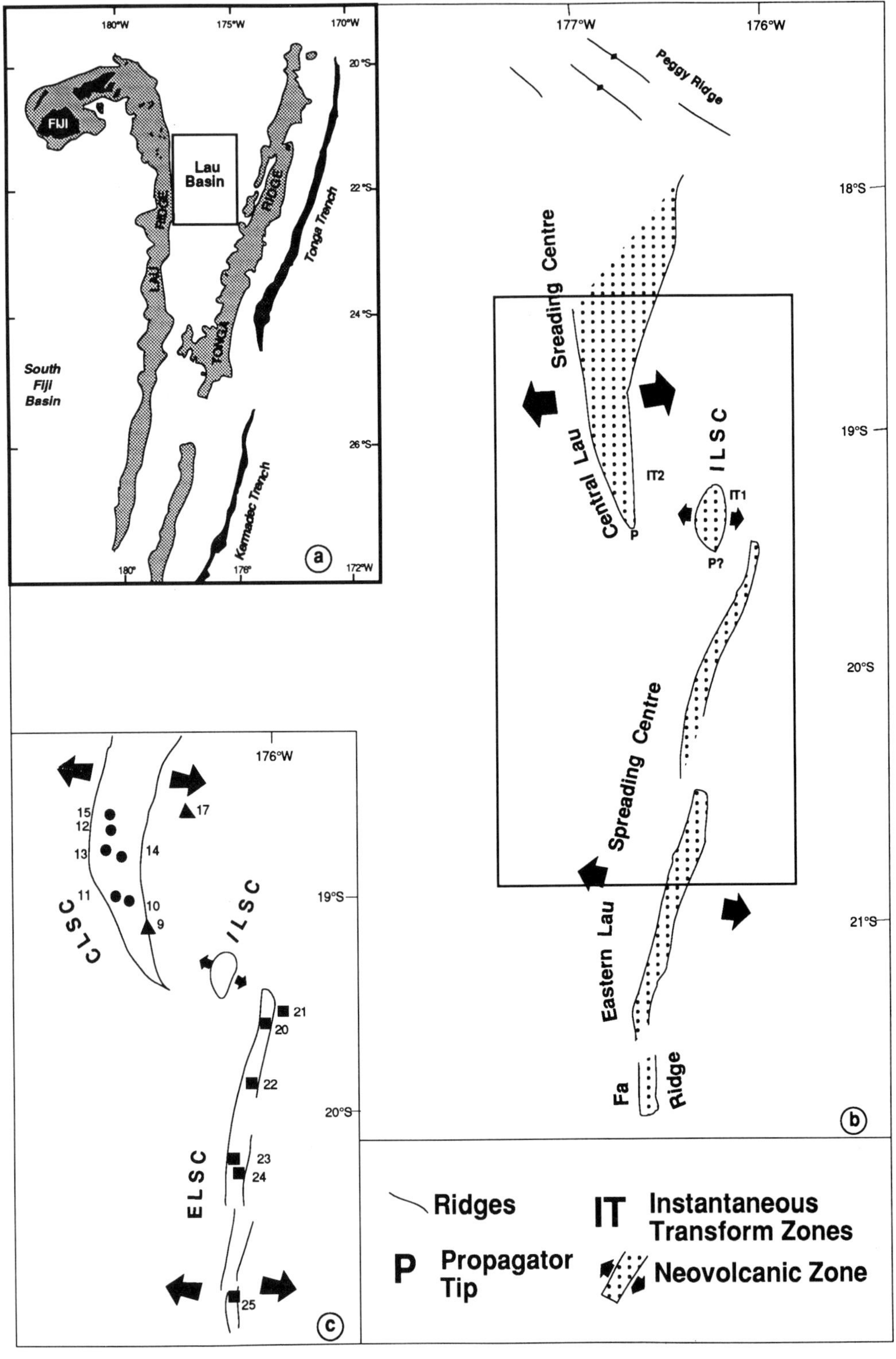

180°W
175°W
170°W
20°S
FIJI
Lau
Basin
22°S
LAU RIDGE
TONGA RIDGE
Tonga Trench
24°S
South
Fiji
Basin
26°S
Kermadec Trench
180°
176°
172°W
a

177°W
176°W
Peggy Ridge
18°S
Sreading Centre
Central Lau
IT2
P
ILSC
IT1
P?
19°S
Spreading Centre
20°S
Eastern Lau
21°S
Fa
Ridge
b

176°W
15
12
17
13
14
11
10
9
CLSC
ILSC
19°S
21
20
22
20°S
ELSC
23
24
25
c

Ridges
IT   Instantaneous
      Transform Zones
P   Propagator
     Tip
Neovolcanic Zone

1986). The first aim of this paper is to investigate the relationship between lava geochemistry and segmentation of the Lau Basin spreading axes.

Many lavas dredged from active back-arc basin spreading centres closely resemble mid-ocean ridge basalts (MORBs), although basalts transitional between MORBs and island-arc tholeiites (defined in terms of their intermediate level of enrichment in large ion lithophile (LIL) with respect to high field strength (HFS) elements) have been reported from the Mariana Trough (Hart *et al.* 1972; Fryer *et al.* 1981; Hawkins & Melchior 1985; Sinton & Fryer 1987; Volpe *et al.* 1987) and east Scotia Sea (Saunders & Tarney 1979; Tarney *et al.* 1981).

Geochemical investigations in the Lau Basin have suggested that at least its central parts are floored with MORB-type oceanic crust with some hot-spot influence in the northern reaches of the basin near the Samoa hot-spot (Hawkins 1974, 1976; Gill 1976; Hawkins & Melchior 1985; Volpe *et al.* 1988). Transitional characteristics were first identified in samples of old Lau Basin crust in the western part of the basin (Hawkins & Melchior 1985) and recent ODP Leg 135 drilling has extended the recognition of transitional and arc-like crust to other, older parts of the basin (Shipboard Scientific Party 1992). Of the active spreading centres, only the Valu Fa Ridge at the southern end of the ELSC at the extreme SE edge of the basin (Fig. 1b) has so far been shown to have transitional affinities (Jenner *et al.* 1987). From a geochemical point of view, therefore, the area between the Central Lau Basin and the northern part of Valu Fa Ridge appears to be a key zone where changes of source composition from MORB-type to transitional type may occur. The second aim of this paper is therefore to examine the relationship between lava chemistry and arc-spreading ridge distance.

Dredge site location and sample description

The samples used for this study were dredged in the central part of the Lau Basin between the latitudes of 18°50'S and 20°35'S during Cruise 33 of the RRS Charles Darwin (CD33) (Parson *et al.* 1990). The dredge sites (Fig. 1c) are concentrated along the neovolcanic zones identified from GLORIA images. Six of the dredge sites (10–15) lie between 18°50'S and 19°10'S along the floor or walls of the well-defined axial graben in the southern part (propagating tip) of the CLSC; six of the sites (20–25) lie between 19°25'S and 20°35'S along the ELSC, on the small ridges or troughs that form the less clearly defined neovolcanic zone in this area; two sites (18 and 41) lie on the ILSC; and two others (17 and 9) lie on small off-axis seamounts, the first east of the CLSC and the second near the southernmost extension of the CLSC. We have divided the rocks from each dredge haul into sub-groups according to flow type and petrography. We summarize the principal characteristics of the dredge hauls in Table 1.

Pillow and/or sheet flow fragments, usually with glassy rims, are the main constituents of all the dredge hauls, although dredge sites D10, D13 and D15 provided exclusively glass fragments. Many of the samples are remarkably fresh and most others are only slightly altered with Mn- and/or Fe-oxy-hydroxide rims coating surfaces and fractures; pervasive alteration, in the form of oxy-hydroxide coatings and secondary minerals (zeolites and clays) in the vesicles and groundmass, is restricted to samples from dredge site D14 and the seamount sites D9 and D17. It is likely that these sites represent older, rather than neovolcanic, crust.

Almost all the CLSC and ILSC samples are aphyric to sparsely plagioclase-phyric, fine-grained to glassy lavas, commonly with a few percent of small vesicles. The plagioclase microphenocrysts or phenocrysts are found together with clinopyroxene microphenocrysts in D10 and D11, and with scarce pyroxene and olivine phenocrysts in D15. Lavas from D14 are the exception, being characterized by olivine (with Cr-spinel) and plagioclase phenocrysts and microphenocrysts set in a cryptocrystalline to fine-grained, sometimes altered, groundmass of plagioclase and clinopyroxene. The D14 samples also have higher vesicularity, 10–15 vol.% on average, and some samples (in sub-group 14–4) are scoriaceous.

The ELSC samples also range in texture from fine-grained to glassy. They include aphyric to plagioclase- and/or olivine-phyric lavas. Clinopyroxene is present in small amounts in most of the samples. Vesicularity is moderate (up to 10 vol.%) in lavas from D20 to D22, but becomes important in lavas from D23 to D25 which typically contain 30–40 (sometimes up to 50) vol.% of micro- and macrovesicles. Pillow fragments from the seamounts are vesicular (20–30 vol.%) and contain olivine, plagioclase and minor clinopyroxene phenocrysts in a finely crystalline plagioclase-clinopyroxene matrix.

Fig. 1. Location of the studied area and the dredge sites. (a) Lau Basin location map with box showing location of sample area. (b) Central Lau Basin spreading ridge geometry from Parson *et al.* (1990). (c) Location of CD33 dredge stations.

Table 1. *Summary of dredge locations and sample descriptions*

Dredge	Lat. (S)	Long. (W)	Depth (m)	Location	Sample description
CLSC					
15	18°53.0	176°33.7	2280–2310	axial graben	Pl-phyric sheet flow glass
12	18°55.3	176°33.2	2320–2370	rift slope	glassy aphyric to sparsely Pl-phyric pillow-basalt; slight alteration in some samples
14	18°58.4	176°31.8	2200–2250	scarp on rift flank	vesicular, moderately Ol- and Pl-phyric basalt; moderately altered; 14–4 is scoriaceous
13	18°58.4	176°34.0	2240–2270	axial graben	aphyric glass fragments
11	19°06.9	176°33.2	2250	rift near tip	aphyric to sparsely Pl-Px-phyric glassy pillow and glass fragments
10	19°08.5	176°31.7	2260	rift near tip	sparsely Pl-Px-phyric glass
ILSC					
18	19°04.3	176°13.0	2650	ILSC rift	aphyric altered basalt; glass fragments
41	19°16.6	176°10.4	2980–3120	ILSC rift	moderately Pl-phyric pillow fragments and glassy sheet flow fragments
ELSC					
21	19°26.4	175°55.6	3000	rift flank	moderately Pl-Ol-phyric glassy basalt; 21–4 is Pl-phyric glass; 5–8% vesicles; Mn-ox rim to crust.
20	19°29.3	175°57.8	2640	central part of axis	moderately Pl-Ol-phyric sheet flow and pillow basalt with less than 5% vesicles.
22	19°41.3	175°59.6	2730	central part of axis	aphyric to sparsely Pl-phyric glassy sheet flow; pillow basalt; 22–6 and –8 are aphyric glass fragments; up to 10% vesicles.
23	20°03.4	176°08.0	2680–2620	graben	highly vesicular (up to 50%) aphyric to sparsely Ol-phyric glassy to fine-grained sheet flow and pillow fragments; 23–3 and some glass fragments are massive.
24	20°05.5	176°05.2	2700	graben	Pl-phyric glass fragments.
25	20°35.0	176°08.6	2620	central graben and wall	vesicular (30–40%) aphyric to moderately Pl-phyric pillow and sheet flow fragments with glassy rims
Seamounts					
9	19°18.6	176°27.6	1430	seamount SE of CLSC tip	vesicular (~20%) Pl-Ol-Px-phyric pillow basalts; 9–2 is scoriaceous and aphyric; thin Mn-ox rims.
17	18°51.2	176°16.9	2300	seamount E of CLSC	vesicular (20–30%) moderately Ol-phyric pillows with glassy rims; moderate alteration.

Rock composition

We have analysed about 90 representative dredge samples (whole rock and glass, all crushed in an agate mortar) for major, minor and trace element (Cr, Ni, Co, Cu, Zn, V and Sc) contents by atomic absorption spectroscopy and colorimetry (P_2O_5 only) at Newcastle upon Tyne University (see Pearce *et al.* 1986 for analytical details). We then analysed a subset of these samples by ICP-MS (VG Plasmaquad) at Durham University for the incompatible trace elements, Rb, Sb, Y, Zr, Nb, Cs, Ba, the rare earth elements (REE), Hf, Ta, Pb, Th and U. To prepare samples for analysis, 0.1 g of powder was dissolved by standard acid attack (using HF and HNO_3), spiked by Rh and Re internal standards and run in a nitric acid matrix at a $0.1 \, \text{g.l}^{-1}$

concentration of total dissolved solids with a dwell-time of $250 \, \mu s$ for each element analysed. The correction procedure includes blank-subtraction, drift-monitoring and correction for oxide/hydroxide interferences and isotopic overlaps. Calibration is by a set of laboratory and international standards. Detection limits (3 sigma of background) for the elements in Table 3 are 0.01–0.02 ppm for all elements except Zr, Nd, Sm, Gd (.05 ppm), and Ba, La, Pb (0.1–0.2 ppm). The precision (same solution run at different times) typically ranges from about 10% at 10× detection limit to about 3% at 100× detection limit. Calculation of precision using different sample preparations doubles these values. All analyses were duplicated. AA analyses are given in Table 2 and selected ICP-MS data are given in Table 3.

Table 2. *AA analysis of representative Lau Basin Volcanic rocks*

sample no:	SiO$_2$	TiO$_2$	Al$_2$O$_3$	Fe$_2$O$_3$	MnO	MgO	CaO	Na$_2$O	K$_2$O	P$_2$O$_5$	LOI	Total	Cr	Co	Ni	Zn	Cu	Sc	V	Li	Rb	Sr
CLSC																						
15-1-2:	50.6	1.16	15.89	10.74	0.18	7.39	12.12	2.42	0.04	0.07	−0.31	*100.33*	335	38	79	86	106	41	318	7	<2	94
15-1-4:	50.3	1.12	15.85	10.85	0.17	7.32	12.32	2.43	0.04	0.07	−0.62	99.85	337	40	75	86	107	45	332	7	<2	108
15-1-5:	50.7	1.18	15.90	10.80	0.17	7.18	12.19	2.45	0.04	0.07	−0.67	*100.05*	328	40	74	84	107	40	359	7	<2	94
15-1-6:	52.2	1.51	13.61	13.69	0.22	6.57	10.91	2.60	0.13	0.12	−0.41	*101.14*	82	49	52	111	98	40	436	9	<2	81
12-1-1:	51.3	1.64	13.79	13.63	0.23	6.67	10.98	2.55	0.08	0.11	−0.27	*100.76*	154	46	61	111	99	45	481	8	4	75
12-1-2:	51.2	1.59	13.86	13.41	0.23	6.73	10.94	2.54	0.12	0.11	−0.25	*100.53*	153	42	62	109	98	38	443	7	<2	79
12-2-1:	51.5	1.65	13.86	13.64	0.22	6.58	11.00	2.58	0.11	0.11	−0.27	*101.00*	151	45	58	112	96	44	468	7	<2	88
12-2-2:	50.8	1.68	14.02	13.49	0.22	6.75	11.01	2.56	0.09	0.12	−0.44	*100.34*	147	42	57	110	116	41	435	7	<2	73
12-3-3:	50.1	1.62	14.08	13.41	0.22	6.71	11.08	2.55	0.10	0.11	−0.34	99.64	158	44	62	114	109	43	444	7	2	80
12-4-3:	51.1	1.65	13.76	13.42	0.21	6.68	11.02	2.56	0.10	0.12	−0.15	*100.50*	159	42	57	111	101	45	469	7	<2	74
12-5-1:	58.6	1.54	12.69	12.55	0.20	2.63	6.48	4.17	0.31	0.32	1.17	*100.64*	14	19	13	125	49	25	213	21	8	73
12-5-3:	58.9	1.57	12.67	12.55	0.20	2.63	6.42	4.24	0.30	0.34	1.19	*100.98*	17	25	10	128	47	24	224	22	6	73
14-1-2:	50.1	0.84	16.09	9.25	0.15	8.76	12.47	2.29	0.13	0.07	0.33	*100.48*	460	40	139	77	107	38	248	4	2	166
14-1-4:	49.8	0.85	16.38	9.31	0.15	8.62	12.41	2.21	0.14	0.06	0.41	*100.35*	440	42	132	79	122	40	263	7	3	165
14-2-1:	50.8	0.86	15.85	9.71	0.16	8.43	11.96	2.37	0.11	0.07	0.32	*100.66*	390	39	124	81	113	38	260	6	3	175
14-4-1:	50.3	0.83	15.84	9.70	0.19	7.79	12.47	2.23	0.10	0.06	0.11	99.61	278	39	87	76	113	41	298	4	2	166
14-5-1:	50.2	0.93	16.12	9.66	0.17	8.36	11.54	2.42	0.04	0.06	0.61	*100.11*	350	37	131	77	110	34	270	4	<2	149
13-1-1:	60.5	1.20	12.80	12.00	0.22	1.48	5.42	4.36	0.39	0.35	1.11	99.83	10	18	5	169	33	18	67	27	10	73
13-1-2:	60.2	1.28	12.70	11.90	0.21	1.46	5.42	4.35	0.38	0.38	1.36	99.64	9	17	4	177	33	20	55	27	9	74
13-1-3:	59.9	1.25	12.73	12.41	0.21	1.35	5.28	4.25	0.36	0.38	1.58	99.73	8	17	4	170	32	21	64	28	8	74
13-1-4:	60.9	1.25	12.70	12.48	0.21	1.36	5.34	4.26	0.37	0.38	1.24	*100.49*	5	17	4	175	34	18	61	28	8	80
13-1-6:	60.4	1.27	12.70	12.88	0.22	1.44	5.38	4.31	0.47	0.40	0.62	*100.13*	5	18	6	170	34	15	74	27	9	88
11-1-1:	50.3	2.77	12.06	18.76	0.28	4.61	8.62	2.95	0.11	0.20	−0.63	*100.08*	10	49	25	166	90	43	638	11	4	74
11-1-2:	50.9	2.55	11.91	18.38	0.28	4.64	8.59	2.96	0.11	0.19	−0.55	*100.00*	9	50	20	167	92	41	639	11	4	83
11-1-5:	50.8	2.57	12.14	18.85	0.29	4.64	8.75	2.96	0.12	0.20	−0.57	*100.79*	8	48	21	165	93	41	618	11	<2	73
11-2-1:	51.0	2.50	12.08	18.91	0.28	4.70	8.66	2.99	0.09	0.20	−0.82	*100.62*	8	48	21	165	89	44	611	11	<2	71
11-2-2:	50.9	2.49	12.03	18.93	0.28	4.70	8.72	2.95	0.09	0.20	−0.89	*100.40*	9	50	15	166	89	44	628	12	>2	71
11-2-3:	50.5	2.62	12.12	19.30	0.28	4.64	8.92	2.96	0.10	0.20	−0.91	*100.73*	11	46	16	173	95	39	663	30	6	73
10-1-1:	52.3	2.61	12.06	18.19	0.27	3.24	7.60	3.24	0.22	0.41	−0.68	99.51	7	36	16	185	50	38	305	17	5	78
10-1-2:	51.3	2.67	12.22	18.21	0.27	3.32	7.88	3.32	0.17	0.40	−0.50	99.30	8	39	13	189	51	38	320	17	<2	80
10-1-3:	51.5	2.53	12.27	18.39	0.27	3.30	7.92	3.34	0.17	0.43	−0.47	99.70	7	38	15	187	50	35	288	18	4	80

Table 2. *continued*

sample no:	SiO₂	TiO₂	Al₂O₃	Fe₂O₃	MnO	MgO	CaO	Na₂O	K₂O	P₂O₅	LOl	Total	Cr	Co	Ni	Zn	Cu	Sc	V	Li	Rb	Sr
ILSC																						
18-1-1:	48.0	1.12	16.88	9.84	0.15	10.00	11.82	2.46	0.05	0.03	−0.40	*99.96*	424	49	206	68	114	35	249	7	<2	104
18-1-2:	48.4	1.16	17.00	9.97	0.16	10.23	11.53	2.40	0.02	0.04	−0.60	*100.30*	440	50	212	72	109	38	266	7	<2	101
41-1-1:	51.2	0.85	16.35	8.90	0.15	8.28	12.91	1.88	0.11	0.05	0.04	*100.72*	380	43	90	68	95	41	280	7	<2	80
41-2-1:	50.5	0.82	16.18	8.86	0.15	8.11	12.82	1.84	0.06	0.06	−0.23	*99.16*	399	43	85	67	90	33	305	6	<2	88
41-2-2:	50.2	0.77	16.14	8.83	0.15	8.20	12.71	1.86	0.06	0.06	0.00	*98.98*	393	39	89	69	93	34	290	6	<2	83
41-3-1:	50.1	0.78	16.65	8.60	0.14	7.99	12.84	1.82	0.07	0.05	0.05	*99.14*	385	40	88	66	92	40	278	7	<2	83
41-3-2:	50.9	0.79	16.58	8.79	0.15	8.13	12.95	1.84	0.06	0.05	−0.23	*100.00*	397	41	89	68	93	34	280	6	<2	83
ELSC																						
21-1-1:	51.4	1.02	15.12	10.39	0.16	8.09	12.57	1.98	0.08	0.07	−0.24	*100.67*	437	38	94	80	104	35	326	8	3	65
21-2-1	51.5	0.85	16.20	9.06	0.15	8.06	12.94	1.74	0.07	0.05	0.02	*100.63*	481	37	108	72	92	32	314	7	<2	59
21-3-1:	51.6	0.80	16.61	9.28	0.15	8.22	12.42	1.82	0.07	0.06	−0.02	*101.02*	440	37	135	72	95	34	318	4	<2	64
21-4-1:	51.5	0.80	16.38	9.13	0.15	8.11	12.61	1.80	0.04	0.05	−0.48	*100.14*	480	40	112	73	89	34	288	6	<2	60
20-1-1:	51.9	1.00	14.83	11.23	0.19	7.73	11.61	2.00	0.04	0.06	−0.20	*100.43*	260	41	75	87	104	35	374	7	<2	59
20-1-2:	51.3	0.99	14.79	11.27	0.18	7.76	11.50	2.02	0.05	0.07	−0.12	*99.85*	201	44	81	87	100	39	368	7	<2	60
20-2-1:	51.2	1.03	14.70	11.04	0.18	7.83	11.68	2.03	0.06	0.06	−0.16	*99.68*	232	41	90	86	103	35	354	9	<2	60
20-2-3:	50.9	0.90	15.77	10.17	0.17	8.49	12.22	2.01	0.05	0.06	−0.08	*100.72*	323	42	110	78	99	35	331	7	<2	63
20-3-2:	51.5	0.95	14.82	11.11	0.19	7.98	11.88	2.01	0.05	0.06	−0.25	*100.33*	203	43	85	87	97	38	371	7	<2	61
20-4-1:	51.6	0.90	14.74	10.98	0.18	7.95	11.97	1.92	0.03	0.06	−0.28	*100.09*	213	42	73	87	103	39	365	7	<2	54
20-5-1:	58.8	1.00	12.63	12.37	0.19	2.58	6.42	4.17	0.31	0.07	−0.47	*98.11*	210	40	69	90	85	39	366	7	2	73
20-5-2:	58.1	1.01	12.53	12.61	0.19	2.59	6.36	4.08	0.31	0.06	−0.55	*97.32*	168	44	64	93	88	35	373	7	<2	59
22-1-1:	51.3	1.01	14.87	10.59	0.17	7.72	12.32	2.12	0.07	0.07	−0.03	*100.20*	311	40	70	81	111	36	354	7	<2	70
22-2-1:	51.6	1.22	14.57	11.90	0.19	7.06	11.28	2.43	0.07	0.09	0.00	*100.46*	81	44	47	91	104	36	393	7	2	76
22-3-1:	52.0	1.21	14.59	12.02	0.19	7.04	11.22	2.43	0.07	0.09	0.04	*100.94*	84	41	55	91	109	36	394	7	4	76
22-4-1:	52.0	1.18	14.57	12.01	0.19	7.03	11.31	2.44	0.09	0.09	−0.12	*100.83*	64	40	46	92	109	35	388	8	<2	76
22-5-1:	52.1	1.19	14.61	12.06	0.19	7.02	11.18	2.45	0.07	0.09	−0.03	*100.92*	65	42	49	91	105	35	399	7	2	75
22-6-1:	51.4	1.19	14.53	11.84	0.19	7.06	11.08	2.39	0.06	0.09	−0.47	*99.36*	80	42	50	94	100	39	406	7	3	76
22-6-2:	52.5	1.21	14.43	11.95	0.19	7.05	11.07	2.42	0.06	0.09	−0.39	*100.54*	77	41	48	96	99	40	399	8	2	75
22-8-1:	51.5	1.20	14.44	11.86	0.19	7.08	11.13	2.37	0.06	0.09	−0.34	*99.57*	79	42	48	96	98	40	416	7	<2	73
22-8-2:	51.1	1.24	14.48	11.84	0.19	7.05	11.15	2.40	0.07	0.09	−0.44	*99.19*	77	39	52	109	98	40	388	7	<2	78
23-1-1:	51.4	0.72	15.85	8.62	0.15	8.34	12.57	1.74	0.10	0.07	0.94	*100.45*	387	37	99	71	128	35	305	6	<2	121
23-2-1:	50.8	0.86	15.84	9.29	0.16	8.01	12.43	1.92	0.11	0.08	0.83	*100.32*	271	37	81	72	114	36	325	4	<2	156

Table 2. *continued*

sample no:	SiO$_2$	TiO$_2$	Al$_2$O$_3$	Fe$_2$O$_3$	MnO	MgO	CaO	Na$_2$O	K$_2$O	P$_2$O$_5$	LOI	Total	Cr	Co	Ni	Zn	Cu	Sc	V	Li	Rb	Sr
23-2-2:	50.6	0.83	15.83	9.23	0.16	8.00	12.38	1.87	0.11	0.08	0.94	*100.03*	274	37	84	75	110	35	344	7	2	154
23-3-1:	52.0	1.13	14.53	11.84	0.19	7.54	11.60	2.06	0.04	0.08	−0.17	*100.86*	140	42	64	95	93	34	401	8	3	66
23-3-2:	51.5	1.11	14.47	11.90	0.19	7.45	11.48	2.08	0.05	0.08	−0.35	*99.93*	129	45	61	95	90	39	399	8	2	66
23-4-1:	51.2	0.84	15.52	9.26	0.16	8.07	12.59	1.98	0.13	0.08	0.76	*100.62*	320	36	86	75	132	40	329	7	2	155
23-4-2:	51.1	0.67	16.06	8.39	0.15	8.21	12.71	1.71	0.12	0.06	0.85	*100.06*	343	37	85	65	105	33	288	6	3	123
23-5-1:	51.9	0.71	15.75	8.65	0.15	8.33	12.66	1.69	0.11	0.06	0.66	*100.74*	297	36	89	69	145	33	291	4	3	121
23-6-1:	51.5	0.70	15.75	8.48	0.15	8.72	12.55	1.66	0.12	0.06	0.46	*100.15*	398	39	112	70	128	35	288	4	<2	116
23-7-1:	51.2	0.69	16.02	8.43	0.15	8.45	12.47	1.71	0.11	0.06	0.69	*100.01*	324	31	98	69	157	36	294	6	3	124
23-7-3:	51.8	0.71	15.85	8.61	0.15	8.85	12.58	1.73	0.10	0.06	0.42	*100.89*	390	38	106	69	135	35	295	6	2	119
23-8-1:	51.7	0.96	15.46	10.48	0.17	7.45	11.99	1.93	0.03	0.08	0.10	*100.38*	280	43	70	83	86	36	340	7	<2	80
23-8-2:	51.8	0.77	15.65	9.09	0.16	7.90	12.24	1.87	0.11	0.06	0.93	*100.63*	320	35	81	74	98	36	336	7	<2	156
24-1-1:	51.1	1.00	14.96	10.68	0.17	7.48	12.03	1.92	0.03	0.07	−0.47	*98.95*	246	42	68	87	89	38	365	7	<2	61
25-1-1:	51.1	0.97	15.90	10.38	0.17	6.57	11.25	2.11	0.16	0.08	0.75	*99.84*	114	36	56	85	101	38	371	6	<2	113
25-2-1:	52.6	1.94	13.92	14.54	0.23	4.13	8.30	3.08	0.32	0.17	0.57	*99.83*	7	34	10	126	42	35	514	8	<2	93
25-2-2:	52.3	1.94	13.80	14.20	0.22	4.20	8.38	3.06	0.22	0.17	1.02	*99.51*	9	35	9	125	37	34	529	9	4	98
25-3-1:	52.7	1.93	13.87	14.45	0.22	3.94	8.07	3.02	0.32	0.18	0.84	*99.57*	5	34	8	127	37	38	511	8	7	103
25-4-1:	51.6	1.13	15.01	11.08	0.18	6.57	11.14	2.52	0.14	0.09	0.33	*99.85*	83	42	44	83	80	44	401	6	<2	98
25-5-1:	51.3	0.96	15.93	10.31	0.17	6.50	11.29	2.11	0.17	0.08	0.14	*99.23*	116	35	55	85	90	34	376	6	<2	108
25-7-1:	52.5	1.02	15.81	10.50	0.18	6.42	11.07	2.21	0.13	0.08	0.45	*100.40*	111	38	55	85	98	33	394	6	2	110
Seamounts																						
9-1-1:	50.1	0.89	16.18	10.85	0.17	6.24	13.64	2.00	0.08	0.06	0.09	*100.30*	299	42	90	75	91	45	324	7	<2	269
9-1-2:	49.5	0.78	15.22	9.96	0.16	9.58	13.65	1.91	0.12	0.05	0.28	*101.20*	480	43	115	77	121	45	308	6	3	239
9-1-7:	49.5	0.80	15.81	10.46	0.18	8.51	13.55	2.06	0.10	0.06	0.22	*101.26*	295	45	106	74	205	45	318	4	<2	256
9-2-1:	50.1	0.98	16.02	10.97	0.19	6.93	13.15	1.98	0.13	0.07	0.21	*100.73*	136	38	53	85	160	46	355	6	<2	273
9-3-1:	48.2	0.83	14.71	10.33	0.20	9.05	13.77	1.92	0.13	0.06	1.41	*100.67*	365	43	123	84	244	44	310	8	4	304
9-3-2:	48.5	0.82	16.63	11.10	0.21	7.31	13.61	2.22	0.10	0.07	0.17	*100.68*	133	46	65	80	105	40	330	6	3	271
17-1-1:	50.3	0.89	15.51	9.52	0.16	8.45	13.37	2.16	0.22	0.06	0.32	*101.00*	378	40	88	69	150	35	315	7	4	216
17-1-2:	50.9	0.86	15.73	9.36	0.16	8.09	13.34	2.13	0.13	0.07	0.38	*101.14*	360	39	75	69	148	38	311	4	3	215
17-1-3:	49.6	0.80	15.71	9.41	0.17	8.55	13.22	2.07	0.11	0.07	0.29	*100.01*	390	42	89	71	171	39	315	7	5	249

Table 3. *ICP-MS analyses of selected Lau Basin samples*

	CLSC												ILSC				ELSC														Seamounts				
	15.1.2	15.1.6	12.1.1	12.5.3	14.4.1	13.1.1	13.1.5	13.1.6	11.1.1	11.1.2	10.1.1	10.1.3	18.1.1	18.1.2	41.2.1	41.3.2	21.1.1	21.4.1	20.1.1	20.5.2	22.1.1	22.6.2	23.2.1	23.3.2	23.7.3	23.8.1	23.8.2	24.1.1	25.2.1	25.5.1	9.1.1	9.2.1	9.3.1	17.1.1	17.1.3
Rb	0.77	1.37	1.36	6.66	1.36	9.80	9.30	9.00	1.67	3.60	4.70	6.06	0.88	0.07	1.35	1.33	1.22	1.03	0.93	0.97	1.03	1.70	1.55	0.53	1.53	0.56	1.40	0.68	4.20	2.80	3.80	2.00	1.78	2.65	2.70
Sr	102	87	80	89	188	80	92	87	80	83	89	86	106	113	88	87	65	61	64	63	77	80	172	73	118	66	170	61	113	119	298	287	323	235	239
Y	28.1	43.2	41.7	142.3	21.3	144.5	178.2	157.4	71.0	65.9	118.4	104.7	24.3	25.4	18.6	18.1	25.1	20.1	27.1	27.4	26.9	30.0	20.0	28.6	15.8	25.4	18.0	24.6	44.8	23.1	20.2	21.5	19.3	19.7	20.5
Zr	61.6	98.8	87.0	505.0	39.6	500.1	618.5	533.6	158.6	155.1	299.0	297.4	64.0	66.0	38.6	36.7	48.0	37.5	50.0	54.0	50.9	65.4	36.7	59.0	31.6	52.3	37.9	47.2	114.5	47.2	38.8	40.8	34.8	44.1	51.3
Nb	1.06	1.52	1.32	6.10	0.63	7.62	8.42	8.49	2.66	3.18	6.00	5.95	0.25	0.29	1.28	1.30	0.86	0.76	0.83	0.94	0.86	1.67	0.40	0.95	0.38	0.71	0.54	0.67	2.14	0.69	0.66	0.83	0.69	2.39	2.77
Ba	10.2	16.3	12.8	49.8	18.0	48.7	63.3	55.0	18.3	19.3	32.2	29.2	2.0	1.3	13.1	13.8	12.1	9.6	8.6	9.1	10.3	13.5	17.8	9.5	20.3	6.9	13.5	9.1	34.5	40.7	13.9	14.9	23.5	21.5	29.8
La	1.77	3.00	2.78	14.91	1.97	14.51	17.34	16.06	4.43	4.68	9.30	8.67	1.33	1.57	1.65	1.63	1.66	1.33	1.64	1.72	1.63	2.21	1.49	1.72	1.19	1.52	1.27	1.51	3.70	1.91	2.57	2.36	2.45	3.35	3.60
Ce	6.36	10.03	9.23	47.37	5.48	45.57	55.32	50.77	15.30	15.43	29.29	29.88	5.40	5.62	5.07	5.06	4.95	4.23	5.04	5.37	5.17	7.09	4.64	5.73	3.85	5.01	4.08	5.03	12.91	6.13	6.60	7.37	5.77	8.82	9.32
Pr	1.15	1.71	1.94	7.82	0.91	7.37	9.02	8.47	2.85	2.82	5.09	5.09	1.04	1.13	0.80	0.87	0.99	0.73	0.94	1.02	0.93	1.21	0.78	1.06	0.73	0.91	0.76	0.92	2.21	1.08	1.16	1.24	1.04	1.35	1.38
Nd	7.04	10.61	10.15	41.57	5.10	40.44	50.01	43.50	16.76	15.54	31.20	30.15	6.26	6.77	4.77	4.99	5.62	4.26	5.59	5.69	5.60	7.17	5.05	6.29	4.14	5.38	4.46	5.09	12.44	6.17	6.23	6.47	5.20	6.84	6.89
Sm	2.67	3.94	3.67	13.95	1.87	12.71	16.56	14.30	6.21	5.49	10.40	10.78	2.31	2.54	1.67	1.83	2.04	1.63	2.13	2.11	2.29	2.55	1.90	2.38	1.51	2.08	1.64	2.04	4.58	2.24	1.95	2.04	1.65	2.02	2.15
Eu	0.99	1.34	1.36	3.01	0.77	3.10	4.03	3.83	2.10	2.13	2.87	3.09	1.04	0.97	0.67	0.71	0.82	0.63	0.90	0.89	0.87	1.03	0.76	0.88	0.62	0.81	0.66	0.76	1.58	0.86	0.77	0.80	0.64	0.75	0.78
Gd	3.52	5.39	4.93	17.15	2.84	18.06	21.42	19.30	8.84	7.67	13.73	13.96	3.17	3.14	2.37	2.33	3.13	2.41	3.32	3.27	3.42	3.77	2.64	3.53	2.01	2.94	2.18	2.88	6.36	2.90	2.82	2.60	2.31	2.49	2.70
Tb	0.67	1.04	0.96	3.16	0.53	3.13	3.88	3.66	1.67	1.43	2.43	2.58	0.59	0.61	0.51	0.47	0.58	0.47	0.63	0.64	0.65	0.71	0.47	0.65	0.41	0.58	0.46	0.58	1.14	0.56	0.49	0.50	0.44	0.47	0.51
Dy	4.61	7.17	6.79	22.03	3.65	21.54	27.85	25.20	11.24	11.27	16.29	17.47	4.02	4.22	3.25	3.21	4.00	3.31	4.22	4.46	4.40	4.70	3.18	4.50	2.63	3.87	2.86	4.01	7.50	3.86	3.29	3.47	2.93	3.14	3.34
Ho	1.04	1.61	1.47	4.77	0.80	4.73	6.06	5.65	2.46	2.43	3.93	4.04	0.93	0.94	0.75	0.78	0.93	0.74	0.96	0.99	0.96	1.00	0.72	1.04	0.56	0.86	0.65	0.90	1.71	0.86	0.73	0.82	0.67	0.69	0.75
Er	2.91	4.55	4.46	13.83	2.32	14.16	17.13	16.32	7.55	6.92	10.94	11.58	2.66	2.67	2.21	1.90	2.83	2.15	2.88	2.97	2.95	3.10	2.09	3.12	1.69	2.42	1.88	2.72	5.13	2.32	2.19	2.25	1.93	1.95	2.21
Tm	0.45	0.70	0.69	2.20	0.37	2.10	2.80	2.50	1.22	1.20	1.70	1.81	0.41	0.44	0.36	0.34	0.44	0.35	0.46	0.47	0.48	0.50	0.34	0.44	0.26	0.39	0.29	0.43	0.79	0.36	0.34	0.35	0.31	0.32	0.35
Yb	3.04	4.58	4.51	14.93	2.36	14.13	18.86	16.76	7.69	7.26	11.18	11.70	2.61	2.77	2.10	2.09	2.78	2.28	2.94	3.13	2.86	3.21	2.13	3.04	1.76	2.53	1.95	2.79	5.05	2.65	2.16	2.33	1.97	1.93	2.21
Lu	0.46	0.74	0.70	2.38	0.37	2.11	2.82	2.47	1.17	1.15	1.63	1.74	0.38	0.41	0.35	0.35	0.43	0.36	0.45	0.48	0.46	0.49	0.34	0.46	0.28	0.41	0.31	0.44	0.78	0.44	0.34	0.36	0.30	0.29	0.32
Hf	1.81	2.55	2.46	13.78	0.94	12.38	16.56	13.57	4.05	4.32	7.78	8.12	1.23	1.31	1.03	1.10	1.10	0.94	1.12	1.19	1.20	1.43	0.85	1.45	0.86	1.48	1.05	1.33	3.18	1.41	0.92	1.02	0.82	1.07	1.07
Ta	0.06	0.08	0.10	0.23	0.04	0.53	0.32	0.55	0.22	0.25	0.42	0.47	0.02	0.03	0.09	0.09	0.07	0.05	0.05	0.06	0.06	0.12	0.03	0.08	0.02	0.04	0.04	0.04	0.14	0.04	0.06	0.05	0.05	0.14	0.14
Pb	0.29	0.58	1.00	1.61	1.39	3.76	2.28	3.29	0.84	1.27	1.75	2.04	0.46	0.35	0.21	0.78	19.70	0.65	0.47	0.48	0.57	0.32	1.17	0.40		0.45	0.46	0.55	1.12	1.30	0.52	0.64	1.33	0.96	1.07
Th	0.07	0.12	0.11	1.13	0.12	1.09	1.00	1.11	0.24	0.29	0.57	0.56	0.03	0.02	0.08	0.08	0.09	0.06	0.06	0.06	0.09	0.12	0.05	0.06		0.06	0.04	0.06	0.27	0.08	0.08	0.11	0.08	0.24	0.32
U	0.02	0.04	0.06	0.44	0.09	0.34	0.33	0.36	0.09	0.11	0.18	0.19	0.03	0.01	0.04	0.03	0.08	0.02	0.05	0.02	0.05	0.04	0.04	0.01		0.02	0.03	0.02	0.08	0.08	0.04	0.04	0.04	0.09	0.15
Chondrite normalized																																			
La	5.69	9.66	8.97	48.10	6.35	46.81	55.94	51.81	14.29	15.10	30.00	27.97	4.29	5.06	5.32	5.26	5.35	4.29	5.29	5.55	5.24	7.14	4.81	5.55	3.84	4.90	4.10	4.86	11.94	6.16	8.29	7.61	7.89	10.81	11.63
Ce	7.87	12.41	11.42	58.63	6.78	56.40	68.47	62.83	18.94	19.10	36.25	36.98	6.68	6.96	6.27	6.26	6.13	5.23	6.24	6.65	6.39	8.77	5.74	7.09	4.76	6.20	5.05	6.23	15.98	7.59	8.17	9.12	7.13	10.92	11.53
Pr	9.43	14.02	15.90	64.10	7.46	60.41	73.93	69.43	23.36	23.11	41.72	41.72	8.48	9.26	6.58	7.13	8.07	5.94	7.70	8.36	7.58	9.89	6.39	8.69	5.98	7.46	6.23	7.54	18.11	8.85	9.51	10.16	8.48	11.07	11.31
Nd	11.73	17.68	16.92	69.28	8.50	67.40	83.35	72.50	27.93	25.90	52.00	50.25	10.43	11.26	7.95	8.32	9.36	7.10	9.32	9.48	9.33	11.95	8.42	10.48	6.89	8.96	7.43	8.48	20.73	10.28	10.38	10.78	8.67	11.40	11.48
Sm	13.67	20.18	18.82	71.54	9.59	65.18	84.92	73.33	31.85	28.15	53.33	55.28	11.82	13.03	8.56	9.38	10.46	8.33	10.91	10.84	11.74	13.08	9.74	12.21	7.72	10.65	8.41	10.44	23.49	11.49	10.00	10.46	8.46	10.36	11.03
Eu	13.47	18.16	18.50	40.95	10.48	42.18	54.83	52.11	28.57	28.98	39.05	42.04	14.15	13.13	9.05	9.66	11.16	8.50	12.24	12.11	11.84	14.01	10.34	11.97	8.37	10.98	8.98	10.39	21.50	11.70	10.48	10.88	8.71	10.20	10.67
Gd	13.59	20.81	19.03	66.22	10.97	69.73	82.70	74.52	34.13	29.61	53.01	53.90	12.22	12.12	9.15	9.00	12.08	9.31	12.81	12.62	13.19	14.56	10.19	13.63	7.74	11.35	8.42	11.13	24.56	11.20	10.89	10.04	8.90	9.61	10.42
Tb	14.24	21.94	20.25	67.23	11.18	66.03	82.55	77.22	35.23	30.17	51.27	54.43	12.34	12.76	10.65	10.00	12.24	9.92	13.22	13.55	13.71	14.98	9.92	13.71	8.65	12.24	9.70	12.24	24.05	11.91	10.34	10.55	9.28	9.92	10.79
Dy	14.32	22.25	21.09	68.42	11.34	66.89	86.49	78.26	34.91	35.00	50.59	54.25	12.48	13.11	10.08	9.97	12.42	10.28	13.11	13.84	13.65	14.60	9.86	13.98	8.17	12.02	8.88	12.45	23.29	11.99	10.22	10.78	9.10	9.75	10.37
Ho	14.42	22.35	20.47	66.43	11.14	65.88	84.40	78.69	34.26	33.84	54.74	56.27	12.88	13.09	10.45	10.86	12.88	10.35	13.42	13.79	13.30	13.93	9.96	14.48	7.73	11.93	9.05	12.53	23.82	11.98	10.17	11.42	9.33	9.61	10.40
Er	13.86	21.64	21.24	65.86	11.05	67.43	81.57	77.71	35.95	32.95	52.10	55.14	12.64	12.71	10.50	9.05	13.45	10.24	13.71	14.12	14.02	14.76	9.95	14.86	8.02	11.52	8.95	12.94	24.43	11.05	10.43	10.71	9.19	9.29	10.54
Tm	13.89	21.45	21.30	67.90	11.42	64.81	86.42	77.16	37.65	37.04	52.47	55.86	12.50	13.43	10.96	10.49	13.58	10.65	14.30	14.58	14.81	15.43	10.49	13.58	7.95	12.14	8.95	13.17	24.38	11.11	10.49	10.80	9.41	9.72	10.89
Yb	14.55	21.89	21.58	71.44	11.29	67.61	90.24	80.19	36.79	34.74	53.49	55.98	12.46	13.25	10.05	10.00	13.30	10.91	14.07	14.99	13.68	15.36	10.17	14.55	8.42	12.11	9.33	13.35	24.16	12.68	10.33	11.15	9.43	9.23	10.57
Lu	14.13	22.98	21.74	73.91	11.49	65.53	87.58	76.71	36.34	35.71	50.62	54.04	11.65	12.73	10.71	10.87	13.35	11.02	13.98	14.83	14.29	15.22	10.56	14.29	8.59	12.84	9.63	13.56	24.22	13.66	10.56	11.18	9.32	9.01	9.94

Major element variations

The major element data set for representative samples from the various dredge sites (Table 1) reveals that compositions vary from basalts to ferrobasalts and andesites, the evolved rock types being confined to the CLSC and the southernmost ELSC site. The basalts are all hypersthene- and slightly quartz-normative tholeiites with the exception of those from D14, which are olivine-normative. The quartz-normative basalts typically range from primitive to more evolved tholeiites with Mg-numbers (Mg No. = Mg × 100/(Mg + Fe)) between 70 and 52; a few samples from D25 (25–2 and –3) are more differentiated (Mg No. = 38–40). The olivine basalts from D14 have only primitive compositions (Mg No. = 64.5–68.5). Basalts from the seamounts have hypersthene and olivine in their norms and they range from primitive to slightly more differentiated tholeiites (Mg No. = 57–69). The rocks sampled near the southern tip of the CLSC (D10 and D11) have basic silica contents and contain FeO* (total Fe as FeO) up to 17 wt% and TiO_2 up to 2.8 wt% (Figure 2), thus satisfying the criteria (>12 wt% FeO* and >2 wt% TiO_2) of Byerly *et al.* (1976) and Natland (1980) for classification as 'FeTi basalts' or 'ferrobasalts'. Volpe *et al.* (1988) have already noted the presence of ferrobasalts in this area. Glass fragments from D12 and all samples from D13 are andesites in terms of their silica contents with Mg No. as low as 20 in D13 samples. In both D12 and D13, TiO_2 and FeO* are lower than in the ferrobasalts (1.2–1.6 and 11.4–11.5 respectively), suggesting that Fe-Ti-oxide fractionation has affected these samples.

For consistency with many of the papers cited below, we plot the major elements on oxide–MgO variation diagrams (Fig. 2). Lavas from the CLSC, ILSC, seamounts and ELSC are assigned separate symbols. Note, however, that here, as elsewhere, we have assigned samples from D14 to the seamount group as its vesicularity, phenocryst content, geochemical characteristics and degree of alteration all indicate an affinity with older, transitional crust rather than with the active CLSC. Liquid lines of descent can clearly be recognised, although some of the seamount samples, and rare samples from the spreading axes, plot off these lines on olivine and olivine + plagioclase cumulation trends. No liquid compositions contain more than 9 wt% MgO. The liquid lines of descent are characteristic of the basalt-ferrobasalt-andesite series, with increases in TiO_2, FeO and P_2O_5 accompanying decreases in MgO up to the point of oxide and apatite crystallization. Silica shows a continuous increase with an inflection at the point of oxide crystallization, K_2O and Na_2O both increase with a slight inflection at this point and CaO and Al_2O_3 both decrease.

Compared with CLSC-basalts with similar MgO content, the ELSC-basalts can be distinguished by their overall slightly higher SiO_2 and lower Na_2O contents (Fig. 2). At more evolved compositions, they also exhibit a lesser enrichment in FeO*, TiO_2 and P_2O_5. The seamount and ILSC lavas are difficult to compare with lavas from the CLSC and ELSC because of the effects of olivine and plagioclase cumulation.

Trace elements

Compatible and the less incompatible elements. Figure 3 depicts the variation in compatible trace element contents with MgO. Ni (Fig. 3a) decreases rapidly with decreasing MgO content in predictable response to olivine crystallization. The two olivine-phyric samples from D18 plot towards high Ni and MgO contents because of olivine cumulation. The evolved samples from D20 (20–5) on the ELSC also plot off the fractionation trend to high Ni values. In theory, this could be explained by magma mixing (which would give straight line 'chords' to the curved fractionation trends) as well as crystal cumulation. However, the fact that this effect is mirrored only by other elements compatible with olivine and chrome spinel (i.e. Cr, Co and Zn, but not V or Ti) indicates that cumulation or resorption of phenocrysts or xenocrysts of these two minerals is the likely explanation.

Co, V and Sc (Fig. 3c–e) all show slight increases in concentration with fractionation, reaching peaks at about 5 wt% MgO, after which concentrations fall off rapidly. The increases in concentration can be explained by the high proportion of plagioclase in the crystallizing assemblage which gives bulk distribution coefficients less than unity despite the compatibility of these elements in one or more mafic phase. The sudden decrease is caused by oxide crystallization, all three elements being strongly partitioned into Fe,Ti-oxides, and hence behaving essentially like Fe and Ti (Fig. 2). Ferrobasalts from D10 do not, however, follow this evolutionary trend: despite similar Ti and Fe contents, D10 is depleted in Co, Sc and, particularly V compared to D11 (610–660 ppm in D11, 280–320 ppm in D10). This depletion in V before Fe and Ti has been observed in other tholeiitic basalt suites (e.g. James *et al.* 1987),

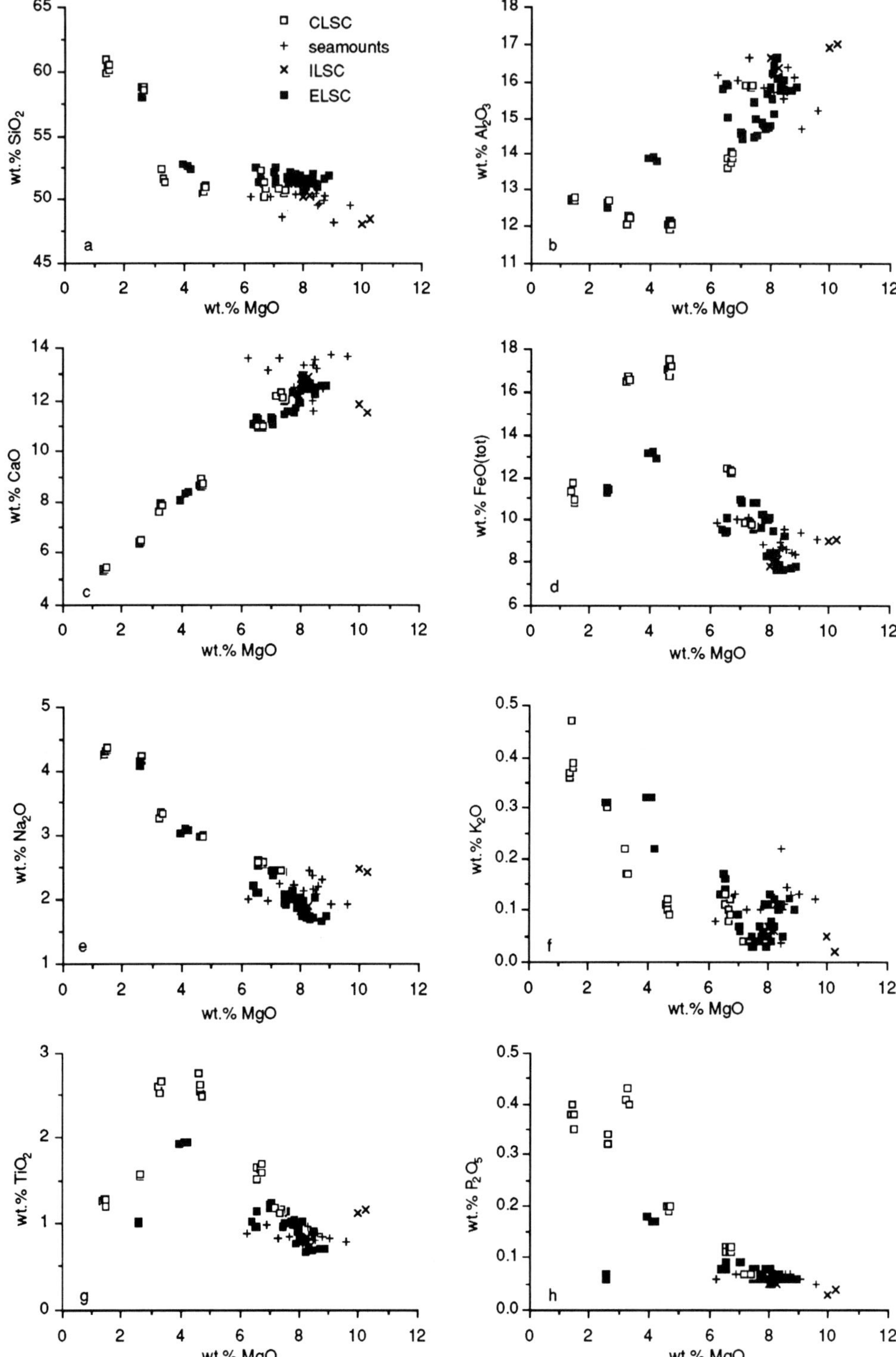

Fig. 2. Oxide-MgO covariation diagrams for CD33 samples.

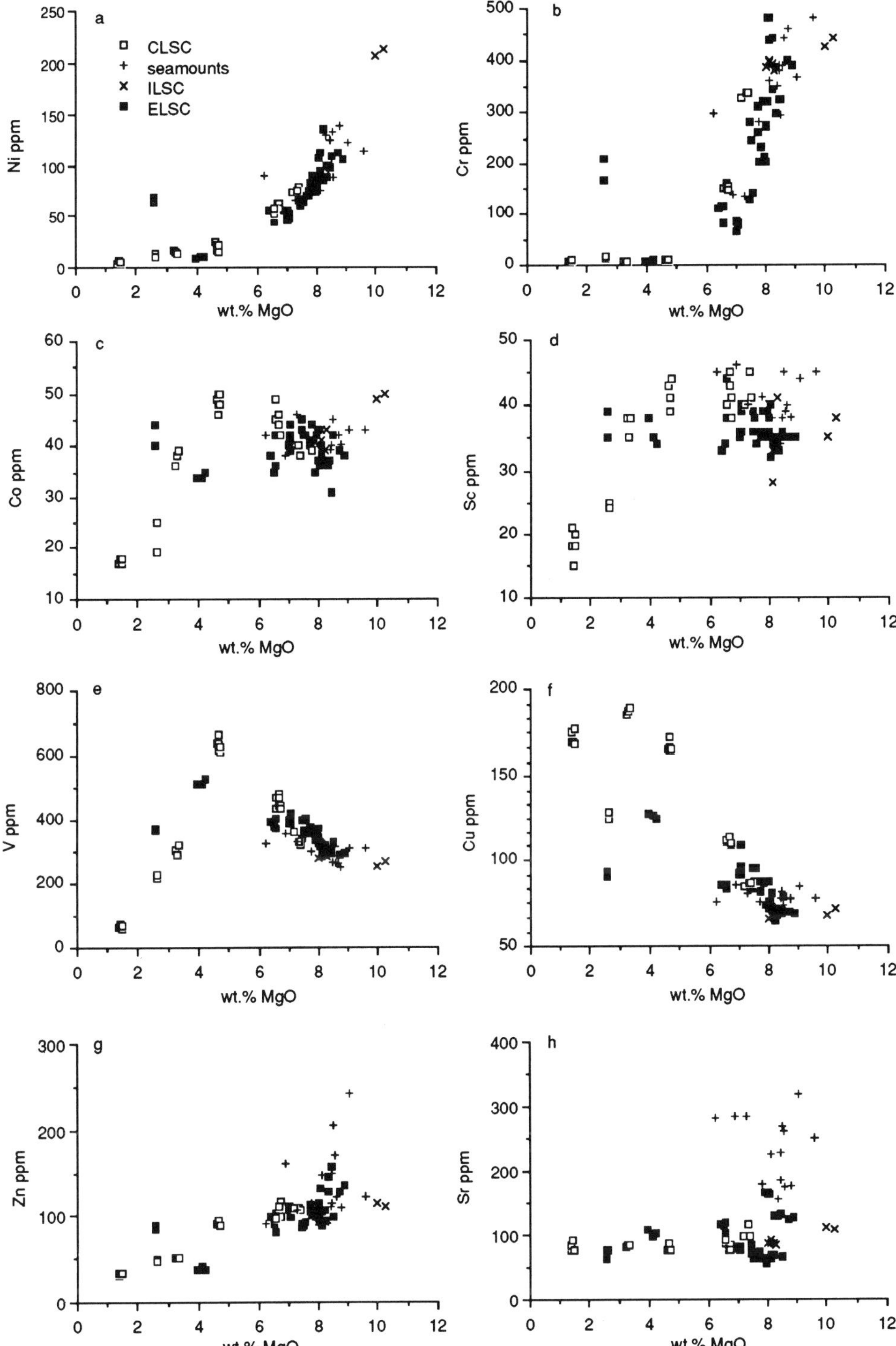

Fig. 3. Trace element-MgO covariation diagrams for CD33 samples.

and can be attributed to crystallization of a V-magnetite prior to ilmenite in response to V-saturation in the melt.

Cu (Fig. 3f) peaks at about 190 ppm at about 3 wt% MgO, perhaps reflecting segregation of a sulphide or volatile phase at this stage of crystallization. Zn (Fig. 3g) continuously decreases in concentration during crystallization (Zn has a higher olivine/liquid partition coefficient than Cu) and therefore gives an inflection, rather than peak, at this point. Some of the seamount lavas also show anomalously high Zn values. Sr (Fig. 3h) has high values in the seamounts and some axis sites (D14, some D23 and D25), an observation attributed in the next section to a subduction component, but otherwise samples plot along an almost horizontal trend with values close to 100 ppm. Given a basalt plagioclase/liquid partition coefficient of about 2, increasing slightly with fractional crystallization, this trend is consistent with about 50 wt% crystallization of plagioclase. In detail, however, the ELSC samples above about 7.5 wt% MgO exhibit decreasing Sr with increase in fractionation, indicating that plagioclase may not have started to crystallize, or was subordinate to mafic minerals, until the magma reached this MgO content.

The trace element variation diagrams also highlight some significant differences between the CLSC and ELSC lava suites. In particular, the CLSC suite contains greater contents of Ni, Cr and Sc for a given MgO content at basic compositions. Of these, the behaviour of Ni is best understood. The explanation must lie either in the partial melting process or in the early, olivine-dominated crystallization that preceeded eruption. During partial melting, Ni is more strongly buffered by residual olivine and thus increases in concentration less rapidly than MgO with increase in degree of partial melting: low Ni at a given MgO content should therefore be a characteristic of high degrees of partial melting (Sato & Tohara 1985). Alternatively, because Ni is partitioned more strongly into olivine than Mg (Kd (Ni:Mg) $\approx$ 3.5: Sato & Banno 1983), enhanced crystallization of olivine, caused for example by expansion of the olivine phase volume due to increased p_{H_2O}, will also lower the Ni content at a given value of MgO. Other processes that might reduce Ni at a given MgO content are separation of a sulphide phase (which strongly concentrates Ni) and an increase in the oxidation state of iron (which reduces the Fo content of olivine). Cr–MgO (Fig. 3b) is similar to Ni–MgO, except that it is the behaviour of chrome spinel rather than olivine that causes decoupling of Cr and MgO.

The mean Sc contents of CLSC- and ELSC-basalts are slightly different (41 and 36 respectively). The distribution coefficients for Sc compiled by Pearce & Parkinson (1993) (olivine/liquid = 0.16, orthopyroxene/liquid = 0.50, clinopyroxene/liquid = 0.85 at 1300°C) demonstrate that Sc will have a bulk distribution coefficient less that unity during melting of spinel lherzolite, and lower Sc contents could therefore be explained by a greater degree of partial melting or by melting of a more depleted source; alternatively, but less likely in this case, enhanced crystallization of clinopyroxene (clinopyroxene/liquid = 2 at 1200°C) prior to eruption could have the same effect.

Rare earth elements. Representative chondrite-normalized REE plots are presented in Fig. 4. Basalts from the CLSC have sub-parallel LREE-depleted patterns (Fig. 4a) similar to those of mid-ocean ridge basalts (N-MORB). With increasing fractionation, a Eu anomaly develops and increases in magnitude, so that highly enriched REE contents (up to 60× chondrite) with marked negative Eu anomalies are characteristic of the ferrobasalts and andesites. Rayleigh fractionation calculations show that 90% or more fractional crystallization of a plagioclase–clinopyroxene–olivine (± Fe–Ti oxide) assemblage from a parental CLSC primitive basalt liquid are necessary to produce evolved liquids with REE contents in the range of those of the CLSC ferrobasalts and andesites. Similarly high degrees of fractional crystallisation have already been invoked to explain the geochemistry of ferrobasalts and andesites sampled in other oceanic areas (e.g. Clague *et al.* 1981; Le Roex *et al.* 1982). Despite nearly identical major element compositions, the two groups of ferrobasalts (D10 and D11) are characterized by very different REE abundances indicating that closed-system fractional crystallization alone cannot account for the diversity observed among the CLSC volcanics. The ELSC patterns (Fig. 4b) lie subparallel to those of the CLSC, but do not attain such highly evolved profiles.

Typical ILSC patterns are shown in Fig. 4c. The D18 pattern falls into the range of CLSC and ELSC patterns, though with a positive Eu anomaly indicative of plagioclase cumulation. The D41 pattern indicates a slightly lower degree of LREE depletion at this site. Seamount and D14 patterns (Fig. 4d) are distinct from the LREE-depleted spreading centre patterns in showing profiles that vary from slight LREE enrichment to slight LREE depletion, reflecting

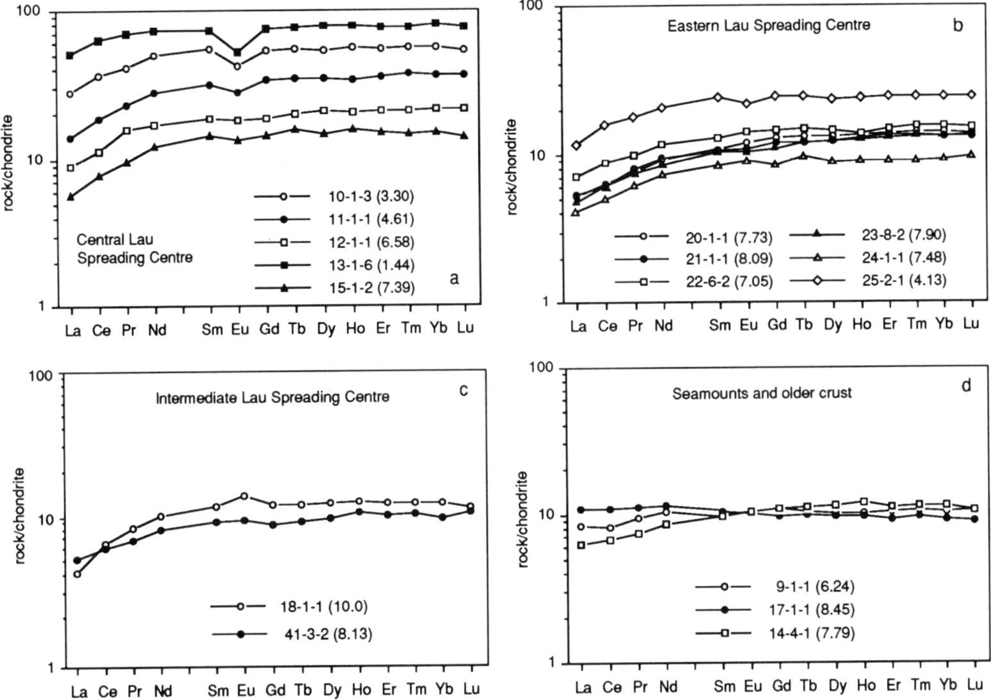

Fig. 4. Chondrite-normalized REE patterns for representative CD33 samples from (**a**) CLSC, (**b**) ELSC, (**c**) ILSC, and (**d**) seamounts.

a more LREE-enriched source compared with the source of the CLSC and ELSC lavas.

Incompatible elements. The incompatible element contents are represented in Fig. 5 as MORB-normalized patterns with elements arranged with increasing incompatibility from right to left but with K and Sr added to the left side (Pearce 1983). The normalizing factors used are taken from the latest N-MORB compilation of Sun & McDonough (1989) which has a more depleted composition than the compilation of Pearce (1983), essentially because the latter included a high proportion of Atlantic samples with a small P-MORB component. The samples chosen to represent the CLSC lavas (the basalts 15–1–2 and 11–1–1, and the ferrobasalt 12–1–1) have patterns with a slight positive slope compared with the MORB normalizing composition and with a small negative Nb anomaly and slight large ion lithophile (LIL) element enrichment (Fig. 5a). The positive slope may indicate a slightly more depleted source than the average N-MORB source, while the negative Nb anomaly may indicate a small subduction component in the mantle source. However, the magnitude of the latter is insufficient to place it

outside the compositional spectrum of N-MORB from ocean basins unrelated to subduction. In particular, we cannot distinguish at this stage between a recently-added subduction component and a long-lived subduction component (the DUPAL component) in the CLSC mantle source.

The samples chosen to represent the southern ELSC (the primitive basalts 20–1–1 and 23–8–2, and the evolved basalt 25–2–1) also have patterns that lie subparallel to the normalizing composition but with a slightly more distinctive negative Nb–Ta anomaly (Fig. 5b). The latter may, however, result in part at least from a greater enrichment in the surrounding LIL elements, notably Ba and Th, rather than an actual depletion of Nb and Ta. The samples chosen are fresh and glassy and these enrichments are unlikely to have resulted from interaction with seawater. These patterns therefore provide evidence for a greater LIL element-rich subduction component in the lavas nearest to the subduction zone.

Patterns for the ILSC basalts are shown in Fig. 5c. The D18 pattern is particularly distinctive in exhibiting strong depletions in the incompatible high-field strength (HFS) elements, Ta and Nb,

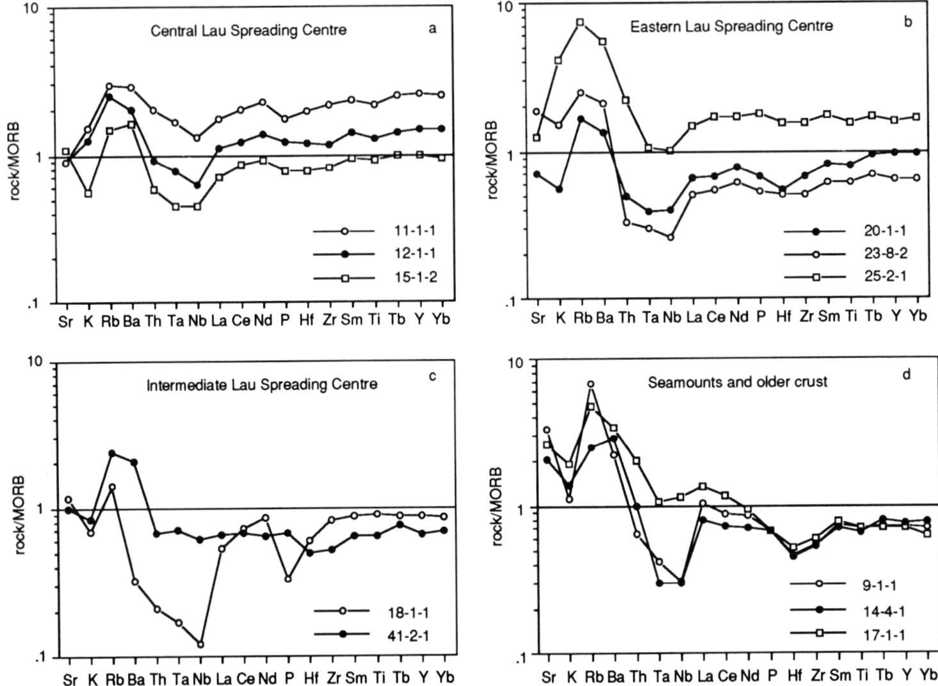

Fig. 5. MORB-normalized incompatible element patterns for representative CD33 samples from (**a**) CLSC, (**b**) ELSC, (**c**) ILSC, and (**d**) seamounts.

and marked enrichment in LIL with respect to HFS elements. This type of pattern is characteristic of addition of a subduction component to a depleted mantle source. By contrast, the D41 pattern is enriched in Rb and Ba with respect to the MORB normalizing value but otherwise has a flat pattern.

The seamounts and older crust also exhibit distinctive patterns (Figure 5d). Seamount D9 and CLSC off-axis site D14 have similar patterns in which the selective enrichment in LIL elements points to a large subduction component. Seamount D17 may contain a small subduction component but is dominated by a component in which enrichment is related to overall incompatibility during mantle melting (the 'within plate' component of Pearce (1983)). The derivation of this within plate component is not apparent from the present-day tectonic setting of the seamount.

Geochemical correlations with ridge segmentation

A striking geochemical feature of the studied area is the presence of highly fractionated rocks among the CD33 samples. Ferrobasalts and

highly differentiated rocks are not common at mid-ocean spreading centres, but have been dredged from the Galapagos Spreading Centre (e.g. Byerly *et al.* 1976; Clague & Bunch 1976; Clague *et al.* 1981), the Southeast Indian Ridge (Anderson *et al.* 1980), and the Southwest Indian Ridge (Le Roex *et al.* 1982) and have been drilled in old East Pacific Rise crust on DSDP Leg 34 (Clague & Bunch 1976) and Leg 92 (Pearce *et al.* 1986). In Fig. 6, we compare the Central Lau Basin samples with the first of these areas on the expanded diagrams (compared with Fig. 2) of FeO*, TiO_2, P_2O_5 and Zr against MgO in order to compare and contrast the phase saturation in the back-arc CLSC lavas with the 'normal ocean' Galapagos Spreading Centre (GSC).

The FeO*- and TiO_2 enrichment trends (Fig. 6a and b) for the CLSC suite overlap those of the GSC (85°W), although the degree of Ti, and to a lesser extent Fe, enrichment is greater in the GSC. Juster *et al.* (1989) use their 1 atmosphere experiments (trends shown in Fig. 6a and b) to demonstrate that the oxygen fugacity for the GSC at the point of oxide precipitation lay between the QFM buffer and NNO buffer +2 log units, probably approximately at the NNO buffer (= QFM + 0.8 at 1200°C). The same

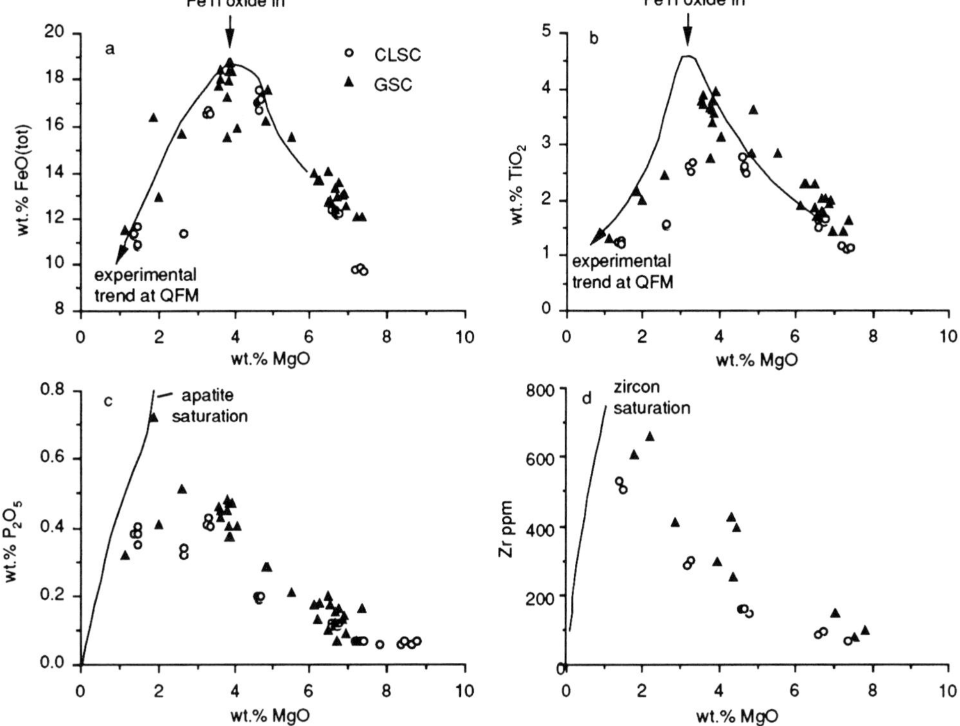

Fig. 6. FeO*–MgO, TiO₂–MgO, P₂O₅–MgO and Zr–MgO plots showing comparison between basalt-ferrobasalt-andesite trends for the Central Lau Spreading Centre (CLSC) and Galapagos Spreading Centre (GSC).

arguments applied to the CLSC suggest oxygen fugacities slightly greater than this. Juster *et al.* (1989) estimated apatite saturation to have been reached at 0.75 wt% P_2O_5 for the GSC (Fig. 6c), corresponding to the saturation curve derived from the solubility data of Harrison and Watson (1984) coupled with their own QFM experiments. Saturation is reached at about 0.4–0.5 % P_2O_5 in the CLSC. This lower value could be explained by the fact that the P_2O_5 content is lower for a given MgO in the CLSC and would thus intersect the saturation curve in Si-Al-alkali space at a lower P_2O_5 value. The Zr-enrichment trend (Figure 6d) shows a continuous increase in Zr with fractionation, indicating that zircon saturation was not reached, even in the most evolved rocks sampled from the GSC (650 ppm Zr) or CLSC (500 ppm Zr). This is consistent with the calculations of DeLong and Chatelain (1990) that zircon saturation of fractionated N-MORB magma should take place at Zr= 700 ppm at 840°C.

Because Zr is incompatible throughout, it can be used to quantify the degree of fractional crystallization, assuming that closed-system

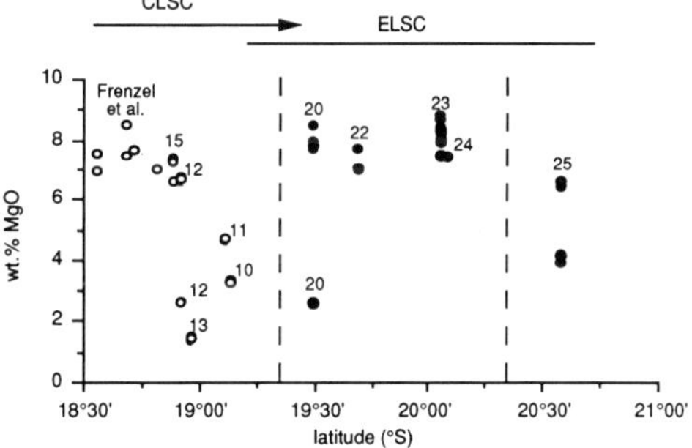

Fig. 7. MgO variations along the Central and Eastern Lau Spreading Centres. Dredge sites identified by numbers are located in Fig. 7. D14 and D21 data have not been used because of their slightly off-axis location. Data from north of our sample area is taken from Frenzel *et al.* (1990).

fractionation operated. The calculation based on Zr using the Rayleigh Fractionation Law for derivation of the andesites from D13 by closed system fractional crystallization from a basalt containing 8 wt% MgO gives a mass fraction of 0.075 of melt remaining (92.5% crystallization).

The MgO profile along the spreading segments of the Central Lau Basin is depicted in Fig. 7 using only data from dredge sites clearly located on the neovolcanic zone. On this diagram, there is an abrupt compositional break between the CLSC and ELSC: large variations in composition within the basalt – ferrobasalt – andesite spectrum are found at and near the propagating tip of the CLSC, whereas lavas from the first segment of the ELSC exhibit 'normal' basaltic compositions. We recovered evolved basalts from the second segment of the ELSC (D25), but no data are available to link these compositions to the well-studied Valu Fa Ridge further south. The discontinuities in Fig. 7 resemble those published from the propagating segments of the Galapagos spreading centre (Christie & Sinton 1981; Schilling *et al.* 1982; Sinton *et al.* 1983), the Spiess Ridge (Le Roex *et al.* 1982) and the Juan de Fuca Ridge (Sinton *et al.* 1983).

In terms of mechanism, the model for magmatism at propagating rifts at the well-studied Galapagos Spreading Centre is equally applicable to the CLSC (Christie & Sinton 1981, 1986; Clague *et al.* 1981; Schilling *et al.* 1982; Sinton *et al.* 1983; Juster *et al.* 1989). As the rift migrates through older and colder lithosphere, the decreasing magma supply from the newly established upwelling zone coupled with high cooling rates allows the development of small magma bodies in which magmatic evolution by fractional crystallization dominates over mixing with basic magma input from below. Fractional crystallization alone cannot produce the whole range of observed lava composition and we can thus infer the involvement in the genesis of the magma of variable degrees of partial melting (Clague *et al.* 1981), variations in depth of melt separation (Christie & Sinton 1986), crystallization at various depths under varying fluid pressure (Juster *et al.* 1989) and/or magma mixing.

Unfortunately, there are insufficient data from the northern part of the CLSC to define fully the extension of the CLSC geochemical anomaly and to ensure that highly fractionated rocks are really restricted to the propagating tip. Further sampling along the tip itself is also necessary to determine whether the spatial distribution of compositional variations along the CLSC is similar to that observed along other propagating segments, namely unfractionated basalts at the tip itself, bimodal distribution with primitive and highly evolved rocks immediately behind the tip and normal basaltic compositions further behind the tip (Christie & Sinton 1981, 1986; Sinton *et al.* 1983).

Geochemical correlations with arc-ridge distance

Any compositional variations linked to arc proximity should appear as differences between CLSC- and ELSC-lavas, given that the latter are at least 75 km closer to the active arc. Moreover, north–south variations along the ELSC could also be related to arc proximity, as the southernmost dredge sites are located where the spreading ridge runs closest to the arc. To investigate subduction zone influence, we focus on differences between the CLSC and ELSC lavas that cannot obviously be related to melting or fractional crystallization. These include: (1) ELSC samples, especially those from D23 and D25 have higher LIL/HFS element ratios than CLSC lavas; (2) at the same Mg-number, ELSC-lavas are SiO_2-enriched and Na_2O-depleted when compared to CLSC-basalts; (3) despite no greater water depths than other sites, the southernmost dredge sites on ELSC yielded the most vesicular rocks, comparable to the Valu Fa andesites further south (Jenner *et al.* 1987); (4) saturation in Fe, Ti and P is reached at lower MgO contents in the ELSC than the CLSC.

We examine the relationship between LIL element enrichment patterns and distance to the trench in Fig. 8. This figure contrasts typical CLSC and ELSC incompatible element patterns with patterns from the Valu Fa ridge (data of Jenner *et al.* 1987 and Davis *et al.* 1990) and Falcon Island in the Tofua arc (unpublished data of JAP). The CLSC sample 12–1–1 has been taken here as the normalizing composition to highlight element enrichments relative to the

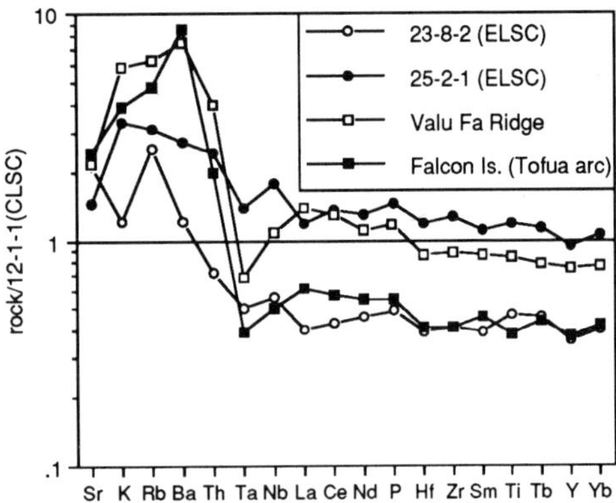

Fig. 8. CLSC-normalized (sample 12–1–1) incompatible element patterns for samples from the southernmost dredge sites on the ELSC (D23 and D25), the Valu Fa Ridge (Jenner *et al.* 1987) and the Tofua arc (J.A.P. unpublished data).

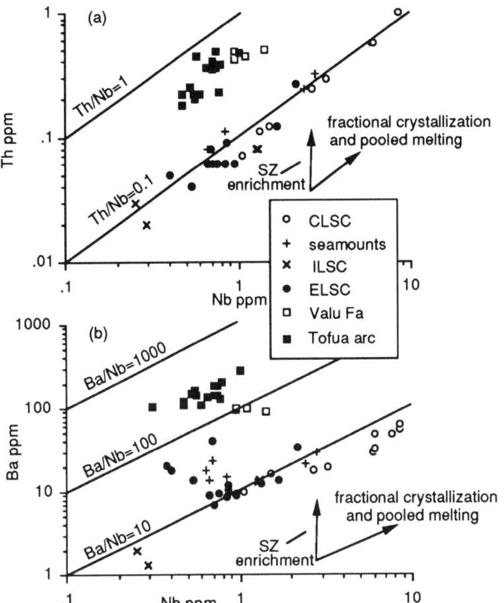

Fig. 9. Th–Nb and Ba–Nb plots used to identify subduction-related inputs of Th and Ba, respectively, in Lau Basin and Tofua arc lavas.

MORB back-arc basin (rather than average MORB) source. From this diagram, it is apparent that the ELSC lavas do not exhibit a Nb–Ta anomaly in the sense that they lack the large enrichments in Th and the LREE compared with the Valu Fa and Falcon Island (Tofua arc) samples. However, there is still a marked enrichment in the alkali and alkali earth elements, plotted on the left side of the pattern. The ELSC can thus be regarded as having, to a first approximation, trace element characteristics intermediate between the other ELSC basalts and the Valu Fa andesites described by Jenner *et al.* (1987). In detail, however, the enrichment in more highly charged LIL elements (Th, LREE) falls off faster than the enrichment in the mono- and divalent LIL elements (Sr, K, Rb, Ba).

Figure 9 examines these processes further with plots of Th and Ba against Nb. The idea of this plot is that Ba, Th and Nb have similar bulk distribution coefficients for mantle melting and fractional crystallization involving olivine, pyroxene and plagioclase. The Th/Nb and Ba/Nb ratios are therefore little affected by these processes, but strongly influenced by subduction which adds Th and Ba, but not Nb, to the mantle wedge. In Fig. 9a , the CLSC, ELSC, ILSC and seamounts all have Th/Nb ratios of 0.1 or just below, similar to the N-MORB ratio of Sun &

McDonough (1989). By contrast, both the Tofua arc and Valu Fa Ridge have ratios of about 0.5. Thus, any subduction-generated Th is at background levels even in the most southerly of our dredge sites on the ELSC. In Fig. 9b, the CLSC has a Ba/Nb ratio of 7–10. This is greater than the N-MORB normalizing value of Sun & McDonough (1989), which has a Ba/Nb ratio of 2.7 but within the range of N-MORB for which values up to 10 have been obtained. Thus the CLSC source may carry a small subduction signature or, as noted earlier, this may be a long-lived characteristic of SW Pacific mantle. By contrast, the ELSC ratios lie slightly above this (9–15) for ELSC sites D20, D21, D22 and D24, and significantly above this (26–60) for some samples from ELSC sites D23 and D25. The Valu Fa ridge has ratios of 80–100 and the Tofua arc has ratios of 200–300. Thus subduction-generated Ba significantly exceeds background only in the two most southerly ELSC sites, D23 and D25. Unlike Th, however, there is a more gradual decline in subduction-generated Ba away from the arc.

Note that the lavas from two CD33 sites with highest Ba/Nb ratios (D23 and D25) are also the most vesicular despite the fact that these sites are at least as deep as the sites with less-vesicular lavas. Figure 10 further highlights the progressive decrease in iron enrichment from the CLSC, through the ELSC (D20-D25) and the Valu Fa Ridge to the Tofua arc. Using the calibration of Juster *et al.* (1989), the difference in oxygen fugacity between the CLSC and Tofua arc is some 2.5 log units at the point of oxide crystallization. Between the CLSC and ELSC, the difference is about half this value. There is thus a clear relationship between oxygen fugacity, water content and enrichment in alkali

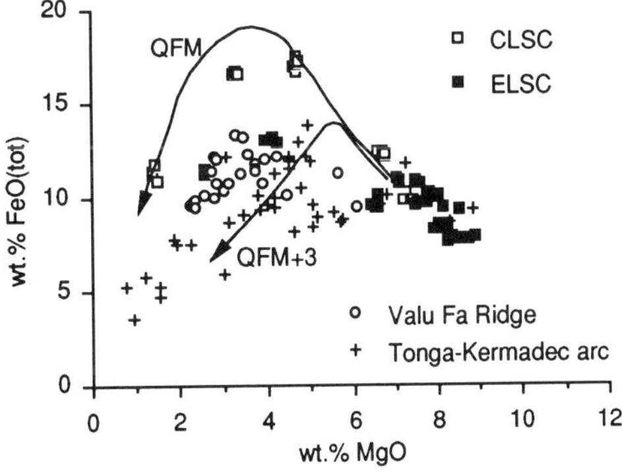

Fig. 10. FeO*–MgO plot used to contrast oxygen fugacities in magmas from the Lau Basin and Tofua arc.

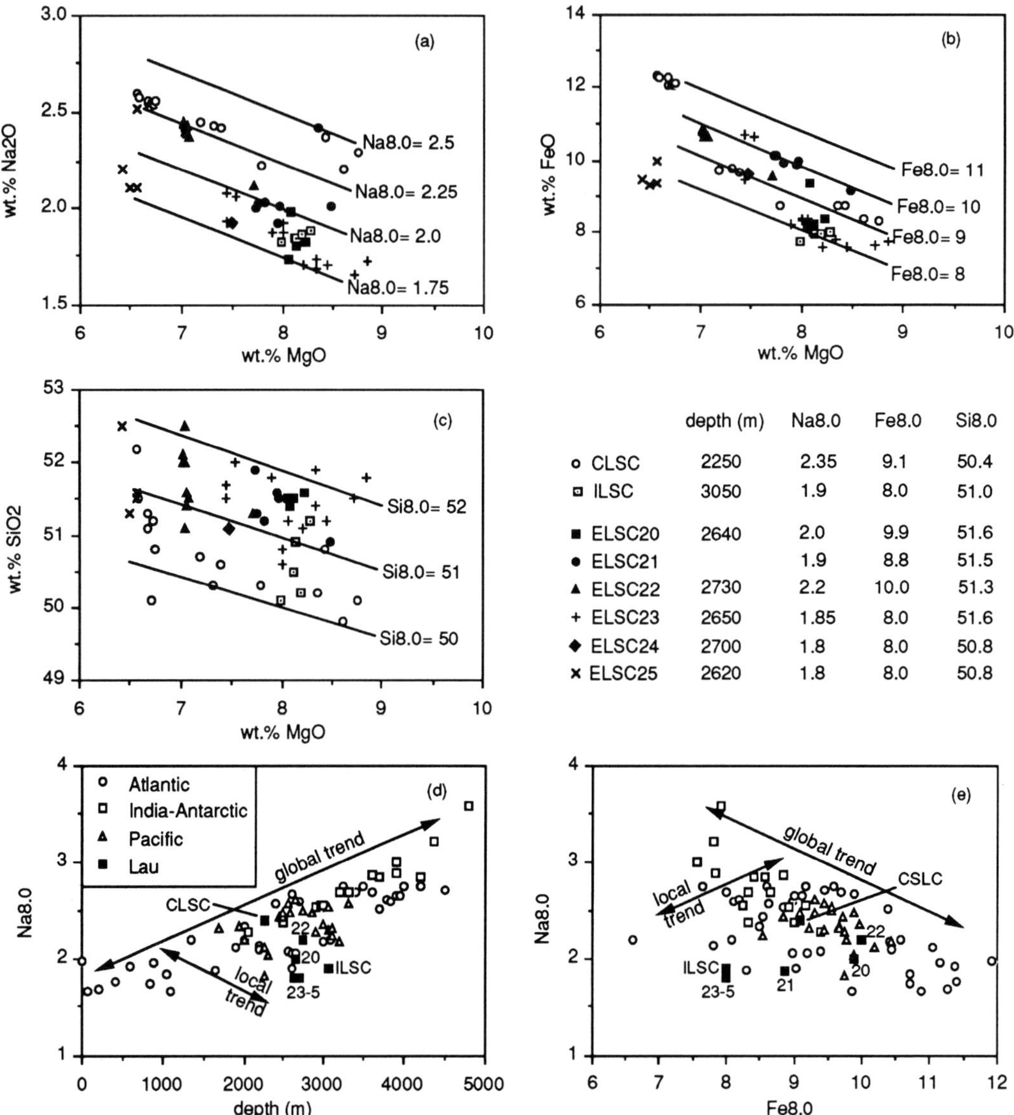

Fig. 11. (a) Na$_2$O–MgO, (b) FeO*–MgO and (c) SiO$_2$–MgO diagrams showing the determination of Na 8.0, Fe 8.0 and Si 8.0 values for the CLSC and ELSC lavas; (d) Na8.0-depth diagram (from Klein & Langmuir 1987) and (e) Na 8.0–Fe 8.0 diagram (from Langmuir *et al.* 1992) showing the Na 8.0 and Fe 8.0 values of the Lau Basin dredge stations with respect to values from the major oceans. Values used are shown in the inset Table.

and alkali earth elements in the magmas of the Tonga arc-basin system. However, the absence of strong iron enrichment even in the intermediate lavas from Site D20 suggests that subducted water may affect mantle well behind the arc, at distances where the enrichment of other elements such as Th and Ba cannot be detected.

We can also identify geochemical differences between the CLSC and ELSC that cannot be linked to shallow-level fractional crystallization or to subduction-induced mantle heterogeneities, i.e. that must be related to the melting process. Na$_2$O, FeO and SiO$_2$ contents, in particular, can be viewed in terms of the global major element correlations identified by Klein & Langmuir (1987, 1989) and Langmuir *et al.* (1992), though with the caveat that these authors did not consider hydrous melting. Figure 11a shows an expanded version of part of the Na$_2$O–MgO diagram in Fig. 3e and enables their

Na8.0 value (the Na_2O content at 8 wt% MgO) to be estimated as 2.35 wt% for the CLSC (mean depth 2300 m), 2.2 wt% for Site D22 on the ELSC (depth 2730 m), and 1.8–2.0 wt% for Sites D20, 21, 23–25 on the ELSC (2620–2750 m depth). Figure 11b enables Fe8.0 values to be estimated as 9.1 for the CLSC, *c.*10.0 for the ELSC Sites D20 and 22, 8.8 for ELSC Site D21 and 8.0 for ELSC Sites D23–25. Figure 11c shows that Si 8.0 values are systematically lower for the CLSC (50.4) than for the ELSC (50.8–51.6).

The CLSC Na 8.0 value plots within the Pacific, Atlantic and Indian Ocean fields on fig. 2 of Klein & Langmuir (1987; reproduced here as Fig. 11d), confirming that the processes involved are essentially MORB-like for at least the southern part of the CLSC. By contrast, ELSC values other than that for D22 plot at the bottom end of the range for the main spreading axes, following a local rather than global trend. The Na8.0–Fe8.0 plot (Fig. 11e) shows the CLSC on the global trend of increasing Fe8.0 with decreasing Na8.0. However, the ELSC data points plot on a local trend of decreasing Fe8.0 with decreasing Na8.0. This local trend intersects the global trend at a lower Na8.0 value than that of the CLSC.

The most obvious explanation for the low Na 8.0 values at the majority of ELSC sites is a high degree of partial melting. This interpretation is consistent with the higher water content at the ELSC compared with the CLSC, in that addition of subducted water to the mantle should increase the degree of melting by lowering the mantle solidus (Davies & Bickle 1991). It is also consistent with the lower contents of Ni, Cr, Sc and Co (at a given MgO content) shown by Figure 3. However, a greater degree of melting beneath the ELSC should generate thicker crust and hence shallower bathymetry, assuming a comparable mantle source and melting régime. Given that the ELSC is typically deeper than the CLSC, we can infer that mantle source and melting régime are not comparable.

The Na8.0–Fe8.0 plot provides a further insight into the cause of these major element variations. The displacement from the CLSC to ELSC Site D22 lies along the global trend toward lower Na8.0 and higher Fe8.0. This could be explained by enhanced melting beneath the ELSC. Note that the global trend as defined by Langmuir and co-workers is related to global variations in temperature, in which higher potential temperatures lead to deeper initiation of melting and greater degrees of partial melting. This will not be precisely the same as a global trend resulting from variable subducted

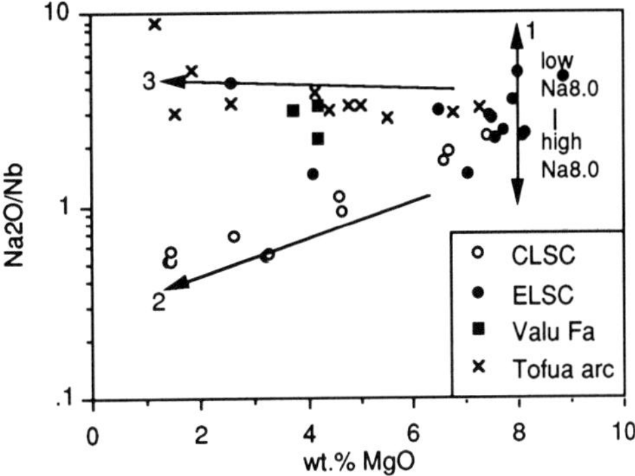

Fig. 12. Plot of Na_2O/Nb against MgO showing a spread of Na_2O/Nb ratios at MgO = 8.0 wt%. ELSC samples with low Na 8.0 also have high Na_2O/Nb ratios indicating that low Na 8.0 in the more arc-proximal samples can be explained by mantle depletion or loss of a low melting fraction in the melting column.

water as the depth function for melt generation will be different (Pearce & Parkinson 1993). However, the slope will still be negative.

The local trend from ELSC Site D22 to ELSC Sites D23–25 thus needs to be explained by variations in source and melting régime. Batiza *et al.* (1988), Klein & Langmuir (1989) and Langmuir *et al.* (1992) all noted that this trend is characteristic of slow-spreading ridges, especially parts of the Mid-Atlantic Ridge. However, they have no satisfactory explanation for the trend despite a much larger data set than our own. The ELSC is a slow-spreading ridge (unlike the CLSC) and the same, as yet unexplained, mechanism may therefore apply. The alternative explanation for the Lau Basin is that there is some relationship to arc proximity, as the Sites closest to the arc (D23–25) plot furthest from the global trend.

To explore the local trend further, we examine the Na_2O/Nb ratio as a function of MgO content (Fig. 12). The principle is that Nb is more incompatible than Na and the assumption is that there is not significant addition of Na from the subduction zone. The Na_2O/Nb ratio is thus very sensitive to mantle source and melting régime as loss of small melt fractions will significantly increase the ratio. It is not, however, sensitive to variations in degree of melting provided the melt fractions are pooled.

Figure 12 highlights three trends: a trend (Trend 1) of variable Na_2O/Nb ratio for a given MgO content that includes all basic lavas; a trend (Trend 2) to low Na_2O/Nb ratios and low MgO contents that includes all CLSC lavas and rare ELSC lavas; and a trend (Trend 3) to constant Na$_2$O/Nb ratios and low MgO contents that includes all Tofua arc, most Valu Fa and rare ELSC lavas. Trends 2 and 3 clearly relate to fractional crystallization, extensive open-system fractionation probably causing the reduction in the Na$_2$O/Nb ratio with decreasing MgO in Trend 2. Trend 1 requires variations in the mantle source and melting régime. The CLSC lavas intesect Trend 1 at low-intermediate Na$_2$O/Nb ratios; the Tofua arc and Valu Fa lavas intersect Trend 1 at intermediate ratios. ELSC Sites D20–22 have low-intermediate ratios on Trend 1, Sites D23–25 having intermediate-high ratios. Thus, the displacement along the local trend to low Na 8.0 in Fig. 11e is accompanied by an increase in the ratio of the highly incompatible element (Na) to a very highly incompatible element (Nb). This is consistent with a more depleted source (or a loss of a small melt fraction from the melting column) beneath the arc-proximal parts of the ELSC. Note, however, that some samples from Sites D23–25 require even more depleted sources (or loss of a greater melt fraction) than beneath the arc itself.

The higher Si8.0 of the ELSC compared with the CLSC may be explained by a greater contribution of melt increments from shallower melting or by greater water content, both of which enhance the effect of incongruent melting of orthopyroxene. Given that the high Si8.0 value is a characteristic of all ELSC Sites, rather than just those with low Na$_2$O/Nb ratios (Sites D23–25), a high water content is the more likely explanation.

Overall, a consistent story can be extracted from the observations and interpretations. Moving north from the Valu Fa Ridge along the ELSC, Th and LREEs exhibit little enrichment by Site D25, although Ba, Rb and H remain at high levels, and only H may be noticeably enriched by Site D20. Deep release of Ba, Rb and H from hydrous phases such as phlogopite could account for this observation; variable diffusion of elements in the mantle wedge away from the subducted slab may also provide an explanation. The major element systematics is the likely product of three factors: the mantle wedge may become more depleted with arc proximity, tending to decrease the degree of melting beneath the ELSC; the water content of the mantle may increase with arc proximity, tending to increase the degree of melting in the ELSC compared with the CLSC; and the dynamics of melt extraction may be different beneath the slow-spreading ELSC compared with the intermediate-fast spreading CLSC. A full explanation must, however, await a full, detailed sampling programme along the ELSC.

The ILSC and Seamounts

Both the seamount, D9, and older crustal site, D24, have patterns indicative of a significant subduction component in their source, and their Ba/Nb ratios between 20 and 50 support this conclusion (Fig. 10b). Given that the lavas from the two sites had ferromanganoan crusts, it is probable that these sites both represent old, rifted, transitional crust similar to that at many of the ODP Leg 135 sites (Shipboard Scientific Party 1992). Seamount D17 has a more intra-plate character, with high Nb/Yb ratios but low Ba/Nb ratios (Fig. 10). How such a composition fits into the evolution of the Lau Basin is not, however, clear.

The two ILSC samples have markedly different patterns, one (D18) indicating a strongly depleted mantle source, the other (D41) indicating a slightly enriched source. As with the seamounts, these compositions may be best explained by rifting of metasomatised lithosphere, and neither may contribute to the neomagmatic story.

Conclusions

(1) Both the segmentation of the spreading axis and the proximity of the Tonga subduction zone influence the petrography and geochemistry of the active ridge lavas in the Central Lau Basin.

(2) The highly fractionated compositions of the Fe-basalts and andesites sampled near the southern end of the Central Lau Spreading Centre are best explained by the evolution of magmatic liquids at the tip of a propagating rift in response to a high cooling rates and low rates of magma supply.

(3) Many of the geochemical characteristics (oxygen fugacity, water content LIL/HFS element ratios) of samples from the Eastern Lau Spreading Centre are intermediate between the more MORB-like Central Lau Spreading Centre basalts and the more arc-like Valu Fa samples, suggesting that the subduction component in the mantle source changes progressively with arc proximity beneath the Lau Basin.

(4) In detail, however, subducted H appears to have influenced the mantle throughout the ELSC, to at least 75 km behind the arc. By contrast, Ba and Rb have influenced mantle only as far as the southernmost ELSC sites dredged, to about 60 km behind the arc. Th and LREEs have influenced mantle only to between 40 and 60 km behind the arc. Phlogopite breakdown at depths greater than the breakdown of amphibole could explain the persistence of Ba, Rb and H, but differential diffusion of elements within

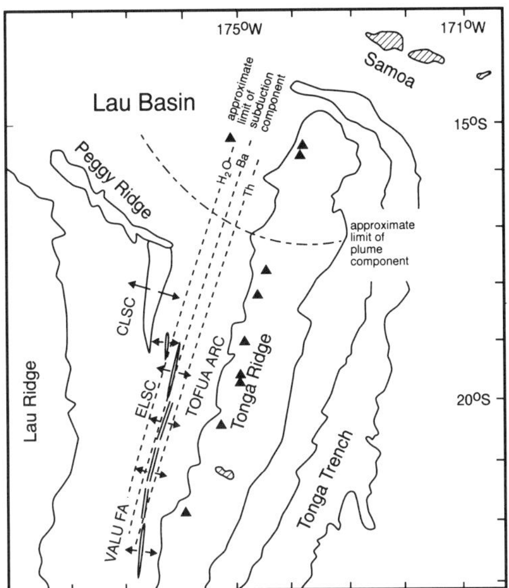

Fig. 13. Map of the Lau Basin showing the approximate limits of plume-influenced and subduction-influenced mantle in the Lau Basin.

the mantle wedge may also be important. This observation gives an updated geochemical map of the Lau Basin as shown in Fig. 13, although it must be emphasized that more detailed sampling is still required along the ELSC.

(5) Na8.0 and Fe8.0 values in the ELSC follow a local, rather than global trend, although the CLSC does lie on the global array. There is no simple explanation for the observed variations in these and other indicators of the melting process. Increasing arc proximity may be accompanied by a combination of a more depleted mantle, greater water content and more complex mantle dynamics. More detailed sampling may be needed to resolve these variables better.

We are grateful to R. Hodkinson, Q. Huggett, S. Miller, P. Rodda, S. Helu and the officers and crew of the RRS Charles Darwin for their shipboard assistance. M.E. thanks the British Council for funding her post-doctoral work in Newcastle. J.A.P. thanks R. Jackson for analytical assistance. J.A.P. and L.M.P thank the Natural Environment Research Council (UK) for funding the CD33 cruise (Grant GST/02/389) and the geochemical analyses.

References

ANDERSON, R.N., SPARIOSU, D.J., WEISSEL, J.K. & HAYES, D.E. 1980. The interrelation between variations in magnetic anomaly amplitudes and basalt magnetization and chemistry along the Southeast Indian Ridge. *Journal of Geophysical Research*, **85**, 3883–3898.

BATIZA, R., MELSON, W.G. & O'HEARN, T. 1988. Simple magma supply geometry inferred beneath a segment of the mid-Atlantic ridge. *Nature*, **335**, 428–431.

BYERLY, G.R., MELSON, W.G. & VOGT, P.R. 1976. Rhyodacites, andesites, ferrobasalts and ocean tholeiites from the Galapagos Spreading Center. *Earth and Planetary Science Letters*, **30**, 215–221.

CHRISTIE, D.M. & SINTON, J.M. 1981. Evolution of abyssal lavas along propagating segments of the Galapagos spreading center. *Earth and Planetary Science Letters*, **56**, 321–335.

——,—— & —— 1986. Major element constraints on melting, differentiation and mixing of magmas from the Galapagos 95°W propagating rift system. *Contributions to Mineralogy and Petrology*, **94**, 274–288.

CLAGUE, D.A. & BUNCH, T.E. 1976. Formation of ferrobasalts at East Pacific mid-ocean spreading centers. *Journal of Geophysical Research*, **81**, 4247–4256.

——, FREY F.A., THOMPSON, G. & RINDGE, S. 1981. Minor and trace element geochemistry of volcanic rocks dredged from the Galapagos spreading center: role of crystal fractionation and mantle heterogeneity. *Journal of Geophysical Research*, **86**, 9469–9482.

DAVIS, A.S., CLAGUE, D.A. & MORTON, J.L. 1990. Volcanic glass compositions from two spreading centers in Lau Basin, South West Pacific Ocean. *Geologische Jahrbuch*, **D92**, 481–501.

DAVIES, J.H. & BICKLE, M.L. 1991. A physical model for the volume and composition of melt produced by hydrous fluxing above subduction zones. *Philosphical Transactions of the Royal Society of London*, **A335**, 355–364.

DeLONG, S.E. & CHATELAIN, C. 1990. Trace-element constraints on accessory-phase saturation in evolved MORB magma. *Earth and Planetary Science Letters*, **101**, 206–215.

FOUCHER, J.-P. 1986. Summary report of the SEAPSO Leg IV cruise, Suva -Nuku'alofa, 30 Dec. 1985–11 Jan. 1986. *Ifremer-Orstom, Technical Report*, 8p.

FRENZEL, G., MÜHE, R. & STOFFERS, P. 1990. Petrology of the volcanic rocks from the Lau Basin, Southwest Pacific. *Geologische Jahrbuch*, **D92**, 395–479.

FRYER, P., SINTON, J.M. & PHILPOTTS, J.A. 1981. Basaltic glasses from the Mariana Trough. *In*: HUSSONG, D.M. & UYEDA S. *et al. Initial Reports of the Deep Sea Drilling Project*, **60**, 601–609.

GILL, J.B. 1976. Composition and age of Lau Basin and ridge volcanics: implications for evolution of an interarc basin and remnant arc. *Geological Societ of America Bulletin*, **87**, 1384–1395.

HARRISON, T.M. & WATSON, E.B. 1984. The behavior of apatite during crustal anatexis: equilibrium and kinetic considerations. *Geochimica et Cosmochimica Acta*, **48**, 1467–1477.

HART, S.R., GLASSLEY, W.E. & KARIG, D.E. 1972. Basalts and sea floor spreading behind the Mariana Island arc. *Earth and Planetary Science Letters*, **15**, 12–18.

HAWKINS, J.W. 1974. Geology of the Lau Basin, a marginal sea behind the Tonga Arc. *In*: BURK, C. & DRAKE, C. (eds) *Geology of Continental Margins*. Springer Verlag, Berlin, 505–520.

—— 1976. Petrology and geochemistry of basaltic rocks of the Lau Basin. *Earth and Planetary Science Letters*, **28**, 283–297.

—— & MELCHOIR, J.T. 1985. Petrology of Mariana Trough and Lau Basin basalts. *Journal of Geophysical Research*, **90**, 11431–11468.

JAMES, S.D., PEARCE, J.A. & OLIVER, R.A. 1987. The geochemistry of the Lower Proterozoic Willyama Complex volcanics, Broken Hill Block, New South Wales. *In*: PHARAOH, T.C., BECKINSALE, R.D. & RICKARD, D. (eds) *Geochemistry and Mineralization of Proteozoic Volcanic Suites*. Geological Society, London, Special Publication, **33**, 395–408.

JENNER, G.A., CAWOOD, P.A., RAUTENSCHLEIN, M. & WHITE, W.M. 1987. Composition of back-arc basin volcanics, Valu Fa Ridge, Lau Basin: evidence for a slab-derived component in their mantle source. *Journal of Volcanological and Geothermal Research*, **32**, 209–222.

JUSTER, T.C., GROVE, T.L. & PERFIT, M.R. 1989. Experimental constraints on the generation of TiFe basalts, andesites, and rhyodacites at the Galapagos spreading center, 85°W and 95°W. *Journal of Geophysical Research*, **94**, 9251–9274.

KARIG, D.E. 1970. Ridges and basins of the Tonga-Kermadec island arc system. *Journal of Geophysical Research*, **75**, 239–254.

—— 1971. Origin and development of marginal basins in the western Pacific. *Journal of Geophysical Research*, **76**, 2542–2561.

KLEIN, E.M. & LANGMUIR, C.H. 1987. Global correlations of ocean ridge basalt chemistry with axial depth and crustal thickness. *Journal of Geophysical Research*, **92**, 8089–8115.

—— & —— 1989. Local versus global variation in ocean ridge basaltic composition: a reply. *Journal of Geophysical Research*, **94**, 4241–4252.

LANGMUIR, C.H., KLEIN, E. & PLANK, T. 1992. Petrological systematics of mid-ocean ridge basalts: constraints on melt generation beneath ocean ridges. *In*: PHIPPS MORGAN, J., BLACKMAN, D.K. & SINTON, J.M. (eds) *Mantle flow and melt generation at mid-ocean ridges*, American Geophysical Union, Washington DC, 183–280.

LAWVER, L. & HAWKINS, J.W. 1978. Diffuse magnetic anomalies in marginal basins, their possible petrologic and tectonic significance. *Tectonophysics*, **43**, 323–339.

LE ROEX, A.P., DICK, H.J.B., REID, A.M. & ERLANK, A.J. 1982. Ferrobasalts of the Spiess Ridge segment of the Southwest Indian Ridge. *Earth and Planetary Science Letters*, **60**, 437–451.

MORTON, J.L. & SLEEP, N.H. 1985. Seismic reflections from a Lau Basin magma chamber. *In*: SCHOLL, D.W.& VALLIER, T.L. (eds) *Geology and offshore resources of Pacific island arcs -Tonga region*. American Association of Petroleum Geologists, Tulsa, 441–453.

NATLAND, J.H. 1980. Effect of axial magma chambers beneath spreading centers on the compositions of basaltic rocks. *In*: ROSENDAHL, B.R., HEKINIAN, R. *et al*. *Initial Reports of the Deep Sea Drilling Project*, **54**, 833–850

PARSON, L.M., PEARCE, J.A., MURTON, B.J., HODKINSON, R.A. & THE RRS CHARLES DARWIN SCIENTIFIC PARTY. 1990. Role of ridge jumps and ridge propagation in the tectonic evolution of the Lau back-arc Basin, southwest Pacific. *Geology*, **18**, 470–473.

PEARCE, J.A. 1983. Role of the sub-continental lithosphere in magma genesis at active continental margins. *In*: HAWKSWORTH, C.J. & NORRY, M.J. (eds) *Continental basalts and mantle xenoliths*. Shiva Publishing, Nantwich, 230–249.

PEARCE, J.A. & PARKINSON, I.J. 1993. Trace element models for mantle melting: application to volcanic arc petrogenesis. *In*: PRICHARD, H.M., ALABASTER, T., HARRIS, N.B.W. & NEARY, C.R. (eds) *Magmatic Processes and Plate Tectonics*. Geological Society, London, Special Publications, **76**, 373–403.

PEARCE, J.A., ROGERS, N., TINDLE, A.J. & WATSON, J.S. 1986. Geochemistry and petrogenesis of basalts from Deep Sea drilling Project Leg 92, Eastern Pacific. *In*: LEINEN, M., REA, D.K. *et al.*, *Initial Reports of the Deep Sea Drilling Project*, **92**, 435–457.

SATO, H & BANNO, S. 1983. NiO-Fo relation of magnesian olivine phenocryst in high magnesian andesite and associated basalt-andesite-sanukite from northeast Shikoku, Japan. *Bulletin of the Volcanological Society of Japan, Series 2*, **28**, 141–156.

SATO, H. & TOHARA, T. 1985. Geochemical characteristics of back-arc basin basalt. *In*: NASU, N. *et al.* (eds) *Formation of active ocean margins*. Terrapub, Tokyo, 399–410.

SAUNDERS, A.D. & TARNEY, J. 1979. The geochemistry of basalts from a back-arc spreading center in the East Scotia Sea. *Geochimica et Cosmochimica Acta*, **43**, 555–572.

SCHILLING, J.-G., KINGSLEY, R.H. & DEVINE, J.D. 1982. Galapagos hot spot-spreading center system. 1. Spatial petrological and geochemical variations (83°W–101°W). *Journal of Geophysical Research*, **87**, 5593–5610.

SCHOLL, D.W. & VALLIER, T.L. 1985. *Geology and offshore resources of Pacific island arcs-Tonga region*. American Association of Petroleum Geologists, Tulsa, Circum-Pacific Council for Energy and Mineral Resources, Earth Science Series, **2**.

SCLATER, J.G., HAWKINS, J.W., MAMMERICKS, J. & CHASE, C.G. 1972. Crustal extension between the Tonga and Lau ridges: petrologic and geophysic evidence. *Geological Society of American Bulletin*, **83**, 505–517.

SINTON, J.M. & FRYER, P. 1987. Mariana Trough lavas from 18°N: implications for the origin of back-arc basin basalts. *Journal of Geophysical Research*, **92**, 12782–12802.

——, WILSON, D.S., CHRISTIE, D.M., HEY, R.N. & DELANEY, J.R. 1983. Petrologic consequences of rift propagation on ocean spreading ridges. *Earth and Planetary Science Letters*, **62**, 193–207.

SHIPBOARD SCIENTIFIC PARTY. 1992. Introduction, background, and principal results of Leg 135, Lau Basin. *In*: HAWKINS, J., PARSON, L. ALLAN, J. *et al. Proceedings of the Ocean Drilling Program, Initial Reports*, **135**, 5–47

SUN, S.-S. & MCDONOUGH, W.F. 1989. Chemical and isotopic systematics of oceanic basalts: implications for mantle composition and processes. *In*: SAUNDERS, A.D. & NORRY, N.J. (eds) *Magmatism in the Ocean Basins*. Geological Society, London, Special Publications, **42**, 313–345.

TARNEY, J., SAUNDERS, A.D., MATTEY, D.P., WOOD, D.A. & MARSH, N.G. 1981. Geochemical aspects of back-arc spreading in the Scotia Sea and western Pacific. *Philosophical Transactions of the Royal Society of London*, **A300**, 263–285.

VOLPE, A.M., MCDOUGALL, J.D. & HAWKINS, J.W. 1987. Mariana Trough basalts (MTB): trace element and Sr-Nd isotopic evidence for mixing between MORB-like and arc-like melts. *Earth and Planetary Science Letters*, **82**, 241–254.

——, —— & —— 1988. Lau Basin basalts (LBB): trace element and Sr-Nd isotopic evidence for heterogeneity in back-arc basin mantle. *Earth and Planetary Science Letters*, **90**, 174–186.

VON STACKELBERG, U. & THE SHIPBOARD SCIENTIFIC PARTY. 1988. Active hydrothermalism in the Lau Back-Arc Basin (SW Pacific). First results from the Sonne 48 Cruise (1987). *Marine Mining*, **7**, 431–442.

WEISSEL, J.K. 1977. Evolution of the Lau Basin by the growth of small plates. *In:* TALWANI, M. & PITMAN, W.C. (eds) *Island arcs, deep-sea trenches and back-arc basins*. Maurice Ewing Series 1 American Geophysical Union, Washington, DC, 429–436.

Basaltic volcanism associated with extensional tectonics in the Taiwan–Luzon island arc: evidence for non-depleted sources and subduction zone enrichment

ULRICH KNITTEL[1] & DIETMAR OLES[2]

[1] *Department of Geology, University of Melbourne, Parkville, Victoria 3052, Australia, present address: Institut für Mineralogie und Lagerstättenlehre, Wüllnerstr. 2, D–52056 Aachen, Germany*

[2] *Institut für Mineralogie und Lagerstättenlehre, Wüllnerstr. 2, D–52056 Aachen, Germany; present address: Departamento de Geologia, Universidade Eduardo Mondlane, Caixa Postal 257, Maputo, Mozambique*

Abstract: Primitive basalts were erupted from more than 100 monogenetic eruption centres within the Macolod Corridor, an extensional structure which cross-cuts the Taiwan–Luzon volcanic arc. The composition of early phenocrysts, HFSE systematics and relatively high Al_2O_3, TiO_2 and Na_2O contents of these rocks suggest that they are derived from sources comparable to the sources of MORB and back-arc basin basalts. This is in contrast to many other arc magmas which appear to be derived from more depleted sources. High LILE abundances and high LILE/HFSE ratios indicate the presence of slab-derived components in the magmas, although the Macolod Corridor is no longer underlain by an actively subducting slab.

Magmatic arcs, developed above subduction zones, usually evolve in tensional tectonic settings because the trenches tend to retreat oceanward, as the subducting lithosphere sinks into the mantle (Molnar & Atwater 1978; Hamilton 1979, 1989; Seno & Maruyama 1984). Extension ultimately may result in the arc splitting along its axis and the formation of a back-arc basin.

The magmas of volcanic arcs are generally characterized by enrichment in large ion lithophile elements (LILE) and depletion of high field strength elements (HFSE) relative to mid-ocean ridge basalts (MORB) (e.g. Pearce 1983). In contrast, the basalts of back-arc basins (BABB) have HFSE abundances similar to MORB and show moderate LILE enrichment. Woodhead *et al.* (1993) interpreted the HFSE systematics of island arc basalts (IAB) and BABB to indicate that BABB are derived from sources similar to the sources of MORB, whereas IAB are derived from more depleted sources. McCulloch & Gamble (1991) suggested that this depletion of the IAB sources might be caused by the extraction of BABB.

In the southern part of the Taiwan–Luzon volcanic arc, an extensional zone called the Macolod Corridor (Defant *et al.* 1988; Förster *et al.* 1990), has developed obliquely rather than parallel to the arc (Fig. 1). Within this zone, primitive basalts (6–11 % MgO) were erupted in the past <1 Ma from more than 100 small, largely monogenetic eruption centres, located behind the volcanic front. In addition, evolved magmas were erupted from three large stratovolcanoes and two caldera complexes. While volcanism of the Taiwan–Luzon arc is considered to be related to the eastward subduction of the oceanic South China Sea crust (e.g. de Boer *et al.* 1980; Cardwell *et al.* 1980), the relationship between subduction and volcanism in the Macolod Corridor is less clear because, due to the southward steepening of the Benioff-Wadati zone (Hamburger *et al.* 1983), the volcanic fields in this area are not underlain by subducted lithosphere. It is suggested that an extensional tectonic regime controls volcanism.

From our study of the primitive basalts of the Macolod Corridor, it is possible to address several questions of critical importance for understanding arc magmatism.

(1) What is the composition of the mantle wedge? In particular, is it generally more depleted than the MORB source?

(2) To what extent are subduction-zone signatures discernible in magmas derived from regions not directly underlain by the subducting slab?

From Smellie, J.L. (ed.), 1995, *Volcanism Associated with Extension at Consuming Plate Margins*, Geological Society Special Publication No. 81, 77–93.

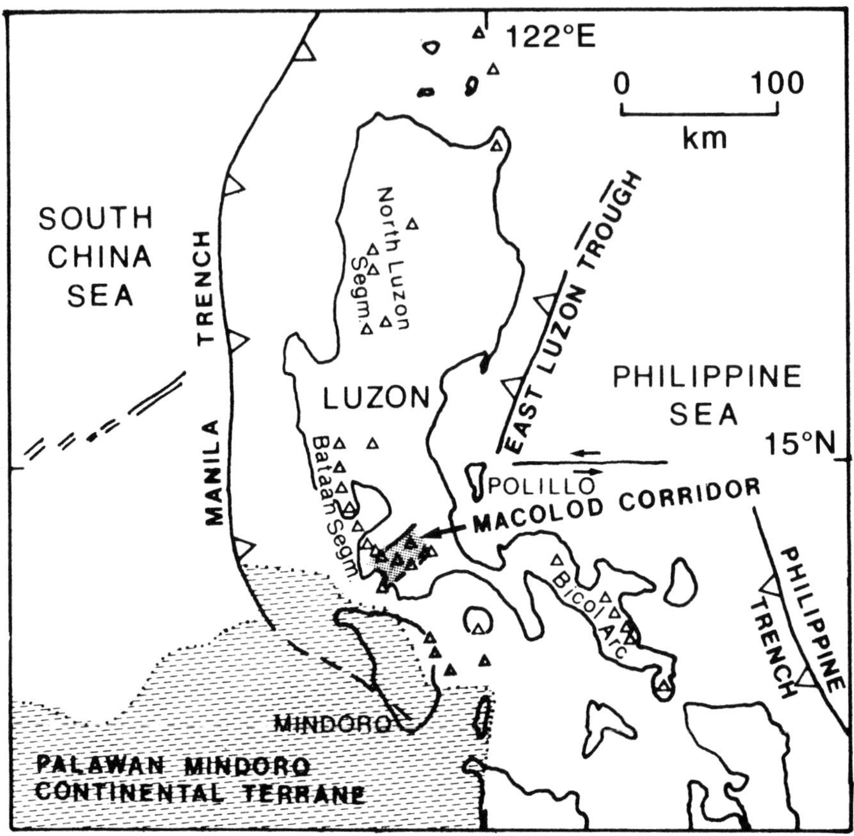

Fig. 1. Simplified tectonic map of Luzon, showing the setting of the Macolod Corridor. Major volcanic centres are indicated by open triangles (note that the volcanic centres in the north Luzon segment are up to 6 Ma old and presently extinct, while most other centres are younger than 2 Ma). The north Luzon and Bataan segments are parts of the Taiwan–Luzon arc. Modified after Rangin *et al.* (1988).

Tectonic setting

Along the western coast of Luzon island, a volcanic arc (the Taiwan–Luzon arc) has developed in response to the eastward subduction of the oceanic South China Sea crust at the Manila Trench. Within central Luzon, the Bataan segment of this arc consists of a well-defined volcanic front located approximately 100 km above the east-dipping subduction zone (Hamburger *et al.* 1983). Only a few volcanoes are located behind the front (Fig. 1).

To the south, the dip of the subduction zone steepens to near vertical below southwestern Luzon (Hamburger *et al.* 1983), possibly in response to the collision of the arc with the North Palawan continental terrain (McCabe *et al.* 1982), which resulted in the slowing or cessation of subduction. Above the near vertical Benioff–Wadati zone, several stratovolcanoes

dated at 3.4–1.3 Ma (De Boer *et al.* 1980; Oles 1988), form the volcanic front along the western coast of southwestern Luzon. In contrast to the Bataan segment to the north, intense volcanism extends far behind the volcanic front in the Macolod Corridor. Three large stratovolcanoes (Mt Makiling, Mt Malepunyo, Mt Banahaw), two large caldera complexes (Lake Taal, Laguna de Bay) and more than a hundred small, largely monogenetic eruption centres are located within a NE–SW-trending zone, which measures 30 × 60 km (Fig. 2; Wolfe & Self 1983; Oles 1988; Defant *et al.* 1988; Förster *et al.* 1990). The majority of the small eruption centres are younger than 1 Ma, possibly younger than 0.5 Ma (Wolfe & Self 1983; Oles 1988), whereas the major stratovolcanoes are mostly older (K–Ar ages range from 1.6–1.3 Ma for Mt San Cristobal, the western cone of the Banahaw complex, 0.9–1.0 Ma for Mt Malepunyo and 0.5–

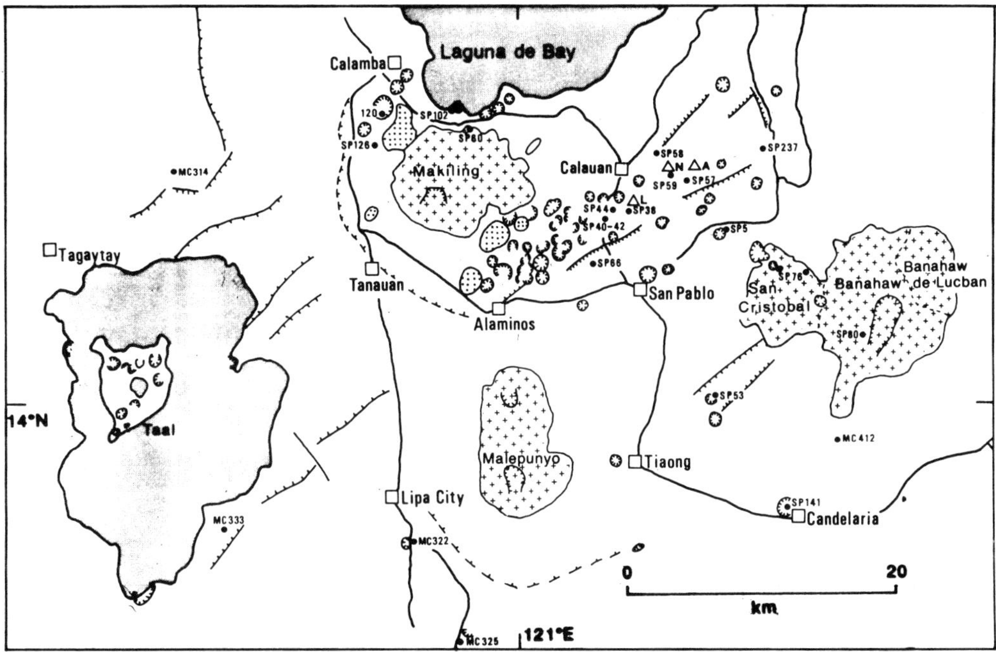

Fig. 2. Map of the Macolod Corridor (based on an unpublished map by Oles, a copy of which may be requested from the senior author). The distribution of andesitic lavas and pyroclastic flow deposits of the major stratocones is indicated by the cross ornament and some dacite domes are indicated by stippled patterns. The location of the summit regions of the polygenetic cones Lagula (L), Nagcarlang (N), and Atimbia (A) are indicated by open triangles. Small filled circles show the sampling location of the samples in Tables 1 and 2.

0.2 Ma for Mt Makiling; de Boer *et al.* 1980; Oles 1988). However, Mt Banahaw may have been active historically.

The known monogenetic eruption centres are concentrated in the San Pablo maar field and in the Taal caldera (Fig. 2), but others may be buried by ignimbrite sheets. They produced scoriae and lavas, which are unique among the volcanic sequences in the southern part of the Taiwan–Luzon arc in that they are SiO_2-poor, MgO-rich basalts, ranging from near-primary magmas to evolved, high-Al basalts (Oles 1988; Knittel & Oles 1991). These basalts are collectively referred to here as the Macolod basalts.

Three small complex cones, Lagula, Nagcarlang and Atimbia, are located at the northeastern end of the San Pablo maar field. These three polygenetic cones attain heights of 480–650 m and have diameters of 1.5–3 km. Their eruption products are referred to as the Calauan basalts.

Basalts, basaltic andesites and andesites were erupted from the Taal complex. A detailed account of their petrology is to be found in Miklius *et al.* (1991).

The huge stratovolcanoes are mainly composed of andesites; basaltic andesites also occur at Mt Banahaw and dacite domes are present on the flanks of Mt Makiling.

Volcanism in the Macolod Corridor appears to be related to a system of young, NE–SW-trending horsts and grabens, which is superimposed on an older, roughly N–S-trending structure (Voss 1971; Oles 1988; Förster *et al.* 1990). Large eruption centres are located at the intersections of major faults. The older N–S system is well developed north of the Macolod Corridor, especially in the northern part of the Laguna de Bay caldera (see Förster *et al.* 1990, fig. 2b), but it has been largely obliterated by younger tectonic activity or buried by ignimbrite sheets to the south.

Several attempts have been made to interpret the tectonics of the Macolod Corridor. Divis (1980) suggested that this zone was a 'leaky' transform fault connecting the Taiwan–Luzon and Bicol arcs. Defant *et al.* (1988) noted that the Macolod Corridor contains a complex fault system and proposed an origin as a 'pull-apart rift zone' between the Manila and Philippine

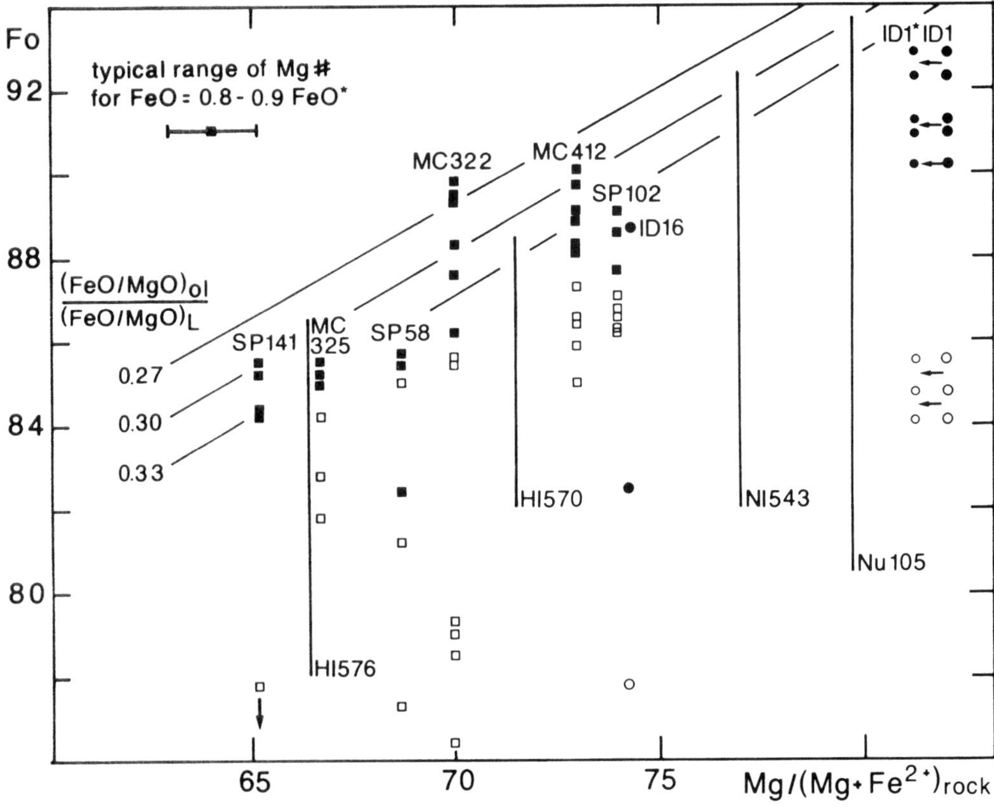

Fig. 3. Fo-contents of olivine plotted versus the $Mg/(Mg+Fe^{2+})$ ratios of the whole rocks, calculated for $Fe^{2+}/(Fe^{2+}+Fe^{3+}) = 0.85$. Uncertainty in $Mg/(Mg+Fe^{2+})$ due to uncertainty in $Fe^{2+}/(Fe^{2+}+Fe^{3+})$ is indicated in upper left corner. Squares are olivines from Macolod basalts, circles are olivines from Okmok picrite and basalt (Nye & Reid 1986; ID1* is ID1 minus about 4% cumulus olivine). Closed symbols refer to core compositions; open symbols are rim compositions. The range of olivine compositions in four basalts from Oshima-Oshima (Yamamoto 1988) is indicated by bars. Note the rather restricted range in olivine compositions in samples SP102 and MC412.

faults. In their model, the 'pull-apart' rifting would have been caused by the relative northward motion of the Philippines north of the Macolod Corridor and east of the Philippine Fault (Defant *et al.* 1988, fig. 7).

We agree that the Macolod Corridor forms a link between the westward-dipping subduction zone associated with the Philippine trench and the eastward-dipping subduction system, developed along the Manila trench (Fig. 1). Subduction along the Manila trench causes the northern part of the Philippine archipelago to move westward, because at present no new crust is formed in the South China Sea. On the other hand, Seno & Maruyama (1984) argued that the Philippine trench is migrating oceanward, hence the southern part of the Philippines should undergo an eastward translation. As there appears to be no single major transform fault accommodating these relative movements, they may have been compensated by internal deformation resulting in the S-shape of the island of Luzon (Hamilton 1979; east of Luzon, a left-lateral transform fault exists north-east of Polillo Island (Fig. 1), which, however, seems to connect only the Philippine trench and the north Luzon trough subduction systems; Lewis & Hayes 1983). The Macolod Corridor is located above the southern end of the Manila trench subduction system and just north of the northern terminus of the Philippine trench subduction system, at the locus where the island of Luzon bends round to the south-east. This suggests to us that the tectonic evolution of the Macolod Corridor is related to the internal deformation of the arc. The strike-slip movements take up the westward and eastward movement of north and south Luzon, respectively. The extensional

NW–SE component may be caused by the fact that north Luzon, at least until fairly recently, was part of the northwestward moving Philippine Sea plate (Lewis & Hayes 1983).

Petrography and mineralogy

Since a detailed description of the petrography and mineralogy of the basalts is beyond the scope of this paper and will be presented elsewhere, only the most important features are summarised in this section.

The most primitive basalts were erupted from the Alligator Lake maar (samples SP102 and MC354) and the Mayabobo cone (sample MC412). They contain about 7% olivine and clinopyroxene microphenocrysts (<1 mm) in a matrix composed of plagioclase microlites and glass. The olivine crystals frequently contain numerous chromite inclusions. The more evolved basalts contain fewer olivine crystals (largely free from chromite inclusions); plagioclase appears as phenocryst; clinopyroxene and plagioclase phenocrysts exhibit increasingly complex zoning patterns. Glomeroporphyritic textures are common in evolved basalts.

Olivine in the most primitive basalts has a very small range of core (Fo_{90-87}) and rim (Fo_{87-86}) compositions (Fig. 3). Core compositions are in equilibrium with whole rock $Mg/(Mg+Fe^{2+})$ ratios assuming $Kd(Fe/Mg)_{ol}/(Fe/Mg)_{liq} = 0.3 \pm 0.03$ (Roeder & Emslie 1970) and FeO to be $0.80–0.90\,FeO_{tot}$ (Nicholls & Whitford 1976). These features suggest that significant accumulation of olivine did not occur. High Ni contents of the Fo-rich olivines suggest that they may have equilibrated with residual olivine in the mantle (Sato 1977).

Chromite inclusions are found in virtually all Fo-rich olivines ($Fo_{>85}$). They are usually characterized by $Cr/(Cr+Al) = 0.55–0.60$ (Fig. 4) and $Mg/(Mg+Fe^{2+}) = 0.55–0.67$, which is consistent with their coexistence with olivine Fo_{90} (Dick & Bullen 1984, fig. 6a). Higher $Cr/(Cr+Al)$ ratios (0.63–0.72) are observed for chromites from Anilao Hill (MC322) and for a sample from the Taal complex. $Fe^{3+}/(Fe^{3+}+Cr+Al)$ ratios of 0.10–0.16 are slightly high, suggesting more oxidizing conditions compared to MORB. TiO_2 contents are mostly low (0.50–0.85 %). $TiO_2 < 1\%$ and $Fe^{3+}/(Fe^{3+}+Cr+Al) > 0.1$ are typical features of chromites from IAB and were used by Arai (1992) to distinguish them from spinels with other origins.

The clinopyroxene crystals are Al-rich diopsides, crystallised at low pressures as indicated by low Al^{VI}/Al^{IV} (Wass 1979). The groundmass

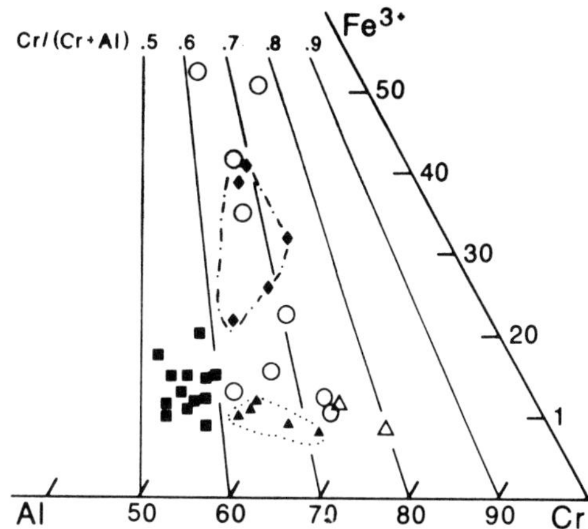

Fig. 4. Spinel compositions plotted in the Cr-rich portion of the Al-Cr-Fe^{3+} triangle. Filled symbols: chromites in Macolod basalts; squares: primitive basalts SP102, MC412, and MC325; triangles: Anilao Hill (MC322); diamonds: Taal main crater. Data for primitive basalts from Okmok (open circles; Nye & Reid 1986) and picrite from Oshima-Oshima (open triangles; Yamamoto 1984) are shown for comparison. Light lines represent constant $Cr/(Cr+Al)$ ratios.

plagioclase in the most primitive basalts is relatively sodic (An_{74-76}), compared to plagioclase phenocrysts in the more evolved basalts (An_{80-88}). Most of the plagioclase crystals analysed by electron microprobe show normal zoning.

Geochemistry

Analytical procedures

Whole-rock major and trace element compositions were determined by XRF at Aachen University on fused discs (sample:flux = 1:10) and undiluted pressed powder pellets, respectively. Six samples were re-analysed by XRF at Melbourne University using the low-dilution method described by Thomas & Haukka (1978). Except for Zr, the results were comparable when the analyses were carried out on aliquots of the same powder (MC322, MC325, MC331, MC333) but gave slightly different results where new powders were produced from the same sample (SP102 and MC412, Table 1). Zr values determined in Aachen are systematically higher than those determined in Melbourne. This discrepancy is not easily resolved, as we obtained nearly identical Zr values of 28–30 ppm for the standard granite MA-N (recommended value: 27 ppm; Govindaraju 1989). In Aachen, Zr concentrations of 109 ppm were determined for basalt BM (recommended value: 110 ppm) and 127 ppm Zr for diorite DR-IV (recommended value: 125 ppm) and, in Melbourne, Zr concentrations

Table 1. *Chemical composition of basaltic rocks (MgO > 6%) from the Macolod Corridor*

Sample	102M	102A	354A	412M	412A	322M	322A	329A	325M	325A	46A	222A	141A	53A	42A	40A	41A	44A	66A	126A
SiO_2	49.44	49.47	50.24	47.73	47.60	51.34	51.30	51.30	51.65	51.04	51.61	49.14	50.97	49.95	52.67	52.55	52.66	51.92	52.07	51.12
TiO_2	0.84	0.83	0.85	0.97	1.00	0.85	0.88	0.95	0.93	0.94	0.99	1.12	0.91	1.00	1.07	1.03	1.05	1.07	1.00	0.95
Al_2O_3	14.87	14.95	15.18	15.84	16.53	16.55	16.36	15.69	16.69	16.40	15.91	16.94	17.07	17.31	17.06	16.86	16.97	17.08	17.44	17.35
FeO^*	8.00	8.16	8.07	8.54	8.69	7.60	7.54	7.94	8.23	8.09	8.57	8.71	8.62	8.39	8.41	8.24	8.37	8.50	8.00	8.95
MnO	0.16	0.16	0.15	0.17	0.17	0.15	0.15	0.17	0.16	0.15	0.16	0.16	0.17	0.17	0.16	0.16	0.16	0.16	0.16	0.17
MgO	10.88	12.17	11.21	11.00	11.39	8.79	8.03	8.00	8.07	7.54	8.59	7.85	7.70	7.23	6.81	6.66	6.68	6.68	6.62	6.17
CaO	10.17	10.49	10.37	8.91	9.36	9.30	9.36	9.53	8.98	8.98	9.93	11.19	10.41	10.16	10.28	9.94	9.95	10.31	10.10	9.56
Na_2O	2.58	2.48	2.48	2.27	2.41	3.02	2.47	2.95	3.07	2.93	2.66	2.82	2.35	3.14	3.16	2.78	3.03	3.23	2.92	2.78
K_2O	0.87	0.83	0.80	0.69	0.70	0.93	0.97	0.98	0.79	0.81	0.80	0.90	0.75	1.23	0.97	0.90	0.90	0.98	1.01	0.83
P_2O_5	0.22	0.20	0.21	0.29	0.29	0.23	0.21	0.21	0.19	0.19	0.21	0.31	0.11	0.34	0.24	0.23	0.24	0.23	0.29	0.10
LOI	0.17	0.16	0.01	1.90	1.79	0.17	0.00	0.01	0.29	0.13	0.24	0.03	0.23	–	0.10	0.15	0.06	0.04	0.28	0.33
Total	98.20	99.90	99.57	98.31	99.93	98.93	97.27	97.73	99.05	97.20	99.67	99.17	99.29	98.92	100.93	99.50	100.07	100.20	99.89	98.31
Rb	14	18	19	5	8	17	21	19	14	16	19	19	15	28	19	19	20	18	23	19
Sr	505	467	478	556	555	455	449	400	450	444	603	654	407	771	636	648	634	635	663	403
Ba	277	285	266	542	614	209	215	202	225	209	312	365	206	561	313	326	334	330	389	257
Th	2	–	–	6	7	7	–	–	4	–	–	–	2	–	–	–	–	–	–	–
Ga	9	13	15	15	17	16	–	–	15	–	15	16	17	19	16	17	16	17	18	18
Nb	6	6	7	9	8	6	8	–	4	–	6	9	5	8	6	6	6	6	7	6
Zr	84	85	93	112	127	93	110	112	93	109	95	102	88	126	101	103	100	95	111	93
Y	13	19	15	20	20	16	18	20	22	23	21	18	21	18	23	23	24	22	19	18
Cr	535	550	–	543	580	365	350	350	295	200	–	–	–	–	–	–	–	–	–	–
Ni	223	231	192	224	196	149	122	103	132	109	78	78	69	51	32	38	33	31	50	55
$^{87}Sr/^{86}Sr$	–	–	–	–	–	0.70448	–	–	0.70423	–	–	–	–	–	–	–	–	0.70471	–	–

Sample numbers 1–299 have the prefix SP and samples 300–412 have the prefix MC. The letters A and M after the sample numbers indicate whether the analysis was obtained in Aachen (A) or Melbourne (M).
SP102, MC354, Alligator Lake. MC412: Mayabobo. MC322, MC329: Anilao Hill. MC325: Rosario Hill. SP46: Imoc Hill. SP222 Samire. SP141 Malasanya. SP53 Dagatan. SP42, SP40, SP41, SP44: Imoc Hill. SP66: Mapula. SP126: Camotes.

of 165 ppm were obtained for basalt BHVO–1 (recommended value: 179 ppm) and of 226 ppm Zr for andesite AGV–1 (recommended value: 227 ppm). We have plotted both sets of Zr data obtained at Aachen and Melbourne in Figs 6–8 to take into account the uncertainty of the Zr concentrations.

Sr isotopes were determined at LaTrobe University, Bundoora (Melbourne, Australia) following procedures described by Gray (1987). For NBS SRM 987 a value of 0.71027 2̌ was obtained for unspiked samples (based on 60 determinations). No age corrections were applied in view of the young age and low Rb/Sr ratios of the samples.

Results

Altogether, 16 Macolod basalts with $MgO > 6\%$ were analysed for major and trace elements (Table 1). They have relatively high SiO_2 contents of 51–53% (calculated anhydrous) except for the most primitive samples, whereas the Calauan basalts (Table 2 and Defant 1985) have SiO_2 contents of around 50%. Al_2O_3 increases fairly regularly from about 15% in the most primitive basalts to about 17.5% at 6% MgO. The Taal basalts differ from the Macolod basalts by lower Al_2O_3 and Na_2O, and higher CaO contents (see below).

Comparison of the Macolod basalts with primitive basalts erupted in other island arcs (Aleutians: Nye & Reid 1986; Vanuatu: Barsdell & Berry 1990; Eggins 1993; Marianas: Woodhead 1989; Bloomer *et al.* 1989) shows that the Macolod basalts are characterised by relatively higher TiO_2 and Na_2O and lower CaO contents (Fig. 5a,b).

Comparison of the two most primitive basalts with experimentally produced melts (Falloon & Green 1988; Falloon *et al.* 1988) suggests that they were in equilibrium with a clinopyroxene-bearing residue at 15–17 kbar (Knittel & Oles, unpublished data).

The Macolod basalts are enriched in the LILE K, Rb, Ba, and Sr relative to MORB, which is a typical feature of magmas erupted at convergent plate margins. However, in contrast to many other arc basalts, the HFSE Nb and Zr are also enriched, though only to a moderate extent. This feature is typical for some continental arcs (e.g. Pearce 1983), but there is no evidence that the Taiwan–Luzon arc is underlain by continental crust (e.g. Karig 1983).

Relatively few $^{87}Sr/^{86}Sr$ data are available for the basaltic rocks of the Macolod Corridor. The basalts of the Taal complex have $^{87}Sr/^{86}Sr$ = 0.7044–0.7047 (Knittel *et al.* 1988; Miklius *et al.* 1991). A similar value was obtained for a basalt from Mt Macolod at the southern margin of the Taal complex, while a basalt of Mt Sungay on the northern rim of the complex has a distinctly

Table 2. *Chemical composition of basaltic rocks of the polygenetic cones Nagcarlang (SP58, SP59), Lagula (SP38) and Atimbia (SP57, SP236)*

Sample	SP58	SP59	SP38	SP57	SP236
SiO_2	51.85	50.60	49.03	50.57	49.39
TiO_2	0.84	0.88	0.93	0.96	1.10
Al_2O_3	15.27	18.09	17.53	17.98	19.39
FeO^*	9.44	9.10	9.24	9.75	10.66
MnO	0.16	0.17	0.18	0.21	0.18
MgO	9.86	6.48	5.91	5.48	4.69
CaO	9.78	11.12	11.04	10.84	10.80
Na_2O	2.44	2.43	2.74	2.57	2.60
K_2O	0.55	0.84	1.15	1.04	0.73
P_2O_5	0.17	0.18	0.26	0.23	0.21
Total	100.36	99.90	98.01	99.70	100.31
Rb	11	20	24	22	18
Sr	402	566	599	547	538
Ba	169	253	300	259	226
Ga	15	16	17	17	18
Nb	6	4	5	3	4
Zr	74	71	97	71	66
Y	19	19	21	26	19
Cr	–	92	57	55	–
Ni	199	27	21	12	15

All data were obtained at Aachen

lower value of 0.70404 (Table 2). Basalts erupted from the monogenetic eruption centres outside the Taal complex yielded values ranging from 0.70423 (Rosario Hill) to 0.70471 (Imoc Hill; Table 1).

Previous studies have established a general southward increase in $^{87}Sr/^{86}Sr$ from north Luzon (Defant *et al.* 1990, 1991; Knittel *et al.* 1990) to Mindoro (Knittel & Defant 1988), accompanied by a decrease in $^{143}Nd/^{144}Nd$ (Knittel *et al.* 1988; Defant *et al.* 1991). Volcanic rocks north and south of the Macolod Corridor (the Bataan- and Mindoro-segments of the Taiwan–Luzon arc, respectively) are distinguished by $^{87}Sr/^{86}Sr$ <0.7050 and >0.7050, respectively (Knittel & Defant 1988).

The volcanoes forming the volcanic front at the western margin of the Macolod Corridor show a comparable transition from 'low' to 'high' $^{87}Sr/^{86}Sr$, from 0.70399 at Palay-Palay in the north, to 0.70500 at Panay in the south (Table 3). Sr isotopic data obtained for stratovolcanoes within the Macolod Corridor again show low $^{87}Sr/^{86}Sr$ for Mt Makiling in the north (0.7043, Knittel & Defant 1988) and high values for San Cristobal and Banahaw in the south (0.70490 and 0.70473, respectively, Table 3). Mt Malepunyo, located in the centre of the

U. KNITTEL & D. OLES

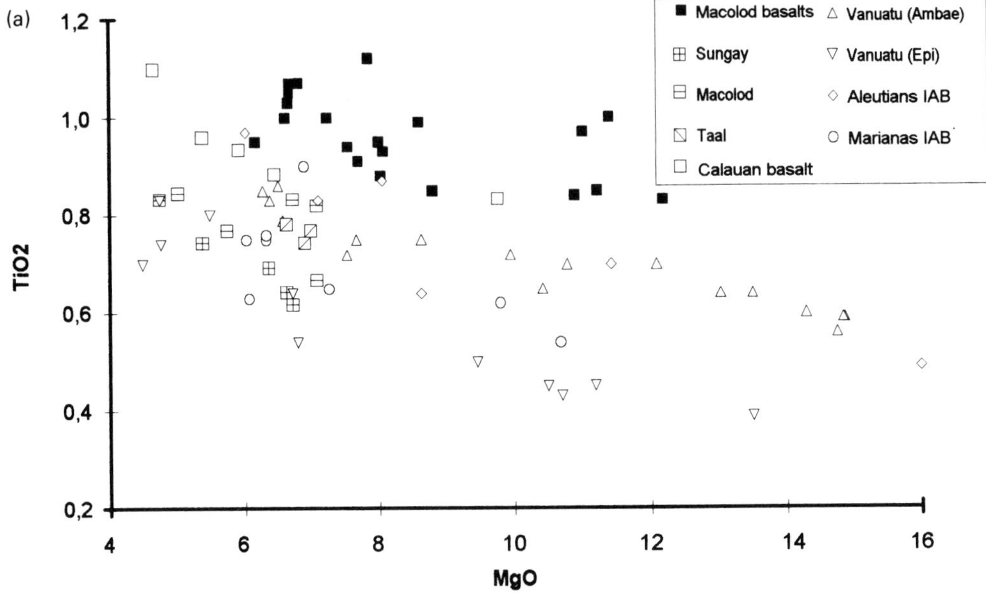

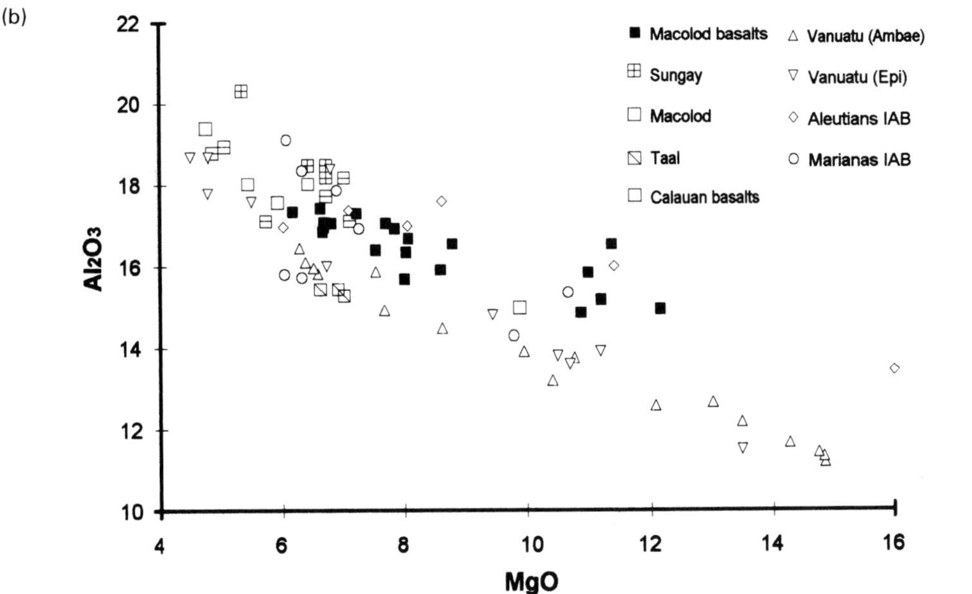

Fig. 5. (**a**) TiO_2, (**b**) Al_2O_3, (**c**) CaO and (**d**) Na_2O variations versus MgO for basalts from the Macolod Corridor (Macolod basalts; Sungay, Macolod, and 1969 Taal basalts of the Taal system), Ambae and Epi (Vanuatu), Okmok (Aleutians), and the Mariana Islands illustrating the major element characteristics of these basalts. Data sources: this paper; Miklius *et al*. 1991; Eggins 1993; Barsdell & Berry 1990; Nye & Reid 1986; Woodhead 1989; Bloomer *et al*. 1989).

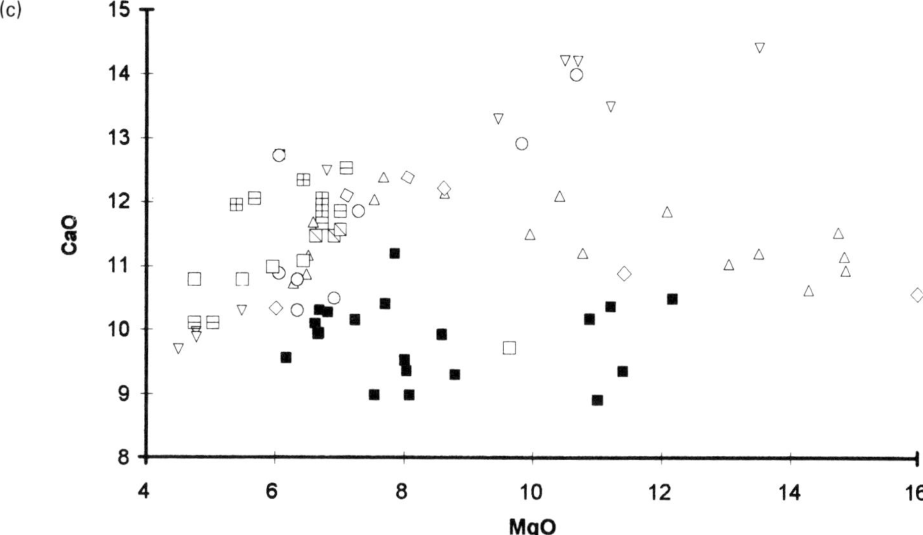

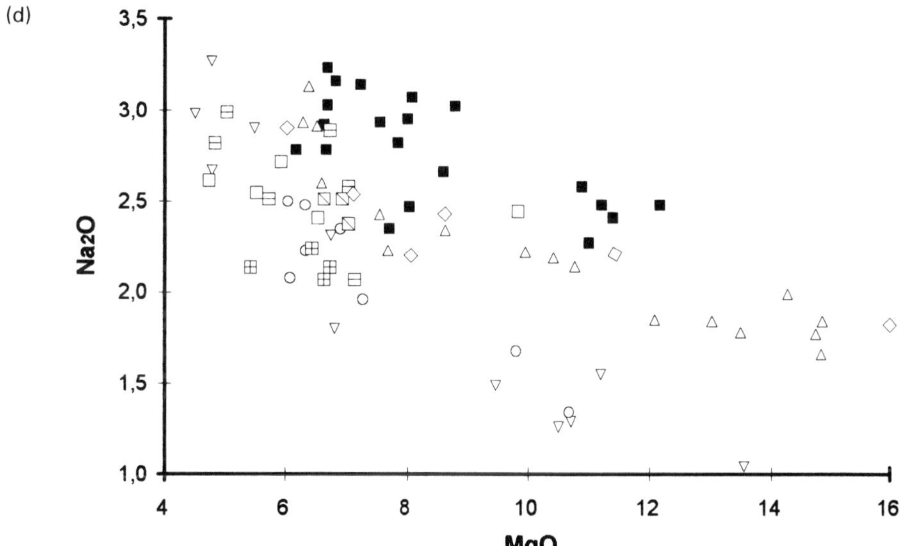

corridor, has extreme Sr isotopic compositions (0.70578–0.70593; Oles 1988 and Table 3).

Thus, consistent with their geographical location, the basalts of the Macolod Corridor have Sr isotopic compositions that fall within the range of values typical for the Bataan- and Mindoro-segments. However, $^{87}Sr/^{86}Sr$ values for centres within the Macolod Corridor are unsystematic; for example, the highest and lowest $^{87}Sr/^{86}Sr$ are found at the northern and southern margins of the corridor, in contrast to the general regional trend.

Discussion

Composition of the mantle wedge

Magmas erupted at destructive plate boundaries are commonly considered to be derived from sources initially more depleted than the sources

Table 3. *Chemical composition of selected samples from the Volcanic front along the western margin of the Macolod Corridor and stratovolcanoes within the Macolod Corridor*

Sample	338A	375A	314M	314A	333M	333A	60A	331M	331A	80A
SiO_2	56.93	56.87	48.83	49.33	50.91	50.74	60.38	57.82	56.93	56.69
TiO_2	0.69	0.69	0.62	0.64	0.82	0.84	0.67	0.75	0.76	0.81
Al_2O_3	17.56	18.22	18.50	18.18	18.21	17.71	16.28	17.53	17.35	18.47
FeO*	7.12	5.74	9.18	9.14	8.14	7.89	5.36	6.70	6.60	6.47
MnO	0.15	0.13	0.17	0.17	0.16	0.14	0.12	0.15	0.15	0.14
MgO	3.59	3.18	6.71	6.67	7.04	6.70	2.64	3.25	3.15	3.77
CaO	7.67	7.93	11.88	12.04	11.86	11.60	5.75	7.16	7.21	7.97
Na_2O	3.40	2.93	2.15	2.07	2.60	2.89	3.66	3.68	3.43	3.39
K_2O	1.30	1.70	0.53	0.55	0.58	0.58	2.47	1.83	1.81	2.00
P_2O_5	0.13	0.18	0.11	0.10	0.15	0.13	0.12	0.20	0.17	0.30
LOI	0.48	1.04	0.21	0.20	0.00	0.21	0.67	0.07	–	0.21
Total	99.02	98.61	98.89	99.09	100.47	99.43	98.12	99.14	97.56	100.22
Rb	39	68	7	12	6	11	63	44	47	50
Sr	359	476	288	303	323	319	370	529	532	731
Ba	294	565	154	156	162	166	322	664	614	582
Th	–	–	2	–	2	–	–	11	–	–
Ga	–	–	15	–	13	–	16	17	–	16
Nb	5	6	2	4	3	5	8	5	6	7
Zr	98	122	43	53	54	65	167	125	150	158
Y	24	19	12	12	16	18	22	22	23	21
Cr	–	–	93	–	109	–	–	10	–	–
Ni	33	<10	27	27	32	35	<10	12	<10	<10
$^{87}Sr/^{86}Sr$	0.70399	0.70500	0.70407		0.70455		0.70422	0.70593		0.70473

Sample number 1–299 have the prefix SP and samples 300–412 the prefix MC. The letters A and M after the sample numbers indicate whether the analysis was obtained in Aachen or Melbourne. Prefix SP for sample nos <300 and MC for sample nos >300 have been omitted.
SP383: Palay-Palay. SP375: Panay. MC314: Mt Sungay. MC333: Mt Macolod. SP60 Mt Makiling. MC331 Mt Malepunyo. SP80 Mt Banahaw.

of mid-ocean ridge basalts (MORB) and subsequently selectively enriched by components derived from subducted crust. The assumption of initially depleted sources is supported by the presence of harzburgitic lithologies at the inner trench walls of several island arcs (e.g. Bonatti & Michael 1989) and the eruption of boninites, which unquestionably are derived from very depleted sources (e.g. Crawford *et al.* 1989). Magnesian olivine (Fo_{92-94}) and Cr-rich chromite ($Cr/[Cr+Al] >0.65$) in primitive IAB (Nye & Reid 1986; Yamamoto 1988; Eggins 1993; Figs 3 & 4) and the low abundances of HFS elements, which are probably not replenished in the sub-arc mantle during subduction zone metasomatism, are additional evidence for initially depleted sources (e.g. Woodhead *et al.* 1993).

Compared to olivine compositions reported for near-primary magmas from other arcs (see above), olivines in the Macolod Corridor basalts are noteworthy for their relatively low Fo-contents, which even in near primary compositions do not exceed 90 mole% (Fig. 3). The

high NiO contents in these olivines, comparable to the NiO contents of olivines in xenolithic mantle fragments (e.g. Sato 1977), indicates that they are not evolved crystals but have been in equilibrium with the residual olivine. The chromites also have $Cr/(Cr + Al)$ ratios lower than those in primitive lavas from the Aleutians, NW Japan and Vanuatu (Fig. 4). Since studies of oceanic mantle xenoliths have shown systematic relationships between the olivine contents of peridotites and the Fo-contents and $Cr/(Cr+Al)$ ratios of the constituent olivines and chromites, respectively (e.g. Dick *et al.* 1984; Michael & Bonatti 1985), it is suggested that the less Fo- and Cr-rich compositions of olivine and chromite in the Macolod basalts reflect a more fertile residue and consequently their derivation from less depleted sources relative to those of primitive magmas in the Aleutians, NW Japan and Vanuatu, provided that the degrees of partial melting do not differ very much. The high Na_2O and low CaO contents of the Macolod basalts (Fig, 5a,b) are compatible with this interpretation. Falloon & Green (1988) observed that

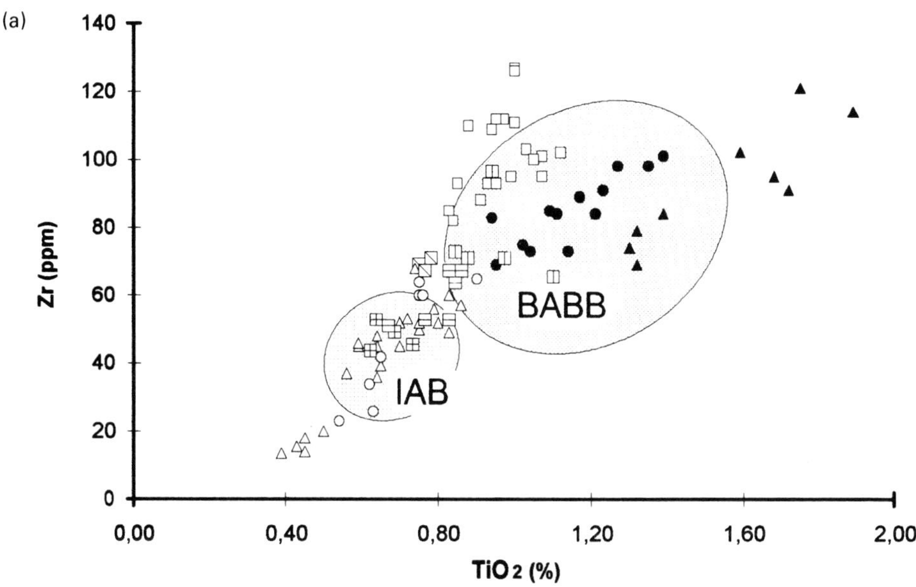

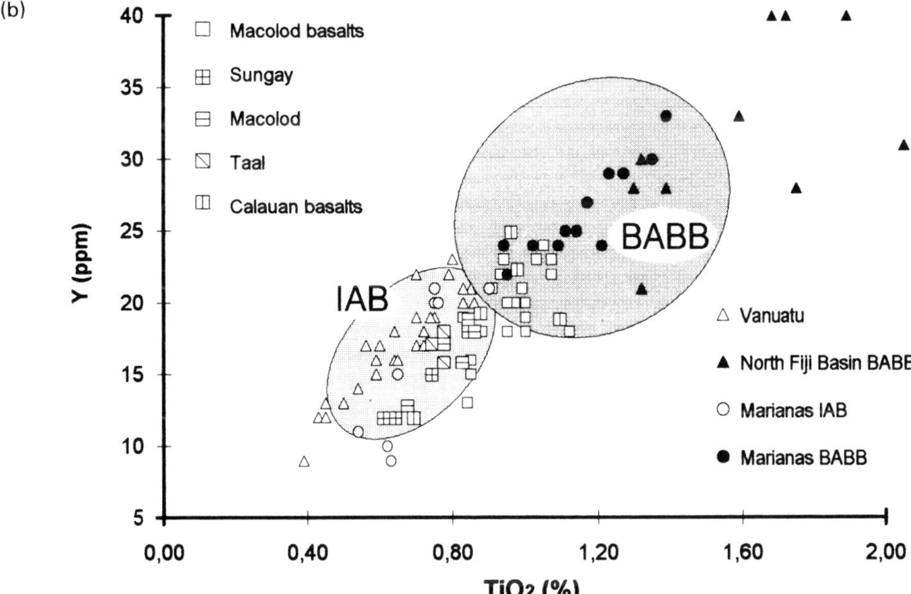

Fig. 6. (**a**) Zr and (**b**) Y plotted *versus* TiO₂ for Macolod basalts and Mt. Sungay, Mt. Macolod and 1969 Taal basalts of the Taal system. Data for the arc–back-arc pairs Mariana arc–Mariana Basin and Vanuatu arc–North Fiji Basin are shown for comparison (data sources: Woodhead 1989; Bloomer *et al*. 1989; Hawkins *et al*. 1990; Eggins 1993; Barsdell & Berry 1990; Price *et al*. 1990). Generalized fields for IAB and BABB are based on the average values for samples from the western Pacific in Woodhead *et al*. (1993).

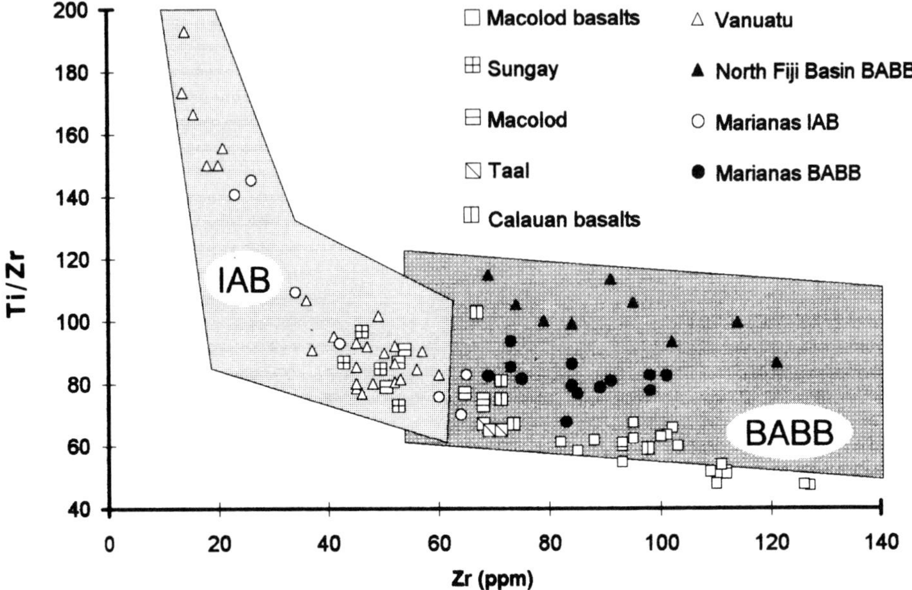

Fig. 7. Ti/Zr v. Zr for relatively primitive IAB and BABB (MgO > 5.5%). Fields for IAB and BABB are based on data from the western Pacific given in Woodhead *et al.* (1993).

melts derived from MORB pyrolite have higher TiO_2, Al_2O_3, and Na_2O and lower CaO contents than melts derived from the more depleted Tinaquillo lherzolite.

Additional support comes from the HFSE systematics of the basalts. Figure 6 shows that Zr (regardless of whether data obtained in Aachen or in Melbourne are considered), TiO_2 and Y contents in Macolod basalts are higher than in average IAB from the western Pacific, but comparable to abundances in associated BABB (Woodhead *et al.* 1993). This implies that the Macolod basalts could be derived from similar sources as other IAB by smaller degrees of partial melting. However, most IAB are also characterized by high Ti/Zr ratios, a feature not observed for the Macolod basalts (Fig. 7; note that the differences in Ti/Zr ratios between analyses obtained in Aachen and Melbourne, respectively, are of the same magnitude as the overall scatter of the individual data sets). In terms of Ti/Zr v. Zr, the Macolod basalts plot in the field of western Pacific BABB (Fig. 7), which are considered to be generated by melting of MORB-like sources (Woodhead *et al.* 1993). The low Zr contents and the steep trends in the Ti/Zr v. Zr diagram shown by IAB from the western Pacific are interpreted to be the signature of their derivation from MORB sources, previously depleted by extraction of 5–10% melt (Woodhead *et al.* 1993). By comparison, the

similar Ti/Zr ratios, as well as similar Zr and Y abundances suggest that the Macolod basalts were derived from sources comparable to MORB and BABB sources.

In contrast, basalt erupted from Taal in 1969 and basalts from Mt Sungay and Mt Macolod, located at the periphery of the Taal system (Fig. 2), are characterized by lower TiO_2, Al_2O_3, Na_2O, Y and Zr and higher CaO contents (Figs 5a,b and 6) and Ti/Zr ratios (Fig. 7). In the Ti/Zr v. Zr plot, they extend the trend defined by the BABB-like Macolod basalts to typical IAB compositions. In addition, Cr-spinels in a Taal basalt have higher Cr/(Cr+Al) ratios than spinels in most Macolod basalts (Fig. 4). Therefore, the basalts of the Taal system appear to be derived from more depleted sources than the Macolod basalts, suggesting heterogeneity of the mantle below the Macolod Corridor. Heterogeneity on a much smaller scale has already been suggested by Miklius *et al.* (1991) for the mantle sources of Taal magmas.

The identification of non-refractory source compositions for arc basalts from the Taiwan–Luzon arc has several implications. Since the Macolod basalts, which are derived from less depleted sources, as argued above, and contain significantly more Zr, TiO_2, Y, and Nb than most arc basalts from the western Pacific, we suggest that HFSE variations in IAB reflect *variable degrees of previous depletion* of the source mantle peridotite rather than the *introduction* of a 'within-plate component', as is commonly suggested (e.g. Pearce 1983).

Furthermore, Sr and Nd are almost completely incompatible in olivine, orthopyroxene, and spinel (see distribution coefficients in Kelemen *et al.* 1990), but moderately compatible in clinopyroxene. Hence, if a slab derived component is superimposed on *depleted* peridotite (harzburgite) containing no or only very small amounts of clinopyroxene, this component will dominate the Sr–Nd isotopic systematics no matter how small the slab contribution is. Conversely, if a small slab-derived component is superimposed on *fertile* peridotite (lherzolite), any pre-existing isotopic signatures are unlikely to be completely obliterated. For the Taiwan–Luzon arc, the possibility of an initially not strongly depleted mantle source was considered by Knittel & Defant (1988), based on the observation that mid-Tertiary plutonic rocks in the Philippines, regardless of their level of LILE enrichment, as well as young volcanic rocks from the Bicol arc in southern Luzon, and basalts from the Philippine Sea and Sanghie arc south of the Philippines all have $^{87}Sr/^{86}Sr$ ratios in the narrow range of 0.7035–0.7038.

In summary, the compositions of early-crystallized olivine and chromite, as well as relatively high HFSE contents in the Macolod basalts argue for their derivation from sources less depleted than the sources of many other IAB. Sub-arc mantle is probably depleted by the extraction of a melt fraction(s) prior to their involvement in IAB genesis (possibly by extraction of BABB; McCulloch & Gamble 1991). In contrast, the sub-Philippine mantle appears to have largely escaped previous extensive melt extraction, possibly because the Philippine arc system is not associated with a back-arc basin.

The presence of a subduction zone component

It has been long recognized that magmas erupted at destructive plate boundaries are characterised by marked enrichment in LILE and have low abundances of HFSE relative to MORB (e.g. Pearce 1983). High LILE/HFSE ratios have been attributed to the selective enrichment in LILE of a depleted source (e.g. Pearce 1983; Woodhead *et al.* 1993). The 'excess' LILE are commonly thought to be derived from subducted lithosphere, a view recently supported by Plank & Langmuir (1993) who found that the overall LILE systematics in arc magmas reflect the composition of the local, subducted sediments. Alternative explanations for the low HFSE abundances in IAB and the LILE/HFSE fractionation include the assumed presence of a phase retaining these elements in the source, but subsequent experimental work (Green & Peason 1986; Ryerson & Watson 1987) and theoretical considerations (Arculus & Powell 1986) have provided no evidence for the presence of such a phase. Kelemen *et al.* (1990) suggested that reaction of the melt with clinopyroxene within the mantle is responsible for the low HFSE abundances in IAB.

If LILE enrichment depends on the composition of locally available sediments (Plank & Langmuir 1993), LILE/HFSE fractionation needs to be evaluated on a regional basis. In the present case, the primitive basalts are compared with evolved, andesitic rocks of the stratovolcanoes. For this purpose, ratios of incompatible elements such as Rb/Zr and Ba/Zr are plotted in Fig. 8 because these elements are probably not significantly fractionated in basalts and andesites (note that the differences in Ba/Zr and Rb/Zr of analyses obtained for a particular sample in Aachen and Melbourne, respectively, are much smaller than the overall scatter). Comparison of the Macolod basalts and the eruption products of Mt Makiling and Mt Banahaw (Fig. 8) shows that the primitive basalts have lower Rb/Zr and Ba/Zr ratios than the andesites of the stratocones, although some overlap exists in terms of Ba/Zr. In these plots, local components, not related to subduction, are represented by the post-spreading olivine tholeiites extruded along the South China Sea Basin spreading centre (Tu *et al.* 1992) and basalts from the floor of the Philippine Sea (Meijer, in Knittel-Weber & Knittel 1990). Comparison of these basalts with the Macolod basalts reveals that the latter are characterised by slightly higher Rb/Zr, and distinctly higher Ba/Zr ratios. Another feature which clearly distinguishes the South China Sea and Philippine Sea basalts from the Macolod basalts are the much higher absolute abundances of TiO_2 (*c.* 2–3% v. 1%), Nb (20–30 ppm v. 5–9 ppm) and Zr (160–200 ppm v. 80–120 ppm) in the sea-floor basalts. Thus it appears that, relative to local sea-floor basalts, the Macolod basalts are characterised by selective enrichment in LILE relative to HFSE. This enrichment is more pronounced in the magmas composing the large stratovolcanoes.

Arc volcanoes are usually considered to be supplied with magma from mantle diapirs rising from near the surface of the subducting slab and melting progressively due to decompression (Marsh 1979; Sakuyama 1983; Plank & Langmuir 1988). The volcanic fields of the Macolod

(a)

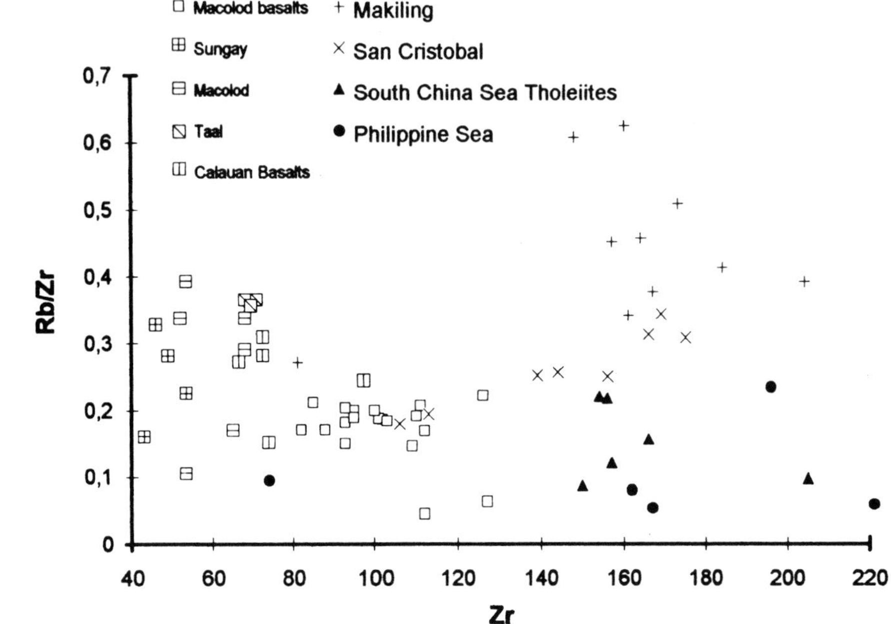

(b)

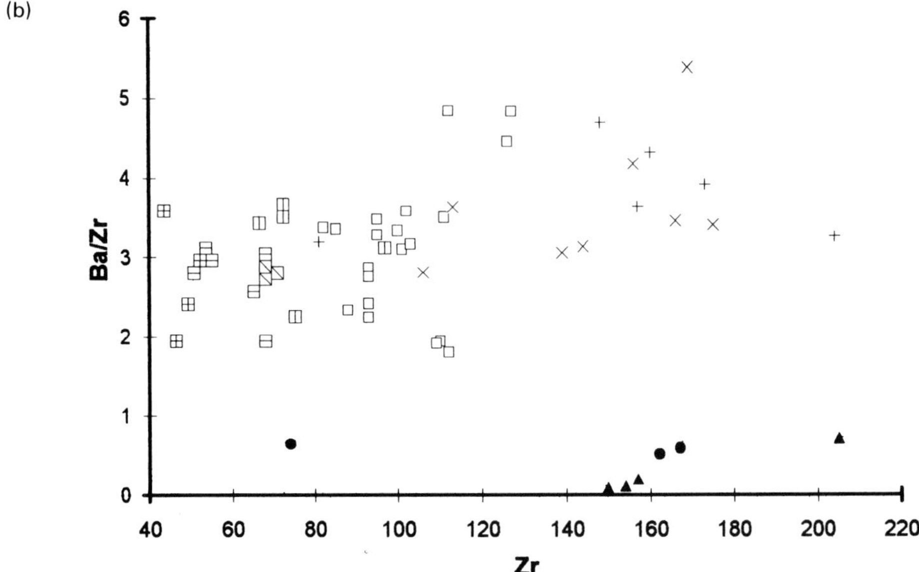

Fig. 8. (a) Rb/Zr and (b) Ba/Zr plotted versus Zr as a measure for LILE/HFSE fractionation commonly thought to be related to the presence of a 'subduction zone component'. Various basalts from the Macolod Corridor are compared to andesites from the stratovolcanoes Makiling and San Cristobal, and ocean floor basalts from the South China Sea and the Philippine Sea. Data sources: this paper; Defant 1985; Meijer *in* Knittel-Weber & Knittel 1990; Miklius 1991; Tu *et al*. 1992 (Note that for a number of samples no Ba data are available).

basalts are not at present underlain by an active Benioff–Wadati zone, although they may have been underlain formerly by a slab, if the southwards-steepening of the Manila trench subduction system is due to a collision in the south. Hence, volcanism in the Macolod Corridor probably is not directly related to present subduction. Moreover, the geographical scatter of the basaltic eruption centres indicates that the magma sources are not focussed. Rather we suggest that extension has decreased the crustal thickness and induced a small degree of melting at shallow levels. These melts were then tapped by trans-lithospheric fractures (Takahashi 1992).

Nevertheless, the elevated Rb/Zr and high Ba/Zr ratios of the Macolod basalts are indicative of the presence of slab-derived components in their sources, suggesting that such components are not exclusively contained and transported by rising mantle diapirs. Slab-'contaminated' sources might be 'extinct' diapirs, but this is not supported by the relatively fertile source composition noted above. Another possible source for the slab-derived components are recrystallized partial melts which escaped melt extraction. Such 'refrozen' partial melts, possibly forming veins, may store slab-derived components for long times after cessation of subduction, if termination of subduction also results in the end of convective overturn in the mantle wedge. Melting of such mantle peridotite may produce magmas that still reflect old subduction zone enrichments rather than the current tectonic regime (e.g. in Papua New Guinea; Johnson *et al.* 1978).

Conclusions

The study of relatively primitive, MgO-rich basalts from the Macolod Corridor, which probably erupted as a consequence of crustal extension, has revealed several unusual features:

(1) Relatively non-refractory olivine ($Fo_{\leqslant 90}$) and chromite phenocrysts, as well as relatively high HFSE abundances in the basalts suggest that they are derived from sources compositionally comparable to MORB sources. This contrasts with several other arcs, where olivine and chromite compositions in primitive basalts as well as low HFSE abundances argue for more depleted sources. As a corollary we note that major element relationships, which indicate a systematic relationship between sodium and calcium contents of arc magmas and crustal thickness (Plank & Langmuir 1988) may not be used to infer crustal thicknesses or depths of

melt segregation unless independent data have established that the magmas are derived from initially depleted sources, because these parameters may reflect the fertility of the source (Falloon & Green 1988).

(2) Though the Macolod Corridor is presently not underlain by an actively subducting slab, the Macolod basalts show distinct LILE enrichment relative to HFSE. Since this area conceivably was formerly underlain by a Benioff–Wadati zone, this suggests that the mantle may retain the 'subduction signature' after cessation of subduction. Reactivation of such mantle produces magmas retaining the 'subduction signature' though the generation of the magmas may be unrelated to subduction.

Supported by Deutsche Forschungsgemeinschaft (grants Fo53/16–1, /16–2 to Förster) and Gesellschaft für Technische Zusammenarbeit (GTZ Grant 85.2522.2–06.100). UK acknowledges the permission to make extensive use of the microprobe of the Geology Department of the University of Melbourne and the help provided by D. Sewell, P. Kelly and M. Haukka. Suggestions made by A. Cundari, J.G. Brophy, two anonymous reviewers and the editor, John Smellie, helped to significantly improve both prose and science of the manuscript. M. Halter kindly had a good look at the revised version.

References

ARAI, S. 1992. Chemistry of chromian spinel in volcanic rocks as a potential guide to magma chemistry. *Mineralogical Magazine*, **56**, 173–184.

ARCULUS, R.J. & POWELL, R. 1986. Source component mixing in the regions of arc magma generation. *Journal of Geophysical Research*, **91**, 5913–5926.

BARSDELL, M. & BERRY, R.F. 1990. Origin and evolution of primitive island arc ankaramites from Western Epi, Vanuatu. *Journal of Petrology*, **31**, 747–777.

BLOOMER, S.H., STERN, R.J., FISK, E. & GESCHWIND, C.H. 1989. Shoshonitic volcanism in the northern Mariana Arc. 1. Mineralogic and major and trace element characteristics. *Journal of Geophysical Research*, **94**, 4469–4496.

BONATTI, E. & MICHAEL, P.J. 1989. Mantle peridotites from continental rifts to ocean basins to subduction zones. *Earth and Planetary Science Letters*, **91**, 297–311.

CARDWELL, R.K., ISACKS, B.L. & KARIG, D.E. 1980. The spatial distribution of earthquakes, focal mechanism solutions, and subducted lithosphere in the Philippine and northeastern Indonesian islands. *In*: HAYES, D.E. (ed.) *The tectonic and geologic evolution of southeast Asian seas and islands*. American Geophysical Union Monographs, **23**, 1–35.

CRAWFORD, A.J., FALLOON, T.J. & GREEN, D.H. 1989. Classification, petrogenesis and tectonic setting of boninites. *In*: CRAWFORD, A.J. (ed.) *Boninites*. Unwin Hyman, London, 1–49.

DE BOER, J.Z., ODOM, L.A., RAGLAND, P.C., SNIDER, F.G & TILFORD, N.R. 1980. The Bataan Orogene: Eastward subduction, tectonic rotations, and volcanism in the western Pacific (Philippines). *Tectonophysics*, **67**, 251–282.

DEFANT, M.J. 1985. *The potential origin of the potassium depth relationship in the Bataan Orogene, The Philippines*. PhD thesis, Florida State University.

——, DE BOER, J.Z. & OLES, D. 1988. The western Central Luzon Arc, the Philippines: two arcs divided by rifting? *Tectonophysics*, **145**, 305–317.

——, MAURY, R.C., JORON, J.-L., FEIGENSON, M.D., LETERRIER, J., BELLON, H., JAQUES, D. & RICHARD, M. 1990. The geochemistry and tectonic setting of the northern section of the Luzon arc (Philippines and Taiwan). *Tectonophysics*, **183**, 187–205.

——, ——, RIPLEY, E.M., FEIGENSON, M.D. & JAQUES, D. 1991. An example of island-arc petrogenesis: geochemistry and petrology of the southern Luzon arc, Philippines. *Journal of Petrology*, **32**, 455–500.

DICK, H.J.B. & BULLEN, T. 1984. Chromian spinel as a petrogenetic indicator in abyssal and alpine-type peridotites and spatially associated lavas. *Contributions to Mineralogy and Petrology*, **86**, 54–76.

——, FISHER, R.L., & BRYAN, W.B. 1984. Mineralogic variability of the uppermost mantle along mid-ocean ridges. *Earth and Planetary Science Letters*, **69**, 88–106.

DIVIS, A.F. 1980. The petrology and tectonics of recent volcanism in the central Philippine islands. *In*: HAYES, D.E. (ed.) *The tectonic and geologic evolution of southeast Asian seas and islands*. American Geophysical Union Monographs, **23**, 127–144.

EGGINS, S.M. 1993. Origin and differentiation of picritic arc magmas, Ambae (Aoba), Vanuatu. *Contributions to Mineralogy and Petrology*, **114**, 79–100.

FALLOON, T.J. & GREEN, D.H. 1988. Anhydrous partial melting of peridotite from 8 to 35 kb and the petrogenesis of MORB. *Journal of Petrology, Special Lithosphere Issue*, 379–414.

FALLOON, T.J., GREEN, D.H., HATTON, C.J. & HARRIS, K.L. 1988. Anhydrous melting of a fertile and depleted peridotite from 2 to 30 kb and application to basalt petrogenesis. *Journal of Petrology*, **29**, 1257–1282.

FÖRSTER, H., OLES, D., KNITTEL, U., DEFANT, M.J. & TORRES, R. 1990. The Macolod Corridor: A rift crossing the Philippine island arc. *Tectonophysics*, **183**, 265–271.

GOVINDARAJU, K. 1989. 1989 COMPILATION OF WORKING VALUES AND SAMPLE DESCRIPTION FOR 272 GEOSTANDARDS. *Geostandards Newsletter*, **13** (Special Issue), 1–113.

GREEN, T.H. & PEARSON, N.J. 1986. Ti-rich accessory phase saturation in hydrous mafic-felsic compositions at high P,T. *Chemical Geology*, **54**, 185–201.

GRAY, C.M. 1987. Strontium isotopic constraints on the origin of Proterozoic anorthosites. *Precambrian Research*, **37**, 173–189.

HAMBURGER, M.W., CARDWELL, R.K. & ISACKS, B.L. 1983. Seismotectonics of the northern Philippine island arc. *In*: HAYES, D.E. (ED.) *The tectonic and geologic evolution of southeast Asian seas and islands, part II*. American Geophysical Union Monograph, **27**, 1–22.

HAMILTON, W. 1979. *Tectonics of the Indonesian Region*. US Geological Survey Professional Paper, **1078**.

HAMILTON, W. 1989. CONVERGENT-PLATE TECTONICS VIEWED FROM THE INDONESIAN REGION. *In*: SENGÖR, A.M.C. (ed.) *Tectonic evolution of the Tethyan region*. Kuwer, Dordrecht, 655–698.

HAWKINS, J.W., LONSDALE, P.F., MACDOUGALL, J.D. & VOLPE, A.M. 1990. PETROLOGY OF THE AXIAL RIDGE OF THE MARIANA TROUGH BACKARC SPREADING CENTER. *Earth and Planetary Science Letters*, **100**, 226–250.

JOHNSON, R.W., MACKENZIE, D.E. & SMITH, I.E.M. 1978. Delayed partial melting of subduction-modified mantle in Papua New Guinea. *Tectonophysics*, **46**, 197–216.

KARIG, D. 1983. Accreted terranes in the northern part of the Philippine archipelago. *Tectonics*, **2**, 211–236.

KELEMEN, P.B., JOHNSON, K.T.M., KINZLER, R.J. & IRVING, A.J. 1990. High-field-strength element depletions in arc basalts due to mantle-magma interaction. *Nature*, **245**, 521–524.

KNITTEL, U. & DEFANT, M.J. 1988. Sr isotopic and trace element variations in Oligocene to Recent igneous rocks from the Philippine island arc: evidence for recent enrichment in the sub-Philippine mantle. *Earth and Planetary Science Letters*, **87**, 87–99.

—— & OLES, D. 1991. Near primary magmas from the Philippine island arc. *European Journal of Mineralogy*, **3**, 139.

——, DEFANT, M.J. & RAZECK, I. 1988. Recent enrichment in the source region of arc magmas from Luzon island, Philippines: Sr and Nd isotopic evidence. *Geology*, **16**, 73–76.

——, TRUDU, A.G., WINTER, W., GRAY, C.M. & STEELE, D.A. 1990. Preliminary petrochemical results from the central segment of the Luzon volcanic arc (Philippines): the Mankayan Mineral District. *Proceedings of the Pacific Rim Congress*, **2**, 217–223.

KNITTEL-WEBER, C. & KNITTEL, U. 1990. Petrology and genesis of the volcanic rocks on the eastern flank of Mount Malinao, Bicol arc (Southern Luzon, Philippines). *Journal of Southeast Asian Earth Sciences*, **4**, 267–280.

LEWIS, S.D. & HAYES, D.E. 1983. The tectonics of northward propagating subduction along eastern Luzon, Philippine islands. *In*: HAYES, D.E. (ed.) *The tectonic and geologic evolution of southeast Asian seas and islands, part II*. American Geophysical Union Monographs, **27**, 57–78.

MARSH, B.D. 1979. Island arc development: some observations, experiments, and speculations. *Journal of Geology*, **87**, 687–713.

McCabe, R., Almasco, J. & Diegor, W. 1982. Geologic and paleomagnetic evidence for a possible Miocene collision in western Panay, central Philippines. *Geology*, **10**, 325–329.

McCulloch, M.T. & Gamble, J.A. 1991. Geochemical and geodynamical constraints on subduction zone magmatism. *Earth and Planetary Science Letters*, **102**, 358–374.

Michael, P.J. & Bonatti, E. 1985. Peridotite composition from North Atlantic: regional and tectonic variations and implications for partial melting. *Earth and Planetary Science Letters*, **73**, 91–104.

Miklius, A., Flower, M.F.J., Huijsmans, J.P.P., Mukasa, S.B. & Castillo, P. 1991. Geochemistry of lavas from Taal Volcano, southwestern Luzon, Philippines: evidence for multiple magma supply systems and mantle source heterogeneity. *Journal of Petrology*, **32**, 593–627.

Molnar, P. & Atwater, T. 1978. Interarc spreading and cordilleran tectonics as alternatives related to the age of subducted oceanic lithosphere. *Earth and Planetary Science Letters*, **41**, 330–340.

Nicholls, I.A. & Whitford, D.J. 1976. Primary magmas associated with Quarternary volcanism in the western Sunda arc, Indonesia. *In*: Johnson, R.W. (ed.) *Volcanism in Australasia*. Elsevier, Amsterdam. 77–90.

Nye, C.J. & Reid, M.R. 1986. Geochemistry and least fractionated lavas from Okmok Volcano, central Aleutians: implications for arc magma genesis. *Journal of Geophysical Research*, **91**, 10271–10287.

Oles, D. 1988. Das San Pablo Vulkangebiet: Grabenvulkanismus im Luzon-Inselbogen. PhD thesis RWTH Aachen.

Pearce, J.A. 1983. Role of the sub-continental lithosphere in the magma genesis at continental margins. *In*: Hawkesworth, C.J. & Norry, M.J. (eds) *Continental basalts and mantle xenoliths*. Shiva, Nantwich, 230–249.

Plank, T. & Langmuir, C.H. 1988. An evaluation of the global variations in the major element chemistry of arc basalts. *Earth and Planetary Science Letters*, **90**, 349–370.

—— & —— 1993. Tracing trace elements from sediment input to volcanic output at subduction zones. *Nature*, **362**, 739–743.

Price, R.C., Johnson, L.E. & Crawford, A.J. 1990. Basalts of the North Fiji Basin: generation of back arc basin magmas by mixing of depleted and enriched mantle sources. *Contributions to Mineralogy and Petrology*, **105**, 106–121.

Rangin, C., Stephan, J.F., Blanchet, R. *et al.* 1988. Seabeam survey at the southern end of the Manila trench. Transition between subduction and collision processes, offshore Mindoro Island, Philippines. *Tectonophysics*, **146**, 261–278.

Roeder, P.L. & Emslie, R.F. 1970. Olivine–liquid equilibrium. *Contributions to Mineralogy and Petrology*, **29**, 275–289.

Ryerson, F.J. & Watson, E.B. 1987. Rutile saturation in magmas: implications for Ti-Nb-Ta depletion in island-arc basalts. *Earth and Planetary Science Letters*, **86**, 225–239.

Sakuyama, M. 1983. Petrology of arc volcanic rocks and their origin by mantle diapirs. *Journal of Volcanology and Geothermal Research*, **18**, 297–320.

Sato, H. 1977. Nickel content of basaltic magmas: identification of primary magmas and a measure of the degree of olivine fractionation. *Lithos*, **10**, 113–120.

Seno, T. & Maruyama, S. 1984. Paleogeographic reconstruction and origin of the Philippine Sea. *Tectonophysics*, **102**, 53–84.

Takahashi, N. 1992. Evidence for melt segregation towards fractures in the Horoman mantle peridotite complex. *Nature*, **359**, 52–55.

Thomas, I.L. & Haukka, M.T. 1978. XRF determination of trace and major elements using a single fused disc. *Chemical Geology*, **21**, 39–50.

Tu, K., Flower, M.F.J., Carlson, R.W., Xie, G., Chen, C.-Y. & Zhang, M. 1992. Magmatism in the South China Sea Basin 1. Isotopic and trace-element evidence for an endogenous Dupal mantle component. *Chemical Geology*, **97**, 47–63.

Voss, F. 1971. Die geologische und morphologische Struktur des aktivsten Vulkangebietes der Philippinen. *Die Erde*, **102**, 307–316.

Wass, S.Y. 1979. Multiple origins of clinopyroxene in alkali basaltic rocks. *Lithos*, **12**, 115–132.

Wolfe, J.A. & Self, S. 1983. Structural lineaments and Neogene volcanism in southwestern Luzon. *In*: Hayes, D.E. (ed.): *The tectonic and geologic evolution of southeast Asian seas and islands, part II*. American Geophysical Union Monographs, **27**, 157–172.

Woodhead, J. 1989. Geochemistry of the Mariana arc (western Pacific): source composition and processes. *Chemical Geology*, **76**, 1–24.

——, Eggins, S. & Gamble, J. 1993. High field strength and transition element systematics in island arc and back-arc basin basalts: evidence for multi-phase melt extraction and a depleted mantle wedge. *Earth and Planetary Science Letters*, **114**, 491–504.

Yamamoto, M. 1984. Origin of calc-alkaline andesite from Oshima-Oshima volcano, north Japan. *Journal of the Faculty of Science, Hokkaido University, Series IV*, **21**, 77–131.

—— 1988. Picritic primary magma and its source mantle for Oshima-Oshima and back-arc side volcanoes, Northeast Japan arc. *Contributions to Mineralogy and Petrology*, **99**, 352–359.

Volcanism associated with extension in an Oligocene–Miocene arc, southwestern Viti Levu, Fiji

M.R. WHARTON,[1] B. HATHWAY[2] & H. COLLEY[3]

[1] *Department of Geological Sciences, Durham University, Durham DH1 3LE, UK.*

[2] *Mineral Resources Department, Private Mail Bag, Suva, Fiji.*
Present address: British Antarctic Survey, High Cross, Madingley Road,
Cambridge CB3 0ET, UK

[3] *School of Construction & Earth Sciences, Oxford Brookes University, Oxford OX3*
0BP, UK

Abstract: The Wainimala Group rocks of southwestern Viti Levu, Fiji, represent part of an Oligocene–Miocene island arc. Some of the early volcanism in this arc took place on a substrate of Eocene–Oligocene frontal-arc crust. Mainly andesitic, transitional calc-alkaline lavas were erupted from major subaerial volcanic edifices, while elsewhere tholeiitic magmas formed low-lying basaltic lava fields and small felsic volcanic centres on the seafloor. The intrusion of a dense bimodal basalt–dacite dyke swarm into the underlying frontal-arc crust indicates that significant crustal extension accompanied the eruption of the inter-edifice lavas.

Geochemical variations within the Wainimala volcanic suite are thought to have been generated through fractional crystallisation processes. While relatively slow ascent and mixing of magmas took place beneath the major edifices, efficient shallow-level fractionation and rapid ascent of magmas occurred in the areas between them. The contrasting styles of volcanism may reflect the interaction of ascending magmas with pre-existing heterogeneities in the frontal-arc substrate.

The western Pacific is now widely accepted as a type-locality for the study of convergent plate margin processes. Extensive programmes of ocean drilling, dredging and surveying have greatly improved our knowledge of the tectonic and geochemical evolution of the intraoceanic arc–back-arc systems in the region. While crustal extension similar to that observed at oceanic spreading centres is recognized as fundamental in producing new crust in the back-arc, accretionary processes within the arc itself are less well understood, because important crustal sections are often buried by the products of later arc volcanism (Stern *et al.* 1989).

Some western Pacific convergent plate margins, in particular those which lack a well developed accretionary wedge (e.g. Izu–Bonin–Mariana, Tonga–Kermadec) may be subject to extensional forces during much of their evolution as a result of subduction zone retreat and plate decoupling (e.g. Hawkins *et al.* 1984). Although there does not appear to be a universal temporal relationship between arc volcanism, rifting and back-arc spreading (Taylor 1992), a broad evolutionary sequence for the initial 25 Ma of this type of plate margin may be summarized as follows.

(i) Establishment of a proto-arc (nascent arc) following the initiation of subduction, with characteristic eruption of depleted (tholeiitic) and ultra-depleted (boninitic) magmas.

(ii) Rifting of the proto-arc to form a trenchward frontal arc, a remnant arc, and an intervening back-arc basin where tholeiitic magmas are erupted during normal asthenospheric upwelling.

(iii) Establishment of a second phase of arc volcanism at the zone of crustal weakness between the rifted margin of the frontal-arc and the back-arc basin. The volcanic rocks produced in the arc may be tholeiitic or calc-alkaline. The system eventually stabilises to become a mature arc, as accretion in the back-arc wanes.

This study of the Lower Oligocene to Middle Miocene Wainimala Group of Viti Levu, Fiji, documents tectonic and geochemical processes occurring at stage (iii) of the evolutionary sequence. The second phase of arc volcanism (the Wainimala arc) began after a short (2–3 Ma) hiatus in arc activity during which the magmatic locus of the system was concentrated in the back-arc (South Fiji Basin). Initially, the

From Smellie, J.L. (ed.), 1995, *Volcanism Associated with Extension at Consuming Plate Margins*, Geological Society Special Publication No. 81, 95–114.

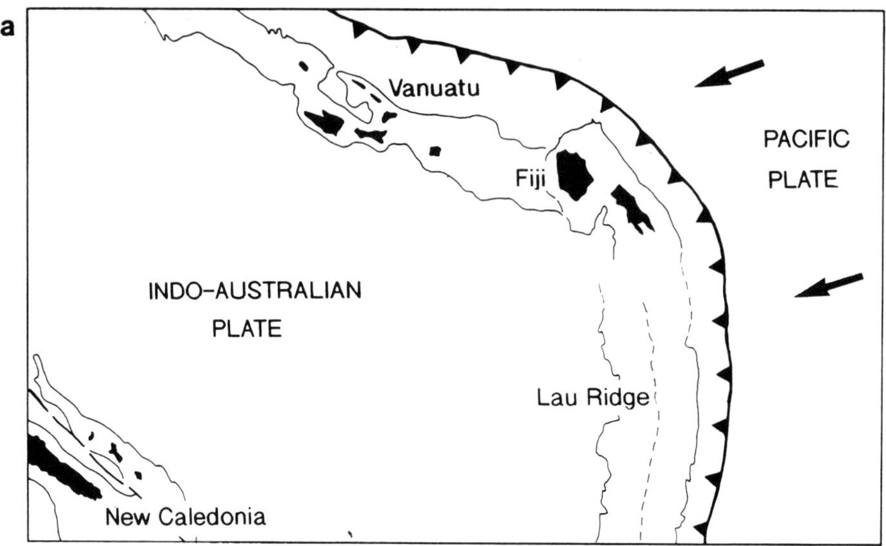

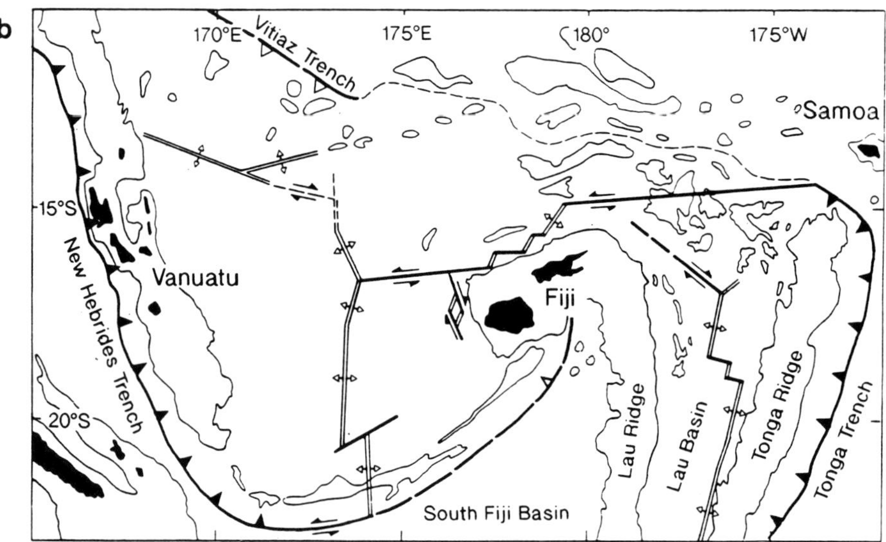

Fig. 1. (**a**) Regional plate tectonic reconstruction for the Late Oligocene–Early Miocene. Plate convergence vectors inferred from poles of rotation published by Packham & Andrews (1975). (**b**) Present-day regional tectonic setting. Modified from Hathway (1993*a*).

volcanic front was marked by major volcanic edifices and inter-edifice lava flows that developed on a substrate of frontal-arc crust. Eventually the zone affected by arc volcanism narrowed and the volcanic front retreated towards the frontal arc–back-arc transition. There, the arc stabilited to become the mature Wainimala arc. This is thought to have formed part of a continuous, northeast-facing Vanuatu–Fiji–Lau Ridge arc (Fig. 1a; sometimes referred to as the Vitiaz arc) until this broke up following

arc-reversal in Vanuatu during the Late Miocene (Carney *et al.* 1985; Hathway 1993*a*). Figure 1b shows the present tectonic setting of the Fiji group within the complex arc-arc transform zone between the opposite-facing Vanuatu and Tonga arcs.

Geological framework

The oldest known rocks in Fiji are assigned to the Yavuna Group (Hathway 1992). They are

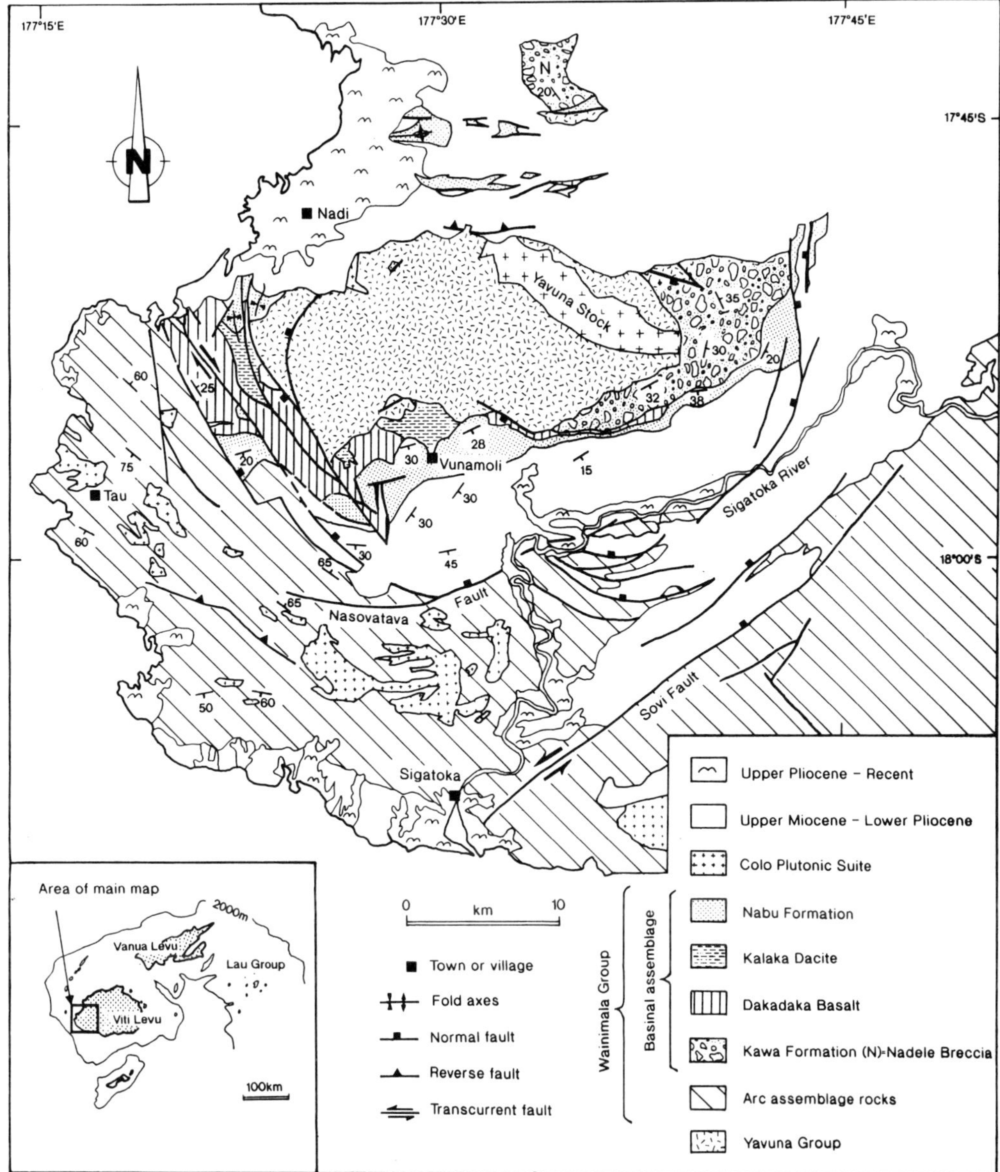

Fig. 2. Geology of southwestern Viti Levu, based on mapping by Hathway (1993*b*) and Houtz (1959, 1960).

exposed in the southwest of the principal island of Viti Levu and record Late Eocene to Early Oligocene volcanism in the Fiji proto-arc (the Yavuna arc). These rocks comprise a sequence of basic lavas interbedded with minor Upper Eocene shallow-water limestones and overlain by Lower Oligocene volcaniclastic rudites and limestones. The lavas show geochemical signatures ranging from island-arc tholeiite to transitional-boninite types (Gill 1987), and are thought to represent crust formed during or shortly following the initiation of subduction in the Middle Eocene. Similar lithologies are preserved in 'Eua, Tonga (Cunningham & Ascombe 1985), and as clasts in sedimentary sequences in Vanuatu (Carney *et al.* 1985), attesting to volcanism in other parts of the SW Pacific at this time. The Yavuna Group volcanic rocks are intruded by a Lower Oligocene trondhjemite pluton, the Yavuna Stock. Radiometric dates on this body range from 36.6 ± 0.8 Ma (U–Pb zircon; I. Williams, RSES, pers.

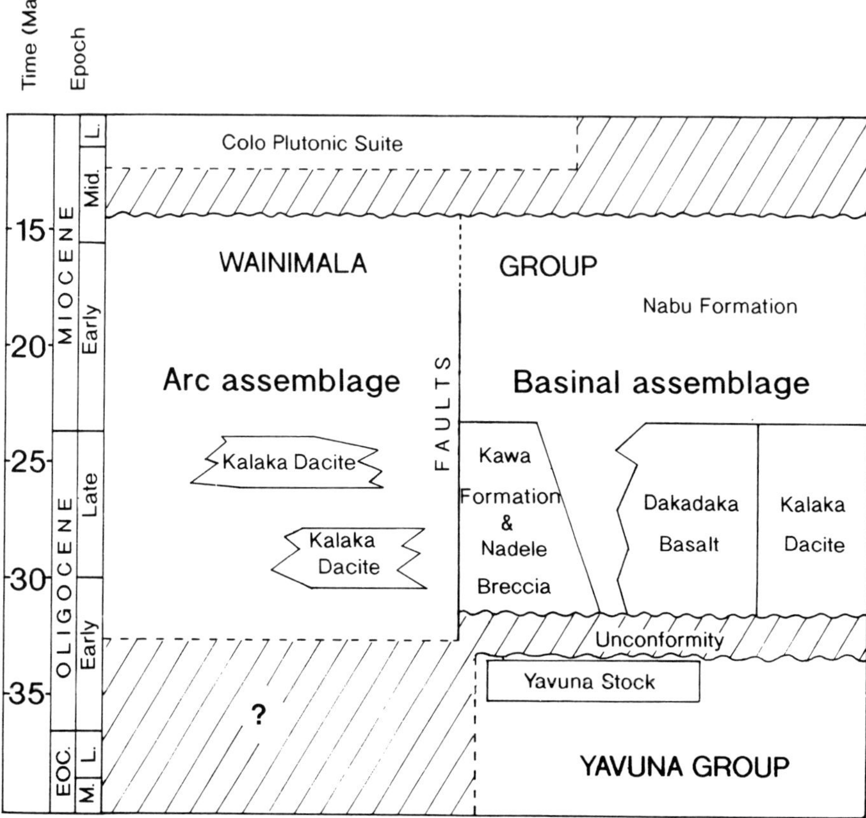

Fig. 3. Eocene to Miocene stratigraphical relationships in SW Viti Levu.

comm. 1991) to 34.9 ± 0.9 Ma (K–Ar on hornblende; Rodda 1983 recalculated from McDougall 1963) and 31.0 ± 1.2 Ma (^{40}Ar/^{39}Ar isochron on hornblende; Whelan *et al.* 1985). Its intrusion represents the final magmatic episode in the proto-arc and may have coincided with proto-arc rifting, as it slightly predates the onset of sea-floor spreading in the northern part of the South Fiji back-arc basin as recorded by magnetic anomaly 12 (Malahoff *et al.* 1982; dated at 31.5 Ma by Harland *et al.* 1990). The remnant arc produced by rifting of the proto-arc has not been located, but it may lie as a submerged ridge far to the SW in the New Caledonia region.

The Wainimala Group is extensively exposed in southern Viti Levu. In the southwest of the island the lull in volcanic activity between the latest magmatism in the proto-arc and the earliest magmatism in the Wainimala arc is marked by a major unconformity that may reflect tectonism during rifting of the proto-arc. At least some of the early Wainimala volcanism in this area took place on a substrate formed by

the Yavuna Group. Middle to Upper Miocene tonalite and gabbro plutons of the Colo Plutonic Suite delineate the axis of the mature Wainimala arc, which extends across southern Viti Levu from WSW to ENE, and across southwestern Viti Levu from NW to SE (Fig. 2).

Palaeomagnetic studies indicate that Viti Levu has rotated in an anti-clockwise direction since the breakup of the continuous Vanuatu–Fiji–Lau Ridge arc in the Late Miocene (Falvey 1978). Recent results have constrained the amount of rotation to approximately 100° (G.K. Taylor pers. comm. 1992), enabling a simplified reconstruction of the elements of the arc system to be made for the Late Oligocene (Fig. 1a).

The Wainimala Group

The Wainimala Group rocks of southwestern Viti Levu can be divided into two assemblages (Fig. 3, Hathway 1993c). One consists mainly of volcaniclastic rudites with subordinate lavas and is interpreted as representing the magmatic axis

Table 1. *New K–Ar dates on plagioclase augite-phyric andesite clasts from the Kawa Formation*

Sample	Location	K_2O Wt%	^{40}Ar $(10^{-4}mmg^{-1})$	Atmospheric content (%)	Age $(\pm 1\sigma)$
VL033	177°39.3'E, 17°49.9'S	0.589 ± 0.001	5.38 ± 0.12	80.4	28.1 ± 0.6
BH197	177°41.1'E, 17°51.5'S	0.709 ± 0.004	5.43 ± 0.12	81.5	23.6 ± 0.5

Conventional whole-rock K–Ar ages were determined at the Department of Geophysics at the University of Newcastle in collaboration with J. Mitchell using a Kratos MS 10 mass spectrometer coupled to a ultra-high vacuum gas extraction line. Isotope dilution analyses were performed using a ^{38}Ar 'spike' calibrated against standard micas P-207 and B2303. Typical sample weights for Ar analysis were 300-mg aliquots. K was determined using a Corning EEL 450 flame photometer with lithium internal standard. The quoted values are the mean of two Ar analyses, and the reported errors take into account both random effects (discrepancies between duplicates) and systematic effects (from spike calibration).

of the Wainimala arc. This assemblage is separated by major faults from a basinal assemblage that crops out further north, around and overlying an uplifted portion of Yavuna Group frontal-arc basement (Fig. 2).

The base of the arc assemblage is not exposed but the assemblage is at least 4 km thick. It is extensively intruded by Colo plutons (Fig. 2). Volcanic rocks from the arc assemblage were analysed by Gill (1987), and include strongly depleted arc volcanic rocks corresponding to the classic 'island arc tholeiite' series of Jakes & Gill (1970). Interbedded shallow-water limestones indicate a Late Oligocene to early Mid-Miocene age range for the exposed portion of the arc assemblage.

This paper focuses on the lower part of the basinal assemblage. This comprises the Kawa Formation and Nadele Breccia, which both consist mainly of volcaniclastic rudites; a bimodal suite of basic and felsic volcanic rocks assigned to the Dakadaka Basalt and Kalaka Dacite, respectively; and the lowermost part of the Nabu Formation. The Nabu Formation consists of volcaniclastic turbidite sandstones and mudstones, hemipelagic limestones and acid tuffs, and it also includes all of the wholly sedimentary upper part of the basinal assemblage. The underlying Yavuna Group rocks are cut by a pervasive bimodal dyke swarm which documents significant crustal extension during eruption of the basinal assemblage volcanic rocks.

The Kawa Formation and Nadele Breccia

These formations are lithologically similar and are thought to be laterally equivalent. They consist mainly of redeposited rudites in which the dominant clast type is a characteristic grey, porphyritic andesite. Clasts of basalt and dacite also occur. The polymict nature of the rudites,

the presence of rounded as well as angular clasts and the presence of sparse shallow-water limestone clasts indicates that they were deposited on the dispersal aprons of major, in part subaerial, volcanic edifices (Hathway 1994). Plagioclase laths up to 3 mm in length are the dominant phenocryst phase in the andesite clasts, comprising up to 40% of the mode. They are usually clouded by alteration and show resorbed cores, secondary overgrowths and zoning, indicating disequilibrium processes. Other phenocryst phases include pristine augite (up to 5%), green-brown hornblende (up to 5%) and magnetite (1–3%). Although fresh orthopyroxene has not been observed, an altered pyroxene phenocryst phase accompanies fresh augite in many samples. Most samples have a fine-grained glassy groundmass which includes small grains of magnetite and secondary clay minerals.

The Nadele Breccia (Rao in press), includes basalt pillow lavas and massive acid andesite lavas as well as volcaniclastic rudites, and is interpreted as part of an early Wainimala arc edifice constructed upon a substrate of Yavuna frontal-arc crust. The Kawa Formation may represent the dispersal apron of a similar edifice situated closer to the arc assemblage (Fig. 2). The Kawa Formation rests unconformably on the Yavuna Group, whereas the base of the Nadele Breccia is not exposed. Whole-rock K–Ar dates of 28.1 ± 0.6 Ma and 23.6 ± 0.5 Ma on two andesite clasts from the Kawa Formation (Table 1) are consistent with late Early Oligocene to Early Miocene biostratigraphical ages for these rocks (Fig. 3, Hathway 1994).

The Dakadaka Basalt

This formation crops out along the western and southern margins of the Yavuna Group outcrop (Fig. 2), and has a maximum exposed thickness

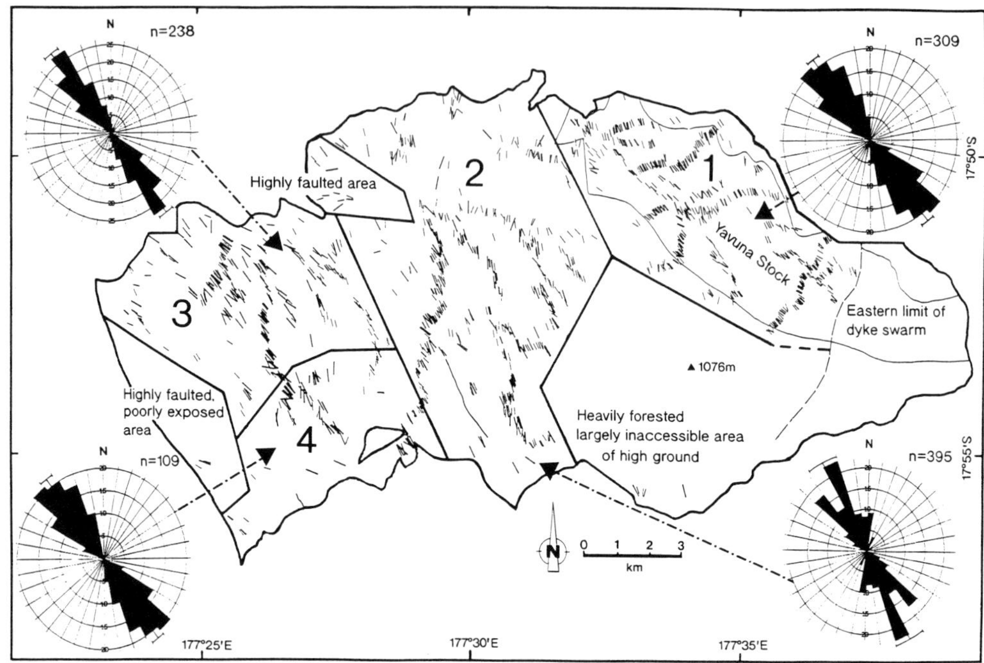

Fig. 4. Distribution of dykes intruding the Yavuna Group. Measured dyke orientations for each sub-area are displayed on rose-diagrams. Figures give data percentage, bar indicates statistical mean and standard deviation. Mean dip for dykes in each sub-area are: Area 1, 77° SW, area 2, 73° SW, area 3, 76° NE, area 4, 44° NE.

of 300–400 m. It consists mainly of massive and pillowed lavas with subordinate pillow breccias. The freshest lava surfaces are mid-grey to grey-green while weathered surfaces are deep maroon in colour. Pillow lavas contain up to 50% by volume of amygdales filled with zeolites or calcite. The lavas are commonly cut by networks of small mm-scale veins bearing the same secondary minerals. Massive basalts are generally more sparsely vesicular. Rare basalt dykes intruding the lavas are interpreted as feeders. Pink pelagic limestone is commonly present as an inter-pillow matrix or as fragments caught up in lava flows.

The basalts may be porphyritic or aphyric and alteration to zeolite-facies mineral assemblages is ubiquitous. Porphyritic samples contain up to 10% by volume of small (sub–1 mm) olivine phenocrysts which are generally pseudomorphed by smectite and other clay minerals, up to 30% plagioclase phenocrysts which are always clouded by alteration, and up to 15% augite phenocrysts which are generally unaltered. Where crystallization orders are discernible, augite generally follows plagioclase as a liquidus phase. Groundmass minerals include microlites of plagioclase and intergranular augite and

magnetite. These phases are generally interspersed with zeolites, clay minerals and secondary Fe-oxides.

No age data are available for the basalts themselves, but nannofossils in the Nabu Formation strata immediately overlying them indicate ages ranging from late Early Oligocene to earliest Miocene (Fig. 3).

The Kalaka Dacite

This formation consists of monomict dacite breccias, massive dacite flows and tuffs. Within the basinal assemblage, these rocks mainly occur along the southern margin of the Yavuna Group outcrop (Fig. 2), where they are interpreted as representing a series of small volcanic centres (Hathway 1994). Up-section or lateral transitions from lava and breccia to thick-bedded dacite tuff are commonly seen around these centres. The dacites are flow-banded and usually contain between 5 and 15% by volume of plagioclase phenocrysts. Some small quartz and green-brown hornblende phenocrysts also occur, but opaque minerals are rare. The groundmass may be either trachytic or felsitic. Secondary minerals include zeolites and patches

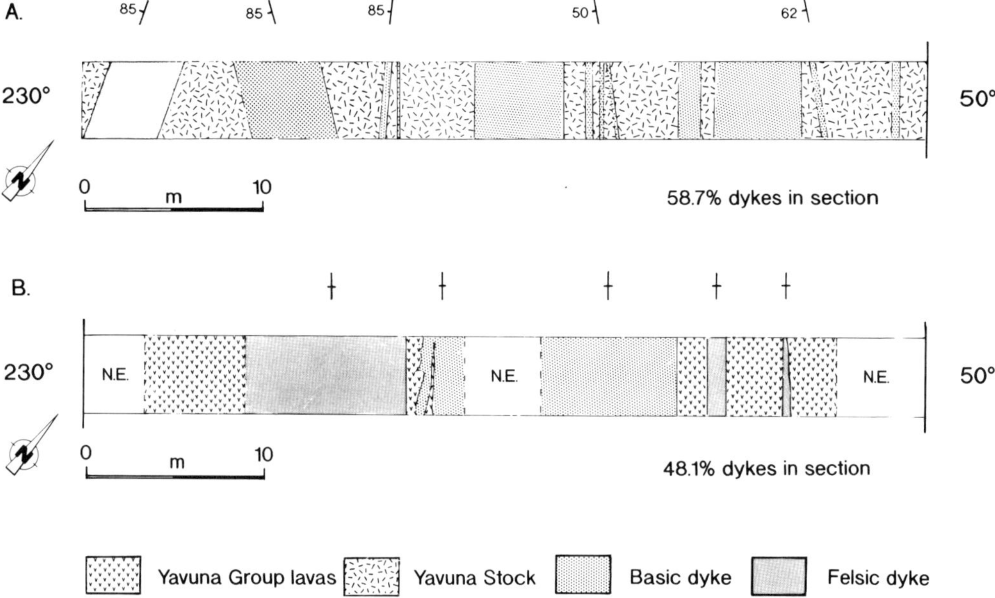

Fig. 5. Representative measured sections through the bimodal dyke swarm cutting the Yavuna Group. (**a**) Namosi Creek (17°50.7′S, 177°33.7′E). (**b**) Masi Creek (17°51.8′S, 177°26.2′E).

of redistributed quartz in the most altered samples.

The NW–SE dyke swarm intruding the Yavuna Group

This pervasive bimodal dyke swarm is widespread in extent across the Yavuna Group outcrop (Fig. 4). Dark blue-grey basic dykes are common throughout the area, but pale grey-fawn felsic dykes are more common in the western part of the swarm, where they are seen to coalesce as feeders to at least one of the Kalaka Dacite centres. Measured sections through the Yavuna Group outcrop demonstrate that the dykes form up to 60% of the crustal section (Fig. 5). Cross-cutting relationships are rare due to the consistent strike of the swarm across the area (Fig. 4). At most of the locations where age relations are apparent, basic dykes cut or chill against felsic dykes; only at one location was a felsic dyke seen to cut and chill against a basic dyke. From these contact relationships, it is concluded that the two end-member compositional groups are of similar age, but that most of the felsic dykes were emplaced prior to the majority of the basic dykes. This conclusion is supported by the presence of NW-striking basic dykes in parts of the Kalaka Dacite.

Basic dykes may be up to 11 m wide but are more commonly between 0.5 and 5 m wide, with centimetre-scale chilled margins. The dykes are generally steeply dipping, apart from in the SW of the area (Fig. 4). Dyke margins are sharp and xenoliths of country rock are rare. Little disruption of the surrounding country rock appears to have taken place during dyke emplacement, and in areas of low concentration, the dykes often terminate in small, cm-scale, finger-like projections. Most of the basic dykes are aphyric, although some are plagioclase-phyric. Along with the Yavuna Group country rocks and the felsic members of the swarm, they generally show greenschist-facies mineral assemblages. They typically consist of plagioclase laths, intergranular augite, interstitial chlorite or fibrous amphibole, and Fe-oxide granules. Plagioclase is invariably turbid and albitised, and augite is commonly partly or completely replaced by chlorite, epidote and/or green fibrous amphibole. Secondary calcite replaces plagioclase and forms irregular patches in some samples.

The felsic dykes may be up to 30 m across. However, they more commonly show similar widths to the basic dykes. In the west, the felsic dykes occur locally as compositionally homogeneous multiple intrusions. At the southern margin of the Yavuna Group outcrop, one

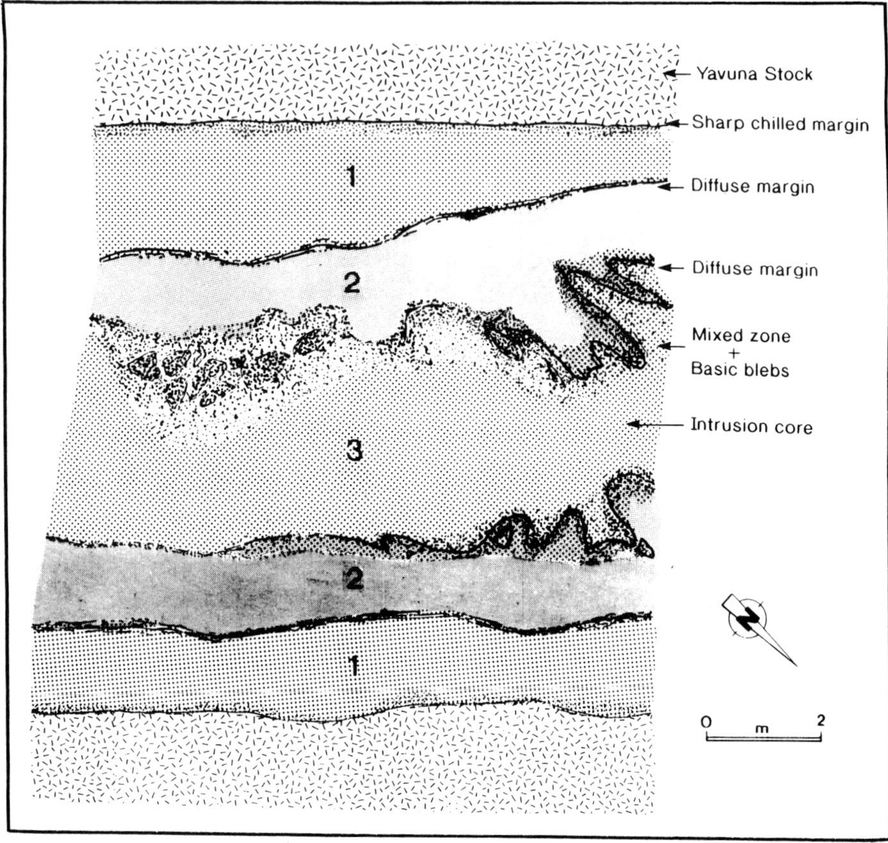

Fig. 6. Relationships between basic and felsic lithologies in a composite dyke at Namosi Creek (17°50.7'S, 177°34.1'E). Numbers indicate order of intrusions. See text for full description.

interval of sheeted felsic dykes can be traced up section into a NW–SE elongate ridge of Kalaka Dacite lavas and breccias (Hathway 1994). The dykes invariably contain plagioclase phenocrysts up to 2 mm in length (<10% of mode). Altered mafic phenocrysts are rare. Plagioclase and quartz occur in the groundmass and the dykes show felsitic or spherulitic textures. Alteration characteristics are similar to those of the basic dykes. Plagioclase is invariably turbid and albitised, and chlorite, secondary quartz and epidote form irregular patches throughout the most altered rocks. In some cases, their intrusive style differs from that of the basic dykes. Although contacts are sharp and well chilled, disruption of the surrounding country rock is more common, and the felsic dykes are seen locally to cut through the Yavuna Group as sill-like apophyses that connect adjacent sub-vertical dykes.

The swarm also includes composite dykes. Although these are rare, they provide important evidence for the synchronous generation of basic and felsic magmas in the basinal assemblage. A typical example about 11 m wide occurs in the upper reaches of Namosi Creek (Fig. 6). The initial intrusion was of the common basic dyke type and has a sharp 3–5 cm wide chilled margin against the trondhjemite country rock. It was subsequently bisected by a felsic dyke with a 7–10 cm wide diffuse and gradational contact, indicating that the earlier basic intrusion had not completely cooled. The felsic dyke is itself bisected by another basic dyke which forms the core of the composite intrusion. The margin of the central dyke is complex and lobate in form and is characterised by the development of mixed zones up to 1 m across that contain basic blebs surrounded by intermediate composition dyke material. These blebs are interpreted as remnants of the core material that have remained isolated from any mixing and re-equilibration with the earlier felsic magma.

Small intrusive pods are occasionally encountered in areas of high dyke concentration. An ovate example exposed in Masi Creek in the

Table 2. *Representative major and trace element data*

Group	Dakadaka Basalt					Dyke swarm cutting Yavuna Group				Kawa Formation		Kalaka Dacite	Highly altered	
Sample	VL036	VL035	BH487	BH173	BH191	BH119	BH126	VL027	BH125	VL034	VL033	VL014	BH045	BH058
SiO$_2$	45.01	47.07	47.79	48.78	49.40	47.73	53.65	54.49	71.83	58.27	59.16	79.32	56.83	56.04
TiO$_2$	1.00	0.72	1.09	0.77	1.06	0.64	1.59	1.42	0.32	0.91	1.01	0.17	1.20	1.11
Al$_2$O$_3$	18.41	19.92	17.61	20.37	17.62	19.20	14.56	14.81	13.36	18.02	16.97	10.61	14.16	18.32
FeO*	10.48	9.48	10.15	9.48	11.85	9.64	13.97	13.76	4.23	5.31	5.72	2.36	9.77	10.05
MnO	0.18	0.17	0.17	0.16	0.23	0.18	0.23	0.22	0.15	0.14	0.25	0.02	0.10	0.16
MgO	11.01	7.45	8.50	5.04	4.98	7.75	4.28	4.26	0.34	1.97	1.91	0.14	2.09	6.01
CaO	11.22	11.53	9.90	10.06	9.52	9.11	5.83	6.17	2.19	8.30	6.79	0.84	8.68	1.50
Na$_2$O	2.71	2.03	2.41	3.86	3.63	2.92	4.91	5.92	5.96	3.90	4.61	4.69	2.87	4.25
K$_2$O	1.02	0.82	1.34	0.56	0.62	1.30	0.03	0.04	0.85	0.53	0.56	1.09	0.71	1.65
P$_2$O$_5$	0.26	0.13	0.27	0.15	0.27	0.05	0.17	0.12	0.06	0.16	0.23	0.03	1.80	0.15
LOI	5.17	7.62	5.36	5.57	6.10	4.00	2.50	3.52	2.56	2.89	3.07	2.62	8.00	13.46
Total	101.29	99.37	99.30	99.28	99.17	98.52	99.20	101.28	99.28	100.48	100.34	99.27	98.20	99.26
ICP-MS														
Rb	14.3	24.4	17.1	9.9	15.7	27.6	2.9	2.1	5.5	7.6	6.7	7.1	7.9	15.9
Sr	306	400	385	483	162	202	54	38	107	291	253	49	209	100
Y	19.3	10.7	17.9	14.2	23.5	11.2	32.2	26.9	49.2	32.7	37.4	116.2	103.8	16.6
Zr	63.4	26.8	72.2	43.0	74.7	27.4	81.2	85.6	120.2	86.2	94.8	309.6	44.2	91.0
Nb	4.42	0.76	3.94	0.86	0.50	0.46	1.21	0.71	2.30	1.14	1.23	4.01	0.42	0.68
Cs	0.04	0.06	0.07	0.06	0.48	0.35	0.14	0.01	0.01	0.40	0.13	0.14	0.02	0.15
Ba	81.1	312.4	126.5	130.0	22.3	73.0	9.0	7.0	104.0	72.1	85.2	81.5	35.1	91.3
La	8.81	4.38	9.53	4.47	2.63	1.33	4.81	3.09	6.83	3.66	4.01	23.15	8.81	3.49
Ce	16.19	8.38	20.66	8.50	6.50	3.31	10.98	7.85	15.00	12.71	9.55	35.22	7.95	8.71
Pr	2.50	1.42	2.65	1.51	1.19	0.57	1.95	1.54	2.79	1.83	2.11	11.28	2.86	1.37
Nd	12.08	7.35	13.51	7.86	6.61	3.33	9.69	8.31	14.42	10.35	12.57	56.81	14.97	6.67
Sm	2.92	2.16	3.28	2.14	1.85	1.12	2.75	2.27	4.31	3.67	3.30	16.14	4.25	2.02
Eu	0.95	0.67	1.14	0.72	0.65	0.47	1.12	0.89	1.33	1.20	1.20	2.40	1.62	0.63
Gd	3.61	1.75	3.62	2.71	2.70	1.49	4.48	3.57	6.39	4.79	5.09	20.15	6.58	2.43
Tb	0.61	0.30	0.57	0.45	0.53	0.29	0.83	0.74	1.19	0.84	0.92	3.54	1.17	0.49
Dy	3.35	1.95	3.27	2.51	3.74	1.88	5.21	4.65	7.82	4.93	6.17	19.41	7.59	3.06
Ho	0.64	0.43	0.68	0.54	0.83	0.45	1.09	0.99	1.67	0.94	1.38	3.79	1.85	0.72
Er	1.92	1.17	1.79	1.57	2.43	1.15	3.27	2.84	4.87	2.84	3.99	11.66	5.78	1.94
Tm	0.27	0.17	0.29	0.26	0.35	0.19	0.51	0.48	0.76	0.45	0.60	1.81	0.92	0.34
Yb	1.86	1.12	1.83	1.54	2.40	1.10	3.35	3.07	5.40	3.03	3.97	10.94	5.50	2.16
Lu	0.30	0.17	0.29	0.25	0.38	0.19	0.50	0.45	0.81	0.46	0.59	1.63	1.06	0.37
Hf	1.55	0.74	1.55	1.23	1.73	0.77	2.19	2.16	3.21	2.00	2.82	7.74	1.07	2.48
Ta	0.25	0.05	0.30	0.07	0.07	0.08	0.12	0.06	WC	0.09	WC	0.33	0.03	0.04
Pb	0.72	1.08	0.70	1.03	0.80	1.57	1.50	1.09	1.26	1.69	1.99	2.42	0.73	7.17
Th	0.69	0.26	0.90	0.41	0.13	0.08	0.32	0.11	0.35	0.16	0.23	0.84	0.04	0.47
U	0.28	0.12	0.39	0.18	0.14	0.01	0.12	0.05	0.15	0.10	0.10	0.45	0.44	0.14
XRF														
Zr	60	29	70	51	87	35	109	86	170	89	94	347	60	96
Y	18	10	19	15	26	14	32	24	52	34	37	99	100	17
Sr	283	393	383	509	203	200	52	36	107	256	241	51	219	122
Ga	14	13	15	19	18	15	16	17	17	16	18	20	16	20
Zn	59	60	55	66	106	70	256	98	100	49	73	77	87	80
Cu	13	90	59	18	61	60	155	105	1	19	31	12	20	145
Ni	135	52	74	24	11	47	13	10	1	16	12	1	8	9
Co	46	43	45	38	45	38	41	49	9	15	16	1	28	30
Cr	509	97	149	36	35	129	3	3	2	14	3	2	9	12
V	263	256	299	290	354	242	344	409	14	145	152	4	760	310
Rb	14	24	19	8	14	27	3	1	7	10	8	12	8	12
Ba	97	378	152	168	32	98	8	3	134	74	90	103	32	115
Sc	28	26	29	25	41	31	29	36	12	27	28	7	31	29

ICP-MS: analytical procedure follows Murton *et al.* (1992).

WC: elevated Ta concentration due to tungsten carbide contamination.

Sample locations: VL036, Dakadaka Basalt, Qaraqara Creek, 18°00′S, 177°37.6′E; VL035, Dakadaka Basalt, Qaraqara Creek, 18°00′S, 177°37.6′E; BH487, Dakadaka Basalt, Togotogi Creek, 17°56.3′S, 177°31.4′E; BH173 Dakadaka Basalt dyke, Rewasali Creek, 17°55.2′S, 177°38.35′E; BH191, Dakadaka Basalt, Savulevu Creek, 17°55.6′S, 177°37.0′E; BH119, Basic dyke cutting Yavuna Group, Nasolo Creek, 17°52.1′S, 177°35.9′E; BH126, Basic dyke cutting Yavuna Group, Qalilevu Creek, 17°49.9′S, 177°33.9′E; VL027, Basic dyke cutting Yavuna Group, Namosi Creek, 17°49.0′S, 177°29.7′E; BH125, Felsic dyke cutting Yavuna Group, Qalilevu Creek, 17°49.9′S, 177°39.3′E;VL034, Clast from Kawa Formation, ridge near Wailulu Creek, 177°39.3′S, 17°49.9′S;VL033, Clast from Kawa Formation, ridge near Wailulu Creek, 177°39.3′E, 17°49.9′S; VL014, Kalaka Dacite flow, Legalega, Nadi Basin, 17°45.5′S, 177°29.3′E; BH045, Dakadaka Basalt, Vunabulubulu Creek, 17°55.9′S, 177°34.0′E; BH058, Basic dyke cutting Yavuna Group, Buseta Creek, 17°55.4′S, 177°33.2′E.

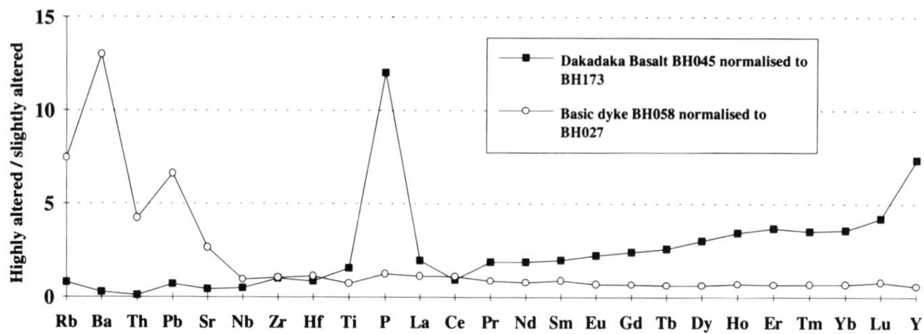

Fig. 7. Trace-element concentrations of highly-altered Dakadaka Basalt and basic dyke samples normalised to less-altered samples from the same groups.

central-western part of the Yavuna Group outcrop is 2–3 m across and was fed from beneath by a NW–SE basic dyke. The transition from basic dyke to pod is marked by a gradational change in texture from aphyric basalt to microgabbro.

A number of points regarding the basinal assemblage volcanic sequence in SW Viti Levu can be summarized as follows.

(i)   At least one major subaerial volcanic edifice, surrounded by a coarse epiclastic debris apron, developed on a substrate of Yavuna Group frontal-arc crust.

(ii) Elsewhere in the basinal assemblage basaltic magmas were erupted to form low-lying lava fields on the seafloor. Subordinate felsic lavas were erupted from fissures in the substrate to form small submarine volcanic centres.

(iii) The stress regime in the early Wainimala arc was extensional. Exposed portions of deeper arc basement reveal that much of the resulting strain was accommodated by the intrusion of dykes.

(iv) The lower part of the Wainimala Group basinal assemblage includes volcanic rocks and volcaniclastic rudites. The upper part of the group is entirely sedimentary and is dominated by mainly fine-grained volcaniclastic turbidites and hemipelagic carbonates. This facies contrast reflects southerly migration of the trenchward limit of arc volcanism, from the area now exposed around the Yavuna Group outcrop, to that represented by the Wainimala arc assemblage to the south.

Geochemistry

Seventy samples from the Wainimala Group have been analysed for major and trace elements using XRF. Major elements were determined on fused glass beads prepared using a lithium tetraborate flux and trace elements were determined on pressed powder pellets prepared using a small quantity of Moviol binding agent. Analyses were carried out on a Philips PW1400 machine at the University of Durham with calibration and corrections made using Philips software. In addition, 12 samples were analysed at Durham for a wider range of trace elements using inductively coupled plasma mass spectrometry (ICP-MS). Results for these representative samples arc listed in Table 2. A complete XRF data set can be obtained from H. Colley on request. During sample preparation, great care was taken in the selection of the least-altered material for analysis. Rocks were crushed to 1–5 mm chips which were then hand-picked under a microscope to exclude obvious vesicle-infill matter and weathered surfaces. The chips were washed in de-ionised water and dried before grinding in agate ball mills. Loss-on-ignition (LOI) values were determined for all samples and range from 1.5% to 8%. In order to study the effects of secondary alteration, two highly altered samples were analysed for the full range of trace elements.

Alteration effects

The most common type of element mobility, which affects samples from each of the formations studied, involves the redistribution of alkaline and alkaline-earth elements. In highly altered samples, this type of alteration has produced variable amounts of leaching and enrichment of the elements K, Rb, Ba, Pb, Sr and Th, such that the concentration of these elements may be unrepresentative of primary magmatic values. Enrichment of these elements in the highly-altered basic dyke sample BH058 is demonstrated by plotting element abundances of the sample normalised to those of a less-altered basic dyke sample VL027 (Fig. 7). A second style of alteration is less common, and affects only a limited number of samples from the Dakadaka Basalt and Kawa Formation and Nadele Breccia groups. It is characterized by

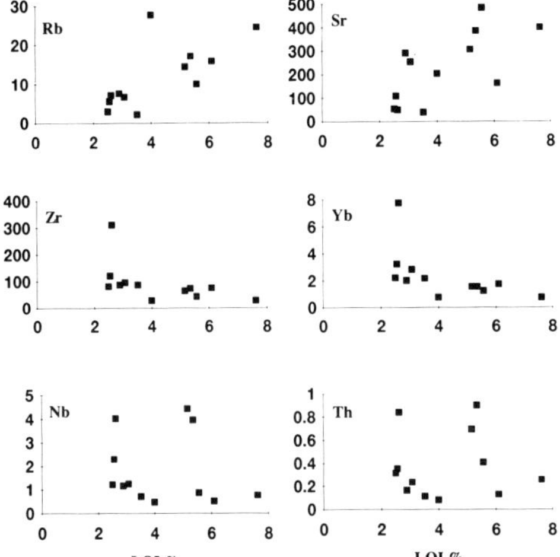

Fig. 8. Selected trace-element concentrations (ppm) plotted against LOI as index of alteration for representative samples in Table 1. Rb and Sr show broad positive correlations with LOI consistent with secondary enrichment. The moderately incompatible elements Zr and Hf show flat patterns for all samples except the highly evolved sample VL014. In contrast, highly incompatible elements Nb and Th show selective and broadly correlated enrichments across the full range of LOI.

redistribution of P, Y and the rare-earth elements (REE) and is highlighted in Fig. 7 in which the multi-element profile for the highly-altered basalt sample BH045 is shown normalised to that of a less altered basalt sample BH173. Similar alteration has been documented during the breakdown of volcanic glass to smectite-group clay minerals in boninite-series volcanic rocks (e.g. Taylor *et al.* 1992; Murton *et al.* 1992). Samples that show this type of enrichment are easily distinguished and are omitted from further discussion.

To illustrate the effects of alteration on the representative samples (Table 2), trace element data are plotted against LOI as an index of alteration (Fig. 8). These plots reveal three groups of apparently similar trace element variation. Alkaline and alkaline-earth elements such as Rb and Sr that are commonly regarded as being mobile during hydrothermal alteration show a broad positive correlation with LOI, suggesting that element enrichment was secondary. The other two groups show contrasting behaviour for elements that are traditionally regarded as less-mobile. One group, which includes the moderately incompatible high-field-strength elements (e.g. Ti, Zr, Hf) and the heavy rare-earth elements (e.g. Yb), shows a

relatively flat trend with the only significantly elevated abundance occurring in the highly evolved Kalaka Dacite sample VL014 (Fig. 8). In contrast, the most incompatible elements including Nb, Th and Ce, show selective, broadly correlated enrichments across the full range of LOI. The pattern of enrichment for the immobile HFS-element Nb as well as for Th and Ce is significant, and indicates that for some of the representative samples, the enrichment in highly-incompatible elements relative to moderately-incompatible elements is a primary magmatic feature.

Classification

A plot of SiO_2 versus K_2O (Fig. 9) illustrates the wide compositional range shown by samples from the Wainimala Group. Although samples plot within each of the low, medium and high-K fields, much of the scatter to high and low-K values occurs within the most altered greenschist-facies samples from the basic and felsic dyke groups. The samples from the other groups cluster mainly along the low-K and medium-K field boundary on this diagram. SiO_2 values for the Kalaka Dacite show a compositional range that may be enhanced by silica mobility, a feature which was noted for these samples in thin-section. The most evolved samples are thus termed high-Si dacites (following Gill & Stork 1979) rather than rhyolites in order to retain their field nomenclature. When plotted on a classic Miyashiro-type diagram (Fig. 10), the Wainimala Group samples collectively define a tholeiite series, with increasing FeO^*/MgO over the range 45–65% SiO_2. Although secondary chlorite has developed in the groundmass and primary olivine and augite have been replaced in some samples, these alteration processes do not appear to have caused a significant change to bulk-rock FeO^*/MgO ratios. In terms of Ni/MgO ratios, most of the samples fall within the mainly tholeiitic low Ni/MgO field A of Gill (1981) rather than the field defined by most orogenic basalts and andesites (Fig. 11). Samples from the Dakadaka Basalt are the least evolved geochemically, although none have Ni contents high enough to represent liquids in equilibrium with mantle peridotite (Fig. 11). Figure 11 also illustrates that the more evolved samples from the Kawa Formation and Nadele Breccia plot in the generally calc-alkaline high Ni/MgO field B of Gill (1981). Thus, while the majority of basinal assemblage Wainimala Group rocks may be broadly classified as low or medium-K arc tholeiites, samples from the major volcanic edifices represented by the Kawa

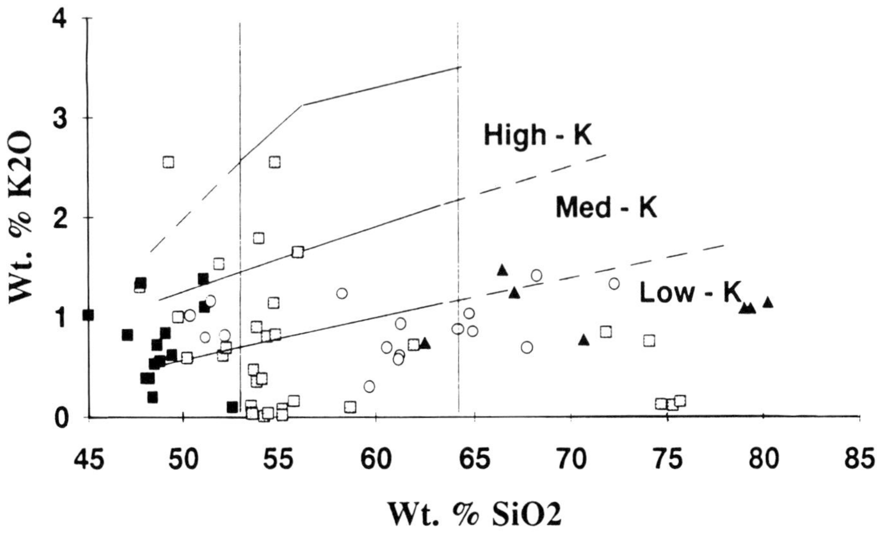

Fig. 9. SiO$_2$–K$_2$O diagram (fields modified from Cole *et al.* 1985).

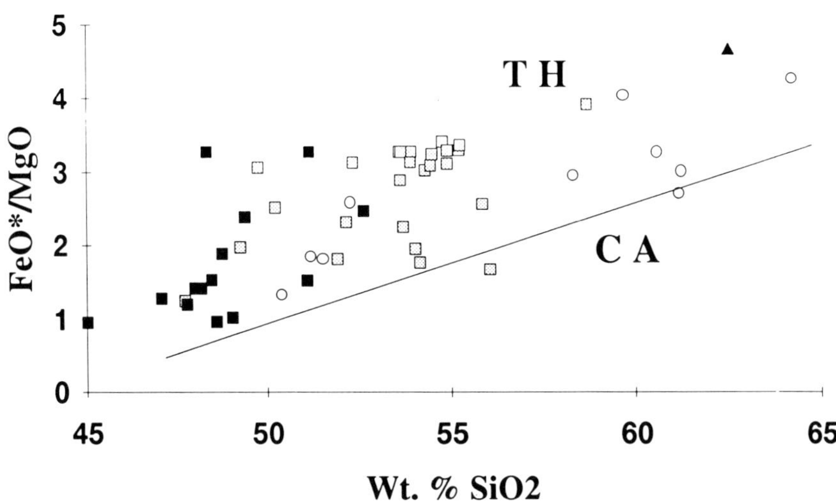

Fig. 10. SiO$_2$–FeO*/MgO diagram, after Miyashiro (1974). Symbols as for Fig. 9.

Formation and Nadele Breccia show geochemical features transitional towards a calc-alkaline volcanic series.

Relationships between volcanic groups

In order to study relationships between the volcanic groups in detail, selected major elements have been plotted against MgNo. as an index of differentiation (Fig. 12). The plot of MgNo. versus Al$_2$O$_3$ defines a broad negative trend from the Dakadaka Basalt samples to the basic dykes group suggesting early plagioclase control during fractionation. Samples from the Kawa Formation and Nadele Breccia show generally higher Al$_2$O$_3$ than other groups at equivalent MgNo. A broad trend of decreasing CaO across

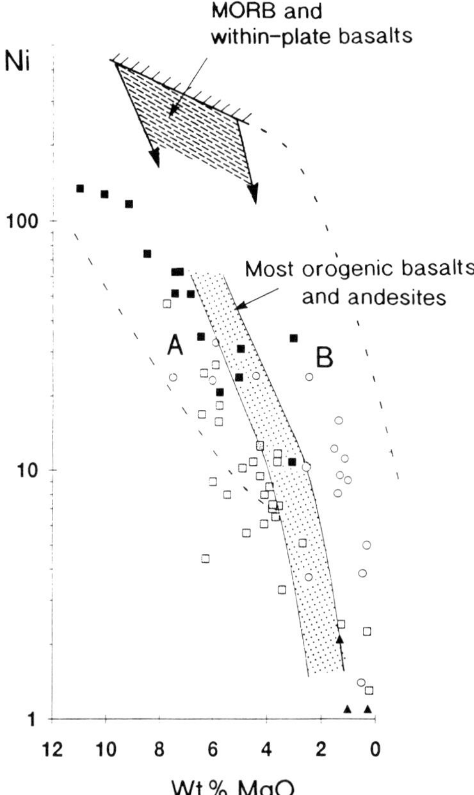

Fig. 11. Ni (ppm) versus MgO diagram after Gill (1981). Field A-mainly tholeiitic and field B-mainly calc-alkaline; heavy line represents liquids in equilibrium with mantle peridotite. Symbols as for Fig. 9.

the range of MgNo. is consistent with fractionation of plagioclase and augite. However, within-group scatter is enhanced both by crystal accumulation and by development of secondary calcite in altered rocks. The trends for FeO* and TiO$_2$ increase from the Dakadaka Basalt to the basic dykcs group at MgNo. − 40 before falling away steeply to lower values in the felsic dykes and the Kalaka Dacite. Values for the Kawa Formation and Nadele Breccia samples are mostly displaced to lower TiO$_2$ at intermediate MgNo. and higher TiO$_2$ at low MgNo. when compared to the fractionation trend defined by the other groups.

Trace element variations between the groups are highlighted by plotting element abundances against Zr (Fig. 13). This allows a clearer distinction to be made between magmatic and secondary processes. For the compatible elements Ni and Cr, the highest values occur in

samples from the Dakadaka Basalt. Lower Ni values in the remaining groups are consistent with fractionation of mafic phases. Basic dykes cutting the Yavuna Group generally have lower Ni and Cr contents than the Dakadaka Basalt. Sr abundances show more obvious scatter both between and within the volcanic groups. Elevated Sr contents in the Dakadaka Basalt samples (up to 826 ppm) are attributed to high (up to 30 %) modal plagioclase. In contrast, basic dyke samples containing similar amounts of plagioclase contain as little as 33 ppm Sr, indicating significant leaching of this element during greenschist-facies alteration. Least-altered, non-porphyritic samples define a trend of decreasing Sr with differentiation across the range of Zr. The least differentiated of these samples have Sr/Zr ratios of 5–6, similar to values of Sr/Zr ratios of 3.8–7.8 for arc tholeiites of the arc assemblage (Gill 1987) but higher than values of 1.0–1.7 for South Fiji Basin basalts (Gill 1987). In the plot of Zr versus Y, the majority of samples define a positive slope towards more evolved compositions. Deviations from this trend at evolved compositions may reflect fractionation of minor phases. Least-differentiated Dakadaka Basalt samples have a MORB-like average Zr/Y ratio of 3.3, which is significantly greater than the Zr/Y ratios of 1–2 for the depleted arc tholeiites of Gill (1987). Sc values increase from non-porphyritic Dakadaka Basalt samples to the basic dykes group. Displacements to high Sc reflect accumulated modal clinopyroxene in samples from both of these groups. Sc values decrease with fractionation through intermediate compositions to values of <10 ppm in evolved samples. Wide within-group variations in V abundance can be attributed to accumulation of Fe–Ti oxides. This feature is most pronounced in the samples from the basic dykes group which contain 5–30% modal Fe-Ti oxide. A distinct compositional gap is noted in V contents between the basic and felsic members of the dyke swarm. The gap is accompanied by a decrease in the values for FeO* and TiO$_2$, suggesting control through concomitant fractionation of Fe–Ti oxide. Early fractionation of this phase may explain the low V contents of the principal edifice samples when compared to similarly differentiated samples in the other magmatic groups.

The between-group trace element variations are consistent with phases observed in thin section having controlled much of the geochemical variation through crystal fractionation and accumulation processes. The Dakadaka Basalt lavas and dykes form the least geochemically differentiated group. Fractionation towards

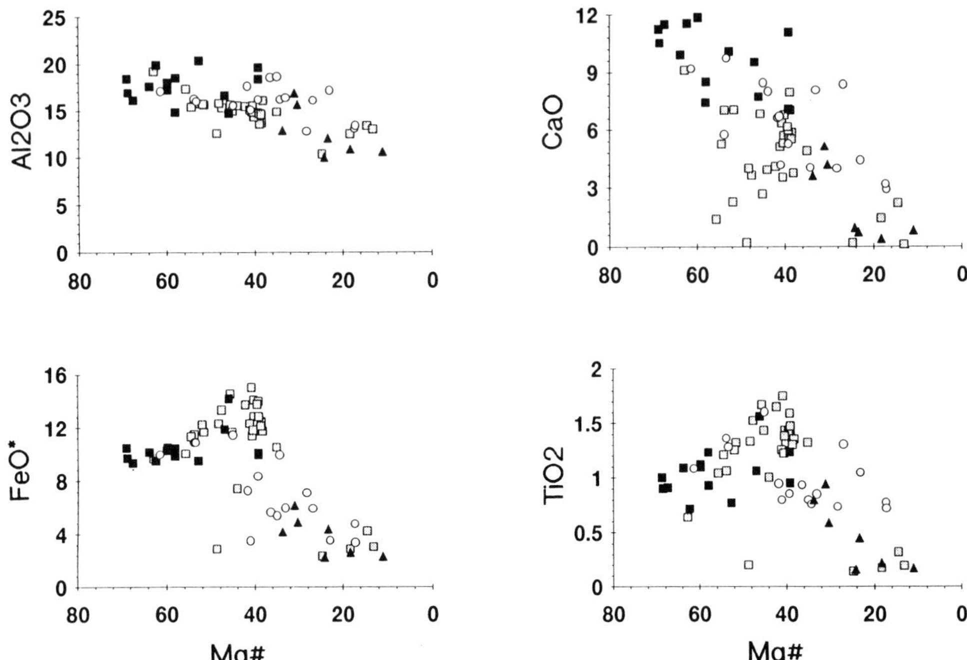

Fig. 12. Major element oxides (wt%) plotted against Mg# ($=100*$Mg/Mg+Fe$_{TOT}$). Symbols as for Fig. 9.

more evolved compositions involved early crystallisation of olivine and plagioclase, whereas fractionation of clinopyroxene was delayed until intermediate compositions. These observations contrast with the results of Perfit *et al.* (1980) who found that plagioclase control was lacking in many island arc suites and that olivine + clinopyroxene + spinel controlled geochemical evolution in the early stages of fractionation. The intermediate samples from the Kawa Formation and Nadele Breccia show significant displacements from the fractionation trends followed by the inter-edifice magmatic groups. Their higher Al$_2$O$_3$ and lower TiO$_2$ and V contents at moderate MgNo. confirm their transitional calc-alkaline nature and suggest that there were fundamental differences in fractionation history or source between the edifice-building and the inter-edifice magmas.

MORB-normalized multi-element profiles for least-differentiated samples (Fig. 14) allow possible source-related variations in the suite to be assessed. Samples from the Dakadaka Basalt define two end-member compositional groups which can be distinguished by their ratios of alteration-resistant incompatible elements. The more-depleted group includes samples BH191 and BH173, which show a negative Nb–Ta

anomaly and a slight enrichment in LREE relative to the HFSE. The second group is represented by samples VL036 and BH487 and shows a contrasting enrichment in Nb–Ta relative to other HFSE, and a more clearly defined enrichment of LREE over HFSE. While mobility and secondary enrichment of the LREE may occur in some highly altered volcanic rocks (e.g. Humphries *et al.* 1978), evidence for significant LREE mobility has not been found in this study and the high LREE/HFSE ratios are considered to be primary. In both of the end-member groups, ratios among moderately incompatible HFSE are similar to MORB (e.g. Zr/Y = 3.0–3.3, Ti/Zr = 94–107, Zr/Hf = 35–41) although absolute levels of abundance are slightly lower (Fig. 14). The two groups are unlikely to be related by the crystal fractionation processes described above because the samples enriched in highly-incompatible elements are less evolved than the depleted samples (MgNo. = 69–64 and 53–47 for the two groups).

Multi-element profiles are also presented for representative Kawa Formation samples and a sample of similar MgNo. from the dyke swarm. Trace-element profiles for these samples are similar for all alteration-resistant elements with the exception of Ti. The lower Ti values in the

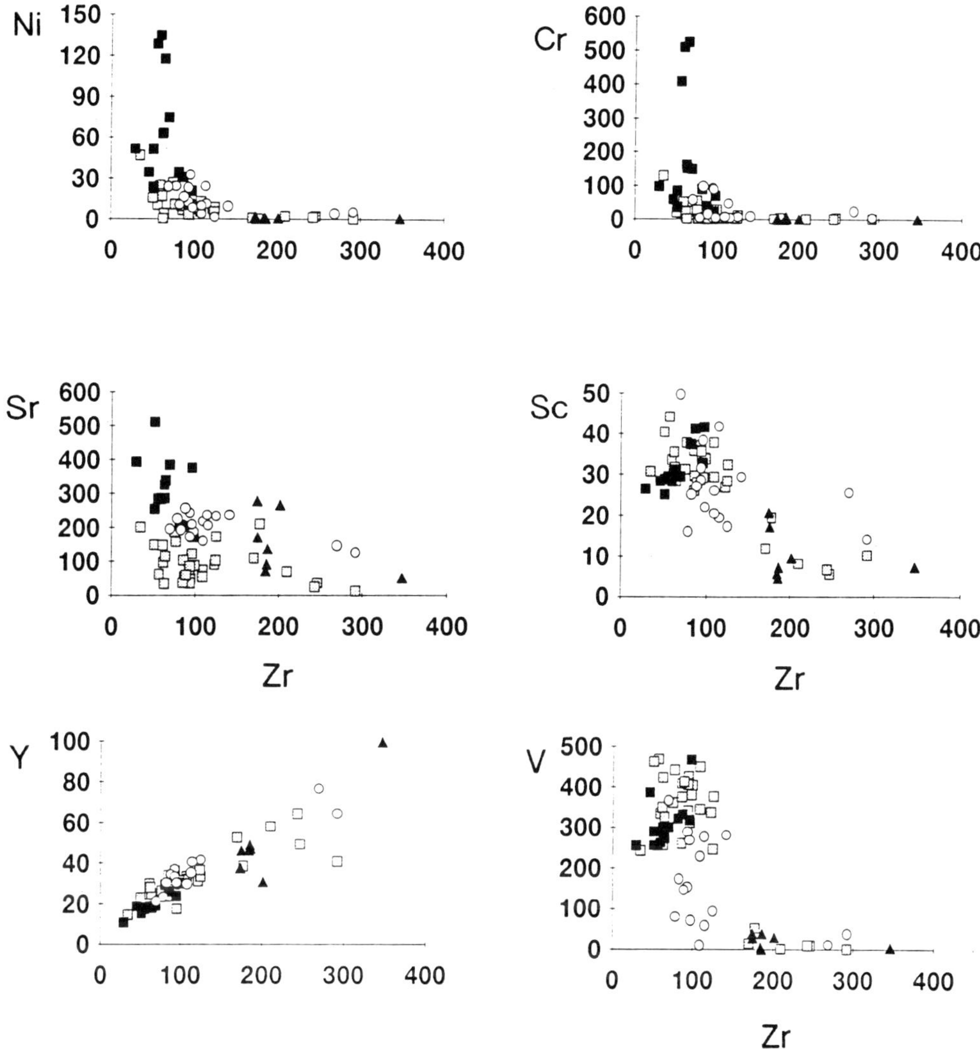

Fig. 13. Trace element concentrations (ppm) plotted against Zr (ppm). Symbols as Fig. 9.

Kawa Formation samples are thus attributed to early fractionation of magnetite in the edifice-forming lavas, a common feature of the calc-alkaline volcanic series (Gill 1981). The source-related trace element variations within the early Wainimala Group samples are summarized in Fig. 15. All the samples contain a subduction zone component with higher than mantle-array Th/Yb ratios. Non-subduction zone-related heterogeneity is most readily identified by Nb–Ta variations relative to the moderately incompatible HFSE and HREE, although this selective trace element enrichment occurs in only the most primitive Dakadaka Basalt lavas. While the majority of samples from the edifice and inter-edifice groups show low Nb/Yb ratios characteristic of other arc tholeiite suites, selective incorporation of a non-subduction zone component, rich in highly incompatible elements, may be required in addition to a conventional high Th/Nb subduction zone component, in order to explain the petrogenesis of the enriched lavas.

Discussion

The edifice-forming and inter-edifice volcanism seen in the Wainimala Group basinal assemblage resembles that described by Stern *et al.* (1989) in their discussion of the volcanic sections

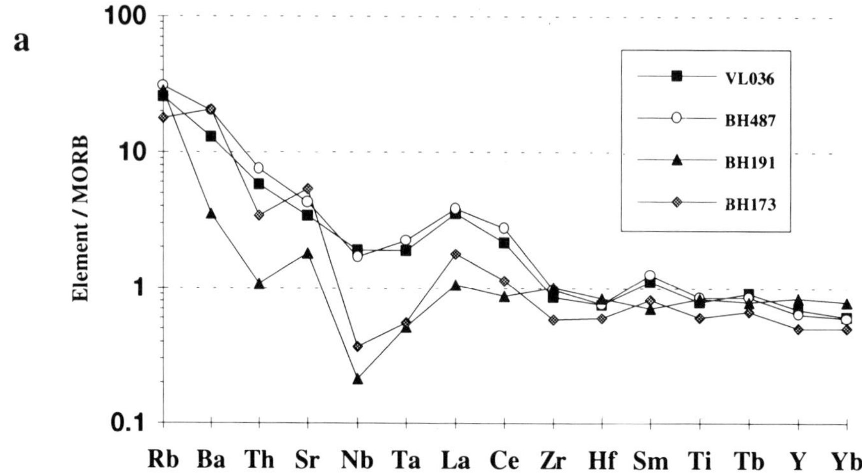

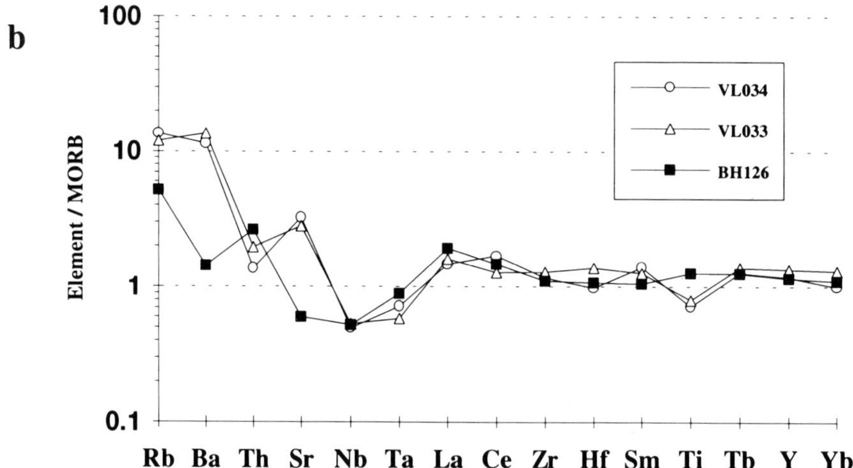

Fig. 14. MORB-normalized multi-element diagrams. (**a**) Least differentiated samples from the Dakadaka Basalt. (**b**) Basic dyke sample BH126 (MgNo. = 39) and two samples from the Kawa Formation (MgNo. = 44 and 41). Diagram illustrates similarity in profiles for alteration-resistant elements except for lower Ti values in the edifice-forming lavas. Normalising factors from Sun & McDonough (1989).

of supra-subduction zone ophiolites. They suggest that fissure-fed lavas may be important, but largely unstudied, components of intra-oceanic arcs, making up much of the seafloor between the major volcanic edifices. They predict that the inter-edifice lavas should be slightly less fractionated but otherwise geochemically similar to those erupted from the major edifices. The present study supports the hypothesis of Stern *et al.* (1989) and allows further detail to be added to their model.

Relationships between the Dakadaka Basalt, the Kalaka Dacite and the bimodal dyke swarm cutting the underlying Yavuna Group provide some insight into part of the magmatic feeder system beneath the arc. None of the basic dykes intruding the Yavuna Group have compositions appropriate to be considered as feeders for the most geochemically primitive Dakadaka lavas. These dykes are therefore considered to represent more evolved equivalents of the Dakadaka Basalt that were emplaced at deeper levels in the arc crust. It is suggested that the Dakadaka Basalt lavas reached the seafloor

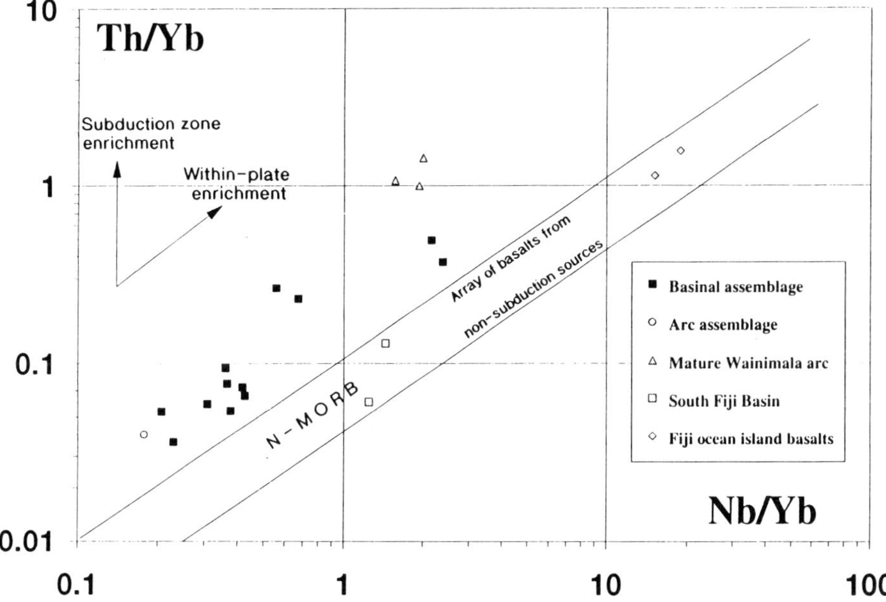

Fig. 15. Th/Yb versus Nb/Yb plot for basinal assemblage volcanic rocks. Diagram adapted from Cole *et al.* (1990) with data from the literature utilized by assuming Nb/Ta = 15. Additional data for arc assemblage: mature Wainimala arc (Namosi Andesite Formation) and South Fiji Basin from Gill (1987), Fiji oceanic island basalts from Gill & Whelan (1989).

through deep-seated fractures in the Yavuna Group substrate. These conduits are not exposed but may lie beneath the present outcrop of the Dakadaka Basalt. Such deep-seated fractures would have allowed primitive magmas to ascend from depth with only limited fractionation *en-route* (Fig. 16).

Field relationships indicate that basic and felsic dykes were emplaced in the arc crust broadly contemporaneously. There is little evidence for magma-mixing in the inter-edifice volcanic rocks and dyke swarm and compositional gaps were generated in tholeiitic trends during fractionation of pyroxene and Fe–Ti oxides. These factors suggest that the dykes were emplaced in the arc crust following efficient segregation of basic and felsic magmas in compositionally stratified magma reservoirs. Magmatic overpressure generated through fractionation and melt vesiculation (e.g. Turner & Campbell 1986) may have been the main factor allowing the most evolved, felsic members of the dyke swarm to reach the sea-floor and form the Kalaka Dacite centres (Fig. 16). This is supported by fracturing and veining of the country rock seen close to some felsic dykes and by the tendency for felsic sills as well as dykes to occur in the arc crust. Magmas with compositions corresponding to the majority of the basic dykes

did not erupt on the seafloor in the basinal assemblage. These magmas either ponded as small gabbroic intrusions in the crustal substrate or terminated in cm-scale finger-dykes. Further upward progress towards the surface occurred only rarely, when more differentiated magma was fed into the same conduit.

The feeder system to the major volcanic edifices is not exposed. However, trace-element profiles indicate that the source for the edifice-forming lavas was similar to that for the inter-edifice lavas. Different fractionation histories in the two regimes and a greater role for disequilibrium processes inferred for the major edifice lavas thus reflects control operating at shallow levels. In their study of tectonic controls on tholeiitic and calc-alkaline volcanism in the Aleutian arc, Kay & Kay (1982) relate tholeiitic volcanoes to the edges of major arc crustal segments where magmas can more easily reach the surface and undergo shallow, closed system fractionation. The calc-alkaline centres occur in the middle of arc crustal segments where transit through the upper plate is more difficult and differentiation takes place at deeper levels. Although the scale of the Aleutian study is much greater than that of the southwestern Viti Levu example, similar inferences regarding the role of the crustal substrate can be made. A greater

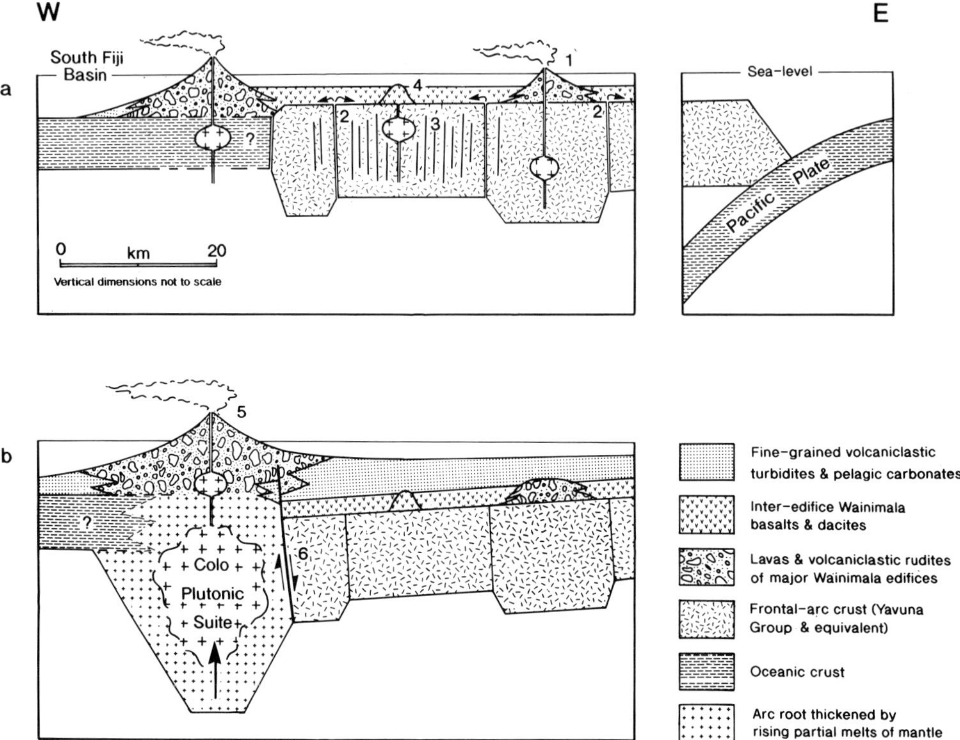

Fig. 16. Schematic cross sections through the southwestern Viti Levu arc and forearc at early (Late Oligocene) and mature (early Middle Miocene) stages of arc evolution. Processes illustrated are: (**a**) 1, Construction of subaerial volcanic edifices marking trenchward limit of volcanic front. Differentiation processes occur at depth and are complicated by magma-mixing; 2, eruption of primitive Dakadaka Basalt through deep-seated fractures in the Yavuna Group substrate to form low-lying mafic lava fields on the seafloor; 3, shallow-level differentiation and pervasive emplacement of a bimodal tholeiitic dyke swarm in the inter-edifice terrane; 4, localized eruption of evolved magmas at submarine Kalaka Dacite centres; (**b**) 5, retreat of volcanic front towards frontal-arc–back-arc transition leading to the concentration of arc volcanism in southern Viti Levu; 6, downthrow of basinal assemblage across major faults.

crustal filtering effect for the frontal arc substrate beneath the major edifice volcanoes would lead to slower magma ascent rates. If differentiation occurred at deeper levels, perhaps under more-oxidising conditions (e.g. Arculus & Wills 1980), both the appearance of hydrous phases and earlier Fe–Ti oxide crystallization would be expected, in addition to an increased role for magma-mixing (Fig. 16). In the inter-edifice regime, less efficient crustal filtering would lead to more rapid ascent of magma, shallower levels of crystallization and a lesser probability of magma-mixing (Fig. 16).

The pervasive NW–SE dyke swarm has a remarkably consistent orientation across the Yavuna Group outcrop (Fig. 4). Both compositional end-members of the swarm occur locally in sheeted forms with internal chilled margins, and veining of the country rock is only

rarely observed. Given the large volumes of magma involved in the dyke swarm (up to 60% of the crustal section over an area of $200\,km^2$), tectonic extension must be invoked to explain the passive emplacement of magma into dykes. A lack of basement exposure trenchwards (i.e. towards the north of the study area following the reconstruction of Fig. 1a), precludes an evaluation of the overall extent of the edifice and inter-edifice terranes. Whether volcanism was occurring in a similar regime in the arc assemblage to the south is also unknown. It is clear, however, that the volcanic front as defined by the trenchward limit of major volcanic edifices retreated southwards within 5–10 Ma of the inception of the arc to become concentrated closer to the frontal-arc–back-arc transition (Fig. 16).

If tectonic extension is typical of intra-oceanic

arcs throughout their evolution (e.g. Hawkins *et al.* 1984), then successive phases of arc volcanism should overprint earlier arc crust within similarly orientated stress regimes. In the case of southwestern Viti Levu, this can be demonstrated by the parallel orientations of the Yavuna Stock of the proto-arc stage and the dyke swarm of the incipient second arc stage (Fig. 4), and by the parallelism of the dyke swarm with the mature arc axis (the Wainimala–Colo plutons in southwestern Viti Levu, Fig. 2). The considerable volume of crustal substrate now occupied by intrusions related to later magmatic episodes should be taken into account when estimates of volcanic production rates in older arc terranes are made. Unless direct field observations can be made to substantiate geophysical surveys of submerged volcanic ridges, the widths of volcanic arcs that have been intruded during later extension-related volcanism may be considerably over-estimated. For example, it may be that along-strike equivalents of the Wainimala arc in Vanuatu and the northern Lau Ridge (Fig. 1a) are also considerably attenuated and intruded at depth.

Conclusions

(i) The initiation of the second phase of arc volcanism in Fiji (the Wainimala arc) began in the late Early Oligocene. Major volcanic edifices initially developed trenchward of the frontal-arc–back-arc transition. Low-lying mafic lava fields and sub-ordinate felsic lavas were erupted onto the seafloor between the major edifices.

(ii) Although parental magmas in the two regimes had similar trace-element profiles, the edifice-forming lavas had a transitional calc-alkaline geochemistry, while the inter-edifice lavas were tholeiitic. Heterogeneities in the underlying crustal substrate may have been fundamental in controlling the distribution and subsequent fractionation histories of edifice and inter-edifice lavas.

(iii) Evidence from the feeder-dyke system to the arc indicates that the stress regime in the early Wainimala arc was extensional. Similarities in structural style in the proto-arc, the second arc and the mature arc stages suggest that the same tectonic regime characterised the arc system throughout much of its history.

The authors extend sincere thanks to former colleagues and staff in the Mineral Resources Department, Suva, Fiji. M.R.W. acknowledges NERC studentship GT4/89/GS/026 and J. Mitchell for K-Ar analyses. W.S. Cole, D.W. Peate and J.L. Smellie are thanked for constructive reviews.

References

ARCULUS, R.J. & WILLS, K.A. 1980. The petrology of plutonic blocks and inclusions from the Lesser Antilles Island arc. *Journal of Petrology*, **21**, 742–799.

CARNEY, J.N., MACFARLANE, A. & MALLICK, D.I.J. 1985. The Vanuatu island arc: an outline of the stratigraphy, structure and petrology. *In*: NAIRN, A.E.M., STEHLI, F.G. & UYEDA, S. (eds) *The Ocean Basins and Margins, Volume 7A*. Plenum Publishing Corporation, New York, 683–718.

COLE, J.W., GILL, J.B. & WOODHALL, B. 1985. A petrological history of the Lau Ridge, Fiji. *In*: SCHOLL, D.W. & VALLIER, T.L. (eds) Geology and offshore resources of Pacific island arcs – Tonga region. *Circum-Pacific Council for Energy and Mineral Resources, Earth Science Series*, **2**, 379–414.

——, GRAHAM, I.J. & GIBSON, I.L. 1990. MAGMATIC EVOLUTION OF LATE CENOZOIC VOLCANIC ROCKS OF THE LAU RIDGE, FIJI. *Contributions to Mineralogy and Petrology*, **104**, 540–554.

CUNNINGHAM, J.K. & ASCOMBE, K.J. 1985, Geology of 'Eua and other islands, Kingdom of Tonga: *In*: SCHOLL, D.W. & VALLIER, T.L. (eds) Geology and offshore resources of Pacific island arcs – Tonga region. *Circum-Pacific Council for Energy and Mineral Resources, Earth Science Series*, **2**, 221–258.

FALVEY, D.A. 1978. Analysis of palaeomagnetic data from the New Hebrides. *Australian Society of Exploration Geophysicists Bulletin*, **9**, 117–123.

GILL, J.B. 1981. *Orogenic andesites and plate tectonics*. Springer-Verlag, New York.

—— 1987. Early geochemical evolution of an oceanic island arc and back-arc: Fiji and the South Fiji Basin. *Journal of Geology*, **95**, 589–615.

—— & STORK, A.L. 1979. Miocene low-K dacites and trondhjemites of Fiji. *In*: BARKER, F. (ed.) *Trondhjemites, dacites and related rocks*, Elsevier, Amsterdam.

—— & WHELAN, P. 1989. Postsubduction ocean island alkali basalts in Fiji. *Journal of Geophysical Research*, **94**, 4579–4588.

HARLAND, W.B., ARMSTRONG, R., COX, A., CRAIG, L., SMITH, A. & SMITH, D. 1990. *A geologic time scale 1989*, Cambridge University Press.

HATHAWAY, B. 1992. Definition of formations in southwestern Viti Levu. *In*: RAHIMAN, A. *Mineral Resources Department: report for the year 1988*. [ISSN 0252–2462] Paper 15 of 1991, 83–89.

—— 1993a. The Nadi Basin: Neogene strike-slip faulting and sedimentation in a fragmented arc, western Viti Levu, Fiji. *Journal of the Geological Society, London*, **150**, 563–581.

—— 1993b. *Geology of the Nadi area (1:50 000 map)*. Fiji Mineral Resources Department.

—— 1994. Sedimentation and volcanism in an Oligocene–Miocene intraoceanic arc and forearc, southwestern Viti Levu, Fiji. *Journal of the Geological Society, London* **151**, 499–514.

HAWKINS, J.W., BLOOMER, S.H., EVANS, C.A. & MELCHIOR, J.T. 1984. Evolution of intra-oceanic arc-trench systems. *Tectonophysics*, **102**, 175–205.

HOUTZ, R.E. 1959. *Regional geology of Lomawai-Momi*. Fiji Geological Survey Bulletin, **3**.

—— 1960. *Geology of Sigatoka area*. Fiji Geological Survey Bulletin, **6**.

HUMPHRIES, S.E., MORRISON, M.A. & THOMPSON, R.N. 1978. Influence of rock crystallisation history upon subsequent lanthanide mobility during hydrothermal alteration of basalts. *Chemical Geology*, **23**, 125–137.

JAKES, P. & GILL, J.B. 1970. Rare earth elements and the island arc tholeiite series. *Earth and Planetary Science Letters*, **9**, 17–28.

KAY, S.M. & KAY, R.W. 1982. Tectonic controls on tholeiitic and calc-alkaline magmatism in the Aleutian arc. *Journal of Geophysical Research*. **87**, 4051–4072.

MALAHOFF, A., HAMMOND, S.R., NAUGHTON, J.J., KEELING, D.L. & RICHMOND, R.N. 1982. Geophysical evidence for post-Miocene rotation of the island of Viti Levu, Fiji, and its relationship to the tectonic development of the North Fiji Basin. *Earth and Planetary Science Letters*, **57**, 398–414.

MCDOUGALL, I. 1963. Potassium-argon ages of some rocks from Viti Levu, Fiji. *Nature*, **198**, 677.

MIYASHIRO, A. 1974. Volcanic rock series in island arcs and active continental margins. *American Journal of Science*, **274**, 321–355.

MURTON, B.J. PEATE, D.W., ARCULUS, R.J., PEARCE, J.A. & VCAN DER LAAN, S. 1992. Trace-element geochemistry of volcanic rocks from Site 786: the Izu-Bonin forearc. *In*: FRYER, P., PEARCE, J.A., SROCKING, L.B. *et al.* (eds). *Proceedings of the Ocean Drilling Program, Scientific Results*, **125**, 211–236.

PACKMAN, G.H. & ANDREWS, J.E. 1975. Results of Leg 30 and the geologic history of South-west Pacific arc and marginal sea complex. *In*: ANDREWS, J.E., PACKMAN, G. *et al.* (eds) *Initial Reports of the Deep Sea Drilling Project*, **30**, 691–706.

PERFIT, M.R., GUST, D.A., BENCE, A.E., ARCULUS, R.J. & TAYLOR, S.R. 1980. Chemical characteristics of island-arc basalts: implications for mantle sources. *Chemical Geology*, **30**, 227–256.

RAO, B. in press. *Revised geology of the Lautoka area*. Fiji Mineral Resources Department Bulletin, **5**.

RODDA, P. 1983. *Fiji radiometric dates recalculated*. Fiji Mineral Resources Department note. Unpublished, on file at Fiji Mineral Resources Department, Suva, Fiji.

STERN, R.J., BLOOMER, S.H., PING-NAN LIN & SMOOT, N.C. 1989. Submarine arc volcanism in the southern Mariana arc as an ophiolite analogue. *Tectonophysics*, **168**, 151–170.

SUN, S.S. & MCDONOUGH, W.F. 1989. Chemical and isotopic systematics of ocean basalts: implications for mantle composition and processes. *In*: SAUNDERS, A.D. & NORRY, M.J. (eds) *Magmatism in the Ocean Basins*. Geological Society, London, Special Publications, **42**, 313–345.

TAYLOR, B. 1992. Rifting and tectonic evolution of the Izu-Bonin-Mariana arc. *In*: TAYLOR, B., FUJIKO, K. *et al.* (eds) *Proceedings of the Ocean Drilling Program, Scientific Results*, **126**, 627–652.

TAYLOR, R.N., LAPIERRE, H., VIDAL, P., NESBITT, R.W. & CROUDACE, I.W. 1992. Igneous geochemistry and petrogenesis of the Izu-Bonin forearc basin. *In*: TAYLOR, B., FUJIOKA, K. *et al.* (eds) *Proceedings of the Ocean Drilling Program, Scientific Results*, **126**, 405–430.

TURNER, J.S. & CAMPBELL, I.H. 1986. Convection and mixing in magma chambers. *Earth Science Reviews*, **23**, 255–332.

WHELAN, P.M., GILL, J.B., KOLLMAN, E., DUNCAN, R.A. & DRAKE, R.E. 1985. Radiometric dating of magmatic stages in Fiji. *In*: SCHOLL, D.W. & VALLIER, T.L. (eds) Geology and offshore resources of Pacific island arcs – Tonga region. *Circum-Pacific Council for Energy and Mineral Resources, Earth Science Series*, **2**, 415–441.

Arc volcanism in an extensional regime at the initiation of subduction: a geochemical study of Hahajima, Bonin Islands, Japan.

REX N. TAYLOR & ROBERT W. NESBITT

Department of Geology, The University, Southampton, SO9 5NH, UK.

Abstract: Hahajima represents an Eocene forearc volcanic centre produced during the initiation of subduction within an intra-oceanic domain of the western Pacific. The Eocene–Oligocene Izu–Bonin arc–forearc is envisaged as a broad region of volcanism generated in an extensional tectonic regime. The forearc is likely to have consisted of a series of small rift basins, punctuated and bordered by coherent volcanic edifices. Hahajima represents one such edifice at the southern termination of the Chichijima–Mukojima rift basin.

Major element, trace element and isotopic data are presented for representative volcanic rocks from Hahajima. The volcanic sequence comprises porphyritic basalt and andesite breccias and flows intercalated with tuffaceous sandstones. All lavas contain high proportions of plagioclase phenocrysts (20–36%) and comparatively low mafic mineral contents. The andesite suite is typified by the presence of magnetite and orthopyroxene phenocrysts and pseudomorphs after olivine. Basalts are dominated by an assemblage of plagioclase, clinopyroxene and olivine. Nodules and xenocrysts typical of the andesites are found in the basalts and vice versa, suggesting an intimate relationship between the two magmas during volcanism.

The basaltic suite follows a tholeiitic fractionation path (iron enrichment) while the andesites are more representative of calc-alkaline magma evolution. These trends are found to be clearer when corrections are made for the excess proportions of plagioclase in the whole rock. Trace element differences between the suites can not be reconciled by a straightforward crystallization model, but can be related to a similar parental magma if the shallow level magma evolution involves a combination of open and closed system fractionation processes.

Comparisons of volcanic suites within and between each site sampled along the Izu–Bonin forearc indicate that each eruptive centre has an individual trace element signature. This suggests that variations within each site can be explained by melting and/or shallow-level fractionation processes, while differences between localities are related to heterogeneities in the composition, trace element and isotopic enrichment, and thermal characteristics of the source. The Hahajima volcanic rocks contrast with the more recent arc magmas of the Izu-Bonin system in that they share some features, such as high-Mg and elevated Zr/Y, with the contemporaneous boninitic lavas.

Hahajima is the southernmost island of the Bonin archipelago, located between the active Izu–Bonin arc and the Bonin Trench (Fig. 1). The Bonin Islands are an emergent section of the Izu–Mariana forearc which is uplifted relative to forearc basins in the west and the Pacific trenches to the east. This forearc high (or outerarc high) has been sampled at several other localities along its length. Drilling at DSDP/ODP Sites 459, 782 and 786 and dredging of the inner wall of the Izu-Bonin-Mariana trenches has recovered Eocene volcanic rocks of island-arc tholeiite and boninitic affinity (Hussong *et al.* 1981; Sharaskin *et al.* 1983; Ishii 1985, Bloomer & Hawkins 1987; Fryer *et al.* 1990). The exact age of the outerarc high volcanic rocks is still unclear, but palaeontological and radiometric dating has constrained eruption to between 40 and 48 Ma (Tsunakawa 1983; Kodama *et al.* 1983; Dobson 1986; Mitchell *et al.* 1992). Lavas from Hahajima have given K-Ar ages in the range 9.6–40.1 Ma (Kaneoka *et al.* 1970; Tsunakawa 1983). Of these ages, the 40 Ma date should be treated as a minimum age due to radiogenic argon loss during hydrothermal alteration. Only one age >43 Ma has been measured for the entire Izu–Bonin forearc high (48.1 Ma; Dobson 1986) but could result from trapped excess radiogenic argon (Mitchell *et al.* 1992). Palaeontological evidence from larger foraminiferal species on Hahajima suggests an age of late Middle Eocene (42–46 Ma) for the sedimentary formations intercalated with, and overlying the volcanic rocks (Ujiie & Matsumaru 1977).

The available age information suggests that most of the volcanic rocks on the Bonin island

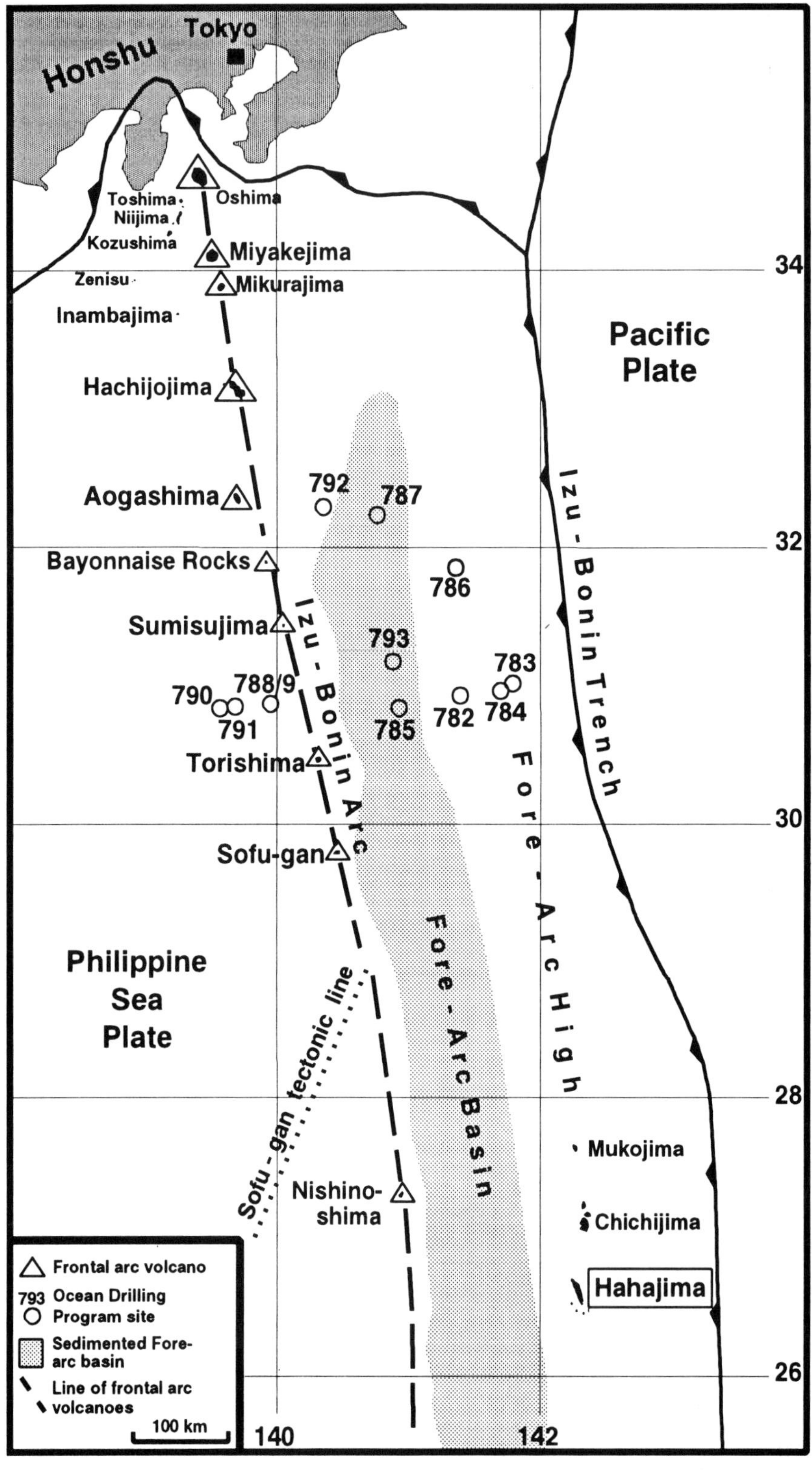

Honshu
Tokyo
Oshima
Toshima
Niijima
Kozushima
Zenisu
Inambajima
Miyakejima
Mikurajima
Pacific
Plate
Hachijojima
Aogashima
792
787
Bayonnaise Rocks
786
Sumisujima
793
783
790
788/9
791
782
784
785
Torishima
Izu - Bonin Trench
Izu - Bonin Arc
Fore - Arc High
30
Sofu-gan
Philippine
Sea
Plate
Sofu - gan tectonic line
Fore - Arc Basin
Mukojima
Nishino-
shima
Chichijima
Hahajima
Frontal arc volcano
793 Ocean Drilling
O   Program site
Sedimented Fore-
arc basin
Line of frontal arc
volcanoes
100 km
140
142
34
32
28
26

Fig. 1. Map of the Izu–Bonin arc–trench region.

group (Mukojima, Chichijima and Hahajima; Fig. 1) were roughly contemporaneous. Clearly, volcanism was spatially and temporally related to the same subduction system. Furthermore, similar associations of middle-late Eocene boninitic and arc lavas are found along the entire forearc high in the Izu-Bonin-Mariana system. This indicates the existence of an elongate terrain with a palaeo-subduction direction roughly parallel to the present Pacific-Philippine plate boundary. In most models, this terrain is the product of initial supra-subduction zone magmatism following the change in Pacific plate motion at around 43 Ma (e.g. Kobayashi 1983; Clague & Dalrymple 1987). The association of boninitic and normal arc lavas provides some important information on magma genesis at the inception of subduction. Boninites are conceived as shallow melts of depleted mantle (e.g. Van der Lann *et al.* 1989), whereas the arc lavas are envisaged as the products of deeper, relatively fertile sources. This incidence raises the problem of the petrogenetic relationship between the two suites. With this association in mind, the geochemistry of Hahajima is documented and contrasted with the products of Cenozoic subduction zone magmatism along the Izu-Bonin margin.

Hahajima volcanic sequence

The Hahajima volcanic succession consists of andesitic and basaltic breccias and massive flows (Fig. 2). Monomict and polymict volcanic breccias with 5–50 cm clasts dominate the succession relative to massive lava flows. Both the andesites and basalts are found throughout the volcanic stratigraphy, although the basaltic rocks are most prevalent at the base of the exposed sequence (Yamaguchi 1985). These

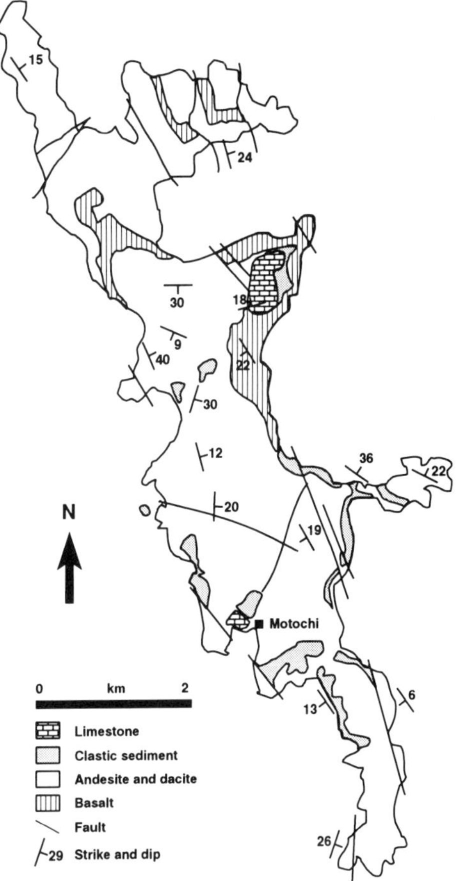

Fig. 2. Geological map of Hahajima (after Yamaguchi 1985)

volcanic rocks are intercalated with tuffaceous sandstones containing the Eocene giant foraminifer *Nummulites boninensis* Hanzawa (Iwasaki & Aoshima 1970) which is characteristic of shallow marine environments (<200 m).

Table 1. *Modal analyses of groundmass and phenocrysts*

Suite Sample no.	Basalt						Andesite		
	780/127	780/31	780/113	780/115	780/119	780/129	780/30	780/130A	780/125I
Groundmass	69.1	62.3	60.4	65.0	56.6	65.0	61.5	59.6	58.8
Plagioclase	19.1	21.8	27.8	30.1	36.2	20.2	25.2	21.7	25.4
Olivine	7.7	6.4	3.1	0.8	5.2	6.0	0.6	1.2	0.1
Clinopyroxene	4.1	8.4	6.3	1.7	1.9	6.6	6.4	10.1	6.7
Orthopyroxene	0	1.0	2.4	1.3	0	2.2	5.1	6.3	8.1
Magnetite	0	0.1	0	1.1	0.1	0	1.2	1.1	0.9

All analyses represent >1000 counts.

R.N. TAYLOR & R.W. NESBITT

Table 2. *Major and trace element analyses of Hahajima volcanic rocks*

	Basalt					Bas/And			Andesite										HCl leach residue
	780/127	780/129	780/31	780/113	780/119	780/115	780/130A	780/130B	780/117	780/126	780/132	780/30	780/128	780/131	780/122	780/124	780/125I split 1	780/125II split 2	780/125II
(wt%)																			
SiO$_2$	51.76	52.13	52.06	52.86	52.03	55.96	59.12	58.62	58.90	58.14	59.62	57.98	58.46	59.42	59.14	60.96	58.65		66.70
TiO$_2$	0.58	0.61	0.62	0.79	0.85	1.00	0.60	0.61	0.69	0.69	0.60	0.67	0.71	0.62	0.68	0.68	0.67		0.78
Al$_2$O$_3$	16.15	16.70	17.24	17.89	18.67	17.06	15.48	15.47	15.63	16.46	15.64	16.60	16.72	15.53	16.12	15.55	16.48		12.61
Fe$_2$O$_3$	9.84	9.74	9.60	9.14	9.90	10.03	7.77	7.81	7.91	8.13	7.91	8.08	7.79	7.77	7.68	7.27	8.19		6.01
MnO	0.18	0.17	0.17	0.16	0.17	0.21	0.14	0.14	0.15	0.18	0.18	0.15	0.15	0.17	0.15	0.12	0.15		0.14
MgO	8.12	7.04	6.57	6.20	4.96	3.76	5.06	4.92	4.74	4.54	4.45	4.43	4.37	4.36	4.34	3.93	4.58		4.51
CaO	10.78	10.72	10.83	10.24	10.68	8.09	8.23	8.18	7.81	8.32	8.02	8.30	8.08	7.81	7.60	6.98	7.85		6.13
Na$_2$O	1.91	2.06	2.15	2.68	2.49	3.34	3.03	3.12	3.06	3.09	3.20	3.20	3.24	3.23	3.15	3.39	3.11		2.72
K$_2$O	0.34	0.19	0.35	0.26	0.29	0.38	0.49	0.48	0.92	0.46	0.77	0.47	0.50	0.77	0.78	1.10	0.43		0.36
P$_2$O$_5$	0.04	0.05	0.05	0.09	0.09	0.10	0.10	0.10	0.12	0.09	0.09	0.10	0.08	0.10	0.11	0.13	0.10		0.06
TOTAL	99.70	99.41	99.64	100.30	100.14	99.93	100.02	99.45	99.92	100.10	100.48	99.98	100.10	98.78	99.76	100.11	100.21		100.02
LOI	0.02	0.25	0.58	0.31	0.15	0.27	0.04	0.22	0.09	0.02	0.18	0.16	0.08	0.28	0.06	0.05	0.45		
(ppm)																			
P	193	204	205	392	378	425	432	432	506	412	404	425	370	439	464	569	430	420	227
Sc			39.8									29.3							
V	287	285	294	301	284	247	200	191	212	230	184	214	225	191	212	182	213	222	148
Cr	240	207	259	102	32	9	129	119	78	72	123	94	73	129	102	100	88	101	60
Co			82									77							
Ni	82	77	77	38	20	6	35	40	27	24	37	26	21	35	34	32	33	33	25
Cu	84	94	80		61	75	29	32	1017	62	29	33	60	28	71	27	37		
Zn	76	72	69	71	102	88	69	69	77	74	73	63	75	63	77	67	116	114	61
Ga	167.0	15.8	16.8	17.0	17.8	18.8	16.4	16.0	16.4	17.5	15.8	17.0	17.2	16.6	16.8	17.1	16.4	16.3	13.6
Rb	6.1	3.2	6.1	3.2	5.8	11.1	6.3	6.1	14.8	7.5	14.9	9.6	9.3	13.9	14.2	20.4	8.1	7.4	7.4
Sr	198	202	200	213	297	235	188	185	217	196	182	193	198	179	212	201	187	184	132
Y	14.9	15.7	16.2	19.9	23.7	28.5	46.1	39.9	26.5	27.0	39.6	25.5	28.9	28.6	27.1	34.7	519	400	33.5
Zr	29.2	29.8	29.4	50.0	52.8	70.9	84.6	83.0	102.7	77.7	83.4	77.2	78.2	83.5	102.6	121.5	80.0	80.9	90.9
Nb	0.5	0.2	0.5	0.8	1.5	2.0	1.3	1.6	2.7	1.5	1.6	1.5	1.3	1.3	2.3	2.7	0.4		
Ba	17	10	26	34	34	47	50	57	91	62	128	54	63	97	71	107	60	61	51
Hf			0.84									1.85							
Pb			0.5									1.1							

Table 2. *Continued*

	Basalt					Bas/And			Andesite								HCl leach residue		
	780/127	780/129	780/31	780/113	780/119	780/115	780/130A	780/130B	780/117	780/126	780/132	780/30	780/128	780/131	780/122	780/124	780/125I split 1	780/125II split 2	780/125II
La			1.93									3.68					51	49	4
Ce			5.36									11.55					45	45	10
Nd			4.34									8.02							
Sm			1.42									2.41							
Eu			0.60									0.85							
Gd			1.75									3.03							
Dy			2.05									3.29							
Er			1.28									2.00							
Yb			1.34									2.15							
Lu			0.22									0.37							

The palaeontological evidence indicates that the volcanic construction(s) which formed the Hahajima volcanic succession were in relatively shallow water or perhaps partially subaerial. Such a palaeoenvironment has also been suggested for other locations along the forearc high. An example is the Eocene silicic volcanic complex forming the basement to Hole 841B in the Tonga forearc, which contains partially welded tuffs overlain by shallow-water volcanic sandstones (Parson *et al.* 1992).

Petrography

The volcanic sequence on Hahajima can be divided on field and petrographical criteria into two broad suites: andesitic and basaltic.

Andesitic suite

Characteristically, these lavas are seriate and porphyritic with 40% modal phenocrysts (Table 1). Plagioclase is the most abundant and largest phase, up to 4 mm long. Oscillatory zoning is conspicuous in most plagioclase crystals, and is frequently picked out by bands of melt and fluid inclusions at various core-rim distances (similar to the Hakone andesite; Mackenzie *et al.* 1982, p. 65). Some examples have wide zone-bands of anhedral, rounded ortho- and clinopyroxene crystals included during crystallization. All andesites contain roughly equal proportions of euhedral or subhedral clinopyroxene and orthopyroxene along with <1 mm grains of a subhedral iron oxide. About half of the andesitic samples examined contain fibrous antigorite pseudomorphs after olivine, representing <1.5% of the mode. All other minerals and the groundmass assemblage are optically fresh.

Basaltic suite

Several differences are apparent in the basaltic suite: magnetite is rare or absent as a phenocryst phase, olivine has a higher modal proportion (2–5% modal) and is fresh away from alteration along cracks, and the proportion of clinopyroxene to orthopyroxene is higher (around 3:1). A further notable difference is that most plagioclase phenocrysts in the basaltic suite lack the oscillatory zoning observed in the andesitic suite, but appear optically to have a continuous normal zoning pattern. One basaltic andesite sample (780/115) is dominated by plagioclase with mafic phases forming <5% of the mode. This sample also has magnetite phenocrysts and glomeroporphyritic aggregates of pyroxene, plagioclase and iron oxide, which closely resemble the assemblage in the andesitic suite.

Two features are characteristic of both suites: high and variable proportions of plagioclase (Table 1), and the mutual occurrence of xenocrysts or aggregates of phases from the other suite. The latter observation suggests an intimate association of the two magma suites during their crystallization/eruption history. This, combined with the reported stratigraphical relationship, is clear evidence that the suites are associated with a single volcanic construction and are connected to the same magmatic plumbing system.

Geochemistry

Major and trace element data are reported for 19 samples (Table 2). Two representative samples, one basaltic and one andesitic were selected for REE and Sr, Nd and Pb isotopic analyses.

Analytical techniques

A Philips PW1400 fully automatic X-ray fluorescence spectrometer fitted with a 3 kW Rh-anode tube was used for the analysis of all samples at the Department of Geology, University of Southampton. Major elements were obtained from fusion beads prepared from 1 g calcined rock powder and 5 g of Spectroflux 100B. Matrix effects were compensated for by using Philips' alphas (influence coefficients). Precision and accuracy are both high and nominally better than 1%. Loss on ignition (LOI) was measured for each sample as the weight lost on samples between 100°C and 1000°C, and is excluded from the totals reported in Table 2.

Trace elements were determined on 40 mm diameter powder pellets pressed to 12 tonnes. Twelve drops of an 8% w/v aqueous solution of polyvinyl alcohol were used as a binder. Corrections for matrix effects for trace elements were made for all wavelength regions, using the Compton scatter technique (full procedural details can be found in Croudace & Gilligan 1990). Modified corrections, necessary when crossing absorption edges, were also applied to Ba, V and Cr. Concentration calibrations were performed using carefully chosen high-quality geological reference samples (basic-intermediate composition). Precision and accuracy are generally better than 3% when the element concentrations are well above their detection limits. Detection limits are approximately 0.5 ppm for Zr and Nb, 1 ppm for Y, Ga, Rb, Sr, Ni, Cu, and Zn, and 6–10 ppm for Cr, V, Ba, Ce and La.

Phosphorus was measured both on fusion beads and powder pellets, although the latter data are preferred because the determinations are considerably precise. Using powder pellets, all samples were counted by recycling the measurements six times, and counting at peak and background positions for 200 and 100 s, respectively. Matrix corrections were applied using mass absorption coefficients calculated from major element compositions. Precision is better than 2% at 100 ppm P.

Hf, Co and Sc were analysed for two samples by instrumental neutron activation analysis at the University of Southampton on a Canberra industries gamma counting system with 2HPGe gamma spectrometers. Samples were irradiated with neutrons (flux = 2.1^{12} n cm^{-2} s^{-1}) for 48 hours, and were counted 40 days after leaving the reactor. Precision is better than 5% for each element.

Rare earth elements (REE) were separated from alkali elements, alkali earths, Fe, Al etc. by mixed-acid elution ion-exchange chromatography, using the method described by Croudace & Marshal (1991). The resulting residue was separated into five fractions (La, Ce, Nd–Sm–Eu, Gd and Dy–Er–Yb–Lu) by stepwise cation exchange on a PTFE column. The REE concentrations were measured by thermal ionisation isotope dilution mass spectrometry on an AEI MS5 instrument at the University of Southampton. Precision is better than 5% for La, Gd and Lu, and better than 2% for all other REE. The blank for all REE analysed is less than 10 ng.

The analytical techniques for Pb isotopes have been described by Vidal and Clauer (1981). Pb was extracted from 0.5–1.0 g of sample as chips previously acid leached. Pb blanks average 1 ng. Isotopic compositions were measured on a Cameca TSN 206 mass spectrometer. Maximum errors estimated from replicate analyses of the standard NBS981 are 0.12%, 0.16% and 0.20% for $^{206}Pb/^{204}Pb$, $^{207}Pb/^{204}Pb$ and $^{208}Pb/^{204}Pb$ respectively. Sr isotopes were normalised to $^{86}Sr/^{88}Sr = 0.1194$ and adjusted to 0.71024 for the $^{87}Sr/^{86}Sr$ ratio of the NBS987 standard. Duplicate analyses indicate that precision is better than ± 0.0001. Nd isotopic ratios were corrected by normalization to $^{146}Nd/^{144}Nd=0.7219$, and $^{143}Nd/^{144}Nd$ ratios are given relative to a value of 0.511860 for the La Jolla isotopic standard. Precision for Nd isotopic ratios is better than ± 0.00003.

Major element geochemistry

The petrographical distinction observed between the basaltic and andesitic suites are reinforced by major element variations. Figure 3 shows MgO variation with other major oxides for samples grouped as basalt and andesite. The basalts show a wide range of concentrations for all elements when compared with the andesites, which cluster at 4–5 wt% MgO. Each suite has a distinct trend for Fe$_2$O$_3$, TiO$_2$ and SiO$_2$. The basalts have constant to increasing Fe$_2$O$_3$ and increasing TiO$_2$ and hence resemble a tholeiitic magma evolution. In contrast the andesites show relative iron depletion at similar MgO contents, and are consequently analogous to a typical calc-alkaline trend. As such, the two suites are also here referred to as calc-alkaline and tholeiitic simply on the basis of Fe–Mg systematics.

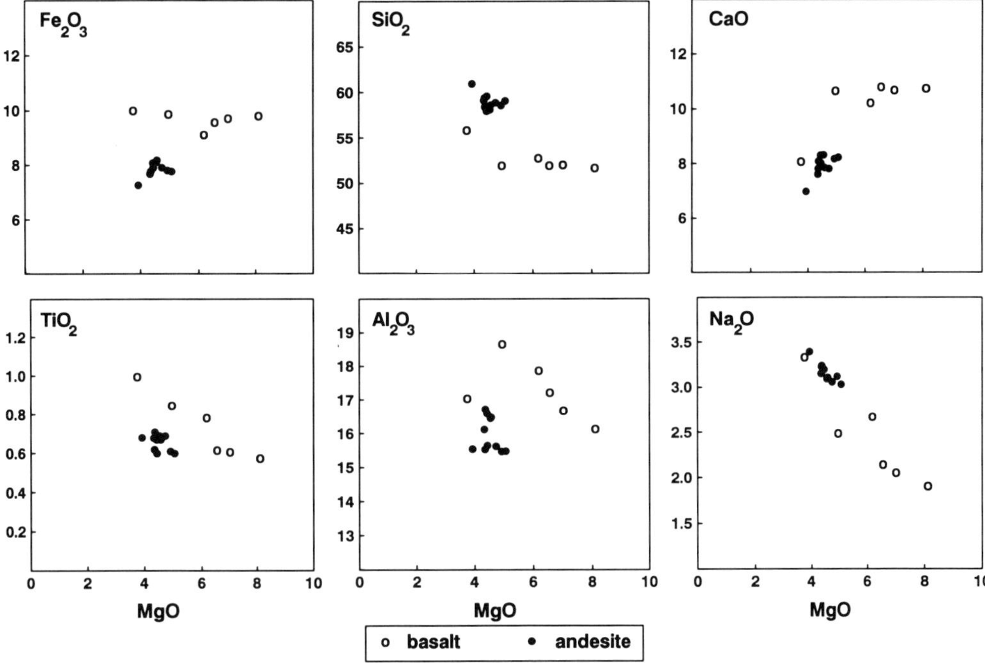

Fig. 3. MgO variation diagrams for the Hahajima volcanic rocks.

Major element trends such as depicted in Fig. 3 are likely to be strongly affected by the quantity and proportions of phenocrysts in these highly porphyritic rocks. In principle, Al contents in the melt should increase during early olivine-dominated fractionation until the onset of cotectic olivine + plagioclase or olivine + plagioclase + pyroxene crystallization. At this point Al concentrations should decline with MgO in accordance with cotectic plagioclase/mafic ratios which should be in the range 1.2 – 1.9 (Nielsen 1988; Langmuir *et al.* in press). Plagioclase is clearly present in excessive proportions relative to mafic phases in Hahajima lavas, with plagioclase/mafic ratios ranging between 1.4 and 6.2. Figure 4a shows that a strong positive correlation exists between whole rock Al$_2$O$_3$ and modal percent plagioclase, suggesting that Al content is dominantly controlled by the quantity of phenocrystic plagioclase. The onset of plagioclase crystallization is suppressed in water-rich magmas, which is typical of arc-related suites such as Hahajima. When plagioclase does crystallize in arc melts, it often appears in unusually high modal proportions. This may be due to selective sorting of felsic–mafic crystals during processes such as non-turbulent convection during magma storage (Brophy 1989), or to straightforward neutrally-buoyant plagioclase flotation and mafic settling in the magma.

To eliminate the effect of excess plagioclase it is possible to recalculate whole rock major or trace element concentrations along a plagioclase control line back to an appropriate liquid line-of-descent (LLD). This is undertaken for the Hahajima samples in a similar fashion to the method described by Langmuir *et al.* in press. This technique utilizes the predicted behaviour of basaltic/andesitic magmas in Al$_2$O$_3$–MgO space. Firstly, a LLD is selected for Al$_2$O$_3$ and MgO appropriate to a water-rich arc magma; in this case the line Al$_2$O$_3$ = 0.6.MgO + 11 was chosen as it qualitatively represents a LLD for a magma with around 0.5 % water (Langmuir *et al.* in press). The amount of excess plagioclase can then be calculated for each sample by extrapolating the tie line between a chosen plagioclase composition and the whole rock through to the chosen LLD. The proportion of excess can then be expressed as the ratio of the distances along the tie line, i.e. sample to LLD : plagioclase to LLD and can be calculated as follows:

$$m^P = (Al^P - Al^R)/(Mg^P - Mg^R)$$
$$c^P = Al^R - (m^P . Mg^R)$$
$$Mg^i = (c^P - c)/(m - m^P)$$
$$\%XP = 100.(Mg^i - Mg^R)/(Mg^i - Mg^P)$$

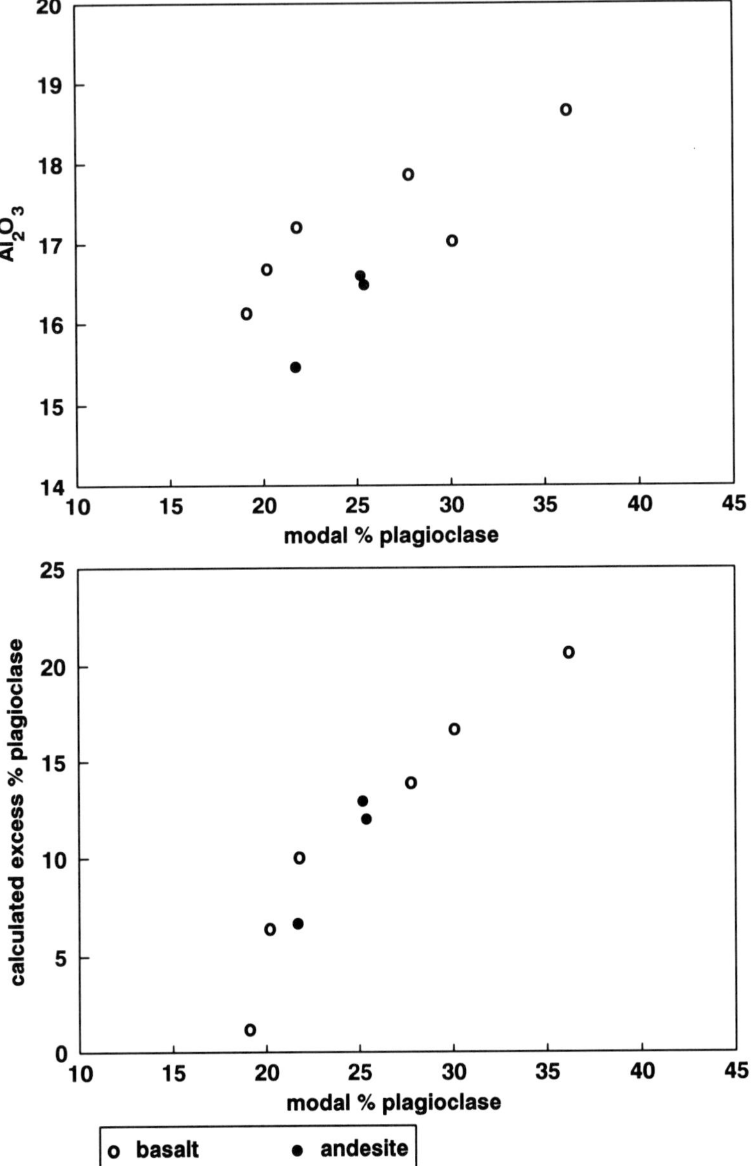

Fig. 4. (**a**) Variation of Al_2O_3 wt% as a function of modal proportion of plagioclase (**b**) Calculated excess plagioclase variation with modal proportion of plagioclase. Proportion of excess plagioclase calculated from the major element concentrations as described in the text.

where:

$Al^R = Al_2O_3$ in sample $Mg^R = MgO$ in sample
$Al^P = Al_2O_3$ in plagioclase $Mg^P = MgO$ in plagioclase
m^P = gradient of plagioclase control line
m = gradient of LLD
c^P = intercept of plagioclase control line

c = intercept of LLD
$Mg^i = MgO$ content of sample extrapolated to LLD
%XP = percentage of excess plagioclase

To simplify the calculations, a plagioclase composition of An_{87} (33.73 wt% Al_2O_3; 0.05 wt% MgO) was used to calculate excess

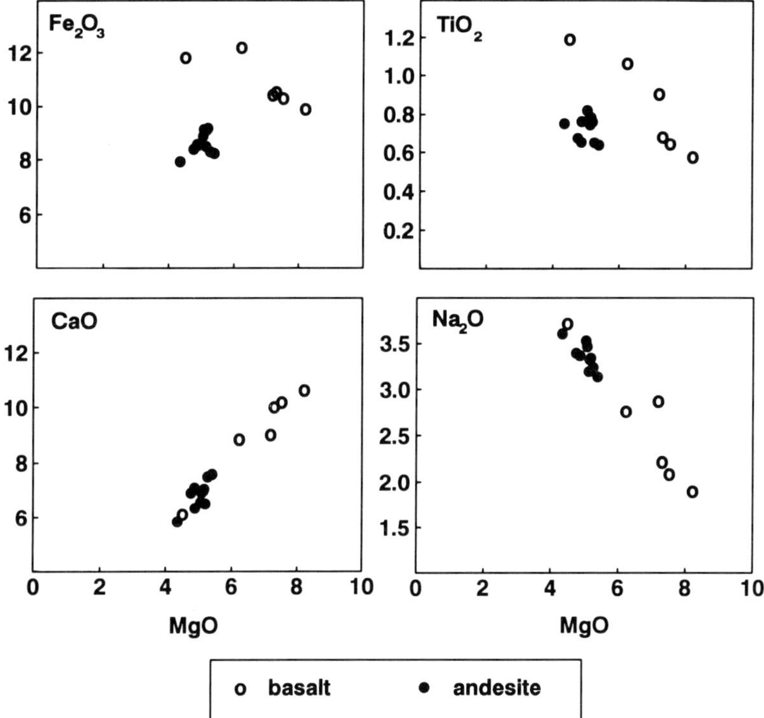

Fig. 5. MgO variation diagrams using data corrected for excess plagioclase.

plagioclase in all the Hahajima lavas. This represents a typical plagioclase composition for Torishima lavas from the active Izu arc (Langmuir *et al.* in press). In reality, the plagioclase composition will vary quite significantly between the zones within each phenocryst and across the range of fractionation that the lavas represent. However, if compositions in the range An$_{67}$ – An$_{97}$ ($\approx$ 30–37 wt% Al$_2$O$_3$) are used in the calculations, the proportion of excess plagioclase is not changed significantly. Once the excess plagioclase is determined for each sample in Al-Mg space, plagioclase can be proportionately subtracted from the whole rock composition. This enables a cotectic-corrected analysis to be derived.

Interestingly there is a remarkable positive correlation between the calculated excess plagioclase and the modal plagioclase contents for the Hahajima samples (Fig. 4b). This trend should intercept at a reasonable estimate of the cotectic proportion of plagioclase phenocrysts ($\approx$12% modal).

The results of correcting Hahajima major element concentrations for excess plagioclase are displayed in the MgO variation diagrams in Fig. 5. Corrected data show more marked increases in Fe$_2$O$_3$ and TiO$_2$ with declining MgO for the basaltic suite, generating a more tholeiitic trend. Furthermore, the CaO and Na$_2$O data form a more coherent relationship with MgO for both suites.

It is clear from these variations that the basaltic (tholeiitic) and andesitic (calc-alkaline) suites of Hahajima are difficult to reconcile by simple fractional crystallization of the same parental magma. As these suites are likely to have originated contemporaneously from the same eruptive system, then either two differing parental magmas were supplying the volcano or a more complex fractionation process was involved in their petrogenesis.

Trace element geochemistry

MORB-normalized incompatible trace element concentrations for representative basaltic and andesitic samples are shown in Fig. 6. Both rocks have sub-parallel profiles and display all the typical characteristics of island-arc magmas. Most notably the low field-strength elements (LFSE) are enriched relative to the rare earth elements (REE) and high field strength elements (HFSE). In turn, REE/HFSE ratios

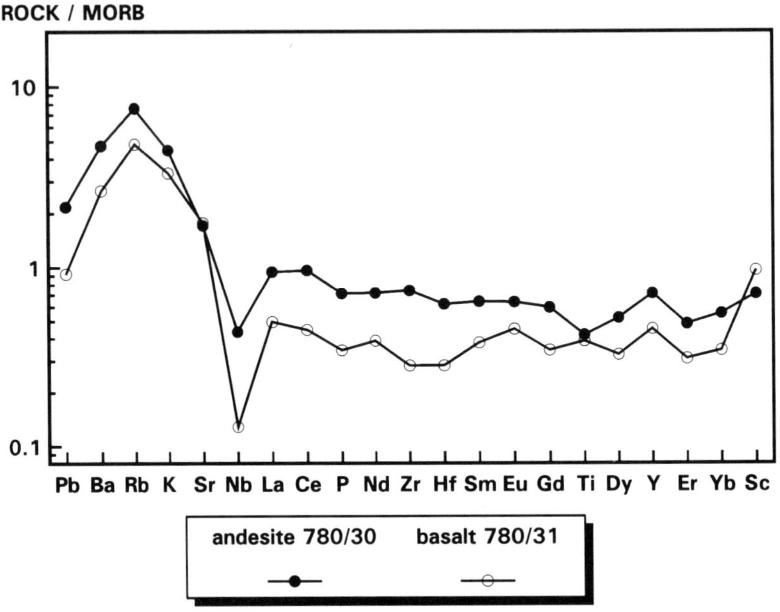

Fig. 6. MORB-normalized plot for representative samples from the andesitic and basaltic suites. Normalising values from Hofmann (1988). Both patterns show the typical arc characteristics of high low-field strength elements and low Nb compared to the REE and high field strength elements.

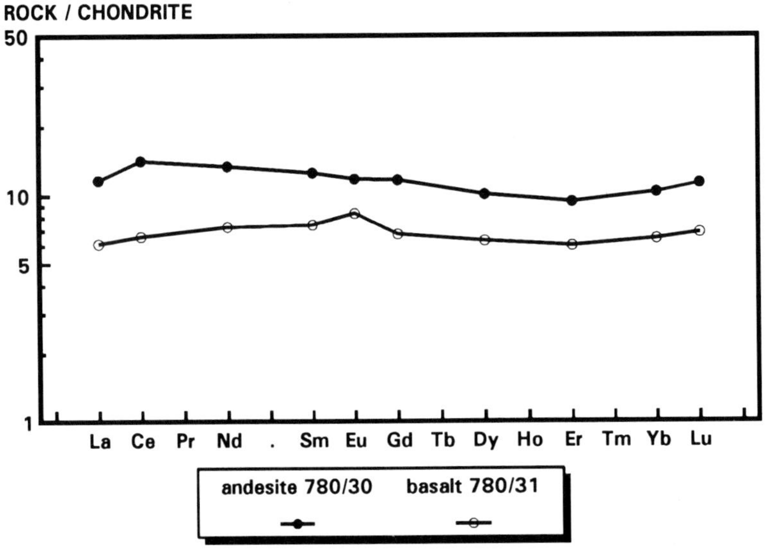

Fig. 7. Chondrite-normalised REE diagram for basalt and andesite from Hahajima.

(e.g. La/Nb, Sm/Zr and Nd/P) are high in basalt 780/31 relative to MORB, which is typical of primitive island-arc tholeiites. However, these ratios are somewhat closer to MORB in andesite 780/30. REE patterns for both samples are similar (Fig. 7) with the andesite slightly more light REE enriched than the basalt, with $(La/Yb)_n$ of 1.16 and 0.97 respectively. $(La/Yb)_n$ ratios ≥ 1 are typical of the Izu–Bonin arc magmas older than 20 Ma (Taylor *et al.* 1992a,

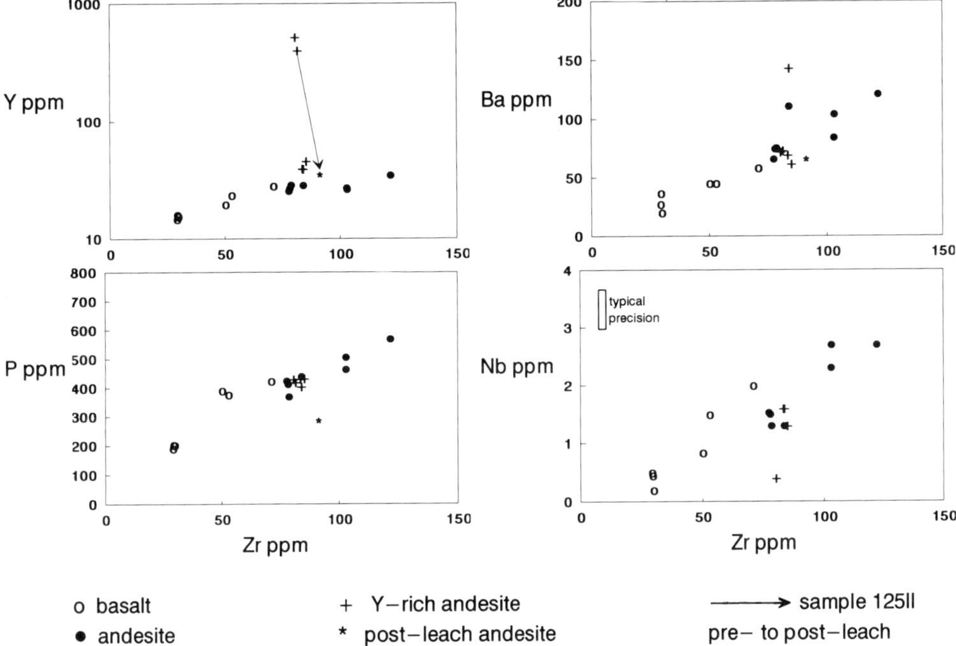

Fig. 8. Variation of trace elements with Zr.

1994; Pearce *et al.* 1992*b*; Gill *et al.* in press), which contrasts with the basalts and andesites from the active Izu-Bonin arc ((La/Yb)$_n$ 0.36–0.79; Taylor, unpublished data).

To illustrate the incompatible trace element variation with magma evolution within and between the two suites, Zr was chosen as a fractionation index (Fig. 8). Y, Ba, P and Nb all show strong positive correlations with Zr, with both petrological suites lying on the same trends. Ba shows a coherent relationship with Zr, but there is a suggestion that some andesites have slightly perturbed Ba concentrations due to secondary mobilization. More of a surprise is the anomalous concentrations of Y in certain andesite samples (up to 519 ppm) which lie well above the general fractionation path. It is now clear from a number of studies on relatively fresh volcanic rocks that Y, the REE and P are extremely mobile under certain low temperature (<150°) aqueous conditions (Price *et al.* 1991; Taylor *et al.* 1992*a*; Kuschell & Smith 1992). In light of this potential for Y–REE mobility, leaching experiments were undertaken on the most anomalous samples. Details of the results of these experiments will be published elsewhere. In brief, leaching samples in <2N HCl removes the excess Y and REE to leave a residue with concentrations similar to the non-anomalous andesites. The change in Y content

following leaching is represented in Fig. 8 as an arrow indicating the compositions of sample 781/125II before and after leaching.

Isotopes

Nd, Sr and Pb isotopes were measured on the two freshest samples representative of the two suites, in which alteration was considered minimal. Radiogenic isotope data are presented in Table 3. This includes a duplicate Pb isotope analysis of andesite sample 780/30 undertaken at the Department of Geology, Royal Holloway and Bedford New College, University of London.

Lead isotopic compositions of the Hahajima volcanic rocks lie above the northern hemisphere reference line (NHRL) in terms of $^{207}Pb/^{204}Pb$-$^{206}Pb/^{204}Pb$ and $^{208}Pb/^{204}Pb$-$^{206}Pb/^{204}Pb$. The andesite and basalt samples have similar Pb isotope ratios, Pb isotopic compositions and $\Delta7/4$ and $\Delta8/4$ values for the Hahajima samples are similar to the boninite series volcanic rocks of Chichijima (Taylor *et al.* 1994). In turn, both Bonin island localities differ from all other Eocene outerarc lavas from the Izu–Bonin–Mariana system, which are clustered around the NHRL (e.g. Site 786; Pearce *et al.* 1992*b*). Hence it would appear that the $\Delta7/4$ and $\Delta8/4$ enrichment of the Bonin islands is a

Table 3. *Radiogenic isotope measurements on Hahajima volcanic rocks*

	780/30 Andesite	780/31 Basalt	780/30 Basalt RHNBC
$^{87}Sr/^{86}Sr_{(m)}$	0.70403	0.70394	
$^{87}Sr/^{86}Sr_{(i)}$	0.70398	0.70386	
$^{143}Nd/^{144}Nd$	0.51304	0.51300	
$\epsilon_{Nd}(40\,Ma)$	7.7	7.1	
$^{206}Pb/^{204}Pb$	18.560	18.501	18.512
$^{207}Pb/^{204}Pb$	15.525	15.550	15.524
$^{208}Pb/^{204}Pb$	38.285	38.350	38.200

780/30 sample denoted RHNBC was analysed for Pb isotopes at Royal Holloway and Bedford New College, London using a VG 354 five-collector mass spectrometer and is corrected for mass fractionation by $c.$ 0.11 %/amu by normalization to SRM981; internal errors are estimated at better than 0.005 %/amu (2 sd).

peculiarity of the mantle source within the forearc high.

The contrast between the Bonin islands and other forearc high localities is somewhat problematical. Most samples measured for Pb isotopes from the Bonin islands (Dobson & Tilton 1989; Taylor *et al.* 1994; this study) give consistent ratios which are displaced relative to the NHRL, the only exception being the Mikazukiyama andesites from Chichijima (Taylor *et al.* 1994). Two possible scenarios could achieve this displacement. First, Pb from pelagic sediment introduced into the mantle wedge. Secondly, a dupal-type anomaly existed beneath the Bonin island region at the time of formation, in a similar fashion to that envisaged for the basinal lavas of the Philippine Sea plate (Hickey-Vargas 1991). In both cases it is possible that slab-derived fluids equilibrated or partially equilibrated with the source of the radiogenic Pb, by passing through either the sediments or anomalous mantle before metasomatizing the source mantle. On the basis of available data it is difficult to discriminate the two models.

$^{143}Nd/^{144}Nd$ (initial) values for the two Hahajima lavas are lower than MORB, with ϵ_{Nd} (40 Ma) values of 7.7 and 7.1 for the basalt and andesite respectively. $^{87}Sr/^{86}Sr$ is more radiogenic than MORB but, like $^{143}Nd/^{144}Nd$, the two suites have very similar values.

Relationship between the Hahajima lavas

On the basis of major element variations, the basaltic (tholeiitic) and andesitic (calc-alkaline) suites of Hahajima have taken different shallow-level fractionation paths. However, the general trace element interrelationships and isotopic characteristics of the two suites are remarkably similar. This suggests that the suites could be related by a similar mantle source or indeed a similar parental magma.

The crystal fractionation relationships are further investigated using trace elements as shown in Fig. 9. The basalts show a marked decline in Ni with increasing Zr (Fig. 9a). In contrast, the andesites have higher and more variable Zr contents compared to basalts with similar Ni contents. A simple fractional crystallization model involving removal of the observed phenocryst phases/proportions from the most primitive (high-MgO) basalt is also shown on Fig. 9a (curve annotated FC). This curve is observed to fit well with the data from the basaltic suite, but deviates significantly from the andesitic data.

One possible explanation of the lack of coherence between the trace element data of the

Fig. 9. Modelled trace element variation for closed system (FC) and open system fractional crystallization (OSF) compared to Hahajima compositions. FC curve is modelled for an assemblage of plag:cpx:olivine = 55:32:13, marked with fraction of crystals formed (F–1). OSF curve is for steady state liquids with a constant fraction crystallized in each cycle (0.02). Phase proportions for the OSF model are plag:cpx:opx:mgt = 55:26:15:4. It is assumed that the mass of primitive magma (of composition equivalent to the most primitive Hahajima basalt) introduced into the system is equal to the mass crystallized (X) + the mass erupted during each cycle (Y). Marks on the OSF curve represent values of Y for the range of steady state liquids.
Pure fractional crystallisation modelled using:
$$C_l/C_o = F^{(D-1)}$$
Open system fractionation modelled using the equation of O'Hara (1977):
$$C_B^S/C_o = (X+Y).((1-X)^{D-1}/1-(1-X-Y)(1-X)^{D-1}$$
where:
D = bulk distribution coefficient for crystallizing assemblage
C_l = concentration in the fractionated liquid
C_o = concentration in the initial primitive liquid
C_{BS} = concentration in the steady state liquid
Kd's used are taken from Irving (1978), Henderson (1982), and Kelemen *et al.* 1990).

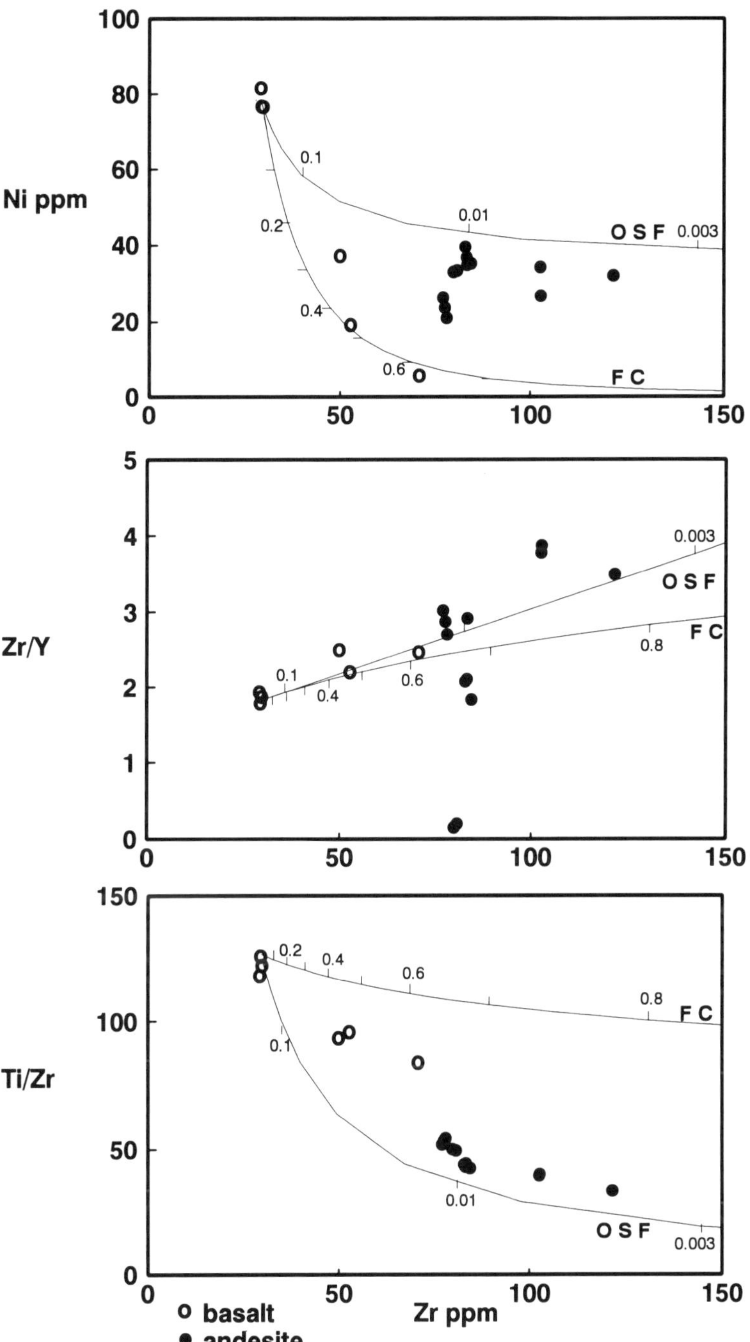
Ni ppm
100
80
60
40
20
0
0.1
0.2
0.4
0.6
0.01
0.003
O S F
F C
Zr/Y
5
4
3
2
1
0
0.1
0.4
0.6
0.8
0.003
O S F
F C
Ti/Zr
150
100
50
0
0.2
0.4
0.6
0.8
0.1
0.01
0.003
F C
O S F
0
50
100
150
Zr ppm
basalt
andesite

two suites is that the andesites and basalts are related to a range of primary liquids, representing various degrees of partial melting. This could potentially generate liquids with a spread of incompatible element concentrations (Zr) at similar levels of compatible elements (Ni). However, this does not explain the coherence of incompatible trace element ratios within and between the two suites or the absence of a series of trends sub-parallel to the FC curve originating from a range of primary basaltic compositions.

In light of the petrographical evidence for magma replenishment and mixing between the two suites (e.g. rhythmically zoned plagioclase phenocrysts and mutually related xenocrysts), it is likely that the andesites were generated in a replenished, tapped and fractionating open magmatic system. In such a system a chamber is periodically supplied at the base with a primitive melt which expels lava at the top. The primitive melt then mixes with the more fractionated liquid and the blended magma continues to crystallise (O'Hara 1977; O'Hara & Mathews 1981). A feature of open system fractionation is that concentrations of the more compatible elements (e.g. Ni, Cr, Mg and to a lesser extent Ti) are buffered to a greater extent compared to closed system fractionation. In contrast, incompatible elements show greater increases in concentration in open systems for similar levels of fractionation. The most extreme case of open system behaviour can be modelled as a steady-state magma reservoir, in which a constant mass of magma is retained between each replenishment-tapping-fractionation cycle. Elements will approach a steady-state concentration after a large number of cycles, the exact number being a function of input, output and crystallization (Z, Y and X) proportions and the relevant bulk distribution coefficient.

Open system steady-state compositions for an assemblage of plagioclase, pyroxenes and magnetite are modelled and shown on Fig. 9 (curves marked OSF). These curves represent constant mass reservoirs with a fixed fraction of crystallization in each cycle (0.02). Constant system mass is preserved by maintaining $Z = Y + X$. Trace element concentrations are formulated by varying parameter Y (output fraction), and values for Y are marked as ticks along the curves. A striking feature of the Hahajima data is that the andesites always lie closer to the open system (OSF) rather than the pure fractional crystallization (FC) trend in contrast to the basalts (Fig. 9). Clearly the open system is more capable of generating the constant Ni concentrations, high Zr/Y, and low Ti/Zr. It is also notable that the basalts lie away from the FC curve on the Ti/Zr vs. Zr plot, despite the use of appropriate Kd values.

To explain the data array in relation to the crystallization models, it is necessary to invoke some interplay between the open and closed systems. This can be achieved by either the magma reservoir not reaching a steady state open system but approaching it between periods of more closed system behaviour, or the andesites being generated by a limited number of reflux events giving rise to mixing trends between primitive liquids and closed system magmas. The latter case seems the least likely as it is not possible to reconcile the low Ti/Zr of the andesites (Fig. 9c) without resorting to excessive quantities of oxide phase precipitation. This in turn, would lead to dramatic reduction in Fe and V with lower Mg, which is not observed.

O'Hara and Mathews (1981) and Nielsen (1988) recognized that it is possible to change element ratios more significantly with an open system process. This is especially true if the element Kd values are close to unity. A consequence of this is that Ti/Zr (and to a lesser extent Zr/Y) will be fractionated more significantly in the open system. Furthermore, it is possible to lower the Ti/Zr (and perhaps buffer Fe/Mg) without recourse to excess magnetite. Hence, it is possible that the Hahajima basalts are an example of a tholeiitic magma suite with some open system tendencies. This is characterized by a lack of magnetite phenocrysts, combined with significantly declining Ti/Zr and only moderate increases in Ti and Fe contents with fractionation.

Arc magmatism in the forearc at the initiation of subduction

Geochemical variation along the forearc

Amidst the more renowned boninite magmatism which proliferated in the Izu-Bonin-Mariana forearc some 40 Ma ago is a significant proportion of 'normal' arc-type volcanic sequences, such as at Hahajima. The non-boninitic magmatism at the initiation of subduction can fall into either of the indefinite categories of calc-alkaline or tholeiitic. Calc-alkaline lavas (in terms of Fe/Mg and phenocryst characteristics at least) are found at all three of the main Izu-Bonin forearc high localities, namely Hole 786B andesite-dacite-rhyolite series; Chichijima Mikazukiyama formation; Hahajima andesite suite. This dispels any notion of calc-alkalinity being a feature of mature arcs.

A noteworthy feature of the normal arc lavas

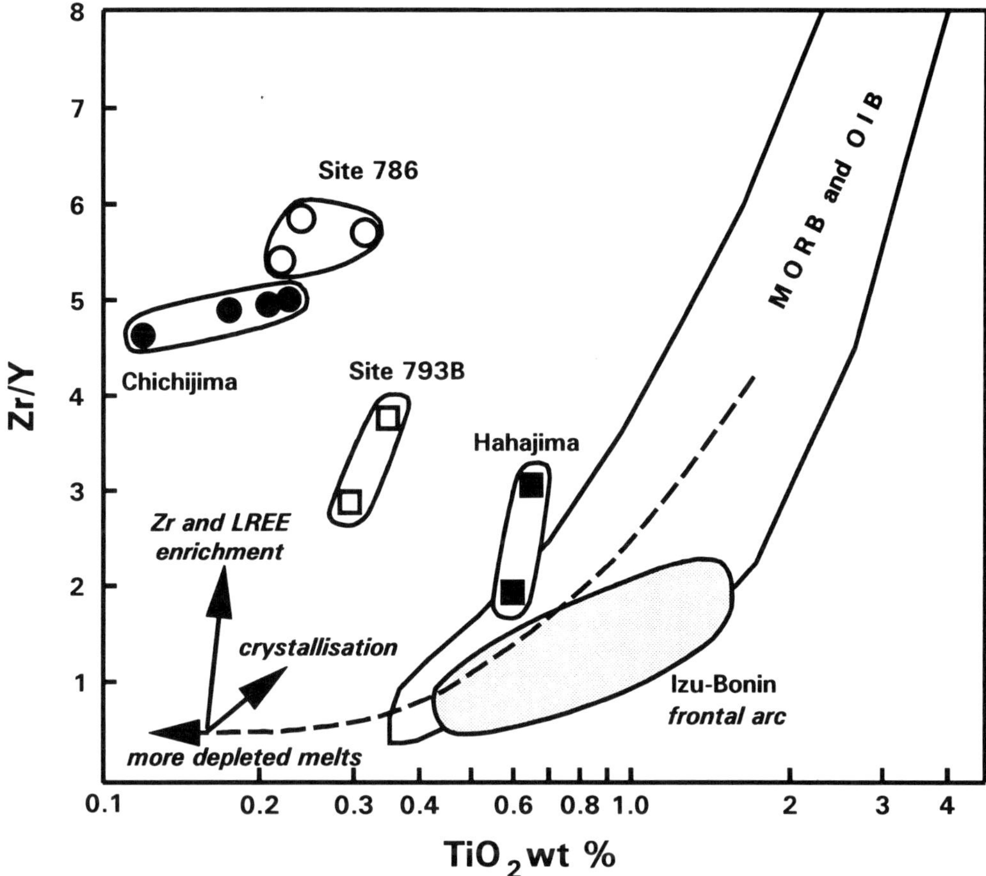

Fig. 10. Variation of Zr/Y as a function of TiO_2 wt% for forearc volcanic suites grouped according to their location. Each point represents an average value for each suite, with samples having anomalous Y and/or P_2O_5 values excluded. Site 786 data points (in order of increasing TiO_2) are LCB series, ICB series, ADR series; Pearce *et al.* (1992*b*) and Murton *et al.* (1992). Chichijima data points are boninite, boninitic andesite, boninitic dacite, Mikazukiyama andesite; Taylor *et al.* (in press). Site 793 data points are basement units 1b–14, unit 1a; Taylor *et al.* (1992*a*). Hahajima data points are for average primitive basalt (samples 780/127, 780129, 780/31) and average andesite.

along the forearc high is their similar trace element characteristics to the contemporaneous boninitic suites at the same locality. This is demonstrated in Fig. 10. On this plot the average values for each distinct suite or series (anomalous Y and P samples excluded) cluster according to their location. This indicates that variation at any particular forearc centre may be partially controlled by shallow level crustal processes. However the inter-locality differences are related to specific characteristics of the mantle which include amount of hydrous flux, temperature and enrichment by exotic components (e.g. low Sm/Zr, ϵ_{Nd} and high $\Delta7/4$).

The Eocene arc in relation to Oligocene–Recent arcs

The existence of boninites in the initial forearc is taken to indicate unusual thermal characteristics in the mantle wedge resulting in tapping of shallow, depleted mantle (e.g. Van der Lann 1989; Taylor *et al.* 1992*b*). Yet, under these same conditions, apparently normal arc volcanic rocks were also generated. Eocene–Oligocene arc magmatism is compared with more recent arcs in Figure 11(a & b). These plots utilize the method described by Plank and Langmuir (1988) of normalizing certain elements (in this case Na_2O

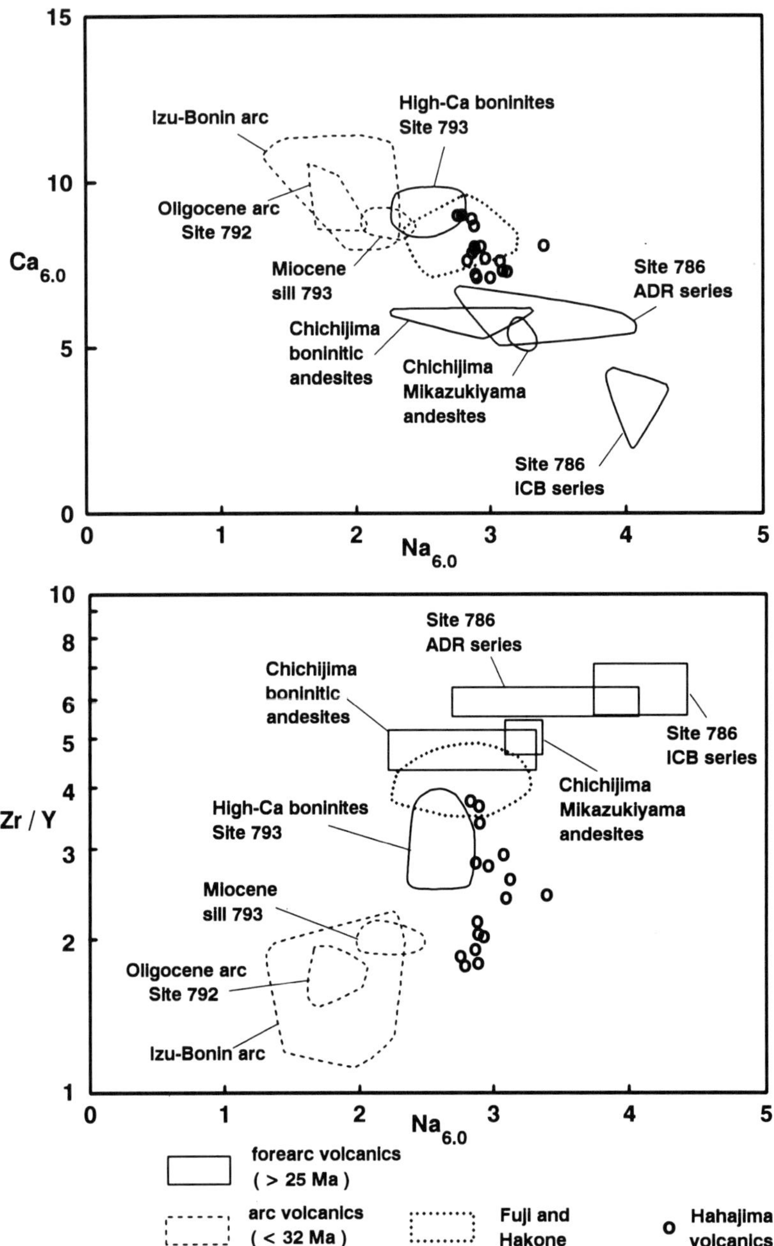

Fig. 11. (**a**) $Ca_{6.0}$ variation with $Na_{6.0}$ for Eocene–Recent Izu–Bonin magmas. Data sources are as for Fig. 10 as well as Miocene sill Site 793 and Oligocene arc Site 792, Taylor *et al.* (1992*a*); Izu–Bonin arc, Taylor, unpublished data; Fuji and Hakone, Arculus *et al.* (1991). (**b**). Zr/Y variation with $Na_{6.0}$ for Izu–Bonin magmas. Data sources as for Fig. 11(a). Square data fields and Site 793 represent the range of $Na_{6.0}$ combined with the average $Zr/Y \pm$ expected error for the particular suite.

and CaO) to a MgO content of 6 wt% ($Na_{6.0}$ and $Ca_{6.0}$). This procedure has the effect of removing most effects of crystal fractionation within any suite, and should allow comparison of major

chemical components with respect to mantle processes.

Figure 11(a) shows $Na_{6.0}$ plotted against $Ca_{6.0}$ for a range of arc and boninitic volcanics rocks from the Izu–Bonin system. As a whole, the data form an array of decreasing $Ca_{6.0}$ with increasing $Na_{6.0}$. The Oligocene–Recent lavas (Site 792 and the Izu-Bonin arc) plot at the low $Na_{6.0}$ end of the array, whereas the boninitic andesites and Hahajima volcanic rocks from the forearc high have significantly higher $Na_{6.0}$. A similar array is also observed on the Zr/Y–$Na_{6.0}$ plot Figure 11(b), with Eocene forearc volcanic rocks displaced to higher Zr/Y.

Originally the $Ca_{6.0}$ and $Na_{6.0}$ normalization procedure was used by Plank & Langmuir (1988) to demonstrate the relationship between elemental abundance and the crustal thickness upon which the arc was generated. These authors demonstrated that volcanic rocks situated on thinner crust (such as Izu-Bonin) have relatively low $Na_{6.0}$ and high $Ca_{6.0}$ compared to continental-based volcanoes. If the forearc data are equated directly to the Plank and Langmuir model, then the crustal thickness at the inception of subduction would have been between 30 and 80 km. This is clearly unreasonable, as it is surmized that the Izu-Bonin system was initiated in an intra-oceanic environment with expected moho depths in the range 7–12 km.

On this basis it is likely that influences other than crustal thickness controlled the chemistry of volcanic rocks in the early forearc. A major control on the composition of boninite series volcanic rocks (which include all Chichijima and Site 786 volcanic rocks) is likely to have been their severely depleted peridotite source (Hickey & Frey 1982; Taylor *et al.* 1994; Pearce *et al.* 1992a). As more depleted lherzolites and harzburgites are involved in melting, a result will be lower Ca in the magma. However, it is not expected that incompatible components such as Na would increase during this process. The answer almost certainly lies in the normalization process itself. It does not take account of the variations in Mg content of primitive liquids leaving the source. For example, if a primitive magma with a high-Mg content evolves, it will produce high-Mg andesitic liquids such as boninitic andesites. When these andesites are normalized to fixed MgO the result will be a higher apparent $Na_{6.0}$.

This high-Mg effect could perhaps be predicted for boninitic suites, but it is also apparent from Fig. 11 that normal arc volcanic rocks from the Eocene forearc (such as Hahajima) are also displaced to significantly higher $Na_{6.0}$ and lower $Ca_{6.0}$. The implication is that the Eocene 'normal' volcanic rocks were significantly different from the Recent arc in having a higher MgO relative to Na and Ca. This could be a product of steeper thermal gradients or melting of shallow, depleted mantle wedge.

Figure 11(b) shows that Hahajima is also discrepant from the Recent arc in having relatively high Zr/Y. The most primitive Quaternary–Recent arc lavas have around 8% MgO and $Zr/Y \approx 1$, which compares with the primitive Hahajima lavas with $Zr/Y \approx 2$ at the same MgO. If Hahajima was supplied from more depleted mantle or by a higher degree of melting or both, this should lead to lower Zr/Y. It is thus more likely that the Hahajima source has been influenced by a similar style of REE and Zr enrichment as observed in the boninitic series (Taylor & Nesbitt 1988; Taylor *et al.* 1994).

Extensional volcanism in the forearc

It has become clear from recent drilling investigations that the Izu-Bonin region was undergoing syn-magmatic extension above a subduction zone during the Eocene–Oligocene. Evidence for extensional magmatism is found on Chichijima and at ODP Site 786, where sheeted dykes occur beneath the boninitic extrusive sequences. Stern and Bloomer (1992) have suggested that the whole arc–forearc crust underwent rapid extension during the Eocene. As a result, coeval magmatism prevailed over the entire supra-subduction terrain. This extension continued into the Oligocene, when the Izu–Bonin forearc basin was rifted giving rise to extrusive boninitic magmatism at Sites 793 and 458 (Taylor *et al.* 1992a).

Within the generally extensional regime of the Eocene forearc it is clear that certain areas were not subjected to the same degree of tension as others. Regions such as Hahajima developed shallow-water (possibly subaerial) volcanoes. In contrast, the dyke evidence from Chichijima and Site 786 suggests volcanic development in small rift basins. As a whole, the Eocene arc–forearc can potentially be viewed as a series of narrow, elongate rift-basins. These basins would be punctuated and bordered by more competent crustal blocks incorporating well-developed volcanoes. A chain of contemporaneous arc volcanoes was developed behind the forearc terrain and is now preserved as the remnant Palau–Kyushu ridge. Such a volcano-tectonic scenario is illustrated in Fig. 12.

The distinction between small rift-basins and volcanoes within the generally extensional forearc regime broadly correlates with lava chemistry. Namely, the normal arc-like magmas (e.g. Hahajima and Mikazukiyama) form volcanic edifices, while boninitic magmas occupy the

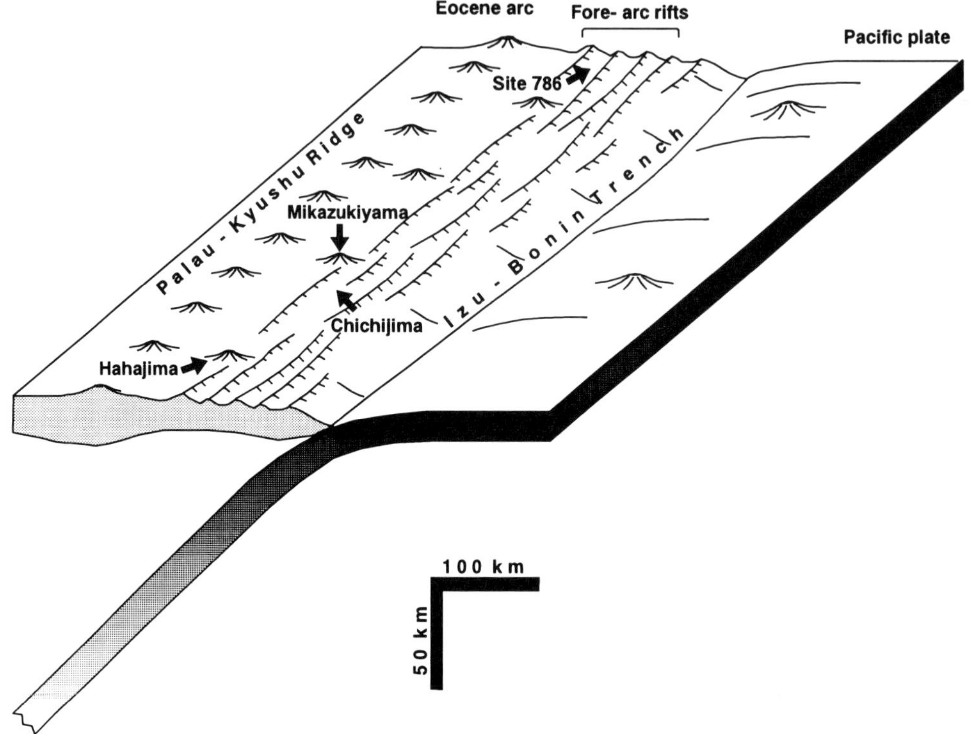

Fig. 12. Schematic model for the Izu–Bonin arc–trench region in the Eocene.

basins. This tectono-magmatic link would correspond with the notion that boninitic magmas are generated at shallow depths (<30 Km), while the arc lavas result from slightly deeper and less extensive melting. Yet in spite of this, differing lava types within particular sections of the forearc retain similar depletion and enrichment characteristics (Fig. 10).

Conclusions

Two lava suites, basaltic and andesitic, are found within the Hahajima succession. These suites contain excess proportions of plagioclase over the mafic phases, which profoundly affects the major element compositions. Correction for this excess produces more systematic trends on variation diagrams which in turn distinguishes the suites as tholeiitic and calc-alkaline in terms of iron enrichment characteristics.

In terms of their trace element and isotopic abundances the suites are remarkably similar, suggesting a common parental magma source. Differences between them can be explained by differing styles of shallow-level fractionation. The basaltic group can be modelled for most

elements as the products of closed system fractional crystallization. In contrast the andesitic group more closely resemble an open magmatic system. These processes may be quite fundamental in explaining the differences between tholeiitic and calc-alkaline magmas.

When typical arc-like lavas from the Eocene forearc are compared with the products of the current Izu–Bonin arc, it is noted that the older arc lavas have higher MgO contents than their modern counterparts. This, together with the coeval occurrence of boninitic magmas, may signify a generally depleted mantle source in the Eocene forearc.

Inspection of lavas from the Oligocene–Eocene forearc terrain shows that the trace element characteristics of lavas are similar at any particular location within the forearc, regardless of their classification as boninitic, tholeiitic or calc-alkaline. However, significant differences in composition are recognized between the localities along the length of the forearc terrain. This spatial variation can be interpreted as an along-arc mantle heterogeneity produced by variable enrichment and earlier melting events, while differences within particular sections can be ascribed to the depth of melting (equating to the amount of local extension) and the nature of the crustal fractionation process.

We thank I. Croudace and J.-A. Barrat for analytical help and discussions during the course of this study. Field and laboratory work was supported by NERC grants GR3/4437, GST/02/416, GST/02/558 and travel grants from the Royal Society.

References

ARCULUS, R.J., GUST, D.A. & KUSHIRO, I. 1991. Fuji and Hakone. *National Geographic Research and Exploration*, **7**, 276–309.

BLOOMER, S.H. & HAWKINS, J.W. 1987. Petrology and geochemistry of boninite series volcanic rocks from the Mariana trench. *Contributions to Mineralogy and Petrology*, **97**, 361–377.

BROPHY, J.G. 1989. Can high-alumina arc basalt be derived from low-alumina arc basalt? Evidence from Kanga Island, Aleutian arc, Alaska. *Geology*, **17**, 333–336.

CLAGUE, D.A. & DALRYMPLE, G.B. 1987. The Hawaiian-Emperor volcanic chain. Part I. Geologic evolution. *In*: DECKER, R.W., WRIGHT, T.L. & STAUFFER, P.H. (eds) Volcanism in Hawaii, Vol. 1. *US Geological Survey Professional Papers*, **1350**, 5–54.

CROUDACE, I.W. & GILLIGAN, J. 1990. Versatile and accurate trace element determinations in iron-rich and other geological samples using X-ray fluorescence analysis. *X-ray Spectrometry*, **19**, 117–123.

—— & MARSHALL, S. 1991. Determination of rare earth elements and Yttrium in nine geochemical reference samples using a novel group separation procedure involving mixed-acid elution ion-exchange chromatography. *Geostandards Newsletter*, **15**, 139–144.

DOBSON, P.F. 1986. *The petrogenesis of boninite: a field, petrologic and geochemical study of the volcanic rocks of Chichi-jima, Bonin Islands, Japan.* PhD. thesis, Stanford University, California.

—— & TILTON, G.R. 1989. Th, U, and Pb systematics of boninite series volcanic rocks from Chichi-Jima, Bonin Islands, Japan. *In*: CRAWFORD, A.J. (ed.). *Boninites and related rocks.* Unwin-Hyman, London, 396–415.

FRYER, P., PEARCE, J.A., *et al.* 1990. *Proceedings of the Ocean Drilling Program, Initial Reports*, **125**.

GILL, J.B., HISCOTT, R.N. & VIDAL, PH. In press. Turbidite geochemistry and evolution of the Izu–Bonin arc and continents. *Lithos*.

HENDERSON, P. 1982. *Inorganic geochemistry.* Pergamon Press, Oxford.

HICKEY, R.L. & FREY, F.A. 1982. Geochemical characteristics of boninite series volcanics: implications for their source. *Geochimica et Cosmochimica Acta*, **46**, 2099–2115.

HICKEY-VARGAS, R.L. 1991. Isotope characteristics of submarine lavas from the Philippine sea: implications for the origin of arc and basin magmas of the Philippine tectonic plate. *Earth and Planetary Science Letters*, **107**, 290–304.

HOFMANN, A.W. 1988. Chemical differentiation of the Earth: the relationship between mantle, continental crust, and oceanic crust. *Earth and Planetary Science Letters*, **90**, 297–314.

HUSSONG, D.M., UYEDA, S., *et al.* 1981. *Initial Reports of the Deep Sea Drilling Project*, **60**, 929pp.

IRVING, A.J. 1978. A review of experimental studies of crystal/liquid trace element partitioning. *Geochimica et Cosmochimica Acta*, **42**, 743–770.

ISHII, T. 1985. Dredged samples from the Ogasawara fore-arc seamount or 'Ogasawara Paleoland' – 'forearc ophiolite'. *In*: NASU, N. *et al.* (eds) *Formation of active ocean margins.* Terra Scientific Publishing Company, Tokyo, 307–347.

IWASAKI, Y. & AOSHIMA, M. 1970. Report on geology of the Bonin Islands. *In: The nature of Ogasawara, report on scientific and natural monuments of the Ogasawara islands.* The Ministry of Education, Japan, 205–220 (in Japanese with English abstract).

KANEOKA, I., ISSHIKI, N. & ZASHU, S. 1970. K–Ar ages of the Izu–Bonin Islands. *Geochemical Journal*, **4**, 53–60.

KELEMAN, P.B., JOHNSON, K.T.M., KINZLER, R.J. & IRVING, A.J. 1990. High-field-strength element depletions in arc basalts due to mantle–magma interaction. *Nature*, **345**, 521–524.

KODAMA, K., KEATING, B.H. & HELSLEY, C.E. 1983. Paleomagnetism of the Bonin Islands and its tectonic significance. *Tectonophysics*, **95**, 25–42.

KOBAYASHI, K. 1983. Fore-arc volcanism and cycles of subduction. *In*: SHIMOSURU, D. & YOKOYAMA, I. (eds) *Arc volcanism, physics and tectonics.* Terra Scientific Publishing Company, Tokyo, 153–163.

KUSCHEL, E. & SMITH, I.E.M. 1992. Rare earth mobility in young arc-type volcanic rocks from northern New Zealand. *Geochimica et Cosmochimica Acta*, **56**, 3951–3955.

LANGMUIR, C., ZHANG, Y., TAYLOR, B., PLANK, T. & RUBENSTONE, J. In press. Petrogenesis of Torishima and adjacent volcanoes of the Izu–Bonin arc: one end member of the global spectrum of arc basalt compositions. *Contributions to Mineralogy and Petrology*.

MACKENZIE, W.S., DONALDSON, C.M. & GUILFORD, C. 1982. *Atlas of igneous rocks and their textures.* Longman, Harlow.

MITCHELL, J.G., PEATE, D.W., MURTON, B.J., PEARCE, J.A., ARCULUS, R.J. & VVAN DER LAAN, S.R. 1992. K–Ar dating of samples from sites 782 and 786 (Leg 125): the Izu-Bonin forearc region. *In*: FRYER, P., PEARCE, J.A. & STOKKING, L.B. *et al. Proceedings of the Ocean Drilling Project, Scientific Results*, **125**, 203–210.

MURTON, B.J., PEETE, D.W., ARCULUS, R.J., PEARCE, J.A. & VAN DER LANN, S. 1992. Trace element geochemistry of volcanic rocks from Site 786: the Izu-Bonin forearc, Leg 125. *Proceedings of the Ocean Drilling Project, Scientific Results*, **125**, 211–235.

NIELSEN, R.L. 1988. A model for the simulation of combined major and trace element liquid lines of descent. *Geochimica et Cosmochimica Acta*, **52**, 27–38.

O'HARA, M.J. 1977. Geochemical evolution during fractional crystallisation of a periodically refilled magma chamber. *Nature*, **266**, 503–507.

O'HARA, M.J. & MATHEWS, R.E. 1981. Geochemical evolution in an advancing, periodically replenished, periodically tapped, continuously fractionated magma chamber. *Journal of the Geological Society, London*, **138**, 237–277.

PARSON, L., HAWKINS, J., ALLAN, J. *et al.* 1992. *Proceedings of the Ocean Drilling Project, Initial Reports*, **135**.

PEARCE, J.A., THIRLWALL, M.F., INGRAM, G., MURTON, B.J., ARCULUS, R.J. & VAN DER LAAN, S.R. 1992*b* . Isotopic evidence for the origin of boninites and related rocks drilled in the Izu–Bonin (Ogasawara) forearc, ODP Leg 125. *Proceedings of the Ocean Drilling Project, Scientific Results*, **125**, 237–261.

——, VAN DER LAAN, S.R., ARCULUS, R.J. *et al.* 1992*a*. Boninite and harzburgite from ODP Leg 125 (Bonin-Mariana forearc): A case study of magma genesis during the initial stages of subduction. *Proceedings of the Ocean Drilling Project, Scientific Results*, **125**, 623–659.

PLANK, T. & LANGMUIR, C.H. 1988. An evaluation of the global variations in the major element chemistry of arc basalts. *Earth and Planetary Science Letters*, **90**, 349–370.

PRICE, R.C., GRAY, C.M., WILSON, R.E., FREY, F.A. & TAYLOR, S.R. 1991. The effects of weathering on rare-earth element, Y and Ba abundances in Tertiary basalts from southeastern Australia. *Chemical Geology*, **93**, 245–265.

SHARASKIN, A.Y., KARPENKO, S.F., LJALIKOV, A.V., ZLOBIN, S.K. & BALASHOV, Y.B. 1983. Correlated $^{143}Nd/^{144}Nd$ and $^{87}Sr/^{86}Sr$ on boninites from Mariana and Tonga Arcs. *Ofioliti*, **8**, 431–438.

STERN, R.J. & BLOOMER, S.H. 1992. Subduction zone infancy: examples from the Eocene Izu-Bonin-Mariana and Jurassic California arcs. *Geological Society of America Bulletin*, **104**, 1621–1636.

TAYLOR, R.N. & NESBITT, R.W. 1988. Light rare-earth enrichment of supra-subduction zone mantle: evidence from the Troodos ophiolite, Cyprus. *Geology*, **16**, 448–451.

——, MURTON, B.J. & NESBITT, R.W. 1992*b*. Chemical transects across intra-oceanic arcs: Implications for the tectonic setting of ophiolites. *In*: PARSON L.M., MURTON, B.J. & BROWNING, P.J. (eds) *Ophiolites and their modern oceanic analogues*. Geological Society, London, Special Publication, **60**, 117–132.

——, LAPIERRE, H., VIDAL, P., NESBITT, R.W. & CROUDACE, I.W. 1992*a*. Igneous geochemistry and petrogenesis of the Izu-Bonin fore-arc basin. *Proceedings of the Ocean Drilling Project, Scientific Results*, **126**, 405–430.

——, NESBITT, R.W., VIDAL, P., HARMON, R., AUVRAY, B. & CROUDACE, I.W. 1994. Mineralogy, chemistry and genesis of the boninite series volcanics, Bonin Islands, Japan. *Journal of Petrology*, **35**, 577–617.

TSUNAKAWA, H. 1983. K-Ar dating on volcanic rocks in the Bonin Islands and its tectonic implication. *Tectonophysics*, **95**, 221–232.

UJIIE, H. & MATSUMARU, K. 1977. Stratigraphic outline of Haha-jima (Hillsborough Island), Bonin Islands. *Memoirs of the National Science Museum, Tokyo*, **10**, 5–22.

VAN DER LAAN, S.R., FLOWER, M.F. & VAN GROOSA F.K. 1989. Experimental evidence for the origin of boninites: near-liquidus phase relations to 7.5 kbar. *In*: CRAWFORD, A.J. (ed.) *Boninites*. Unwin Hyman, London, 112–147.

VIDAL, PH. & CLAUER, N. 1981. Pb and Sr isotopic systematics of some basalts and sulphides from the East Pacific Rise at 21°N (Project RITA). *Earth and Planetary Science Letters*, **55**, 237–246.

YAMAGUCHI, T. 1985. *Geology and petrology of Hahajima, Bonin Islands, Japan*. Graduate thesis, Toyama University, Japan.

Concomitant evolution of tectonic environment and magma geochemistry, Ambrym volcano (Vanuatu, New Hebrides arc)

C. PICARD[1], M. MONZIER[2], J.-P. EISSEN[3] & C. ROBIN[2]

[1] *Université Joseph Fourier, Institut Dolomieu, rue Maurice Gignon, 38031 Grenoble, France*

[2] *UR 1F, ORSTOM, BP 76, Port Vila, Vanuatu*

[3] *UR 1F, ORSTOM, BP 70, 29280, Plouzane, France*

Abstract: Ambrym volcano lies in the central part of the New Hebrides island arc near the latitude where the D'Entrecasteaux Zone is colliding with the arc. It is a 35 km × 50 km, mainly basaltic stratovolcano constructed in four distinct parts: (i) well-preserved remnants of an old edifice in the north; (ii) a N100°-aligned, oval-shaped basal Hawaiian-type shield volcano; (iii) a pyroclastic cone, cut by a concentric 12 km wide caldera; and (iv) post-caldera basaltic suites, both intra- and extra-caldera. Considering K_2O, La and Zr variations, three major trends characterize the geochemistry of Ambrym; one medium-K (MK) to high-K (HK) tholeiitic to calc-alkaline basaltic trend, and two more evolved trends, respectively MK and HK, from basaltic to dacitic and rhyodacitic compositions. These magmatic suites correspond to several volcanic phases: (1) older MK basalts forming the Tuvio–Vetlam–Dalahum edifice; (2) MK to HK basalts forming the basal shield volcano; (3) MK andesites and rhyodacites forming the first pyroclastic sequence of the Ambrym Pyroclastic Series (APS), which probably initiated the formation of the caldera; (4) MK to HK basalts and andesites forming Surtseyan then Strombolian pyroclastic sequences 2, 3 and 4 of the APS; and (5) post-caldera MK to HK basalts forming the recent olivine porphyritic- and plagioclase-porphyritic suites, locally associated with more evolved HK andesitic volcanic rocks in the eastern part of the caldera. (La/Yb)*n* and (La/K)*n* ratios (2–5.9 and 1.3–2.2, respectively) and Zr contents indicate that the parental magmas originated from incremental batch melting or fractional melting (21–25%) during uplift of a single spinel lherzolite source (from at least 60 to 45 km depth) in response to modifications in the tectonic environment due to ongoing collision with the D'Entrecasteaux Zone. However, K and La contents require prior enrichments by possible mantle metasomatism associated with subduction. Fractionation of olivine, plagioclase, clinopyroxene and Fe–Ti oxides explain most of the major and trace element variations in the resulting liquids and interaction between magmas rising from two distinct chambers combined with massive introduction of seawater into the edifice appear to be the major causes of the giant eruption leading to caldera formation. More recent activity of the volcano is principally related to shallow incremental batch melting or fractional melting and to fractionation in magma chambers associated with N100° rifting.

Diverse schemes based on major and/or trace element geochemistry have been used to classify island arc magmas and to propose petrological models for their origin. Among them, K content, LREE/HREE values and isotope data, particularly for Sr and Nd, provide useful information about depth of magma genesis, possible source heterogeneities, degree of partial melting, crustal contamination, and possible participation of subducted sediment during magma genesis, as well as about the differentiation processes operative in island arc magma chambers.

Based on such geochemical studies, models for the spatial and temporal evolution of arc systems have been generally well established in many calc-alkaline provinces. This is especially true for studies of arc development over the lifetime of the arc, or for large parts of an arc system over a reduced period. In contrast, on the scale of a single volcano over a short period (less than 0.1 Ma), examples showing large chemical variations are scarce and most relate to continental calc-alkaline volcanoes where the difficulty of establishing primary magma composition makes the problem of magma genesis even more complicated.

Along the New Hebrides arc, distribution of K_2O reveals important variations which are not well understood (Gorton 1974; MacFarlane *et*

From Smellie, J.L. (ed.), 1995, *Volcanism Associated with Extension at Consuming Plate Margins*, Geological Society Special Publication No. 81, 135–154.

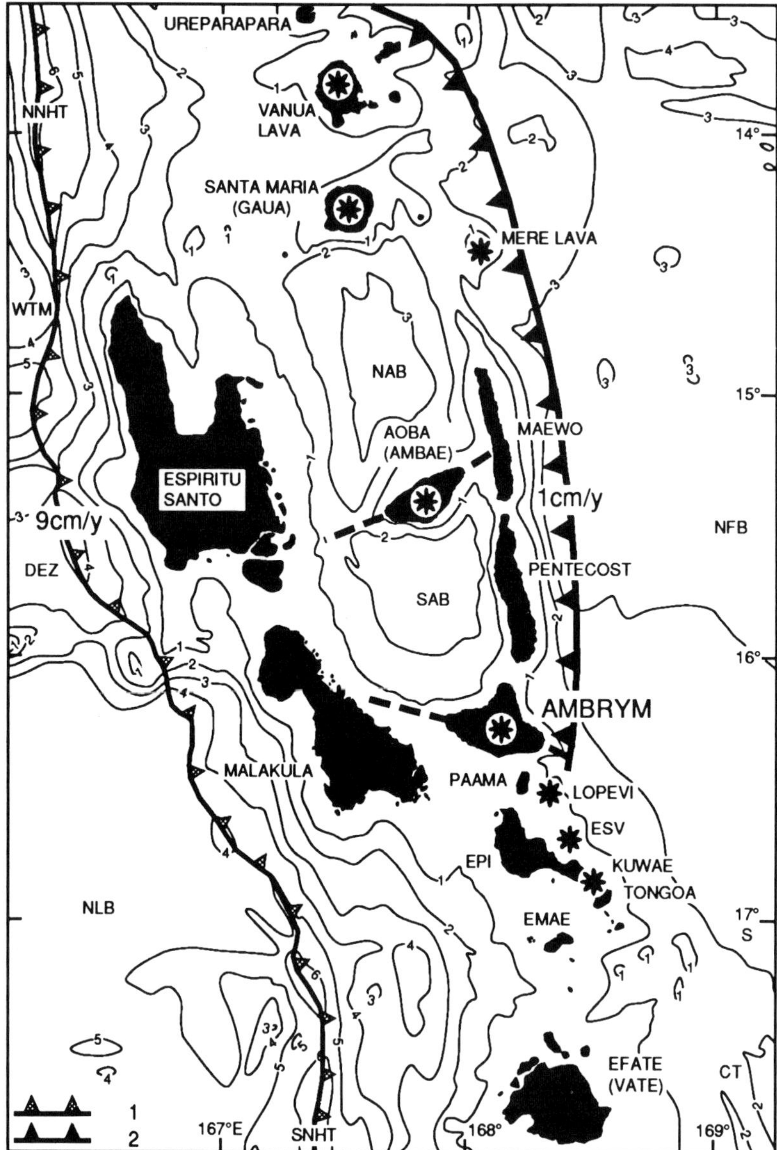

Fig. 1. Map showing central and northern Vanuatu islands (in black). Bathymetry in km from Chase & Seekins (1988). NLB, North Loyalty Basin; DEZ, D'Entrecasteaux Zone; WTM, West Torres Massif; SNHT, Southern New Hebrides Trench; NNHT, Northern New Hebrides Trench; NAB, North Aoba Basin; SAB, South Aoba Basin; ESV, Epi Submarine volcanoes; NFB, North Fiji Basin; CT, Coriolis troughs. New Hebrides convergence (1) and reverse back-arc thrusting (2) emphasized by thick lines and black triangles with relative motions in cm/y (Louat & Pelletier 1989). Volcanoes with activity during the last five centuries are marked by a black star (Simkin *et al.* 1981).

al. 1988), although high-K compositions dominate only in the large volcanoes located at the latitudes where the D'Entrecasteaux Zone collides with the New Hebrides arc (Roca 1978; Eggins 1989). These variations are particularly significant in the Ambrym volcano, in which we draw attention to the coexistence of high K_2O (herein HK) and medium-K_2O (MK) basalts and related andesitic to rhyodacitic lavas.

This paper deals with the petrological characteristics of successive pre-, syn- and post-caldera volcanic suites of Ambrym volcano and focuses especially on the event responsible for caldera formation, in which magmas with both HK and MK compositions took part. Their study allows us to propose a model explaining the chemical evolution of the volcano in relation to its tectonic setting.

Seismo-tectonic setting of Ambrym volcano

Ambrym volcano (168°08′ E; 16°15′ S) is situated in the central part of the New Hebrides (NH) island arc near the latitude where the D'Entrecasteaux Zone (DEZ, Fig. 1) is colliding with the arc. Eastward subduction of the Australian plate occurs along the NH trench. Convergent relative motion between the Australian plate and the NH island arc varies from 15–16 cm/year in the north to 12 cm/year in the south (Louat & Pelletier 1989).

The DEZ is a high-relief, double-spined ridge on the Australian plate (Fig. 1). Collision and subduction of this E–W-trending ridge under the arc was probably initiated near 17°S about 3 Ma ago (Louat *et al.* 1988). According to Roca (1978), Collot *et al.* (1985), Collot (1989) and Eggins (1989), subduction of the DEZ and the progressive northern migration of the collision area (due to the obliquity of the DEZ compared with the general N65–75° azimuth of the relative plate convergence) explain most of the unusual characteristics of the arc between Vanua Lava and Efate islands. Present-day rates of subduction are least in front of the DEZ (9 cm/year) and back-arc thrusting (1 cm/year) occurs at the same latitude instead of the extensional regime characteristic of most of the back-arc areas of the NH (Louat & Pelletier 1989).

Intense tectonic erosion occurs in the fore-arc areas previously affected by the migrating collision zone (Efate–Malakula; Fig. 1), whereas uplift of both forearc (Santo–Malakula) and back-arc (Maewo–Pentecost) areas, together with subsidence of Aoba Basins and major faulting transverse to the arc, occur in front of the present collision zone (Collot 1989). The unusual shallow- and intermediate-depth seismicity between 14° and 18°S (Louat *et al.* 1988), and the intense volcanism along transverse fracture zones in the same area (the most active part of the volcanic arc, containing the large volcanoes of Gaua, Ambae and Ambrym; Macfarlane *et al.* 1988; Greene *et al.* 1988, and where high-K compositions clearly dominate; Roca 1978; Eggins 1989), appear to be strongly associated with past and present-day subduction of the DEZ.

The area between Ambrym and Tongoa islands (Fig. 1) is at present the most active part of the arc (Simkin *et al.* 1981) and includes from north to south: the 12 km wide Ambrym caldera with the two active cones Marum and Benbow (Robin *et al.* 1993), the Lopevi active volcano (Warden 1967), three active submarine volcanoes near Epi island (Exon & Cronan 1983; Crawford *et al.* 1988), the Karua active submarine volcano in the caldera of Kuwae (Crawford *et al.* 1988), and lastly, some thermal springs on Tongoa (Warden *et al.* 1972). In addition, recently extinct volcanoes with well preserved cones are common on Epi, Tongoa and Emae (Warden *et al.* 1972), and Kuwae caldera is only 500 years old (from our unpublished data, and age given by Garanger 1972). Whereas Ambrym and Lopevi are clearly located on transverse fractures, interpreted by Greene *et al.* (1988) as a major active transcurrent wrench fault, other volcanoes do not display such a strong tectonic control.

Active volcanoes of the NH arc are generally located between 100 and 200 km above the Benioff zone (Louat *et al.* 1988). Nevertheless, a conspicuous gap in intermediate-depth seismicity occurs between Malakula and Efate (Marthelot *et al.* 1985; Louat *et al.* 1988), slightly west of the most active volcanic segment of the arc. This seismic gap may correspond to a window in the subducted lithosphere, generated between detached, sinking lithosphere and a short, newly established slab (Louat *et al.* 1988; Collot 1989). The high volcanicity of the area might be related to processes occurring around 200 km depth at the torn upper edge of the detached lithospheric slab.

Detailed present-day seismotectonics of the Ambrym area are shown in Fig. 2. The back-arc, west-dipping thrust zone of the central New Hebrides arc dies out to the south near the active Lopevi volcano, immediately east of Ambrym. The brief (3 days) 1987 superficial seismic crisis beneath eastern Ambrym, the only one recorded since 1977, included shallow earthquakes displaying both strike-slip and underthrust focal mechanisms which strongly support a N100° compressional stress for the crust in this area. Interestingly, the western and eastern rifts of Ambrym, from which numerous lava flows were erupted, display the same N100° trend and, thus, probably correspond to large scale tensional 'cracks' opened by a N100° regional compressive stress, rather than to active transcurrent faults (Greene *et al.* 1988). In addition, most of the recent intra-caldera lava flows were erupted along N100° fractures. As pointed out by Collot (1989), this regional compressive stress in the crust is a consequence of the collision between

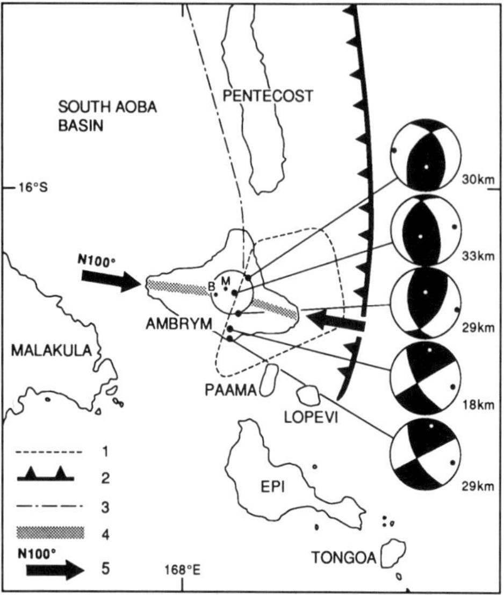

Fig. 2. Present-day seismotectonics of the Ambrym area. (1) Boundary of the seismic crisis area of 26 to 28 November 1987 as recorded by the local seismic network in Vanuatu; CMT focal mechanism solutions are shown on equal-area projections of the lower hemisphere of the focal sphere, with focal depth in km (US Department of the Interior/Geological Survey 1987); (2) central New Hebrides back-arc west-dipping thrust zone (from Collot 1989); (3) western boundary (projected) of the active thrust belt; (4) western and eastern rifts of Ambrym; (5) regional compressive stresses through the crust and Ambrym island. B and M abbreviations correspond to the Benbow and Marum active cones in the Ambrym caldera.

the DEZ and the arc. The underlying west-dipping thrust zone in the crust does not appear to have influenced eruption from the volcano, as the western and eastern rifts have produced identical volumes of compositionally similar basaltic lava.

Characteristics of the Ambrym volcano.

The summit of Ambrym volcano (1270 m) is about 1800 m above the surrounding seafloor (Chase & Seekins 1988). It is a 35 km × 50 km stratovolcano (MacFarlane 1976; Fig. 3) constructed in four distinct parts (Monzier *et al.* 1991; Robin *et al.* 1991, 1993): (i) well-preserved remnants of an old edifice in the north; (ii) a N100°-aligned, oval-shaped basal Hawaiian-type shield volcano; (iii) a pyroclastic cone (the Ambrym Pyroclastic Series or APS: Monzier *et al.* 1991; Robin *et al.* 1993), 24 km in diameter,

cut by a concentric 12 km wide caldera (McCall *et al.* 1970); and (iv) post-caldera volcanic rocks, both intra- and extra-caldera.

The old edifice comprises three cones (Tuvio, Vetlam and Dalahum) aligned along a N10° direction in the northern part of the island, and is composed of subaerial basaltic lava flows and pyroclastic deposits. The lavas are glomeroporphyritic basalts with aggregates of plagioclase (An_{82} cores to An_{55} rims), clinopyroxene (salite and augite), olivine (Fo_{83-79}) and Fe–Ti oxides in a microlitic plagioclase- and pyroxene-rich groundmass.

The gently-dipping flanks (2–3°) of the basal shield volcano are exposed only along the periphery of the island, and its oval shape may reflect incipient activity along N100° rifts. It is principally composed of low-viscosity (pahoehoe type) basaltic flows which typically have a microlitic texture formed by isolated or glomeroporphyritic phenocrysts of plagioclase (An_{91} cores to An_{67} rims), clinopyroxene (salite and augite), olivine (Fo_{84} cores to Fo_{51} rims) and Fe–Ti oxides.

The basal shield volcano is overlain by the APS which consists of: (i) dacitic pyroclastic flow deposits related to plinian and phreatomagmatic eruptions (sequence 1, up to 60 m thick); (ii) well-bedded, sometimes well-sorted, basaltic vitric tuffs with intercalated agglomerates, and ash flow deposits related to hydromagmatic (Surtseyan-type) eruptions and minor Plinian eruptions (sequence 2, which represents the major part of the tuff cone); (iii) basaltic ash flow deposits, essentially composed of highly to extremely vesicular (60–90%) droplets and basaltic pumiceous lapilli formed during a Plinian eruption (sequence 3, 10–25 m thick); and (iv) Strombolian basaltic deposits on the caldera rim (sequence 4, up to 250 m thick around the Woosantapaliplip vent, Fig. 3, Robin *et al.* 1993). Plagioclase and clinopyroxene phenocrysts are abundant and form up to 50 % of some beds as coarse ash or lapilli (sequence 2). The phenocrysts observed within the basaltic to basic-andesitic glasses of the APS are compositionally homogeneous: bytownite–anorthite (An_{80-91}), augite–salite (Wo_{43-44}; $En_{40.5-44}$; Fs_{12-16}), olivine (Fo_{72-80}) and Fe–Ti oxides (4–6.5% TiO_2). The dacitic glasses include rare Fe-olivine microphenocrysts (Fo_{42-44}), two clinopyroxenes (Wo_{41-42}; En_{34-36}; Fs_{21-24}; and Wo_{43-45}; En_{36-40}; Fs_{16-19}), titanomagnetite (12–14% TiO_2) and andesine with reverse zonation (An_{34-45}) in a vitric glass with a few andesine microlites (An_{35}). The thickness of the APS is estimated to be 200–450 m near the caldera edge, which represents a calculated volume of 60–80 km³, which

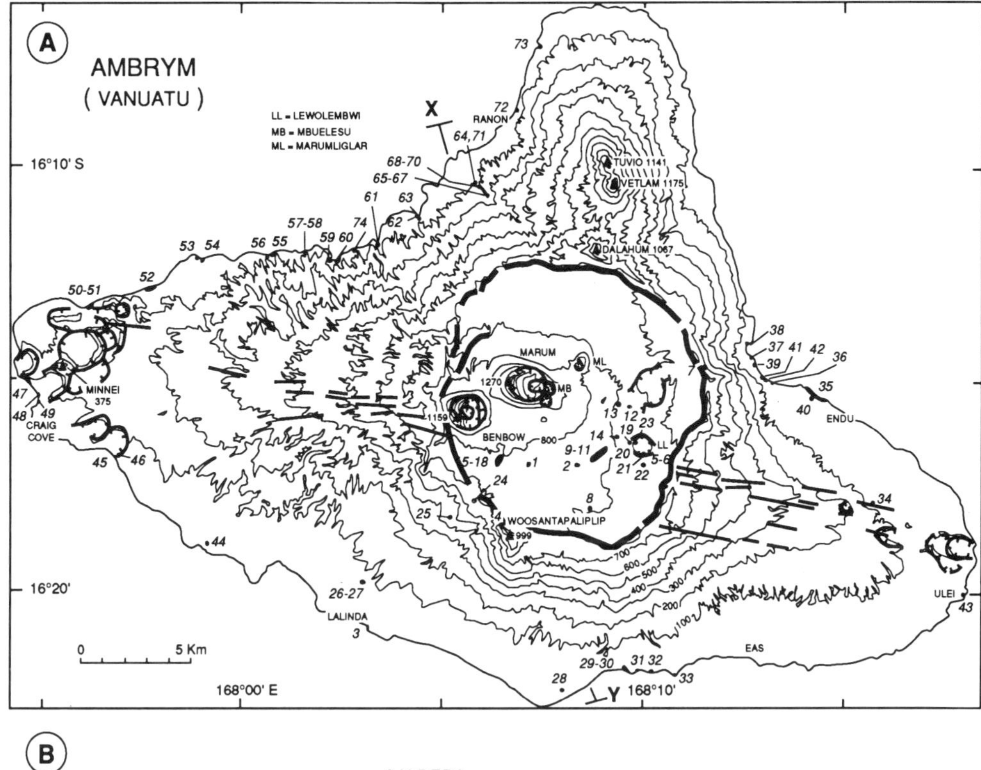

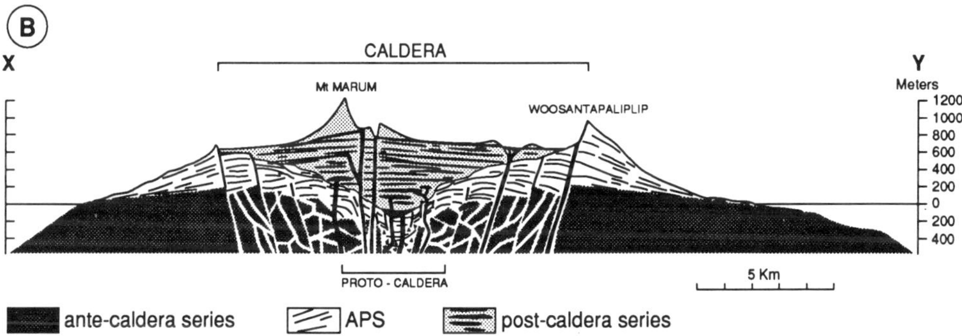

Fig. 3. (**A**) Topographical map of Ambrym island with sample locations (numbered). The main volcanic features (caldera, cones, maars, fissures) are shown in bold lines. Contour interval = 100 m. LL, Lewolembwi maar. (**B**) Composite section of the Ambrym volcano after Robin *et al.* (1993) showing the basal shield volcano (dark grey), tuff cone (white) and the post-caldera volcanic sequences (light grey). Vertical exaggeration ×3. APS, Ambrym Pyroclastic Series.

is more than 20 km³ of dense rock equivalent (Robin *et al.* 1993).

The caldera, which cuts the APS, is almost circular with a continuous scarp a few tens of metres to 450 m high. Interestingly, its centre lies on the southern extension of the older N10° volcanic lineation and the N100° rifts clearly transect south of this centre (Fig. 3). On the north side of the tuff cone, an unusually viscous and glassy andesitic lava flow, 20–25 m thick overlies the APS. This flow was extruded near the caldera edge and flowed as far as the NW coast, forming a large 2 km wide lava tongue. Near the vent, part of this flow remained in the caldera and was intersected by the last collapse, indicating a close relationship to the terminal phase of the syn-caldera episode.

Post-caldera activity, dominantly from Marum (1270 m) and Benbow (1160 m) cones in the western part of the caldera (Fig.3), occurs along N100° fissures inside and outside of the caldera. Episodic lava lakes in the Benbow and

C. PICARD ET AL.

Table 1. *Whole-rock analyses and average compositions of glasses (obtained by microprobe) from volcanic rocks of Ambrym*

N°	Eg.66 MK Bas [1]	Eg.M78 MK Bas [1]	Eg.63 MK Bas [1]	68 MK Bas	72 MK Bas	Eg.74 MK Bas [1]	60D MK Bas	61E MK Bas	43A MK Bas	33 MK Bas	50E MK Bas	32 MK Bas	31 MK Bas	53 HK Bas	50F HK Bas	42 HK Bas	36 HK Bas	48 HK Bas	35 HK Bas	55 MK And	59 MK And
	Tuvio and Vetlam basalts								Ante-caldera basalts												
SiO_2	49.12	49.16	49.29	49.24	48.78	48.50	50.95	47.17	49.90	50.51	50.79	49.98	50.07	51.29	51.23	52.39	53.02	52.64	52.72	55.19	55.46
TiO_2	0.69	0.62	0.73	0.67	0.80	0.78	0.61	0.23	0.86	0.79	0.95	0.69	0.75	0.88	0.89	0.97	0.99	0.92	0.85	0.88	0.90
Al_2O_3	13.53	14.06	14.30	13.97	14.84	15.25	16.84	22.98	16.72	17.86	15.55	18.94	18.34	16.93	17.32	16.73	16.68	18.02	18.88	16.09	16.15
FeO^*	9.44	9.03	9.54	8.85	9.69	10.06	9.29	6.32	12.00	10.09	12.70	9.52	10.31	9.87	10.53	10.10	9.68	8.51	7.93	10.35	10.06
MnO	0.17	0.17	0.17	0.17	0.19	0.18	0.19	0.12	0.23	0.19	0.24	0.18	0.20	0.20	0.20	0.19	0.18	0.16	0.15	0.23	0.23
MgO	12.79	12.24	12.04	11.68	10.22	9.85	6.30	6.20	5.07	4.95	4.89	4.71	4.58	5.24	4.51	4.46	4.42	4.28	4.01	3.73	3.57
CaO	11.40	11.91	10.79	13.16	12.57	11.74	12.54	15.37	11.15	11.29	10.14	12.13	11.78	10.63	10.53	9.35	8.79	9.46	9.26	8.44	8.34
Na_2O	1.96	1.99	2.17	1.74	2.07	2.83	2.22	1.38	2.52	2.53	2.98	2.40	2.43	2.77	2.86	3.08	3.26	3.20	3.21	3.46	3.41
K_2O	0.75	0.63	0.84	0.38	0.64	0.67	0.87	0.12	1.29	1.52	1.47	1.22	1.28	1.84	1.61	2.32	2.55	2.41	2.57	1.32	1.56
P_2O_5	0.15	0.18	0.12	0.13	0.19	0.14	0.20	0.10	0.26	0.28	0.28	0.23	0.25	0.34	0.32	0.41	0.43	0.41	0.42	0.30	0.30
LOI 1050°	−0.16	0.00	−0.09	−0.03	−0.29	0.14	0.02	0.10	−0.12	−0.11	0.32	0.02	−0.11	−0.11	0.29	−0.19	−0.18	0.04	−0.06	−0.26	−0.10
Rb	13.0	10.0	12.0	4.1	11.0	11.0	17.0	2.0	24.0	26.0	22.5	19.0	20.5	32.0	29.0	45.0	48.0	47.0	42.0	15.5	24.0
Ba	210.0	231.0	221.0	142.0	418.0	187.0	244.0	87.0	300.0	323.0	358.0	287.0	300.0	380.0	370.0	450.0	480.0	475.0	457.0	412.0	420.0
Nb	1.0	0.9	1.0	1.3	1.0	1.0	1.3	0.7	2.1	2.0	2.5	1.4	2.3	2.8	2.7	3.9	3.4	3.7	3.0	2.3	5.1
K	6236	5231	6999	3170	5344	5527	7189	1000	10670	12579	12204	10100	10664	15304	13325	19256	212087	19971	21342	10972	12989
La	5.7		4.7	5.0	4.2	4.1	5.8	3.2	8.8	9.0	9.4	7.7	8.1	10.5	9.8	12.0	12.8	13.0	11.7	10.4	23.5
Sr	334.0	366.0	360.0	280.0	304.0	332.0	410.0	430.0	540.0	624.0	528.0	578.0	565.0	620.0	600.0	630.0	638.0	710.0	745.0	400.0	398.0
P	664.5	785.6	531.9	570.1	834.0	616.2	878.8	437.9	1139.3	1234.3	1238.8	1017.6	1103.5	1486.5	1409.7	1775.8	1858.0	1800.9	1840.6	1320.8	1313.1
Nd	8.2		8.0	7.0	7.5	8.6	9.0	3.0	12.5	13.0	14.5	11.0	11.5	14.0	14.0	17.0	17.5	16.5	15.0	15.0	27.5
Zr	39.0	42.0	42.0	36.0	44.0	40.0	45.0	8.5	59.0	64.0	68.0	51.0	55.0	76.0	70.0	85.0	93.0	87.0	82.0	79.0	139.0
Eu				0.8	0.9	0.9	0.8	0.4	1.1	1.0	1.3	1.0	1.0	1.1	1.1	1.3	1.3	1.1	1.2	1.4	1.3
Ti	4138	3717	4384	4036	4824	4656	3682	1384	5177	4723	5713	4133	4487	5285	5326	5794	5956	5491	5117	5262	5411
Dy				2.6	2.9	3.2	3.1	1.1	3.3	3.3	3.5	2.7	2.9	3.3	3.5	3.6	3.5	3.3	3.1	4.5	5.0
Y	15.0	13.0	15.0	15.0	16.0	16.0	16.5	5.5	19.0	17.0	22.0	16.0	17.0	20.0	20.0	20.0	21.0	19.0	18.0	27.5	31.0
Er				1.6	1.6	2.0	1.6	0.7	2.0	1.6	2.1	1.5	1.6	2.2	1.9	2.2	2.1	1.8	1.9	2.9	3.5
Yb				1.4	1.6	1.6	1.6	0.7	1.8	1.6	2.0	1.4	1.5	1.8	1.8	1.8	1.9	1.8	1.6	2.6	2.9
V	289.0	250.0	308.0	245.0	279.0	280.0	310.0	144.0	370.0	312.0	420.0	284.0	323.0	340.0	370.0	365.0	360.0	320.0	297.0	248.0	245.0
Cr	801.0	801.0	547.0	620.0	381.0	324.0	198.0	83.0	16.0	15.0	5.5	65.0	18.3	59.0	20.0	16.0	21.0	70.0	69.0	14.0	15.0
Co				50.0	48.5		40.0	36.0	44.0	39.0	45.0	36.0	38.0	38.0	36.0	35.0	32.0	29.0	26.0	29.6	30.5
Ni	223.0	173.0	211.0	130.0	101.0	97.0	56.0	43.0	26.0	28.0	17.0	27.0	23.0	41.0	28.0	29.5	34.0	48.0	49.0	11.7	13.0
Sc	44.0	40.0	38.0	48.0	42.5	44.0	39.0	28.0	35.0	31.0	35.0	32.0	31.6	30.5	29.0	25.0	24.0	20.0	18.0	29.5	30.0
MgNo.	0.75	0.75	0.74	0.75	0.71	0.69	0.61	0.69	0.49	0.53	0.47	0.53	0.50	0.55	0.49	0.50	0.51	0.53	0.53	0.45	0.45
$(La/Yb)_n$				2.62	1.92	1.89	2.55	3.48	3.59	4.03	3.37	3.89	3.85	4.26	4.00	4.76	4.91	5.31	5.35	2.86	5.81
$(La/K)_n$	2.10		1.54	3.63	1.81	1.71	1.84		1.89	1.65	1.77	1.75	1.74	1.58	1.68	1.43	1.39	1.49	1.26	2.17	4.16
Normative composition																					
Qz	0.00	0.00	0.00	0.00	0.00	0.00	0.00	0.00	0.00	0.00	0.00	0.00	0.00	0.00	0.00	0.00	0.00	0.00	0.00	3.87	4.15
Or	4.43	3.72	4.98	2.25	3.78	3.93	5.11	0.71	7.58	8.94	8.67	7.18	7.58	10.88	9.47	13.69	15.08	14.20	15.18	7.80	9.23
Ab	16.56	16.83	18.37	14.69	17.50	18.28	18.72	11.67	21.29	21.34	25.18	20.31	20.51	23.37	24.14	26.05	27.59	27.06	27.16	29.24	28.82
An	25.88	27.55	26.74	29.15	29.28	26.89	33.40	56.16	30.46	32.88	24.67	37.25	35.30	28.31	29.66	24.94	23.28	27.66	29.47	24.44	24.10
Ne	0.00	0.00	0.00	0.00	0.00	3.07	0.00	0.00	0.00	0.00	0.00	0.00	0.00	0.00	0.00	0.00	0.00	0.00	0.00	0.00	0.00
Di-Wo	12.36	12.66	10.82	14.71	13.27	12.68	11.44	8.11	9.62	8.86	9.89	8.90	8.95	9.24	8.52	7.82	7.31	6.90	5.71	6.44	6.38
Di-En	8.09	8.28	6.96	9.57	8.14	7.60	6.00	4.80	4.09	4.07	4.02	4.09	3.90	4.42	3.68	3.45	3.29	3.26	2.70	2.53	2.50
Di-Fs	3.41	3.50	3.14	4.13	4.37	4.41	5.11	2.91	5.55	4.72	5.96	4.73	5.03	4.68	4.84	4.35	3.98	3.56	2.93	3.98	3.97
Hy-En	3.69	3.49	3.48	5.20	1.45	0.00	7.44	3.13	3.27	3.49	2.68	3.06	3.44	3.10	3.82	3.76	3.80	3.02	2.86	6.75	6.38
Hy-Fs	1.56	1.47	1.57	2.24	0.78	0.00	6.33	1.90	4.43	4.04	3.98	3.53	4.44	3.28	5.03	4.75	4.59	3.30	3.10	10.60	10.15
Ol-Fo	14.04	13.10	13.66	10.01	11.07	11.85	1.57	5.26	3.67	3.33	3.81	3.21	2.84	3.86	2.61	2.71	2.74	3.06	3.09	0.00	0.00
Ol-Fa	6.52	6.10	6.79	4.77	6.55	7.57	1.50	3.52	5.48	4.25	6.24	4.08	4.03	4.51	3.79	3.78	3.65	3.67	3.69	0.00	0.00
Mt	1.81	1.72	1.89	1.69	1.85	1.92	1.78	1.21	2.30	1.93	2.43	1.82	1.97	1.89	2.01	1.93	1.86	1.63	1.52	1.98	1.93
He	0.00	0.00	0.00	0.00	0.00	0.00	0.00	0.00	0.00	0.00	0.00	0.00	0.00	0.00	0.00	0.00	0.00	0.00	0.00	0.00	0.00
Il	1.31	1.18	1.39	1.28	1.52	1.47	1.17	0.44	1.64	1.49	1.80	1.31	1.42	1.67	1.68	1.83	1.88	1.74	1.62	1.66	1.71
Ap	0.36	0.43	0.29	0.31	0.45	0.33	0.48	0.24	0.62	0.67	0.67	0.55	0.60	0.81	0.76	0.96	1.01	0.98	1.00	0.72	0.71

Table 1. *Continued*

		Pyroclastites: APS sequence 1									Pyroclastites: APS sequence 2							
N°	57C And MK	60A And MK	60B1 Dac MK	60C Dac MK	70 Dac Mk	60C Dac MK Ave. n=8	29A2 MK Bas	29A1 MK Bas Ave. n=22	50 HK Bas Ave. n=11	39 HK Bas Ave. n=13	39B HK Bas	39 MK And Ave. n=2	50 MK And Ave. n=3	60I2 MK And	60H MK And	39 HK And Ave. n=6	39A HK And	39 HK Dac Ave. n=18
SiO_2	58.35	60.81	66.56	64.30	65.77	68.52	50.39	51.28	51.35	52.00	51.26	54.35	54.50	54.57	54.89	54.98	56.07	68.39
TiO_2	0.72	0.62	0.43	0.55	0.50	0.33	0.73	0.79	0.94	0.97	0.93	1.09	1.25	0.80	0.87	0.93	0.78	0.41
Al_2O_3	15.44	15.48	14.99	14.94	15.84	15.19	16.20	15.68	15.28	16.05	16.42	15.39	14.11	16.56	16.17	16.80	16.13	15.67
FeO^*	9.51	8.09	5.34	6.92	5.04	4.30	11.59	11.43	12.41	11.21	11.54	11.14	11.90	10.54	10.40	9.92	9.02	3.78
MnO	0.23	0.21	0.16	0.21	0.17	0.10	0.23	0.23	0.27	0.22	0.22	0.28	0.26	0.22	0.22	0.21	0.19	0.13
MgO	3.02	2.49	1.11	1.38	0.91	0.22	5.98	5.88	5.14	5.03	4.96	4.65	4.65	3.93	3.91	3.65	3.67	0.53
CaO	7.04	6.26	3.16	4.41	2.92	2.88	11.37	11.19	10.28	9.79	9.65	8.70	9.32	8.65	8.63	8.15	7.54	1.75
Na_2O	3.64	3.78	5.48	4.43	4.66	5.08	2.31	2.52	2.67	2.68	2.87	3.40	2.95	3.18	3.24	2.84	3.35	3.87
K_2O	1.83	2.07	2.61	2.68	3.99	3.38	0.98	1.01	1.66	2.05	1.81	1.00	1.07	1.29	1.31	2.51	2.96	5.48
P_2O_5	0.21	0.19	0.16	0.19	0.20	–	0.21	–	–	–	0.33	–	–	0.25	0.36	–	0.29	–
LOI 1050°	0.04	0.26	4.34	0.54	0.08	–	2.47	–	–	–	2.29	–	–	−0.03	0.70	–	1.53	–
Rb	32.0	38.0	29.0	51.0	76.0		18.0				31.0			24.0	22.6		54.0	
Ba	525.0	588.0	635.0	725.0	800.0		250.0				358.0			390.0	394.0		522.0	
Nb	3.2	3.2	3.8	3.8	5.5		2.0				2.7			2.0	2.4		4.5	
K	15201	17193	21661	22279	33147		8161				15015			10719	10842		24543	
La	12.3	13.1	16.0	15.8	19.0		7.0				10.3			8.3	9.0		13.5	
Sr	360.0	335.0	255.0	288.0	310.0		445.0				572.0			418.0	416.0		488.0	
P	922.0	841.7	697.1	839.6	877.8		903.1				1443.3			1100.5	1558.3		1245.6	
Nd	17.0	18.0	22.0	19.0	23.0		9.5				15.0			12.0	13.0		17.5	
Zr	105.0	114.0	120.0	142.0	182.0		43.0				69.0			68.0	71.0		112.0	
Eu	1.2	1.2	1.4	1.3	1.2		0.9				1.3			1.2	0.9		1.1	
Ti	4342	3712	2554	3278	3015		4404				5576			4777	5199		4706	
Dy	5.3	5.3	5.7	6.1	5.0		2.9				3.4			3.9	4.2		4.0	
Y	33.5	34.0	32.0	41.0	31.0		17.2				20.5			23.5	24.5		24.0	
Er	3.3	3.5	3.6	4.1	3.2		1.7				1.7			2.9	3.0		2.0	
Yb	3.4	3.5	3.2	4.1	3.3		1.6				1.8			2.5	2.6		2.3	
V	260.0	167.0	29.5	70.0	24.0		325.0				330.0			299.0	244.0		245.0	
Cr	2.0	14.0	1.6	31.0	2.0		204.0				155.0			5.0	14.0		59.0	
Co	26.0	20.0	6.0	12.0	8.0		43.0				41.0			32.0	38.0		29.0	
Ni	9.0	14.0	4.5	17.0	3.0		106.0				100.0			13.0	11.5		20.0	
Sc	26.5	21.7	12.0	17.0	10.5		37.0				29.0			29.0	29.0		22.5	
Mg#	0.42	0.41	0.32	0.31	0.29	0.10	0.54	0.54	0.48	0.50	0.49	0.49	0.47	0.46	0.46	0.46	0.48	0.24
$(La/Yb)_n$	2.59	2.71	3.59	2.78	4.19		3.24				4.15			2.36	2.50		4.27	
$(La/K)_n$	1.86	1.75	1.70	1.63	1.32		1.97				1.57			1.78	1.90		1.26	
Normative composition																		
Qz	7.81	11.34	14.51	14.45	13.70	18.09	0.00	0.00	0.00	0.00	0.00	2.30	4.07	3.67	4.20	2.79	1.95	17.90
Or	10.81	12.23	15.41	15.84	23.58	19.92	5.80	5.97	9.80	12.12	10.67	5.89	6.33	7.62	7.71	14.83	17.45	32.37
Ab	30.78	31.92	46.38	37.41	39.38	42.88	19.49	21.30	22.53	22.67	24.28	28.70	24.95	26.85	27.42	24.03	28.35	32.80
An	20.36	19.15	8.58	12.96	10.52	8.64	30.91	28.44	24.80	25.66	26.53	23.77	22.04	27.08	25.67	25.63	20.19	8.66
Ne	0.00	0.00	0.00	0.00	0.00	0.00	0.00	0.00	0.00	0.00	0.00	0.00	0.00	0.00	0.00	0.00	0.00	0.00
Di-Wo	5.49	4.44	2.53	3.18	1.10	2.35	10.05	11.29	10.92	9.54	7.99	8.07	10.07	5.91	6.16	6.15	6.40	0.00
Di-En	2.00	1.58	0.70	0.86	0.28	0.22	4.69	5.27	4.60	4.22	3.45	3.45	4.19	2.36	2.48	2.47	2.69	0.00
Di-Fs	3.61	2.96	1.95	2.49	0.88	2.39	5.26	5.89	6.35	5.29	4.54	4.63	5.93	3.61	3.74	3.74	3.73	0.00
Hy-En	5.51	4.60	2.05	2.57	1.97	0.34	6.77	6.40	3.94	4.61	4.22	8.11	7.37	7.43	7.24	6.64	6.44	1.31
Hy-Fs	9.95	8.61	5.69	7.43	6.21	3.72	7.59	7.16	5.43	5.79	5.57	10.88	10.42	11.37	10.92	10.06	8.93	5.29
Ol-Fo	0.00	0.00	0.00	0.00	0.00	0.00	2.39	2.06	2.97	2.57	3.27	0.00	0.00	0.00	0.00	0.00	0.00	0.00
Ol-Fa	0.00	0.00	0.00	0.00	0.00	0.00	2.96	2.54	4.51	3.55	4.75	0.00	0.00	0.00	0.00	0.00	0.00	0.00
Mt	1.82	1.55	1.02	1.32	0.96	0.82	2.22	2.19	2.37	2.14	2.21	2.13	2.28	2.02	1.99	1.90	1.73	0.72
He	0.00	0.00	0.00	0.00	0.00	0.00	0.00	0.00	0.00	0.00	0.00	0.00	0.00	0.00	0.00	0.00	0.00	0.00
Il	1.37	1.17	0.81	1.04	0.95	0.63	1.39	1.49	1.79	1.84	1.76	2.06	2.36	1.51	1.64	1.76	1.49	0.77
Ap	0.50	0.46	0.38	0.46	0.48	0.00	0.49	0.00	0.00	0.00	0.78	0.00	0.00	0.60	0.84	0.00	0.67	0.00

Marum craters drained out as basaltic lava flows onto the caldera floor, as occurred in 1988–1989. Magma was also frequently ejected as ash and scoria, mantling the caldera and the western part of the volcano. In the eastern part of the caldera, a recent maar (Lewolembwi crater), located on a N100° fracture, is surrounded by a 2 km wide tuff ring, and older coalescent structures of the same type are infilled by 1986 lava flows. Extra-caldera activity is dominated by lava flows generated along the N100° rifts, and recent maars are present at both extremities of the island, where eruptions interacted with sea-water. During the last two centuries, Ambrym has had a large number of eruptions, often with extra-caldera flows, produced by either over-flowing the caldera margins or by rift activity. These have occasionally been quite destructive (1888, 1894, 1913–14, 1929, 1937, 1942, 1952–53, etc., Fisher 1957; Williams & Warden 1964; Gèze 1966), the latest being in 1986 and 1988–89 (Eissen *et al.* 1991). In 1990–92, no lava flows were erupted and the volcanic activity was moderately explosive and strongly fumarolic, and concentrated in the Mbuelesu area (Fig. 3). Lava flows are of aa or pahoehoe types and they are principally composed of plagioclase-rich basalts with glomeroporphyritic aggregates of zoned plagioclase (An_{92} to An_{65}), Fe–Ti oxides $\pm$ clinopyroxene (salite–augite) and olivine (Fo_{78} to Fo_{55}) in a plagioclase (An_{70-60})- rich microlitic groundmass. Flows of olivine-rich basalts have been observed locally near the floor of the Lewolembwi maar or as widely distributed blocks. The olivine-rich basalts contain olivine (Fo_{79} to Fo_{58}), diopside–augite and rare plagioclase phenocrysts (An_{83} cores to An_{66} rims), in olivine (Fo_{42-45}) and clinopyroxene-rich micro-litic groundmass.

Geochemistry

Sampling and analytical techniques

64 new ICP whole-rock analyses (major, trace and rare earth elements) were carried out for this study (Table 1; Fig. 3): 4 samples from ankaramitic lava flows of the Tuvio-Vetlam-Dalahum suite; 13 samples from coastal lava flows of the pre-caldera shield; 13 samples from juvenile glasses of the APS; and 34 analyses from post-caldera lava-flows. 12 average compositions of the APS vitric clasts obtained from 105 new microprobe analyses and 4 whole-rock analyses of Eggins (unpublished data) from the Tuvio high-MgO basalts have also been selected for this study (Table 1).

Selected rock fragments were ground in agate. Powders were digested with a concentrated acid

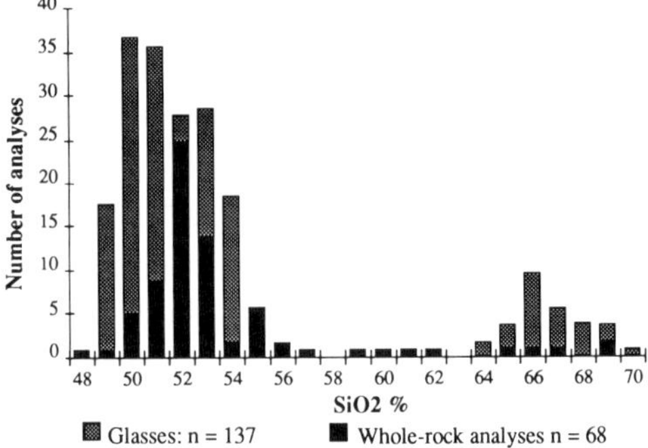

Fig. 4. SiO_2 histogram of 137 microprobe analyses of vitric clasts from the syn-caldera Ambrym Pyroclastic Series (APS) and 68 whole-rock analyses from the entire volcano. Class interval: 1 %.

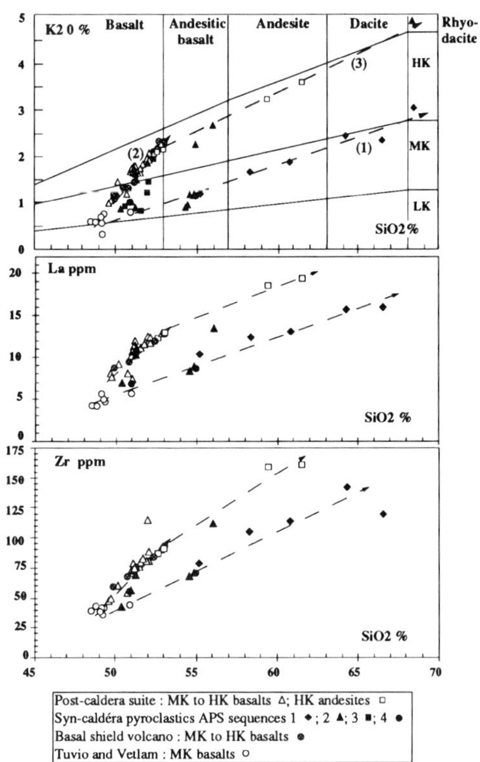

Fig. 5. K_2O, La and Zr versus SiO_2 distribution for the Ambrym volcanic rocks. Major elements are expressed in wt% summed to 100% volatile free.

mixture (1:8 HNO_3:HF). Fluorides were then dissolved and HF neutralized with an H_3BO_3 solution. International standards (JB2, BEN, ACE, GSN, MICA-Fe) were used for calibration. Rb determination was performed by flame Atomic Emission Spectrometry. All other elements were determined by

Inductively Coupled Plasma Emission Spectrometry. Relative standard deviation for major oxides: SiO_2 = 1%; TiO_2 = 3%; Al_2O_3 = 2%; Fe_2O_3 = 2%; MnO = 5%; MgO = 2%; CaO = 2%; Na_2O = 3%; K_2O = 3%; P_2O_5 = 4%. Limits of detection and relative standard deviations for trace elements are as follows: Rb = 1 ppm, 5%; Sr = 0.2 ppm, 5%; Ba = 2 ppm, 5%; Sc = 0.25 ppm, 5%; V = 2 ppm, 5%; Cr = 1 ppm, 5%; Co = 1 ppm, 5 to 10%; Ni = 2 ppm, 5%; Zr = 1 ppm, 5%; Nb = 0.8 ppm, 5 to 10%; Y = 0.5 ppm, 5%; La = 0.8 ppm, 5%; Nd = 2 ppm, 10%; Eu = 0.2 ppm, 5 to 10%; Dy = 0.4 ppm, 5%; Er = 0.8 ppm, 10%; Yb = 0.2 ppm, 5%.

Results

A histogram of SiO_2 content (Fig. 4) for all of the analysed Ambrym samples clearly shows a bimodal magmatic suite, with maxima at 50–53 % (basalts and andesitic basalts) and 66% SiO_2 (dacites). Basaltic compositions are the most abundant (Table 1), and thus Ambrym is principally a basaltic volcano. CIPW norm calculations (Table 1) show that all samples are olivine or quartz normative, and all have a sub-alkaline affinity. On a K_2O v. SiO_2 diagram (Fig. 5), most of the basaltic samples range from MK to HK basalts, whereas the more differentiated samples form two separate MK and HK trends which evolve from basaltic to rhyodacitic compositions. These trends also appear in La v. SiO_2 and Zr v. SiO_2 diagrams (Fig. 5). Al_2O_3 varies from 13 to 17 %, except in some plagioclase-rich samples (cumulates) having 17–19% Al_2O_3 (Fig. 6). MgO decreases rapidly (from 13 to 5%, Fig. 6) in the MK basalts of the Tuvio edifice and in the olivine porphyritic basalts of the post-caldera suite, suggesting important olivine fractionation. In the more evolved samples, the MgO contents decrease more slowly and reveal a second fractionation trend.

FeO and TiO_2 contents increase initially (from 8 to 12% for FeO^t; 0.75 to 1.5 % for TiO_2, Fig. 6), then decrease in the andesitic to dacitic samples indicating a tholeiitic affinity for the Ambrym volcanic suite, although the K_2O behavior rather suggests a calc-alkaline affinity. In N-MORB normalized spider-diagrams (Fig. 7), the basalts appear very similar to N-MORB for the heavy rare earth elements (HREE), Zr and Y, while the light rare earth elements (LREE), Rb, Ba, K, Sr and P are more enriched suggesting an E-MORB or calc-alkaline affinity. These volcanic rocks also show the strong relative Nb depletion which characterizes island arc series. Thus Ambrym volcanic rocks are geochemically intermediate between tholeiitic

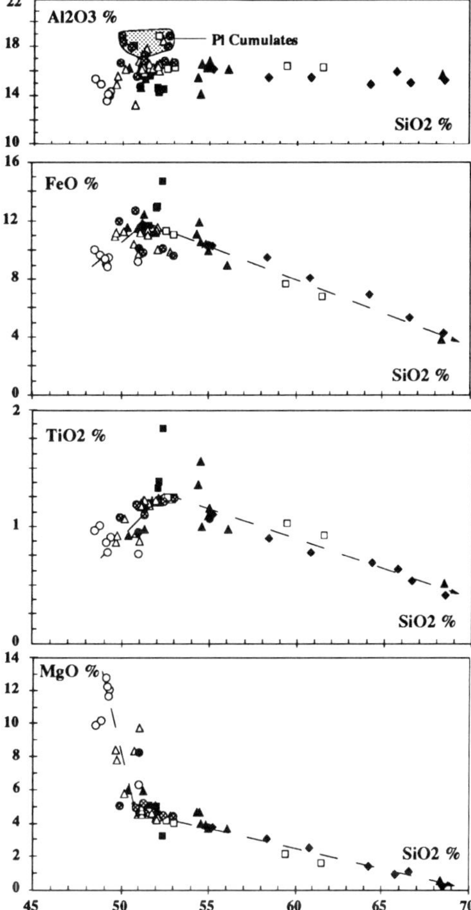

Fig. 6. Al_2O_3, FeO^t, TiO_2 and MgO versus SiO_2 distribution for the Ambrym volcanic rocks. Major elements are expressed in wt % summed to 100% volatile free. FeO^t = total iron as FeO. Same legend as Fig. 5.

and calc-alkaline series, attesting to the difficulty in classifying arc magma (Gill 1981).

The Tuvio and Vetlam volcanic rocks range from LK to MK basalts (K_2O = 0.38 to 0.87 %), show the highest MgO, Cr and Ni contents of the Ambrym volcano (MgO = 9.85 to 12.79 %, Cr = 324 to 801 ppm; Ni = 97 to 223 ppm), the lowest Zr, Y and Nb (Zr = 36 to 45 ppm, Y = 13 to 17 ppm; Nb = 0.9 to 1.3 ppm) contents and have high Mg-numbers ($Mg^{2+}/[Fe_t^{2+} + Mg^{2+}]$ = 0.69 to 0.75). These high-MgO basalts represent the most primitive magmas sampled on Ambrym.

The plagioclase glomeroporphyritic basalts of the basal volcano range from MK to HK basalts, are more evolved with lower MgO, Cr and Ni contents (MgO = 3.57–5.24%, Cr = 5.5–70

Fig. 7. N-MORB normalized incompatible element abundance patterns of Ambrym volcanic rocks. Elements are ordered in a sequence of decreasing incompatibility, from left to right, in oceanic basalts; order and concentrations in N-type MORB are from Sun & McDonough (1989). For comparison, the field covered by all samples (in grey) is reproduced on each diagram.

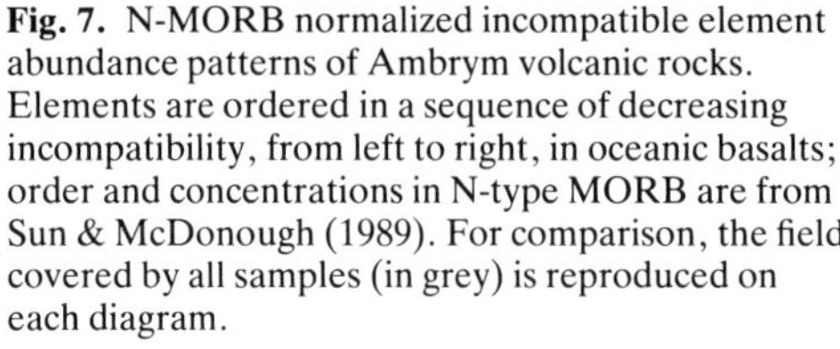

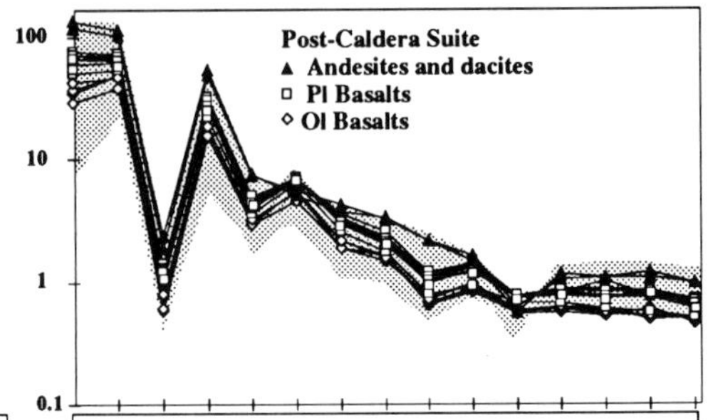

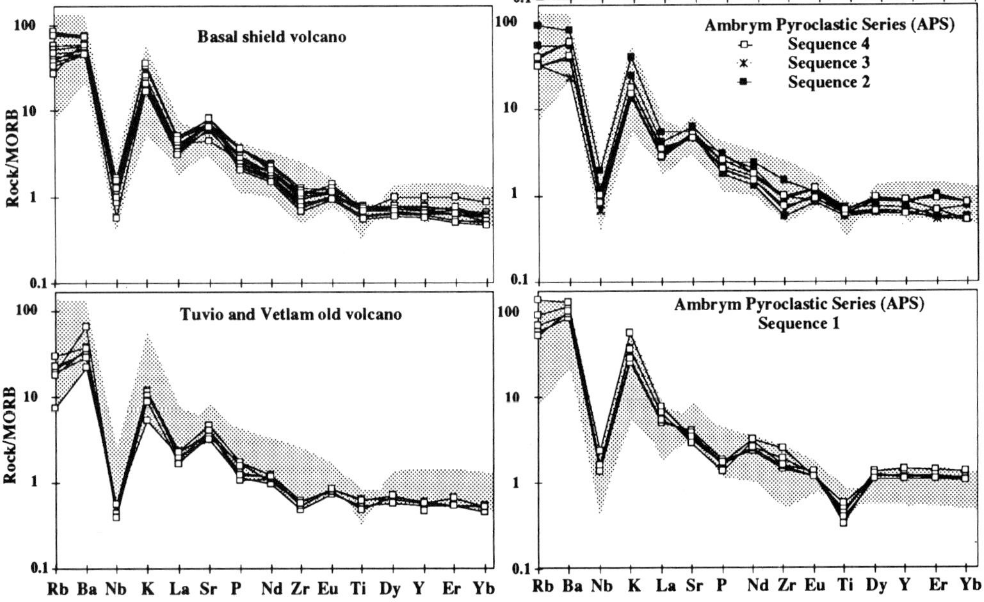

ppm; Ni = 12–49 ppm), lower Mg-number (0.45–0.55) and higher Zr, Y and Nb contents (Zr = 51–87 ppm, Y = 16–27 ppm; Nb = 1.4–3.9 ppm).

Samples of the lower APS (sequence 1) vary from MK andesites to MK rhyodacites. They represent evolved compositions with low MgO, Cr and Ni contents (Table 1), low Mg-number (0.42–0.1) and high Zr contents (Zr = 105–182 ppm). They have low [La/Yb]n values (2.59–4.19), fairly similar to those of the Tuvio-Vetlam volcano (1.89–3.48). Pyroclastic rocks of sequence 2 include MK to HK basaltic compositions and two magmatic trends: one from MK basalt to MK andesite; and another from HK basalt to HK rhyodacite. Sequence 3 is composed of MK to HK basaltic pyroclastics and sequence 4 evolves from MK basaltic to MK andesitic compositions. All these evolved volcanic rocks have low MgO, Cr and Ni contents (Table 1), and low Mg-number (0.54–0.24). In detail, the individual pyroclastic sequences are very heterogeneous, containing vitric clasts which vary from basaltic to rhyodacitic compositions in a single stratum (e.g. sample 39, Table 1). These observations indicate that the APS magmas have formed by mechanical mixing of a range of magmas from the MK and HK series, spanning basalt, andesite and rhyodacite compositions.

Most of the lava and pyroclastic rocks of the post-caldera suite evolved from MK to HK basaltic compositions (K_2O = 1.08–2.49%) in similar fashion to the pre-caldera glomeroporphyritic basalts. Only a few vitric samples from the 1986 lava flow and around the Lewolembwi maar range from HK basalt to HK andesite, indicating the presence of a HK differentiated magma chamber below the eastern part of the caldera. The more primitive olivine and pyroxene porphyritic basalts are exclusively MK compositions. They have relatively high MgO, Cr and Ni contents (MgO = 5.74–9.73 %, Cr = 51–342 ppm; Ni = 35–180 ppm), relatively high Mg-number (0.54–0.69) and low Zr, Y and Nb contents (Zr = 48–60

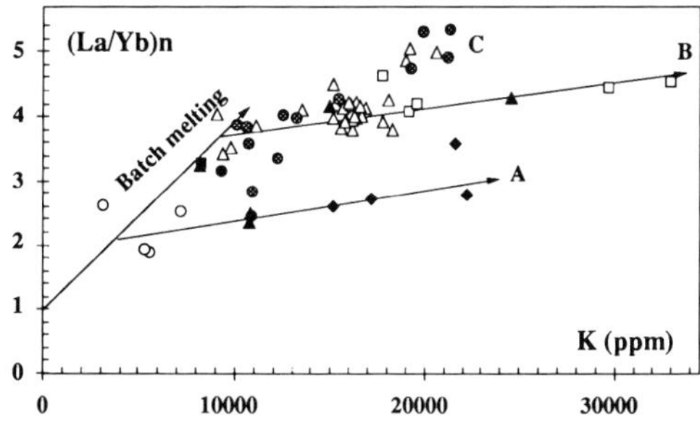

Fig. 8. [La/Yb]$_n$ *versus* K diagram for the Ambrym volcanic rocks. Trends A, B and C correspond to: (**A**) Tuvio–Vetlam–Dalahum high-MgO MK basalts and evolved MK andesites, dacites and rhyodacites of the APS sequence 1 (lowest values of [La/Yb]$_n$); (**B**) MK and HK basalts from the basal shield volcano and from the post-caldera suites, MK – HK basaltic glasses of the APS, HK andesites–rhyodacites of the APS sequence 2, and HK andesites from the Lewolembwi area; and (**C**) several HK basalts from the basal shield volcano and from the post-caldera suites (highest values of [La/Yb]$_n$). Same legend as Fig. 5.

ppm, Y = 15–17 ppm; Nb = 1.4–2.4 ppm). The plagioclase porphyritic basalts of this suite are more evolved, with HK compositions, lower MgO, Cr and Ni contents, higher Zr, Y and Nb contents and low Mg-number (0.45–0.53, Table 1).

Discussion

Evidence for three magma evolution trends and their temporal relationships

Considering the K$_2$O, La and Zr variations versus SiO$_2$ (Fig. 5), three major trends characterize the geochemistry of Ambrym; one MK to HK basaltic trend, and two more evolved trends, respectively MK and HK, from basaltic to dacitic and rhyodacitic compositions. Basaltic volcanic rocks are present in all parts of the volcano, while the more evolved andesitic, dacitic and rhyodacitic rocks form pyroclastic sequences in the APS and locally around the Lewolembwi maar. Thus, the volcano shows several magmatic phases, well illustrated by the K$_2$O and SiO$_2$ variations: (1) older MK basalts forming the Tuvio-Vetlam-Dalahum edifice; (2) MK to HK basalts forming the basal shield volcano; (3) MK andesites and rhyodacites forming the first pyroclastic sequence of the APS, which probably initiated the formation of the caldera; (4) MK to HK basalts and andesites forming Surtseyan then Strombolian pyroclastic

sequences 2, 3 and 4 of the APS; and (5) post-caldera MK to HK basalts forming the recent olivine porphyritic and plagioclase-porphyritic suites, locally associated with more evolved HK andesitic volcanic rocks in the eastern part of the caldera.

In agreement with previous observations, the volcanic rocks form different trends in the [La/Yb]n v. K diagram (Fig. 8). Interestingly, the Tuvio–Vetlam–Dalahum high-MgO MK basalts and the evolved MK andesites, dacites and rhyodacites of the APS sequence 1 fall on the same lowest [La/Yb]n trend (trend A). Most MK to HK basalts from the basal shield volcano and from the post-caldera suites form the second trend (B) which includes MK to HK basaltic glasses and HK andesitic to rhyodacitic pyroclastic rocks of the APS sequence 2, and HK andesites from the Lewolembwi area. Finally, some HK basalts from the basal shield volcano and from the post-caldera suites form a third group (C), characterized by the highest [La/Yb]n.

Genesis of the magmas and origin of the K and La contents

The [La/Yb]n variations defining trends A and B in the more primitive volcanic rocks ([La/Yb]n = 2 to 4, Fig. 8) are principally due to variations in La (La = 4.1–7.4 ppm), whereas Yb remains relatively constant (Yb = 1.4–1.7 ppm). This reflects probable changes in the conditions of melting and possible source effects. On the other hand, the significant increase of K contents over a limited range of [La/Yb]n values illustrates the role of fractionation processes in the magmatic evolution. These processes are discussed below in order to explain the chemical characteristics of the analysed samples.

Primary magmas. The clinopyroxene-rich and olivine-rich high-MgO MK basalts of the Tuvio–Vetlam–Dalahum cones (MgO = 12.79% in Eg.66; 10.22% in Amb72; Table 1), and the olivine-rich basalts of the post-caldera suite (MgO = 9.73% in Amb 27; Table 1) represent the most magnesian lavas sampled from Ambrym volcano. Using relationships between mole % of MgO and FeO at 1 atmosphere pressure (Roeder & Emslie 1970), these highest-MgO whole-rock compositions imply that olivine equilibrium compositions should be Fo$_{89}$ for the Tuvio–Vetlam–Dalahum and Fo$_{85}$ for the post-caldera suite. Actually, the most magnesian olivine crystals are Fo$_{83}$ in the Tuvio-Vetlam-Dalahum edifice, Fo$_{84}$ in the shield volcano, and Fo$_{79}$ in the post-caldera suite,

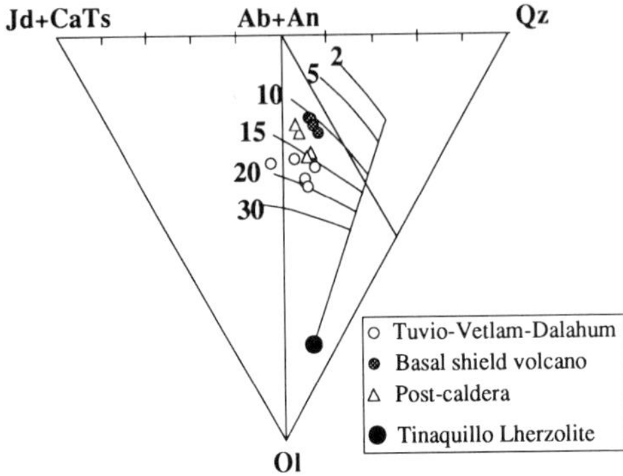

Fig. 9. Plot of the more primitive basalts from the Ambrym volcano on Ol–Jd + CaTs–Qz molecular normative projection following the method of Jaques & Green (1980), Falloon & Green (1988) and Falloon *et al.* (1988). Fine lines are cotectic grids for the Tinaquillo lherzolite after Falloon & Green (1988) and Falloon *et al.* (1988). Numerals indicate the cotectic pressures in kbar.

which reveals a lower MgO content in the equilibrium liquid and a possible effect of cumulates. However, Eggins (1989) reported higher-magnesian olivine phenocrysts (Fo93) in the shield volcano suggesting that the parental magmas had a picritic composition (13–15% MgO, Eggins, unpublished data). Thus, we may conclude that the Ambrym primary magmas were probably high-MgO basaltic liquids with a minimum of 10–11 % MgO.

Sources and melting processes. The flat HREE patterns of the more primitive lavas (Fig. 7) suggest that garnet was not residual during melting and that melting probably occurred at relatively shallow levels in the mantle. In order to estimate depth of melting, we have plotted the most primitive basalts on the Ol–Jd + CaTs–Qz molecular normative projection (Fig. 9) following the method of Jaques & Green (1980), Falloon & Green (1988) and Falloon *et al.* (1988). Thus, the most primitive basalts observed in the Tuvio–Vetlam–Dalahum suite appear to have last equilibrated at pressures between 20 and 15 kbar (60 to 45 km below the surface) during first stages of crystallization, while the most primitive basalts from the chronologically later basal shield volcano and from the post-caldera suites indicate equilibrium pressures between 15 and 10 kbar (45 to 30 km depth). These data suggest that the Tuvio–Vetlam–Dalahum primary melts segregated at a minimum depth of 60 km in the spinel lherzolite zone while the more recent primitive liquids may have segregated at shallower levels (about 45 km), suggesting possible uplift of the mantle source related to the DEZ collision, N100° rifting and doming of the volcano. Such a hypothesis may be supported by the increasingly Fe-rich compositions of the most magnesian olivines in successive volcanic suites of the volcano and by MgO decreasing in the more primitive volcanic rocks (Jaques & Green 1980). If true, such a model could also account for variations of K and La in the parental magmas and therefore some variations of (La/Yn)n ratios (Fig. 8). However, these variations, principally related to variations in La (Yb remains relatively constant), may be also caused by incremental batch melting or fractional melting of the mantle. Such models are strongly supported by the frequent recurrence of primitive basalts in the APS and post-caldera series, which suggest new influxes of parental magmas probably resulting from renewed melting in the mantle. According to Watson & McKenzie (1991) and McKenzie & O'Nions (1991), who showed that >1% melts cannot remain in contact with their source region, such models seem more geologically valid than a simple batch melting model. Nevertheless, using the simple equilibrium batch melting model (Hanson & Langmuir 1978; Steinberg *et al.* 1979; Hanson 1980) and the Zr contents of the high-MgO olivine-rich basalts (samples M78 and Amb 67, Tables 1–2), we calculate that the the the primary melts of the Tuvio–Vetlam–Dalahum and post-caldera suites required 25 and 21% melting, respectively, from an appropriate mantle source.

Source effects. (LaYb)n variations, (La/K)n ratio (1.3–2.2; Table 1) and the observations given above are consistent with variable degrees of melting during possible uplift of a single spinel lherzolite source. However, K and La contents of the more primitive MK basalts (K = 5231 ppm and La = 4.1 ppm in the Tuvio samples; K = 9065 ppm and La = 7.4 ppm in the olivine porphyritic basalts of the post-caldera suite) indicate that degrees of melting are respectively 9% for potassium and 16% for lanthanum (calculated using the simplified equation of melting, C_o/C_i, Kd_{K-La} <0.1 and the mantle composition of Sun & McDonough 1989). These differences with the Zr calculations indicate that the mantle source may have been previously enriched in K relative to La, and in these elements relative to Zr, suggesting possible mantle metasomatism. Assessment of the relative roles of source effect and possible mantle metasomatism or crustal contamination on the K and La variations require further investigation.

Table 2. *melting model using the equilibrium batch melting equation F = [(Co/Cl) − D]/(D −1) of Hanson (1980) and Zr bulk partition coefficient of Pearce & Norry (1979) for the most primitive olivine-rich basalts of the Tuvio–Vetlam–Dalahum and post-caldera suites (sample M78 and Amb 67, Table 1) and for a spinel lherzolite source (Wilson 1989)*

	Olivine	Opx	Cpx	Spinel
Zr bulk partition coefficient	0.01	0.03	0.1	0.1
Source: spinel lherzolite	66% Ol + 24% Opx + 8% Cpx + 2% Sp			
	D(Zr) = 0.0238			
Tuvio olivine porhyritic basalts:				
Higher MgO basalt: sample M78*	Zr = 42.0 ppm	F =	24.88%	
Post-caldera olivine porhyritic basalts:	Zr = 48 ppm	F =	21.46%	
Higher Mgo basalt: sample Amb 67				

* Eggins, unpublished data Zr content in the mantle from Sun & McDonough (1989).

Isotopic studies of these suites are already in progress (Baize 1992).

Crystallization processes. The K and La variations for similar values of [La/Yb]n in the more evolved volcanic rocks (Fig. 8) and the significant changes in SiO_2, MgO, K_2O, FeO_t and CaO distributions (Figs 5, 6, 10) emphasize the important role played by crystallization processes. Using these variations and the least-squares mixing program of Mason (1987), we have modelled fractional crystallization processes for the different periods of the volcanic activity and for each segment of the fractionation path (Table 3).

Thus, the MK and the low [La/Yb]n trends observed from the MK basalts of the Tuvio–Vetlam edifice to the evolved MK andesites, dacites and rhyodacites of the first sequence of the APS (trend 1 on Fig. 5; trend A on Fig. 8) can be explained by two phases of fractional crystallization (trends 1–3 on Fig. 10, Table 3): the first is characterized by a strong decrease of MgO while FeO_t, TiO_2 and K_2O increase in the more primitive basalts (Figs 6, 10) and is related to olivine (16.97% Fo_{79}), plagioclase (21.16% An_{68}) and clinopyroxene (20.6% augite) fractionation; the second is characterized by a more subdued decrease of MgO while FeO_t, TiO_2 and CaO strongly decrease in the more evolved volcanic rocks and is principally controlled by plagioclase (29.61% An_{76}), clinopyroxene (25.77% augite) and Fe–Ti oxide (8.65%) with minor olivine (3.82% Fo_{79}) fractionation.

Due to the observed compositional similarities between the basal shield volcano and the basaltic post-caldera suites (trend 2 on Fig. 5; trends B-C on Fig. 8, and trends 1–2 on Fig. 10), we have modelled the crystallization sequences of these two volcanic phases (Table 3) using the

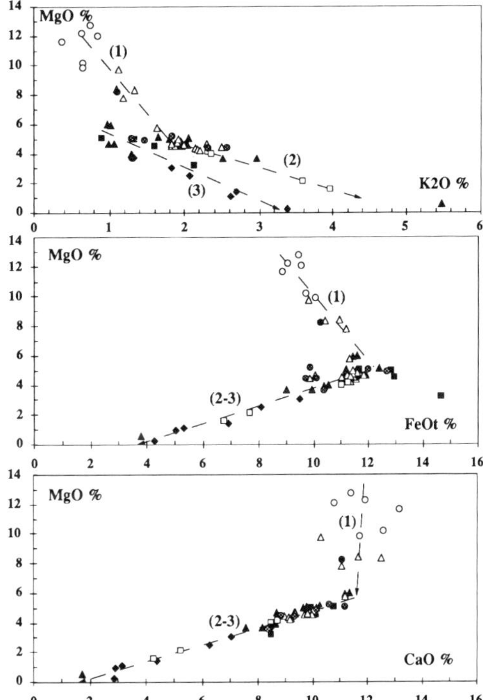

Fig. 10. K_2O, FeO^t and CaO versus MgO distribution for Ambrym volcanic rocks, illustrating crystallization trends described in the text. Major elements are expressed in wt % summed to 100% volatile free. FeO^t = total iron as FeO. Same legend as Fig. 5.

more primitive olivine-rich basalt of the post-caldera suite. These volcanic rocks also show two fractionation paths (Figs 6, 10): the first, characterized by a strong decrease of MgO while FeO^t, TiO_2 and K_2O increase in the more primitive basalts is principally related to olivine

Table 3. *Results of fractional crystallization modelling using the least-squares mixing program of Mason (1987) for different periods during the evolution of Ambrym volcano*

	MK trend : phase 1 – Tuvio-APS sequence 1								MK trend phase 2 : basaltic to rhyodacitic APS seq. 1							
Mass%	Ol basalt M78 100.00 =	Pl basalt 29Al 40.95 +	Fo_{81} 16.00 +	Augite 16.97 +	An_{70} 21.16 +	Fe-Ti Oxide 0.14 =	Calc. parent	Residuals r	Pl basalt 29Al 100.00 =	Rhyodacite 60C 32.27 +	Fo_{79} 3.82 +	Augite 25.77 +	An_{76} 29.61 +	Fe-Ti oxide 8.65 =	Calc. parent	Residuals r
SiO_2	49.16	51.28	42.52	48.63	51.08	0.10	49.18	−0.01	51.28	68.52	38.92	50.45	49.78	0.10	51.33	−0.01
TiO_2	0.62	0.79	0.00	0.62	0.04	6.18	0.45	0.17	0.79	0.33	0.03	0.39	0.04	6.18	0.77	0.01
Al_2O_3	14.06	15.68	0.00	6.63	29.43	5.10	14.05	0.01	15.68	15.19	0.02	4.27	31.46	5.10	15.71	−0.01
FeO	9.03	11.43	17.20	6.52	1.01	80.97	9.05	−0.01	11.43	4.30	18.45	7.17	0.80	80.97	11.44	0.00
MnO	0.17	0.23	0.43	0.15	0.04	0.35	0.21	−0.04	0.23	0.10	0.24	0.10	0.03	0.35	0.11	0.00
MgO	12.24	5.88	39.41	14.83	0.21	3.72	12.22	0.02	5.88	0.22	40.97	14.87	0.17	3.72	5.89	−0.01
CaO	11.91	11.19	0.36	21.19	13.38	0.16	11.90	0.01	11.19	2.88	0.32	21.83	15.67	0.16	11.22	−0.1
Na_2O	1.99	2.52	0.03	0.39	3.76	0.06	1.92	0.07	2.52	5.08	0.00	0.26	2.67	0.06	2.50	0.03
K_2O	0.63	1.01	0.00	0.00	0.30	0.00	0.48	0.15	1.01	3.38	0.02	0.00	0.20	0.02	1.15	−0.14
							$\Sigma r^2=$	0.06							$\Sigma r^2=$	0.02

	Basal shield volcano : phase 1								Basal shield volcano : phase 2							
	Ol basalt Amb_{67} 100.00 =	Pl basalt Amb_{50} 70.62 +	Fo_{79} 7.34 +	Augite 14.97 +	An_{72} 6.09 +	Fe-Ti Oxide 1.13 =	Calc. parent	Residuals r	Pl basalt Amb_{50} 100.00 =	Andesite Amb_{36} 72.61 +	Fo_{75} 0.66 +	Augite 8.50 +	An_{72} 15.50 +	Fe-Ti oxide 3.01 =	Calc. parent	Residuals r
	49.64	51.23	38.92	50.00	49.60	0.14	49.63	0.01	51.23	53.02	38.62	51.98	49.60	0.14	51.20	0.03
	0.70	0.89	0.03	0.55	0.08	4.46	0.77	−0.07	0.89	0.99	0.00	0.33	0.08	4.46	0.90	−0.01
	14.87	17.32	0.02	4.63	30.44	6.65	14.88	−0.01	17.32	16.68	0.01	2.93	30.44	6.65	17.35	−0.02
	10.94	10.53	18.45	7.41	1.06	85.37	10.93	0.00	10.53	9.68	22.19	7.17	1.06	85.37	10.53	0.00
	0.21	0.20	0.24	0.25	0.09	0.34	0.21	0.01	0.20	0.18	0.43	0.19	0.09	0.34	0.18	0.03
	8.42	4.51	40.97	14.34	0.12	4.03	8.43	−0.01	4.51	4.42	38.86	10.47	0.12	4.03	4.54	−0.03
	11.68	10.53	0.32	22.19	14.49	0.05	11.69	−0.01	10.53	8.79	0.20	21.46	14.49	0.05	10.58	−0.05
	2.22	2.86	0.00	0.29	2.89	0.02	2.24	0.02	2.86	3.26	0.00	0.21	2.89	0.02	2.84	0.01
	1.09	1.61	0.02	0.00	0.00	0.00	1.14	−0.05	1.61	2.55	0.00	0.01	0.00	0.00	1.86	−0.25
							$\Sigma r^2=$	0.01							$\Sigma r^2=$	0.07

Postcaldera series : phase 1

Ol basalt Amb_{67} 100.00	=	Ol basalt Amb_{20} 79.90	+	Fo_{79} 6.25	+	Augite 8.79	+	An_{72} 5.34	=	Calc. parent	Residuals parent
49.64		50.17		38.92		50.00		49.60		49.63	0.01
0.70		0.86		0.03		0.55		0.08		0.74	−0.04
14.87		16.06		0.02		4.63		30.44		14.89	−0.01
10.94		11.31		18.45		7.41		1.06		10.91	0.03
0.21		0.21		0.24		0.25		0.09		0.21	0.00
8.42		5.74		40.97		14.34		0.12		8.44	−0.02
11.68		11.18		0.32		22.19		14.49		11.69	−0.02
2.22		2.54		0.00		0.29		2.89		2.21	0.01
1.09		1.63		0.00		0.00		0.00		1.30	−0.21
										$\Sigma r^2=$	0.05

Postcaldera series : phase 2

Ol basalt Amb_{20} 100.00	Pl basalt Amb_{10} 69.24	Fo_{79} 1.58	Augite 13.08	An_{72} 13.59	Fe-Ti Op 2.41	Calc. parent	Residuals r
50.17	52.25	38.92	51.98	49.60	0.14	50.17	−0.01
0.86	0.99	0.03	0.33	0.08	4.46	0.85	0.01
16.06	16.37	0.02	2.93	30.44	6.65	16.05	0.01
11.31	11.41	18.45	7.17	1.06	85.37	11.31	0.00
0.21	0.24	0.24	0.19	0.09	0.34	0.22	0.00
5.74	4.24	40.97	15.47	0.12	4.03	5.74	0.01
11.18	9.20	0.32	21.46	14.49	0.05	11.17	0.01
2.54	3.07	0.00	0.21	2.89	0.02	2.55	−0.01
1.63	2.21	0.00	0.00	0.00	0.03	1.58	0.05
						$\Sigma r^2=$	0.01

APS sequence 2 : HK basaltic to rhyodacitic trend

glass 39A1 100.00	=	glass 39A2 34.20	+	Fo_{79} 2.93	+	Augite 22.57	+	An_{65} 31.32	+	Fe-Ti Op 8.83	=	Calc. parent	Residuals r
52.00		68.39		38.92		50.45		51.29		0.11		52.01	−0.01
0.97		0.41		0.03		0.39		0.02		8.98		1.03	−0.06
16.05		15.67		0.02		4.27		29.80		4.76		16.05	0.00
11.21		3.78		18.45		7.17		1.29		82.15		11.20	0.01
0.22		0.13		0.24		0.10		0.03		0.33		0.11	0.10
5.02		0.53		40.97		14.87		0.07		2.56		5.02	0.00
9.79		1.75		0.32		21.83		13.48		0.08		9.78	0.01
2.68		3.87		0.00		0.26		4.10		0.01		2.67	0.02
2.05		5.48		0.02		0.04		0.32		0.02		1.99	0.07
												$\Sigma r^2=$	0.02

Postcaldera HK basaltic to dacitic Lewolembwi trend

Pl basalt Amb_{10} 100.00	=	Andesite Amb_{14} 48.72	+	Fo_{79} 1.34	+	Augite 16.97	+	An_{68} 25.09	+	Fe-Ti Op 7.60	=	Calc. parent	Residuals r
52.25		61.53		38.92		51.98		50.08		0.18		52.27	−0.02
0.99		0.74		0.03		0.33		0.04		6.75		0.94	0.05
16.37		16.22		0.02		2.93		29.43		5.45		16.36	0.02
11.41		6.73		18.45		7.17		1.01		84.06		11.42	−0.01
0.24		0.21		0.24		0.19		0.04		0.44		0.18	0.06
4.24		1.58		40.97		15.47		0.21		2.86		4.23	0.02
9.20		4.29		0.32		21.46		13.38		0.05		9.17	0.02
3.07		4.29		0.00		0.21		0.76		0.09		3.10	−0.03
2.21		3.97		0.00		0.00		0.40		0.03		2.05	0.16
												$\Sigma r^2=$	0.04

(Fo$_{79}$), plagioclase (An$_{72}$) and clinopyroxene fractionation; and the second, formed by the more evolved basalts and characterized by a more subdued decrease of MgO while FeO^t, TiO$_2$ and CaO decrease, is due to olivine (Fo$_{75-79}$), plagioclase (An$_{72}$), clinopyroxene and minor Fe-Ti oxide fractionation.

The differentiated suites from HK basalts to HK andesites of the Lewolembwi maar and the HK differentiated suite formed by any samples of the APS sequence 2 appear very similar (trend 3 on Fig. 5; trend B on Fig. 8, and trend 2 on Fig. 10) and are characterized by a subdued decrease of MgO while SiO$_2$, K$_2$O, Zr and La increase and FeOt, TiO$_2$ and CaO strongly decrease. For both suites, chemical variations are principally controlled by plagioclase (An$_{65-68}$), clinopyroxene, Fe-Ti oxides and minor olivine (Fo$_{79}$) fractionation (Table 3e–d).

Magmatic evolution of the volcano

Except for some samples characterized by very high [La/Yb]n (C on Fig. 8), the volcanic rocks of Ambrym volcano seem to be derived from two magmatic series. The older MK series ((La/Yb)n = 2; trend 1 in Fig. 5; A in Fig. 8), including the Vetlam–Tuvio–Dalahum and MK volcanic rocks of the APS (sequence 1 and some glasses of sequences 2 and 3), formed by c.25% melting of a spinel lherzolite source at a moderately deep level (at least 60 km). This suite may correspond to an ancient episode of magma generation probably related to normal subduction, which formed a magma chamber below these edifices in which magma progressively differentiated according to the fractionation scheme modelled in Table 3a. The second series ((La/Yb)n = 3–5; trends 2–3 in Fig. 5; B–C in Fig. 8), formed by a minimum of 21% melting of a similar spinel lherzolite source at probably shallower levels (about 45–30 km), includes volcanic suites of the basal shield volcano, some glasses of sequences 2, 3 and 4 of the APS, and volcanic rocks of the post-caldera suite (B on Fig. 8). This second suite may correspond to a more recent magmatic episode mainly related to rifting and possibly responsible for the formation of a second, less evolved magma chamber lying along the rift axis.

Extension and uplift accompanying the formation of the rift system, after the Vetlam–Tuvio and Dalahum MK episode, might explain the compositional evolution of the volcano. The giant eruption responsible for the APS and the formation of the caldera could have been caused by emplacement of hot, primitive basaltic magma of the rift-related phase into the differentiated MK andesitic to rhyodacitic magma of the Tuvio–Vetlam–Dalahum magmatic chamber. The mechanical mixing of mafic and felsic glasses and the range in composition of the juvenile clasts observed in the APS are strong arguments favouring such a model. The intrusion of hot basaltic magma probably took place between the Tuvio–Vetlam–Dalahum edifice and the N100° rift axis, thus explaining the location of the caldera centre half-way between the two volcanic structures. Simultaneous introduction of seawater into the edifice, related to regional doming and fracturing following uprise of the large basaltic magma body to shallow depth and/or seismic events affecting the N100° fracture zone, induced phreatomagmatic activity as demonstrated by Robin *et al.* (1993) and contributed to the paroxysmal eruption. After the violent emptying of the differentiated magma from the old chamber, massive emission of basaltic magma from the replenished reservoir progressively dominated the volcanic activity. Introduction of seawater during this period is responsible for the basaltic surtseyan-like deposits of sequences 2 and 3 of the APS (Robin *et al.* 1993). The decreasing role of water at the end of the caldera episode explains the gradual changes in the dynamics of the eruptions, which became more strombolian with time (APS sequence 4 and post-caldera volcanism). Since the caldera-forming episode, the current magma chamber, located below the rift, appears to have been fed regularly by primitive magma and was responsible for all the recent magmatism. Lastly, the differentiated suite associated with the Lewolembwi maar and the 1986 lava flow in the western part of the caldera indicate that differentiation processes are presently active in small secondary reservoirs of the present chamber.

Conclusions

Sources uplifted from at least 60–45 km depth during extension and doming accompanying rift formation, together with high degrees of melting (about 25%) of spinel lherzolite and possible K–La enrichment by mantle metasomatism or crustal contamination can explain the compositions of the different high-MgO primitive liquids on Ambrym volcano, whereas olivine, plagioclase, clinopyroxene and Fe–Ti fractionation explain the variations of most of the major and trace elements in the associated MK and HK volcanic suites. The tectonic evolution of the region, mainly controlled by the ongoing collision between the DEZ and the arc, seems to be responsible for the geochemical variations of the magmatic series and for the changes in the volcanic regime. Thus, interaction between MK and HK magmas from an ancient and a more recent reservoir was probably the major cause of the giant eruption leading to the caldera formation during the APS episode, assisted by massive introduction of seawater into the system. The more recent volcanism is principally related to rifting. This model, featuring controls by the tectonic setting and interaction between two magmatic series aided by infiltration of seawater, can explain the characteristics of the Ambrym basaltic complex.

This work was financially supported by ORSTOM UR 1F, French Foreign Affairs Ministry (MAE) and the 'Direction de la Recherche et des Etudes Doctorales' (DRED). We thank C. Douglas for useful participation during fieldwork. C. Mortimer, Director of the Department of Geology, Mines and Rural Water Supply of Vanuatu, and C. Reichenfeld, Director of the ORSTOM Centre of Port-Vila, constantly supported our investigations. We also express our thanks to Tony Crawford, S. Eggins, J. Scarrow and J.L. Smellie who reviewed our manuscript and perfected our English. Finally, J. Cotten of the Université de Bretagne Occidentale in Brest is thanked for his dedication to providing excellent ICP analyses.

References

BAIZE, S. 1992. *Déséquilibres U-Th, Géochimie des éléments traces et des isotopes des laves d'Ambrym (Arc des Nouvelles-Hébrides)*. Mémoire de DEA, Géosciences, Université de Clermont-Ferrand, OPGC.

CHASE, T.E. & SEEKINS, B.A. 1988. Submarine topography of the Vanuatu and southeastern Solomon Islands regions. *In*: GREENE, H.G. & WONG, F.L. (eds) *Geology and offshore resources of Pacific island arcs – Vanuatu region*. Circum-Pacific Council for Energy and Mineral Resources Earth Science Series, Houston, Texas, **8**, 35–36.

COLLOT, J.Y. 1989. *Obduction et collision: examples de la Nouvelle-Calédonie et de la zone de subduction des Nouvelles-Hébrides*. Thèse Doctorat d'Etat ès-Sciences Naturelles, Université de Paris-Sud, Orsay.

——, DANIEL, J. & BURNE, R.V. 1985. Recent tectonics associated with the subduction/collision of the D'Entrecasteaux Zone in the central New Hebrides. *Tectonophysics*, **112**, 325–356.

CRAWFORD, A.J., GREENE, H.G. & EXON, N.F. 1988. Geology, petrology and geochemistry of submarine volcanoes around Epi island, New Hebrides island arc. *In*: GREENE, H.G. & WONG, F.L. (eds) *Geology and offshore resources of Pacific island arcs – Vanuatu region*. Circum-Pacific Council for Energy and Mineral Resources Earth Science Series, Houston, Texas, **8**, 301–327.

EGGINS, S.M. 1989. *The origin of primitive ocean island and island arc basalts*. PhD Thesis, University of Tasmania, Australia.

EISSEN, J-PH., BLOT, C. & LOUAT, R. 1991. *Chronology of the historic volcanic activity of the New Hebrides island arc from 1595 to 1991*. Report, ORSTOM, Nouméa, New Caledonia, **69**.

EXON, N.F. & CRONAN, D.S. 1983. Hydrothermal iron deposits and associated sediments from submarine volcanoes of Vanuatu, soutwest Pacific. *Marine Geology*, **52**, M43-M52.

FALLOON, T.J. & GREEN, D.H. 1988. Anhydrous partial melting of peridotite from 8 to 35 Kb and the petrogenesis of MORB. *Journal of Petrology, Special Lithosphere Issue*, 379–414.

——, ——, HATTON, C.J. & HARRIS, L. 1988. Anhydrous partial melting of a fertile and depleted peridotite from 2 to 30 Kb and application to basalt petrogenesis. *Journal of Petrology*, **29**, 1257–1282.

FISHER, N.H. 1957. *Catalogue of the active volcanoes of the world including solfatara fields, Part V. Melanesia*. IAVCEI, Rome.

GARANGER, J. 1972. *Archéologie des Nouvelles-Hébrides*. ORSTOM, Publications de la Société des Océanistes, Musée de l'Homme, Paris, **30**.

GÈZE, B. 1966. Sur l'âge des derniers cataclysmes volcanotectoniques dans la région centrale de l'arc des Nouvelles-Hébrides. *Bulletin de la Société Géologique de France*, **7**, 329–333.

GILL, J.B. 1981. *Orogenic andesites and plate tectonics*. Springer-Verlag, Berlin Heidelberg New York.

GORTON, M.P. 1974. *The geochemistry and geochronology of the New Hebrides*. PhD Thesis, Australian National University, Australia.

GREENE, H.G., MACFARLANE, A., JOHNSON, D.P. & CRAWFORD, A.J. 1988. Structure and tectonics of the central New Hebrides arc. *In*: GREENE, H.G. & WONG, F.L. (eds) *Geology and offshore resources of Pacific island arcs – Vanuatu region*. Circum-Pacific Council for Energy and Mineral Resources Earth Science Series, Houston, Texas, **8**, 377–412.

HANSON, G.N. 1980. Rare earth elements in petrogenetic studies of igneous systems. *Annual Review of Earth and Planetary Sciences*, **8**, 371–406.

—— & LANGMUIR, C.H. 1978. Modelling of major elements in mantle-melt systems using trace element approaches. *Geochimica et Cosmochimica Acta*, **42**, 725–741.

JAQUES, A.L. & GREEN, D.H. 1980. Anhydrous melting of peridotite at 0–15 Kb pressure and the genesis of tholeiitic basalts. *Contributions to Mineralogy and Petrology*, **73**, 287–310.

LOUAT, R. & PELLETIER, B. 1989. Seismotectonics and present-day relative plate motions in the New Hebrides-North Fiji Basin region. *Tectonophysics*, **167**, 41–55.

——, HAMBURGER, M. & MONZIER, M. 1988. Shallow and intermediate-depth seismicity in the New Hebrides arc: constraints on the subduction process. *In*: GREENE, H.G. & WONG, F.L. (eds) *Geology and offshore resources of Pacific island arcs -Vanuatu region*. Circum-Pacific Council for Energy and Mineral Resources Earth Science Series, Houston, Texas, **8**, 329–356.

MCCALL, G.J.H., LEMAITRE, R.W., MALAHOFF, A.,

ROBINSON, G.P. & STEPHENSON, P.J. 1970. The Geology and Geophysics of the Ambrym Caldera, New Hebrides. *Bulletin of Volcanology*, **34**, 681–696.

MACFARLANE, A. 1976. *Geology of Pentecost and Ambrym, 1:100.000. New Hebrides Geological Survey Sheet 6*. Geological Survey Department., Port Vila, New Hebrides.

——, CARNEY, J.N., CRAWFORD, A.J. & GREENE, H.G. 1988. Vanuatu – A review of the onshore geology. *In*: GREENE, H.G. & WONG, F.L. (eds) *Geology and offshore resources of Pacific island arcs – Vanuatu region*. Circum-Pacific Council for Energy and Mineral Resources Earth Science Series, Houston, Texas, **8**, 45–91.

MARTHELOT, J.-M., CHATELAIN, J.-L., ISACKS, B.L., CARDWELL, R.K. & COUDERT, E. 1985. Seismicity and attenuation in the central Vanuatu (New Hebrides) islands: a new interpretation of the effect of subduction of the D'Entrecasteaux Fracture Zone. *Journal of Geophysical Research*, **90**, 8461–8650.

MASON, D.R. 1987. *Least-squares mixing program for MacIntosh*. Glenside, Australia.

MCKENZIE, D. & O'NIONS, R.K. 1991. Partial melt distributions from inversion of rare earth element concentrations. *Journal of Petrology*, **32**, 1021–1091.

MONZIER, M., ROBIN, C., EISSEN, J.-P. & PICARD, C. 1991. Découverte d'un large anneau de tufs basaltiques associé à la formation de la caldeira d'Ambrym (Vanuatu, SW Pacifique). *Compte Rendu de l'Académie des Sciences*, **313**, 1319–1326.

PEARCE, J.A. & NORRY, M.J. 1979. Petrogenetic implications of Ti, Zr, Y and Nb variations in volcanic rocks. *Contributions to Mineralogy and Petrology*, **69**, 33–47.

ROBIN, C., EISSEN, J.-P. & MONZIER, M. 1993. Giant tuff cone and 12km wide associated caldera at Ambrym volcano (Vanuatu, New Hebrides Arc). *Journal of Volcanology and Geothermal Research* (in press).

——, MONZIER, M., EISSEN, J.P., PICARD, C. & CAMUS, G. 1991. Coexistence de lignées HK et MK dans les pyroclastites associées à la caldéra d'Ambrym (Vanuatu, Arc des Nouvelles Hébrides). *Compte Rendu de l'Académie des Sciences*, **313**, 1425–1432.

ROCA, J.-L. 1978. *Contribution à l'étude pétrologique et structurale des Nouvelles-Hébrides*. Thèse Doctorat 3ème Cycle, Université des Sciences et Techniques du Languedoc, Montpellier.

ROEDER, P. & EMSLIE, R.F. 1970. Olivine-liquid equilibrium. *Contributions to Mineralogy and Petrology*, **29**, 275–289.

SIMKIN, T., SIEBERT, L., MCCLELLAND, L., BRIDGE, D., NEWHALL, C. & LATTER, J.H. 1981. *Smithsonian Institution, Volcanoes of the world*. Hutchinson Ross Publishing Company, Stroudsburg, Pennsylvania.

STEINBERG, M., TREUIL, M. & TOURAY, J.C. 1979. *Géochimie: Principes et méthodes -tome 2 – Cristallochimie et éléments en traces*. Doin, Paris.

SUN, S.-S. & MCDONOUGH, W.F. 1989. Chemical and isotopic systematics of oceanic basalts: implications for mantle composition and processes. *In*: SAUNDERS, A.D. & NORRY, M.J. (eds) *Magmatism in the Ocean Basins*. Geological Society, London, Special Publications, **42**, 313–345.

US DEPARTMENT OF THE INTERIOR/GEOLOGICAL SURVEY 1987. *Preliminary Determinations of Epicenters, Monthly listing*. National Earthquake Information Center, November 1987.

WARDEN, A.J. 1967. *The geology of the Central Islands*. New Hebrides Geological Survey, Port Vila, Vanuatu, **5**.

——, CURTIS, R., MITCHELL, A.H.G. & ESPIRAT, J.-J. 1972. *Geology of the Central Islands, 1:100.000*. New Hebrides Geological Survey Sheet 8, Port Vila Vanuatu.

WATSON, S. & MCKENZIE, D. 1991. Melt generation by plumes : A study of Hawaiian volcanism. *Journal of Petrology*, **32**, 501–537.

WILLIAMS, C.E.F. & WARDEN, A.J. 1964. *Progress report of the Geological Survey for 1959–1962*. New Hebrides Geological Survey , Port Vila, Vanuatu.

WILSON, M. 1989. *Igneous Petrology, a global tectonic approach*. Unwin Hyman Ltd, London.

Gamilaroi Terrane: A Devonian rifted intra-oceanic island-arc assemblage, NSW, Australia.

J.C. AITCHISON[1] & P.G. FLOOD[2]

[1] *Department of Geology and Geophysics, University of Sydney, NSW 2006, Australia.*

[2] *Department of Geology and Geophysics, University of New England, Armidale, NSW 2351, Australia.*

Abstract: The Gamilaroi terrane comprises part of the New England orogen of eastern Australia and contains Devonian strata preserving evidence of their development in a variety of arc-related environments. Radiolarian studies provide new and more precise constraints on stratigraphy and together with geochemistry allow development of a new model for evolution of the terrane. A diverse suite of volcaniclastic sedimentary rocks intercalated with regionally extensive extrusive and high level intrusive meta-felsic igneous rocks of low-K and calc-alkaline volcanic affinity is present. These rocks constitute the oldest *in-situ* volcanic rocks in the Gamilaroi terrane and are similar to those found in modern intra-oceanic volcanic island-arc settings. Basaltic intrusions and lavas occur at higher stratigraphic levels. Major and trace element compositions, together with relatively flat chondrite-normalized REE distributions, resemble those from rocks formed in modern intra-oceanic island-arc rift zones.

The close spatial association of these rock types together with an absence of continent-derived detritus is consistent with interpretation of development of the Gamilaroi terrane in an intra-oceanic island arc which experienced local rifting. This interpretation is a departure from previous models for the New England orogen, which interpret the Gamilaroi terrane as a forearc basin that developed above a long-lived (Cambrian to Permian), west-dipping subduction zone fringing the continental margin of Gondwana. By analogy with modern arc–continent collision zones in which crust between colliding continents and island arcs is preferentially subducted beneath the approaching arc, we suggest that oceanic crust intervening between Gamilaroi terrane and Gondwana was subducted eastwards under the western margin of the Gamilaroi terrane arc. During a latest Devonian collision event the Gamilaroi terrane was obducted onto the Gondwana margin. This resulted in subduction flip and subsequent development of an east-facing continental margin arc system on top of the Gamilaroi terrane.

The eastern (Australian) margin of Gondwana was an active zone of terrane accretion during the Palaeozoic to Early Mesozoic (Flood & Aitchison 1988; Coney *et al.* 1990). Accretion events and terrane interactions resulted in the development of a complex tectonic collage known locally as the New England orogen (NEO), that extends over approximately 1300 km along eastern Australia from Bowen (20°S) to Newcastle (33°S). Differences in the tectonic setting and relative motions of the constituent terranes, together with their subsequent post-accretionary interactions, disruption and dispersal, have produced the complex geology now observed. This study addresses aspects of the development of the Gamilaroi terrane, a major tectonostratigraphic entity in the southern NEO.

The southern NEO is subdivided into two zones of inherently different strata across the Peel Manning fault system (PMFS). To the west of the fault lie structurally simple rocks of the Tamworth Belt (Harrington 1974; zone A of Leitch 1974). To the east (zone B of Leitch 1974) is a zone which includes a dismembered Early Cambrian ophiolite (Aitchison *et al.* 1992*a*) but dominated, overall, by complexly deformed strata which appear to have developed in a series of Devonian and Carboniferous subduction complexes (Cawood 1983; Fergusson & Flood 1984; Aitchison *et al.* 1992*b*).

Most plate tectonic models for development of the NEO have inferred a long-lived east-facing volcanic arc, forearc basin, subduction complex geometry (Leitch 1975). Various refinements of this model have been developed (Cawood 1983; Cawood & Leitch 1985; Murray *et al.* 1987). The 'consensus' model for the tectonic evolution of the NEO (Murray *et al.* 1987) interprets rocks in the Tamworth Belt, to the west of the PMFS, as forearc basin strata deposited adjacent to a continental margin arc.

From Smellie, J.L. (ed.), 1995, *Volcanism Associated with Extension at Consuming Plate Margins*, Geological Society Special Publication No. 81, 155–168.

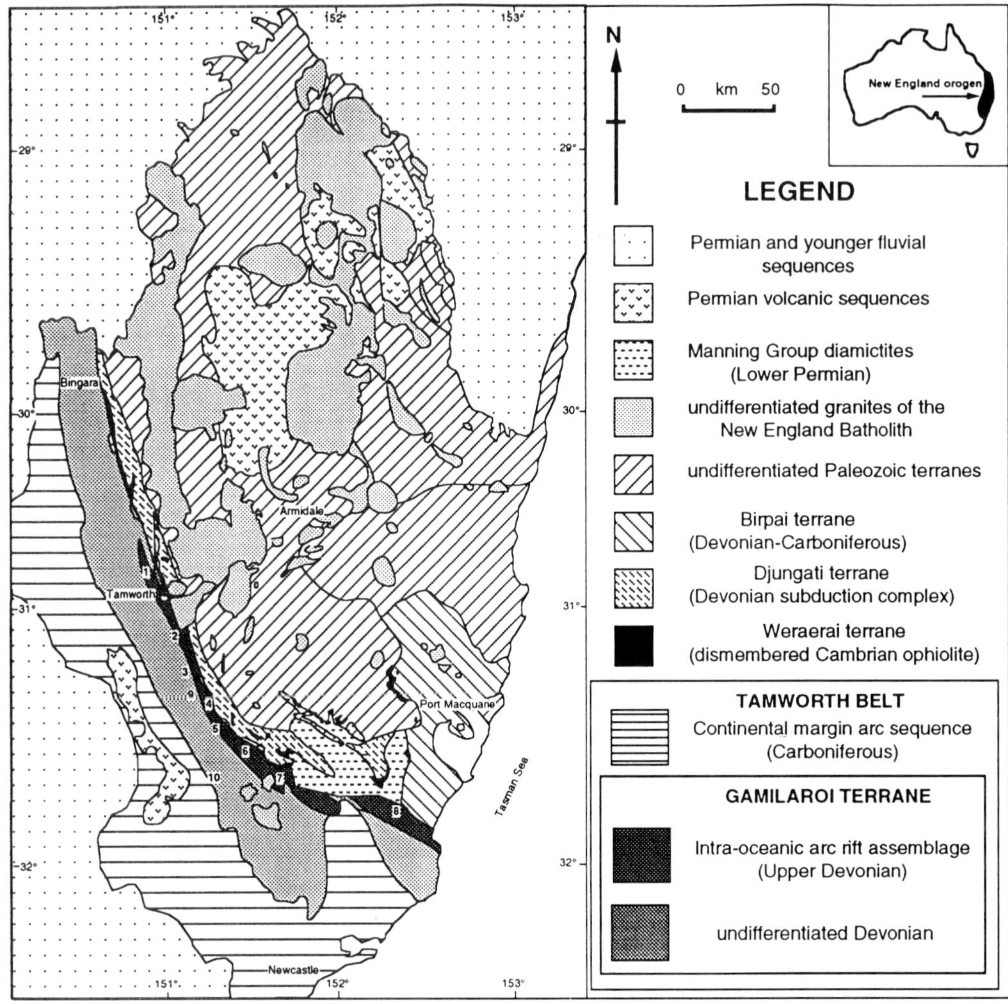

Fig. 1. Map of the of the southern section of the New England orogen indicating distribution of Devonian Gamilaroi terrane intra-oceanic island arc strata and successor Carboniferous continental margin arc-related rocks. Prior to Gamilaroi terrane accretion the eastern limit of Gondwana was the Lachlan Fold Belt. Palaeozoic geology between the Lachlan Fold Belt and the New England orogen is masked by fluvial sediments of Mesozoic-Cenozoic successor basins. Palaeozoic terranes east of the Djungati terrane are not differentiated on this figure. Locations of stratigraphic columns of Fig. 3 are indicated. 1: Yarramanbully; 2: Tamworth; 3: Woolomin; 4: Nundle; 5: Upper Barnard (Barry); 6: Glenrock; 7: Pigna Barney; 8: Bundook; 9: Chaffey Dam; 10: Timor.

This arc is postulated to have developed above a west-dipping subduction zone in association with subduction complex rocks east of the PMFS.

Recent radiolarian (Aitchison 1988*a,b*, 1990; Aitchison *et al.* 1992*a*; Ishiga *et al.* 1988) and radiometric (Aitchison *et al.* 1992*b*) studies have provided the first reliable age constraints for many previously undated rocks in the NEO and have significantly improved our ability to constrain the timing of development of various elements of this tectonic collage. With the benefit of this knowledge we now realize that some of the links previously inferred, in tectonic models, to exist between constituent terranes of the orogen are implausible. Although stratigraphical resolution within terranes has been greatly improved, original relations between terranes remain indeterminate.

The name, Gamilaroi terrane, was introduced for arc-related rocks, in Zone A, west of the PMFS (Flood & Aitchison 1988; Aitchison &

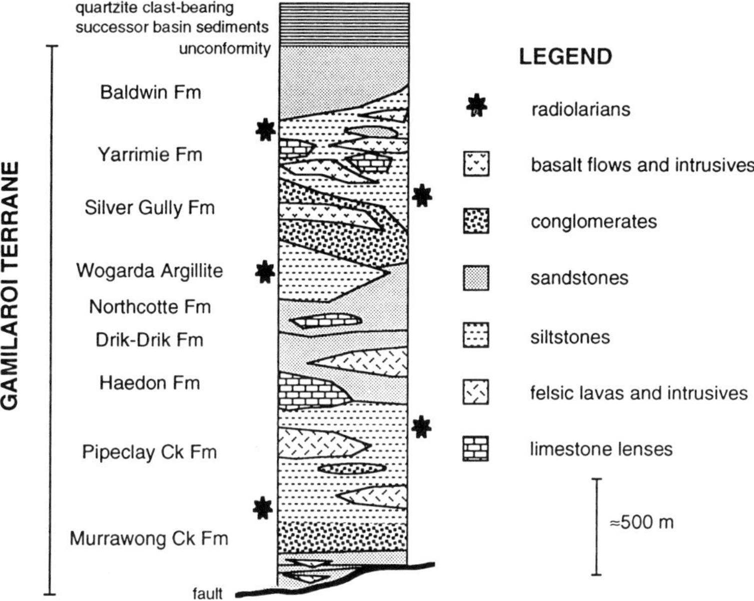

Fig. 2. Composite stratigraphical column of Gamilaroi terrane lithostratigraphy in the Tamworth Nundle area (modified after Cawood 1983; Aitchison *et al.* 1992). Dominant lithologies are indicated but bed/unit thicknesses are approximate.

Flood 1990). Originally, this arc was thought to have influenced the development of the entire Tamworth Belt (Harrington 1974) evolving gradually from an island arc into a continental margin arc (Korsch 1984). However, as a result of further study and improved stratigraphical resolution, two distinct, temporally separate, arc-related sequences are now recognized in the Tamworth Belt: a Devonian intra-oceanic island arc and a Carboniferous continental margin arc (Flood & Aitchison 1992). In the light of increased understanding of these rocks, revision of the definition of the Gamilaroi terrane was made by Aitchison *et al.* (1992a) with restriction of the Gamilaroi terrane to the rocks of the Devonian arc.

Rocks of this terrane appear to have developed in an intra-oceanic arc setting isolated from the continental margin of Gondwana, whereas Carboniferous rocks are part of a continental margin arc (McPhie 1983, 1984, 1986) which developed atop the Gamilaroi terrane subsequent to its accretion to the margin of Gondwana. The eastern boundary of the Gamilaroi terrane is defined by the PMFS and the western boundary of the terrane is obscured by volcano-sedimentary overlap-assemblage rocks of the Carboniferous arc.

Gamilaroi terrane

The Gamilaroi terrane (Fig. 1) contains (?Silurian–) Devonian rocks which preserve evidence of development in a variety of intra-oceanic island-arc-related environments. Volcaniclastic sedimentary rocks, including deep-water marine high-density mass-flow conglomerates, volcaniclastic turbidite sandstones, altered felsic tuffs and resedimented tuff turbidites, dominate the terrane (Crook 1961; Cawood 1983). The volcaniclastic rocks are intercalated with meta-andesites, rhyolites and dacites together with meta-basalts (spilites) originally mapped by Benson (1915, 1918). The lithostratigraphical succession is well established between Tamworth and Nundle (Fig. 2). This succession can be very broadly correlated to other areas throughout the extent of the terrane (Fig. 3). Uppermost Devonian and younger strata unconformably overlie the Gamilaroi terrane and comprise the eastern margin of a foreland basin sequence, which overlaps both the Gamilaroi terrane and the Lachlan orogen (the pre-Carboniferous eastern limit of Gondwana) to the west (Veevers 1984). Initiation of foreland basin development was related to accretion of the Gamilaroi terrane to

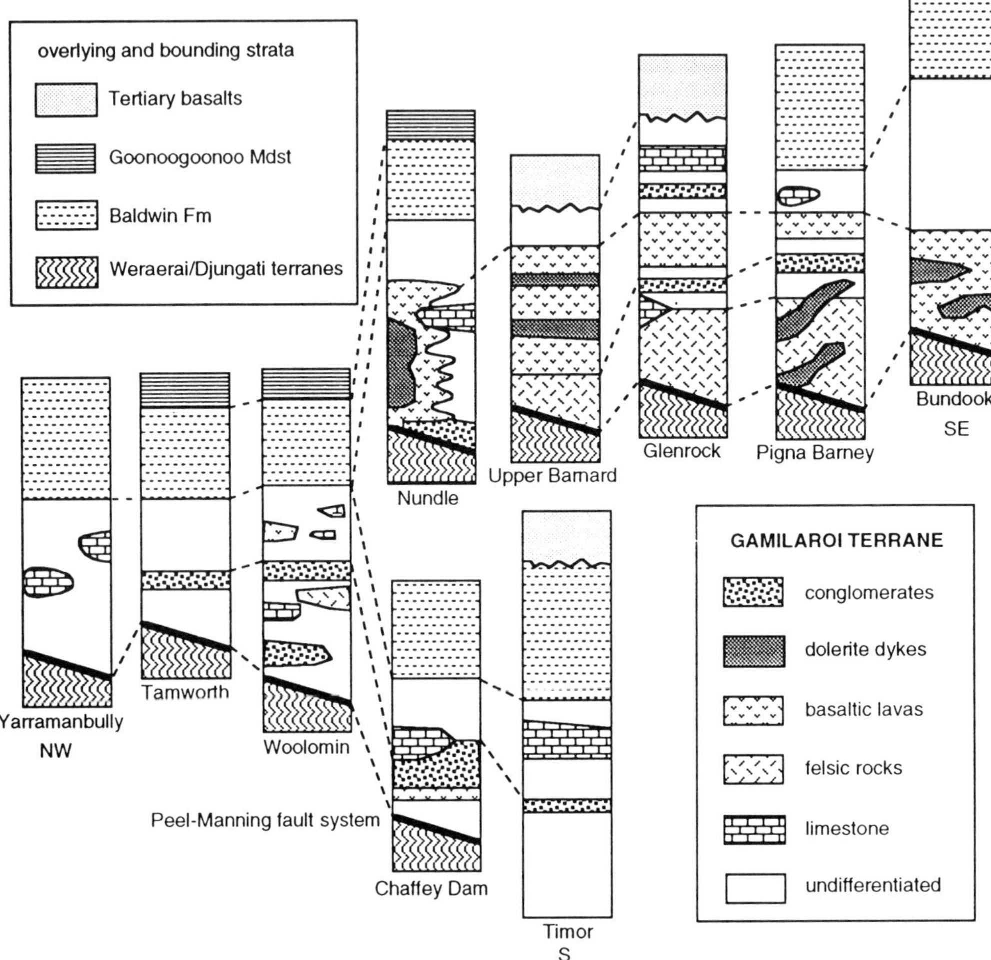

Fig. 3. Schematic stratigraphical columns of Gamilaroi terrane lithostratigraphy indicating correlation between major areas of outcrop indicated on Fig. 1. Dominant lithologies are indicated but bed/unit thicknesses are approximate.

Gondwana. It is marked by unconformity, major facies changes (from shallow marine oolitic limestone to submarine canyon fill) and the first appearance of sediments containing conspicuous, mineralogically-mature, Gondwana-derived quartzite clasts (Flood & Aitchison 1992).

The maximum age of rocks in the Gamilaroi terrane is somewhat controversial. The Tamworth Group (Crook 1961) is the lowermost lithostratigraphical unit within the Gamilaroi terrane. It comprises a sequence of volcaniclastic sandstones, conglomerates and breccias interbedded with voluminous fine-grained felsic tuffs. These strata have been metamorphosed to prehnite-pumpellyite facies. Much of this group is poorly exposed and ages from fossils occurring

in several major Devonian shallow water bioclastic limestone lenses have previously been used as a basis for dating parts of the terrane. However, these limestones occur within deep-marine, mass flow dominated successions and doubt exists as to their autochthoneity (Crook 1961; Pedder 1966). Cawood (1976) reported the discovery of Middle Cambrian and Ordovician fossils in allochthonous limestone clasts within conglomerates originally mapped as part of the Tamworth Group. In subsequent publications these conglomerates have been assigned to the Lower Palaeozoic and have been inferred to underlie strata of a modified Tamworth Group (Cawood 1983; Leitch & Cawood 1987; Cawood & Flood 1989). However, after regional investigation and examination of type sections, we favour an alternative interpretation of the stratigraphy and consider that the original assignment of these strata to a single lithostratigraphical group (Tamworth Group of Crook 1961) remains appropriate. There is no a priori reason why the age of fossiliferous conglomerate clasts should be the same as that of the units into which they were resedimented (Korsch & Harrington 1981; Aitchison *et al.* 1992*a*). Metamorphosed sills and flows of felsic volcanic rock occur throughout the lower half of the stratigraphy in both inferred Lower Palaeozoic and the restricted Tamworth Group of Cawood (1983). Detrital mode compositions of sandstones throughout the entire Tamworth Group (*sensu* Crook 1961) indicate their derivation through the erosion of an undissected magmatic arc (Cawood 1983). In such a dynamic tectonic setting, minor unconformities such as those which occur in the sequence are to be expected. A Devonian age for the entire lithostratigraphical unit (Tamworth Group) was previously inferred by Crook (1961).

Radiolarians occur in abundance in silicic tuffs throughout the terrane (Hinde 1899) and are providing new biostratigraphical evidence as to the age and stratigraphy of the terrane (Aitchison 1990; Aitchison *et al.* 1992*a*). Well preserved Upper Devonian (Frasnian) radiolarians have been extracted from *in situ* sedimentary rocks of the Yarrimie Formation (upper Tamworth Group) (Aitchison 1990). These sedimentary rocks envelope allochthonous blocks of limestone which contain a Middle Devonian fauna (Pedder 1966). Uppermost Devonian (Famennian) radiolarians have also been recovered from overlying sedimentary rocks of the Baldwin Formation (T. Naka, pers. comm.). Wogarda Argillite also contains well preserved entactinid radiolarians with robust tri-bladed spines reportedly characteristic of Middle Devonian or younger faunas (Holdsworth & Jones 1980; Nazarov & Ormiston 1986). Extraction of diagnostic specimens from the Pipeclay Creek Formation, considered to be of Cambrian age by Cawood (1983), has proved difficult and further work is currently in progress. Nevertheless, strong tri-bladed spines are observed in thin sections of radiolarian-rich tuffs from this unit (Aitchison 1990). The relative abundance of radiolarians, which generally have a small shell diameter, and the presence of forms with tri-bladed spines suggests Devonian not Cambrian affinity (Nazarov & Ormiston 1986).

We have traversed the type sections of all formations together with several auxiliary sections. In each area, exposure is poor and stratigraphical relations are uncertain.

However, similar rocks are found throughout the succession. Studies of detrital sandstone petrography and pyroxene geochemistry (Cawood 1983) clearly indicate derivation of clastic sediments from an island-arc source. There is no evidence indicating the presence of detritus sourced in areas of continental crust, as observed higher in the sequence (Leitch & Willis 1977; Flood & Aitchison 1992). Were it not for the limestone clasts in conglomerates, which have yielded Cambrian and Ordovician faunas, there would be no reason to suspect that the succession represents anything other than a pile of volcaniclastic rocks related to a single period of arc development. The varied nature of volcaniclastic sedimentation in the Tamworth Group, including local unconformities, rapid facies changes, lateral thickness variations and the absence of continental detritus (Cawood 1983), is consistent with interpretation as an intra-oceanic island arc sequence. Sedimentation rates in modern island arc environments are typically high and are well documented in DSDP/ODP results (e.g. Taylor *et al.* 1990). By analogy, it is argued that the Tamworth Group also accumulated rapidly. Sedimentation of the entire sequence (less than 5 km thick) is more likely to have occurred in a relatively short time (e.g. ?Silurian–Devonian) rather than over the 150 Ma time span (Middle Cambrian–Late Devonian) inferred by Cawood (1983). There is no evidence which requires separation of the lowermost Tamworth Group (as defined by Crook 1961) into a separate lower Palaeozoic sequence. Many examples of large allochthonous sediment rafts, similar in scale to the Devonian limestones of the Tamworth Group, are known from within fine-grained arc-derived sediments worldwide (Conaghan *et al.* 1976; Ineson 1985). The source of smaller Cambrian and Ordovician limestone clasts is unclear, although recent studies have revealed the existence of Cambrian rocks within other terranes of the NEO (Aitchison *et al.* 1992*b*). Tectonism related to the development of a Devonian island arc may have resulted in the foundering of sections of a pre-existing carbonate reef/platform terrane. The authors consider that lithological affinities, likely sedimentation rates and radiolarian data are compatible with the conclusion that the entire Gamilaroi terrane sequence accumulated in an island-arc setting during the (?Silurian–) Devonian.

Arc volcanic rocks

There are numerous occurrences of arc-related igneous rocks within the terrane. Detailed

Table 1. *Geochemical analyses of felsic igneous rocks of the Gamilaroi terrane from the Pigna Barney (47350–47353) and Barry statiuon (LD5–LD9) areas*

Sample	47350	47351	47352	47353	LD5	LD6	LD7	LD9
SiO_2	54.16	54.81	58.21	59.63	59.15	68.09	65.67	73.91
TiO_2	1.08	1.11	1.19	1.16	0.57	0.30	0.25	0.24
Al_2O_3	16.80	16.85	16.67	17.09	17.40	14.33	12.40	12.04
$Fe_2O_3^T$	10.25	7.76	7.20	6.73	6.18	5.00	5.64	3.67
MnO	0.18	0.15	0.12	0.09	0.00	0.00	0.00	0.00
MgO	3.48	5.52	2.63	2.60	0.09	0.15	0.10	0.12
CaO	5.96	3.99	2.88	1.03	1.05	1.42	5.42	2.26
Na_2O	4.38	4.70	6.66	7.47	2.83	2.47	1.90	0.24
K_2O	1.85	0.88	1.45	1.03	8.73	3.62	4.31	5.22
P_2O_5	0.37	0.16	0.36	0.29	0.78	1.77	0.07	0.09
S	0.00	0.00	0.01	0.00	0.21	0.10	0.09	0.10
LOI	1.27	3.78	1.96	2.35	3.05	2.07	4.00	1.85
Total	99.79	99.73	99.31	99.28	100.04	99.31	99.87	99.74
Nb	11	6	6	6	8	5	4	3
Zr	117	147	140	151	92	39	92	76
Y	28	30	29	30	15	16	9	9
Sr	941	525	718	266	166	124	69	64
Rb	22	14	24	16	9	30	2	1
Th	8	7	9	5	4	3	4	5
Pb	12	4	7	5	6	4	4	6
U	4	3	1	1	2	2	2	2
Ga	17	17	19	16	13	11	10	9
Zn	113	101	87	85	38	65	83	757
Cu	440	35	10	3	114	17	35	99
Ni	14	64	7	7	14	4	65	15
Cr	60	117	6	48	15	7	271	30
Ce	43	22	30	26	25	12	19	20
Nd	28	19	27	23	16	13	12	12
Ba	543	806	548	277	238	2772	<30	56
V	313	149	156	103	153	59	150	67
La	25	13	15	16	12	10	12	12
Sc	29	18	17	15	15	26	22	15

Pigna Barney samples were collected originally by Cross (1983) from the Pitch Creek Volcanics. All samples were analysed by XRF for major and trace element abundances at the University of New England, Armidale, analyst J. Bedford.

descriptions of the mineralogy and geochemistry of the igneous rocks have been provided by Vallance (1969), Offler (1982), Cross (1983), Morris (1988) and Cawood & Flood (1989). Regionally extensive extrusive and high-level intrusive meta-andesite, rhyolite and dacite intrusions, flows and breccias (locally referred to as keratophyres) are intercalated with lower Tamworth Group sedimentary rocks. Geochemical analyses indicate that they are altered subalkaline volcanic arc andesites with subordinate dacites (Cross 1983; Cawood & Flood 1989; Table 1). Their overall low Zr + Nb contents are comparable to low-K and calc-alkaline volcanic rocks elsewhere (Cawood & Flood 1989). Offler *et al* (1988) reported moderate LREE enriched patterns typical of calc-alkaline island arc lavas for felsic volcanic rocks in the Glenrock area.

These rocks constitute the oldest *in-situ* volcanic rocks in the Gamilaroi terrane and they are similar to those found in modern intraoceanic volcanic island-arc settings. They are most common in the Barry, NE Glenrock and

Pigna Barney districts. A sample of tonalite, compositionally similar to lithologies found in the Pitch Creek Volcanics at the base of the Gamilaroi terrane in the Pigna Barney/Glenrock area, has recently been U–Pb dated as Late Silurian at Australian National University, Canberra (unpublished SHRIMP data of Aitchison & Ireland). A similar Early Silurian age (436 ± 9 Ma) has been reported for samples from the Pola Fogal Suite at Pigna Barney (Kimbrough *et al.* 1993). This complex is intrusive into the Pitch Creek Volcanics (felsic lavas in the Gamilaroi terrane) and is geochemically distinct from the nearby Weraerai terrane ophiolite (Cross 1983; unpublished data of Aitchison) which is of Early Cambrian age (Aitchison *et al.* 1992b). Silurian fossils are, as yet, unknown from the Tamworth Belt and radiometric age data may indicate that unfossiliferous silicic igneous rocks of the Gamilaroi terrane comprise part of a Silurian–Devonian succession.

Benson (1915, 1918) described meta-basaltic rocks which crop out in the Tamworth Group

Table 2. *Geochemical analyses of basic igneous rocks of the Gamilaroi terrane from the Pigna Barney (47060–47083), Upper Barnard (47050–47054) and Barry station (LD8) areas*

Sample	47050	47051	47052	47053	47054	47060	47061	47062	47063	47064	47065	47066	47068	47073	47074	47077	47078	47081	47083	LD8
SiO_2	49.92	52.03	50.51	51.47	41.54	47.68	48.53	49.08	50.89	49.91	51.05	53.87	51.23	50.00	48.01	50.32	51.31	52.32	48.45	53.30
TiO_2	2.20	2.17	2.10	1.99	2.89	1.40	1.09	1.43	1.44	1.43	1.55	1.08	1.89	2.10	0.90	1.12	1.28	1.65	1.19	0.29
Al_2O_3	13.58	13.55	13.77	14.30	11.89	16.13	15.41	15.19	14.44	16.48	14.08	16.97	16.16	13.77	16.11	15.05	15.19	14.25	16.10	12.70
$Fe_2O_3{}^{T}$	13.04	13.89	14.71	14.38	19.48	10.65	10.65	12.39	12.84	11.73	13.37	10.36	12.20	14.71	9.81	12.08	12.54	11.27	11.54	12.25
MnO	0.20	0.22	0.18	0.18	0.19	0.16	0.17	0.20	0.17	0.17	0.18	0.16	0.18	0.18	0.16	0.16	0.17	0.17	0.19	0.00
MgO	4.47	4.41	5.62	4.18	6.05	7.91	7.57	6.20	5.99	4.77	4.29	3.46	3.97	4.57	8.21	6.00	4.68	3.64	5.67	0.20
CaO	7.92	5.37	8.85	5.47	8.28	10.63	11.37	7.94	5.55	7.22	7.16	6.79	6.83	6.13	10.22	6.98	6.54	7.77	8.83	7.50
Na_2O •	5.20	5.55	4.32	5.69	2.82	2.32	2.33	4.42	4.50	4.48	4.75	3.41	4.51	5.75	2.62	4.78	5.33	6.04	4.36	4.62
K_2O	0.23	0.75	0.02	0.06	0.33	0.37	0.14	0.10	1.59	0.94	0.32	1.99	0.50	0.06	0.72	0.42	0.25	0.25	0.02	3.59
P_2O_5	0.27	0.33	0.23	0.21	0.17	0.22	0.11	0.19	0.15	0.16	0.17	0.38	0.24	0.20	0.08	0.16	0.14	0.19	0.24	0.50
S	0.24	0.19	0.03	0.06	0.34	0.12	0.14	0.12	0.01	0.16	0.07	0.01	0.04	0.20	0.12	0.00	0.06	0.14	0.03	0.05
LOI	2.47	1.70	3.28	2.10	5.99	2.55	2.51	2.55	2.47	2.44	2.51	1.39	2.04	1.85	2.57	2.54	2.56	2.64	3.36	5.35
Total	99.75	100.18	99.98	100.09	99.96	100.13	100.03	99.80	100.03	99.91	99.50	99.87	99.78	100.03	99.53	99.60	100.04	100.31	99.97	100.35
Nb	4	5	3	3	3	4	1	4	2	3	3	10	3	2	2	2	3	3	6	3
Zr	137	187	97	102	84	101	55	94	75	77	106	116	136	114	41	59	58	108	95	23
Y	51	51	43	39	28	27	22	32	31	29	44	28	43	43	20	27	29	39	24	7
Sr	169	209	143	144	297	303	280	329	159	324	181	777	299	147	479	263	227	141	191	140
Rb	3	10	1	2	6	8	2	2	17	21	6	22	8	3	15	9	7	3	1	8
Th	3	5	2	7	5	5	6	5	5	5	6	7	6	6	3	4	5	5	8	4
Pb	4	5	8	7	5	1	3	5	5	3	6	7	1	4	3	2	3	3	6	5
U	2	1	2	1	3	2	1	2	4	2	0	4	2	1	2	1	0	0	1	3
Ga	17	20	18	17	19	17	14	18	16	20	21	18	21	19	15	14	17	15	16	10
Zn	124	114	132	99	123	82	74	100	88	75	103	105	86	111	68	85	84	100	87	81
Cu	85	30	121	60	273	67	100	110	34	95	73	246	64	78	98	58	85	78	109	108
Ni	29	7	18	13	82	117	76	48	22	23	20	9	14	23	85	36	21	23	31	33
Cr	94	9	30	26	47	224	256	71	18	17	30	10	9	26	240	142	22	29	35	150
Ce	15	26	15	19	13	16	12	15	15	14	13	40	24	14	12	14	10	13	24	14
Nd	16	27	14	18	12	15	12	19	11	15	15	28	19	15	9	11	10	15	19	9
Ba	51	210	47	29	125	78	125	87	126	196	67	618	97	59	151	83	73	19	38	253
V	448	334	541	504	1062	266	310	350	379	379	458	310	317	464	259	325	468	403	294	344
La	12	15	9	7	8	12	4	10	6	6	12	23	11	8	2	8	8	5	18	7
Sc	40	36	38	37	55	34	43	43	34	31	42	25	27	33	46	39	36	32	35	62

Pigma Barney and Upper Barnard samples were collected originally by Cross (1983). All samples were analysed by XRF for major and trace element abundances at the University of New England, Armidale, analyst J. Bedford.

from Nundle to near Taree. These basaltic intrusions and commonly pillowed extrusive rocks mostly occur at stratigraphical levels above the meta-felsic igneous rocks (upper Tamworth Group; Yarrimie Formation and equivalents). Mineralogically, they consist almost entirely of plagioclase + Ca-rich pyroxene + Fe–Ti oxides or low-grade alteration products of these minerals (e.g. albite, chlorite, carbonate, ± epidote and rare pumpellyite). Tamworth Group basalts and dolerites have undergone low-grade (prehnite-pumpellyite facies) metamorphic alteration.

Numerous geochemical analyses are available for Gamilaroi terrane basalts (Offler 1982; Offler *et al.* 1988; Cross 1983; Morris 1988) and, together with the authors' data (Table 2), provide the basis for our interpretation. Cross (1983) originally recognized the similarity of many Tamworth Group basalts to relatively evolved back arc basalts. Major and trace element abundances (Table 2; Cross 1983; Offler 1982; Morris 1988) resemble those from rocks formed in both modern and ancient island-arc rift zones (e.g. Fryer *et al.* 1990*a*; Fryer *et al.* 1992; Ikeda & Yuasa 1989; Pederson & Hertogen 1990; Stern *et al.* 1990; Smellie & Stone 1992; Taylor *et al.* 1992a; Taylor *et al.* 1992b). Si, Ti, Al, Fe and Mg relations are transitional between those of island arc tholeiites and relatively evolved MORB. Ti/V ratios for basaltic rocks are similar to those of back-arc basin basalts (Shervais 1982) and high ($\geqslant$20) Ti/V ratios for the felsic rocks are characteristic of calc-alkaline lavas. Offler *et al.* (1988) reported relatively flat chondrite-normalized REE distributions and noted that LILE and HFSE contents of Gamilaroi terrane igneous rocks are characteristic of tholeiitic and calc-alkaline magmas erupted in island-arc settings. On some basalt discriminant diagrams (e.g. Ti–Zr; Fig. 4), Gamilaroi terrane basalts plot predominantly within the MORB field of Pearce (1980). Associated felsic rocks plot within the field of arc lavas. Conversely, most basalts analyses plot within an island-arc tholeiite field on a Cr–Y discriminant plot (Fig. 4). Notably, no samples plot within the MORB field on this diagram and all samples show characteristics which indicate a considerably more refractory source than would be expected for normal MORB.

Island-arc tholeiitic rocks are more common in the SE of the terrane. Basalts in the Glenrock area, within a sinistrally displaced fault sliver of Gamilaroi terrane rocks located to the NE of the main body of the terrane, have the most depleted characteristics of all rocks analysed

from the terrane whereas basalts in the SW of Glenrock have more MORB-like characteristics (Offler *et al.* 1988). Overall, rocks in the SW Glenrock to Nundle areas are less depleted than those in the Pigna Barney, NE Glenrock and Barry areas. However, stratigraphical resolution is presently insufficient to determine if this is an up-section trend indicating gradual evolution of the arc rift or, if the rocks are coeval, whether or not this reflects relative positions in a transect across a rifted island arc.

Discussion

The close spatial association of meta-felsic igneous rocks, voluminous siliceous tuff, volcaniclastic sedimentary rocks and the absence of continent-derived detritus or contamination in the Gamilaroi terrane clearly indicate an intra-oceanic island-arc setting (Cawood & Flood 1989). Comparison with modern intra-oceanic island arcs and their associated forearc regions indicates that the volcanic rocks were probably erupted in an intra-arc environment and are unlikely to have been erupted within or flowed as far as a forearc basin. The lack of high-Mg series rocks is more compatible with rifting in the axial regions of an intra-oceanic island arc rather than the forearc. Meta-felsic igneous rocks are succeeded up-section by volcaniclastic sedimentary rocks intercalated with island-arc tholeiitic basalts and E-type MORB. This is interpreted as evidence of rifting within the Gamilaroi terrane intra-oceanic island arc during the Middle to Late Devonian (Cross 1983). Modern examples of similar rock associations are known from the Lau Basin (Gill 1976, 1987; Leg 135 Scientific Party 1992), Scotia arc (Saunders & Tarney 1979) and Mariana Trough (Hawkins & Melchoir 1985). Results of recent ODP Legs 125 and 126 drilling in the Izu-Bonin-Mariana system (Fryer *et al.* 1990*b*, 1992; Taylor *et al.* 1990, 1992a) provide excellent modern analogues, including the Sumisu Rift, an actively developing intra-oceanic island-arc rift. In each of these areas, felsic volcanic-arc terrains have rifted with subsequent eruption of island-arc tholeiites and progressively more MORB-like basaltic lavas in the developing rift. Thick sequences of tuffs and volcaniclastic sediments eroded from the flanks of the rifted arc have accumulated in the rift basins.

If, as the evidence suggests, the Gamilaroi terrane developed as a Devonian intra-oceanic island arc which experienced local rifting, then previous interpretation as a forearc basin is incorrect. Indeed, Morris (1988) indicated that compositions of the Tamworth Group basalts

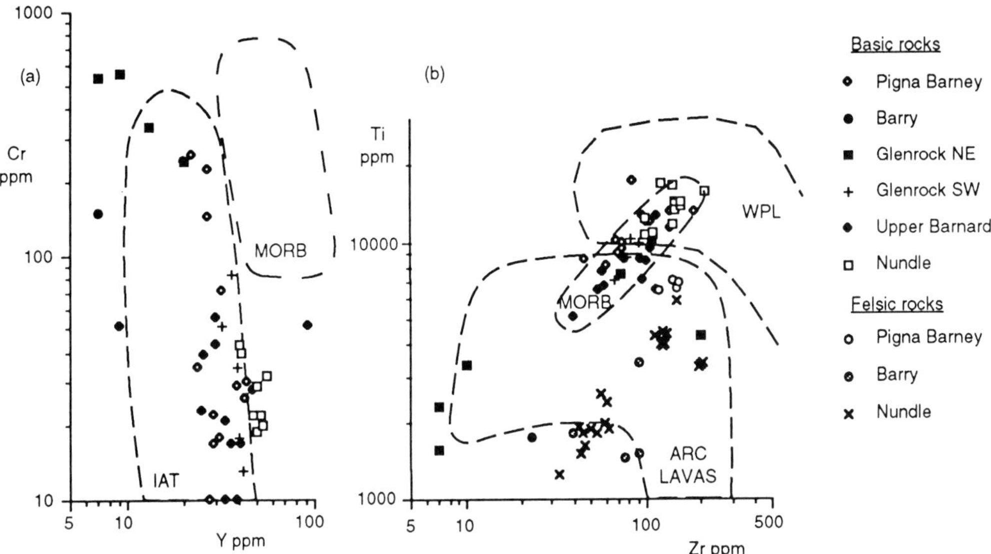

Fig. 4. (a) Cr–Y relations of basaltic rocks in the Gamilaroi terrane (fields for island arc tholeiites and MORB after Pearce 1982; Pearce *et al.* 1984). (b) TiO$_2$–Zr relations of basic and felsic igneous rocks in the Gamilaroi terrane (fields for mid-ocean ridge, within plate and arc-related lavas after Pearce 1980). Analyses used are those published by Offler (1982) for basalt samples in the Glenrock area, Morris (1988) for basalts from the Upper Barnard and Nundle districts, and Cawood & Flood (1989) for felsic rocks in the Nundle district together with those reported in Tables 1 & 2 of this paper.

are anomalous in terms of existing models for the Gamilaroi terrane, in which a forearc setting is inferred. However, caution is required in interpretation, as results of studies of western Pacific modern oceanic island-arc systems show that the volcanic front is imperfectly defined, with volcanism in the Izu-Bonin-Mariana system spanning over 100 km across the strike of the arc system. The location of the new arc which develops after rifting is not necessarily always trenchward of the remnant arc. Nascent rifts may develop in either back-arc or forearc basins. Geochemistry and elemental abundances, often used by those who study ancient rocks as diagnostic clues to tectonic configuration, are no panacea. Although some geochemical patterns appear to exist in data sets from arc transects (eg. Taylor *et al.* 1992b), there are many anomalies and databases for ancient arc rifts are, in many cases, considerably more restricted than those for now well-documented modern systems. Ancient systems are inherently more difficult to interpret accurately because of uncertainties introduced by later tectonic dismemberment and poor exposure. Unless excellent biostratigraphical constraints are available it is easy to confuse spatial variations in geochemistry with those of a temporal nature related to evolution of magmatic systems.

Reservations notwithstanding, the Gamilaroi terrane appears to contain intra-oceanic island-arc rocks and products of rifting. The presence of both felsic and basic lavas within this succession and the geochemical characteristics of these rocks provides evidence for a rifted arc or arc-flank rather than forearc setting at the time of its development. However, from existing knowledge it is not possible to state unequivocally whether rifting took place in a forearc, intra-arc or back-arc setting. Neither is it possible to determine, from studies of the Gamilaroi terrane rocks alone, the polarity of this arc system.

Implications

Reappraisal of Gamilaroi terrane lithostratigraphy, aided by data from radiolarian studies and comparisons with sedimentation rates of modern intra-oceanic arc environments, permits development of a more appropriate model for the development of the terrane. Many aspects of the existing tectonic models for the NEO, while developed on the basis of all available existing data at the time, are simplistic and can be greatly improved as new data and results of studies of analogous rocks elsewhere become available (Flood 1988; Flood & Aitchison 1993). Previous interpretation of the Gamilaroi terrane was based on widespread belief in an intimate relationship between all NEO terranes throughout their evolution. New radiolarian and chronostratigraphical age constraints clearly show that this hypothesis is untenable (Aitchison *et al.* 1992*a*). With improved stratigraphical resolution it is timely to consider evolution of individual terranes prior to development of any orogen-wide model. Previous interpretations of rocks in the Tamworth Belt have considered this zone to represent a long-lived (possibly Cambrian to Permian) forearc basin. However, we can now recognize that the oldest *in-situ* rocks in the belt are those of a Silurian–Devonian rifted intra-oceanic island-arc assemblage (Gamilaroi terrane) which is discrete from younger overlying convergent continental margin arc rocks of Carboniferous age.

Significantly, as the Gamilaroi terrane was an intra-oceanic island arc, it could not have developed as part of the continental margin of Gondwana during the Devonian and thus it must be allochthonous to eastern Australia. By definition, all other terranes of similar, or older, age outboard of the Gamilaroi terrane must now also be regarded as suspect. Subduction complex rocks to the east of the PMFS have been used previously to provide a basis for inferring subduction polarity (e.g. Cawood 1982, 1983). However, these rocks constitute part of a separate terrane which was not necessarily related genetically to the Gamilaroi terrane during the Devonian. Indeed, recently obtained radiolarian data indicate that their development was not coeval (Aitchison 1990). No unambiguous evidence is known, which could indicate the polarity of the Gamilaroi terrane arc.

In an attempt to resolve the question of polarity, the authors have examined all modern settings where continents and arcs are converging, through the subduction of intervening oceanic crust. In such areas, subduction is occurring beneath the arc rather than the continental margin (e.g. Timor and Australia, Karig *et al.* 1987; Taiwan and China, Aubouin 1990; Dorsey 1992). Through analogy we suggest that oceanic crust intervening between the Gamilaroi terrane and Gondwana was subducted eastwards under the western margin of the Gamilaroi terrane arc. This contrasts with most existing models in which various authors (e.g. Leitch 1975; Cawood 1983; Murray *et al.* 1987) have suggested that the geology of the NEO is the product of magmatic arc development associated with a west-dipping subduction zone active along the eastern margin of Australia through the Palaeozoic. Although both

Devonian and Carboniferous volcanic rocks are recognized in the vicinity of the Gamilaroi terrane outcrop, there is a marked change in the nature and locus of volcanic activity. The Gamilaroi terrane, a ?Silurian–Devonian lithotectonic entity, pre-dates younger Carboniferous Gondwana margin calc-alkaline arc volcanic rocks (McPhie 1987). Carboniferous volcanic rocks developed along the Gondwana margin and were associated with west-dipping subduction beneath this margin. However, these rocks post-date Gamilaroi terrane/Gondwana continent collision and amalgamation. The authors suggest that a flip in subduction polarity may have occurred in a manner similar to that associated with the Cenozoic Papua New Guinea/Australian continent collision (e.g. Cooper & Taylor 1987).

Several other factors point to eastward subduction of intervening crust beneath the Gamilaroi terrane island arc. This inferred west-facing arc system would place the Gamilaroi terrane in an upper plate position facilitating its accretion and preservation through obduction during collision with Gondwana. Recent deep seismic surveys carried out by the Bureau of Mineral Resources have revealed the presence of probable Lachlan orogen layered rocks structurally underlying westward overthrust Gamilaroi terrane rocks north of Tamworth (Korsch *et al.* 1992, 1993). Conversely, if Gondwana had been in the upper plate position then we would not expect to find significant evidence of the Gamilaroi arc. Further evidence for Gamilaroi terrane having been thrust over an older basement is becoming available through SHRIMP studies of inheritance in magmatic zircon cystals in igneous rocks of the Carboniferous arc. Whereas there is no evidence for any inherited older cores in zircon crystals from the one analysed Gamilaroi terrane sample (unpublished data of Aitchison & Ireland), zircons from overlying Carboniferous continental margin volcano-sedimentary overlap assemblage rocks contain evidence of both Silurian and Precambrian inheritance (Roberts *et al.* 1991). Further SHRIMP studies on zircons from the Port Macquarie region have revealed a significant component of inheritance probably related to older crust now underlying much of the New England region (unpublished data of Aitchison & Ireland). Both the SHRIMP and deep seismic results support interpretation of the Gamilaroi terrane as having overridden the eastern margin of Gondwana.

We propose a model for development of the Gamilaroi terrane in a west-facing intra-oceanic island arc setting during the (?Silurian–)

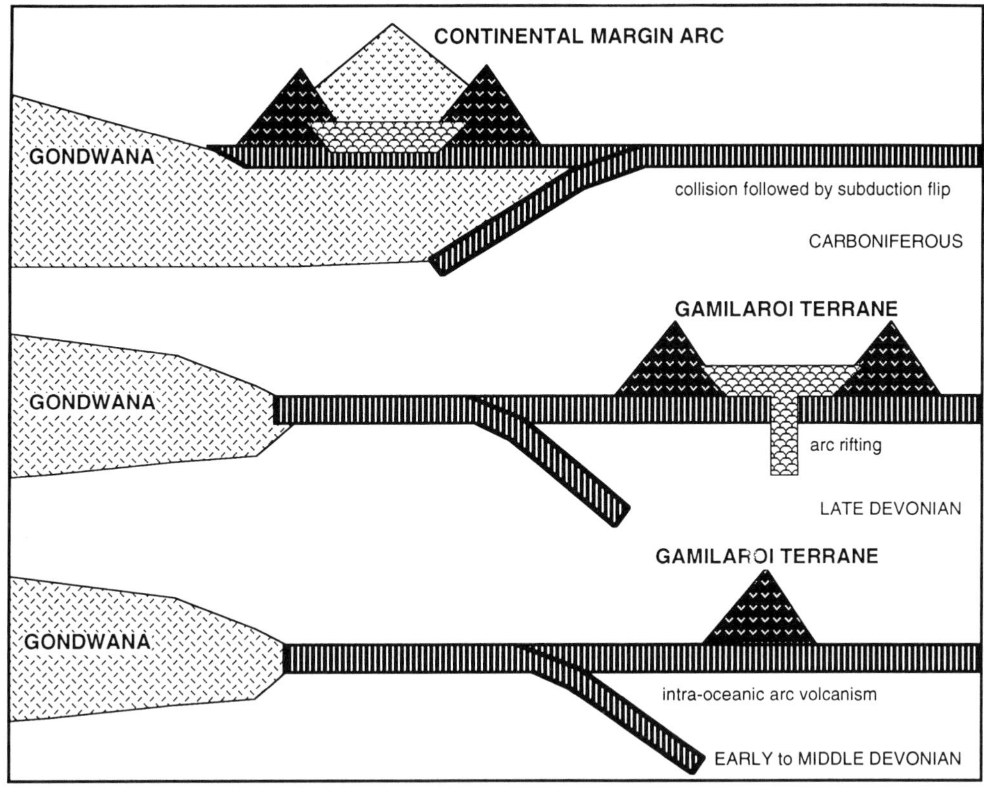

Fig. 5. Cartoon depicting the proposed model for the tectonic evolution of the Gamilaroi terrane. Not to scale.

Devonian in which the arc was over-riding eastward-subducting oceanic crust (Fig. 5). Meta-felsic lithologies represent parts of this island arc. During the Middle to Late Devonian, the arc experienced a period of rifting. Mass wasting of volcaniclastic sediments into arc-rift basins, together with limestone blocks shed off uplifted older basement arc edifices resulted in development of the Tamworth Group. These sediments were intruded by dolerites and are intercalated with erupted basalts. Near the end of the Devonian, the Gamilaroi terrane, which was in an upper plate setting, collided with and overrode, or was obducted onto, the leading edge of older crust rocks of Gondwana. Collision resulted in a subduction flip and during the Early Carboniferous a new east-facing subduction complex developed east of accreted Gamilaroi terrane rocks in association with west-dipping subduction. Collision is marked by a hiatus in volcanic activity followed by a major westward shift in the locus of volcanism.

The authors suggest that, in the light of this model, closer examination of Carboniferous Tamworth Belt rocks will allow determination of the precise timing of collision and recognition of a foreland basin sequence related to this collision event. Any such collision of an intra-oceanic island arc with a continental margin would result in orogenesis followed by extensional relaxation and is likely to be reflected in the sedimentary record. Further examination of the New England region may also lead to the recognition of post-collisional extension-related eruption of subduction-related (shoshonitic) magmas similar to those found in analogous settings elsewhere. Recently published studies (Morand 1993) on rocks of the Late Silurian to Devonian Calliope terrane (Aitchison & Flood 1991) in the northern NEO, Queensland, give details of rocks which we consider likely correlatives of the Gamilaroi terrane. These rocks are closely similar, both in their stratigraphy and geochemistry, to rocks described in this paper. As with the Gamilaroi terrane, they are overlain by latest Devonian to Carboniferous overlap assemblages and continental margin arc rocks (Leitch *et al.* 1992). The Calliope asemblage has also been considered to represent parts of an exotic island arc accreted to the Gondwana margin (Powell 1984). Recognition of such assemblages in the NEO is a major advance and ongoing reappraisal of New England geology in light of new data from modern analogues will eventually lead to an improved understanding of the tectonic evolution of the eastern margin of Gondwana.

We gratefully acknowledge the financial assistance of BHP Utah and the Australian Research Council. K. Cross (Western Mining, Olympic Dam) has graciously allowed us access to valuable data. Reviews and comments from R. J. Arculus, A. J. Crawford and R. J. Korsch on earlier versions of this manuscript and discussions with R. Offler and J. Gamble are appreciated.

References

AITCHISON, J.C. 1988a. Middle–Late Devonian Radiolaria from the Yarrimie Formation, Tamworth Group, northeastern New South Wales, Australia. *In*: SCHMIDT-EFFING, R. & BRAUN, A. (eds) *First International Conference on Radiolaria (EURORAD V), abstracts.* Geologica et Paleontologica, Marburg, Germany, 4.

—— 1988b. Late Paleozoic radiolarian ages from the Gwydir terrane, New England orogen, eastern Australia. *Geology*, **16**, 793–795.

—— 1990. Significance of Devonian–Carboniferous radiolarians from accretionary terranes of the New England orogen, eastern Australia. *Marine Micropaleontology*, **15**, 365–378.

—— & FLOOD, P.G. 1990. Preliminary tectonostratigraphic terrane map of the southern part of the New England Orogen, Eastern Australia. *In*: WILEY T.J., HOWELL, D.G. & WONG, F.L. (eds) *Tectonostratigraphic terranes of the Circum-Pacific region, Circum Pacific Council for Energy and Mineral Resources. Earth Science Series*, **13**, 81–85.

—— & —— 1991. New England Orogen. *In*: MOULLADE, M. & NAIRN, A.E.M. (eds) *The Phanerozoic Geology of the World. I. Palaeozoic. A.* Elsevier, Amsterdam, 301–308.

——, —— & SPILLER, F.C.P. 1992a. Tectonic setting and paleoenvironment of terranes in the southern New England orogen as constrained by radiolarian biostratigraphy. *Palaeogeography, Palaeoclimatology, Palaeoecology*, **93**, 31–54.

——, IRELAND, T.R., BLAKE, M.C. JR. & FLOOD, P.G. 1992b. 530 Ma zircon age for ophiolite from the New England Orogen: Oldest rocks known from eastern Australia. *Geology*, **20**, 125–128.

AUBOUIN, J. 1990. The west Pacific geodynamic model. *In*: ANGELIER, J. (ed.) *Geodynamic Evolution of the Eastern Eurasian Margin. Tectonophysics*, **183**, 1–7.

BENSON, W.N. 1915. The geology and petrology of the Great Serpentine Belt of New South Wales. Part iv. The dolerites, spilites and keratophyres of the Nundle district. *Proceedings of the Linnean Society of New South Wales*, **40**, 121–173.

—— 1918. The geology and petrology of the Great Serpentine Belt of New South Wales. Part viii. The extension of the Great Serpentine Belt from the Nundle district to the coast. *Proceedings of the Linnean Society of New South Wales*, **43**, 593–599.

CAWOOD, P.A. 1976. Cambro-Ordovician strata, northern New South Wales. *Search*, **7**, 317–318.

—— 1982. Correlation of stratigraphic units across the Peel Fault System. *In*: FLOOD, P.G. & RUNNEGAR, B. (eds) *New England Geology.* University of New England, Armidale, 53–61.

—— 1983. Modal composition and detrital clinopyroxene geochemistry of lithic sandstones from the New England Fold Belt (east Australia): A Paleozoic forearc terrane. *Geological Society of America Bulletin*, **94**, 1199–1214.

—— & FLOOD, R.H. 1989. Geochemical character and tectonic significance of Early Devonian keratophyres in the New England Fold Belt, eastern Australia. *Australian Journal of Earth Sciences*, **36**, 297–311.

—— & LEITCH, E.C. 1985. Accretion and dispersal tectonics of the southern New England Fold Belt, Eastern Australia. *In*: HOWELL, D.G. (ed.) *Tectonostratigraphic terranes of the Circum-Pacific region, Circum Pacific Council for Energy and Mineral Resources. Earth Science Series*, **1**, 481–492.

CONAGHAN, P.J., MOUNTJOY, E.W., EDGECOMBE, D.R., TALENT, J.A. & OWEN, D.E. 1976. Nubrigyn algal reefs (Devonian), eastern Australia: allochthonous blocks and megabreccias. *Geological Society of America Bulletin*, **87**, 515–530.

CONEY, P.J., EDWARDS, A., HINE, R., MORRISON, F. & WINDRIM, D. 1990. The regional tectonics of the Tasman orogenic system, eastern Australia. *Journal of Structural Geology*, **12**, 519–543.

COOPER, P. & TAYLOR, B. 1987. Seismotectonics of New Guinea: a model for arc reversal following arc-continent collision. *Tectonics*, **6**, 53–67.

CROOK, K.A.W. 1961. Stratigraphy of the Tamworth Group (Lower and Middle Devonian), Tamworth–Nundle District, N.S.W. *Journal and Proceedings of the Royal Society of New South Wales*, **94**, 173–188.

CROSS, K.C. 1983. *The Pigna Barney ophiolitic complex and associated basaltic rocks, northeastern New South Wales, Australia.* PhD thesis, University of New England, Armidale, Australia.

DORSEY, R.A. 1992. Collapse of the Luzon Volcanic Arc during onset of arc-continent collision: evidence from a Miocene-Pliocene unconformity, eastern Taiwan. *Tectonics*, **11**, 177–191.

FERGUSSON, C.L. & FLOOD, P.G. 1984. A late Palaeozoic subduction complex in the Border Rivers area of southeast Queensland. *Proceedings of the Royal Society of Queensland*, **95**, 47–55.

FLOOD, P.G. 1988. New England Orogen: Geosyncline, mobile belt and terranes. *In*: KLEEMAN, J.D. (ED.) *New England orogen tectonics and metallogenesis.* Armidale, University of New England, 1–6.

—— & AITCHISON, J.C. 1988. Tectonostratigraphic terranes of the southern part of the New England

Orogen. *In*: KLEEMAN, J.D. (ed.) *New England orogen tectonics and metallogenesis*. Armidale, University of New England, 7–10.

—— & —— 1992. Late Devonian accretion of the Gamilaroi terrane to Gondwana: provenance linkage provided by quartzite clasts in the overlap sequence. *Australian Journal of Earth Sciences*, **39**, 539–544.

—— & —— 1993. Understanding New England orogen: the comparative approach. *In*: FLOOD, P.G. & AITCHISON, J.C. (eds) *New England orogen, eastern Australia*. University of New England, Armidale, 1–10.

FRYER, P., TAYLOR, B., LANGMUIR, C.H. & HOCHSTAEDTER, A.G. 1990*a*. Petrology and geochemistry of lavas from the Sumisu and Torishima backarc rifts. *Earth and Planetary Science Letters*, **100**, 161–178.

——, PEARCE, J.A., STOKKING, L.B., *et al*. 1990*b*. *Proceedings of the Ocean Drilling Program, Initial Reports*, **125**, 1092 pp.

——, ——, *et al*. 1992. *Proceedings of the Ocean Drilling Program, Scientific Results*, **125**, 716 pp.

GILL, J.B. 1976. Composition and ages of Lau Basin and Ridge volcanic rocks: implications for evolution of an inter-arc basin and remnant arc. *Geological Society of America Bulletin*, **87**, 1384–1395.

—— 1987. Early geochemical evolution of an oceanic island arc and backarc: Fiji and the South Fiji Basin. *Journal of Geology*, **95**, 589–615.

HARRINGTON, H.J. 1974. The Tasman Geosyncline in Australia. *In*: DENMEAD, A.K., TWEEDALE, G.W. & WILSON, A.F. (eds) *The Tasman Geosyncline – a symposium*. Geological Society of Australia, Brisbane, 383–407.

HAWKINS, J.W. & MELCHOIR, J.T. 1985. Petrology of Mariana Trough and Lau Basin basalts. *Journal of Geophysical Research*, **90**, 11431–11468.

HINDE, G.J. 1899. On the Radiolaria in the Devonian Rocks of New South Wales. *Quarterly Journal of the Geological Society of London*, **55**, 38–64.

HOLDSWORTH, B.K. & JONES, D.L. 1980. Preliminary radiolarian zonation for Late Devonian through Permian time. *Geology*, **8**, 281–285.

IKEDA, Y. & YUASA, M. 1989. Volcanism in nascent back-arc basins behind the Shichito Ridge and adgacent areas in the Izu-Ogasawara arc, northwest Pacific: evidence for mixing between E-type MORB and island arc magmas at the initiation of back-arc rifting. *Contributions to Mineralogy and Petrology*, **101**, 377–393.

INESON, J.R. 1985. Submarine glide blocks from the Lower Cretaceous of the Antarctic Peninsula. *Sedimentology*, **32**, 659–670.

ISHIGA, H., LEITCH, E.C., WATANABE, T., NAKA, T. & IWASAKI, M. 1988. Radiolarian and conodont biostratigraphy of siliceous rocks from the New England Fold Belt. *Australian Journal of Earth Sciences*, **35**, 73–80.

KARIG, D.E., BARBER, A.J., CHARLTON, T.R., KLEMPERER, S. & HUSSONG, D.M. 1987. Nature and distribution of deformation across the Banda Arc-Australian collision zone at Timor. *Geological Society of America Bulletin*, **98**, 18–32.

KIMBROUGH, D.L., CROSS, K.C. & KORSCH, R.J. 1993. U–Pb isotopic ages for zircons from the Pola Fogal and Nundle granite suites, southern New England Orogen. *In*: FLOOD, P.G. & AITCHISON, J.C. (eds) *New England orogen, eastern Australia*. University of New England, Armidale, 403–412.

KORSCH, R.J. 1984. Sandstone compositions from the New England Orogen, eastern Australia: implications for tectonic setting. *Journal of Sedimentary Petrology*, **54**, 192–211.

—— & HARRINGTON, H.J. 1981. Stratigraphic and structural synthesis of the New England Orogen. *Journal of the Geological Society of Australia*, **28**, 205–226.

——, WAKE-DYSTER, K.D. & JOHNSTONE, D.W. 1992. Preliminary geological results from deep seismic reflection profiling in the Gunnedah Basin and New England orogen. *Geological Society of Australia abstracts*, **32**, 27.

——, —— & —— 1993. The Gunnedah Basin – New England orogen deep seismic reflection profile: implications for New England tectonics. *In*: FLOOD, P.G. & AITCHISON, J.C. (eds) *New England orogen, Armidale, eastern Australia*. University of New England, 85–100.

LEG 135 SCIENTIFIC PARTY 1992. Evolution of backarc basins: ODP Leg 135, Lau Basin. *EOS American Geophysical Union Transactions*, **73**, 241–247.

LEITCH, E.C. 1974. The geological development of the southern part of the New England Fold Belt. *Journal of the Geological Society of Australia*, **21**, 133–156.

—— 1975. Plate tectonic interpretation of the Paleozoic history of the New England Fold Belt. *Geological Society of America Bulletin*, **86**, 141–154.

—— & CAWOOD, P.A. 1987. Provenance determination of volcaniclastic rocks: the nature and tectonic significance of a Cambrian conglomerate from the New England Fold Belt, eastern Australia. *Journal of Sedimentary Petrology*, **57**, 630–638.

—— & WILLIS, S.G.A. 1977. Nature and significance of plutonic clasts in Devonian conglomerates of the New England Fold Belt. *Journal of the Geological Society of Australia*, **29**, 83–89.

——, FERGUSSON, C.L. & HENDERSON, R.A. 1992. The intra-Devonian angular unconformity at Mt Gelobera, south of Rockhampton, central Queensland. *Australian Journal of Earth Sciences*, **39**, 121–122.

MCPHIE, J. 1983. Outflow ignimbrite sheets from Late Carboniferous calderas, Currabubula Formation, New South Wales, Australia. *Geological Magazine*, **120**, 487–503.

—— 1984. Permo-Carboniferous silicic volcanism and palaeogeography on the western edge of the New England Orogen, north-eastern New South Wales. *Journal of the Geological Society of Australia*, **31**, 133–146.

—— 1986. Primary and redeposited facies from a large-magnitude, rhyolitic, phreatomagmatic eruption: Cana Creek Tuff, Late Carboniferous,

Australia. *Journal of Volcanological and Geothermal Research*, **28**, 319–350.

—— 1987. Andean analogue for Late Carboniferous volcanic arc and arc flank environments of the western New England Orogen, New South Wales, Australia. *Tectonophysics*, **138**, 269–288.

MORAND, V.J. 1993. Stratigraphy and tectonic setting of the Calliope Volcanic Assemblage, Rockhampton area, Queensland. *Australian Journal of Earth Sciences*, **40**, 15–30.

MORRIS, P.A. 1988. Petrogenesis of fore-arc metabasites from the Paleozoic of New England, eastern Australia. *Mineralogy and Petrology*, **38**, 1–16.

MURRAY, C.G., FERGUSSON, C.L., FLOOD, P.G., WHITAKER, W.G. & KORSCH, R.J. 1987. Plate tectonic model for the Carboniferous evolution of the New England Fold Belt. *Australian Journal of Earth Sciences*, **34**, 213–236.

NAZAROV, B.B. & ORMISTON, A.R. 1986. Trends in the development of Paleozoic Radiolaria. *Marine Micropaleontology*, **11**, 3–32.

OFFLER, R. 1982. Geochemistry and tectonic setting of igneous rocks in the Glenrock Station area, N.S.W. *Journal of the Geological Society of Australia*, **29**, 443–455.

——, GAMBLE, J. & FARDY, J. 1988. Palaeozoic forearc volcanism in the southern New England fold belt, New South Wales. *9th Geological Society of Australia Convention, Brisbane. Abstracts*, 449.

PEARCE, J.A. 1980. Geochemical evidence for the genesis and eruptive settings of lavas from Tethyan ophiolites. *In*: PANAYIOTOU, A. (ed.) *Ophiolites: Proceedings of the International Ophiolite Symposium Cyprus 1979*. Geological Survey Department, Cyprus, 261–272.

—— 1982. Chemical and isotope characteristics of destructive margin magmas. *In*: THORPE, R.S. (ed.) *Andesites*. John Wiley, Chichester, 525–548.

——, LIPPARD, S.J. & ROBERTS, S. 1984. Characteristics and tectonic significance of suprasubduction zone ophiolites. *In*: KOKELAAR, B.P. & HOWELLS, M.F. (eds) *Marginal Basin Geology: Volcanic and associated sedimentary and tectonic processes in modern and ancient marginal basins*. Geological Society, London, Special Publications, **16**, 77–94.

PEDDAR, A.E.H. 1966. The Devonian System of New England, New South Wales, Australia. *International Symposium on the Devonian System. Proceedings Alberta Society of Petroleum Geologists, Calgary*, **2**, 135–142.

PEDERSON, R.B. & HERTOGEN, J. 1990. Magmatic evolution of the Karmøy Ophiolite Complex, SW Norway: relationships between MORB-IAT-boninitic-calc-alkaline and alkaline magmatism. *Contributions to Mineralogy and Petrology*, **104**, 277–293.

POWELL, C. McA. 1984. Silurian to mid-Devonian – a dextral transtensional margin. *In*: VEEVERS, J.J. (ed.) *Phanerozoic Earth History of Australia*. Oxford Geological Science Series 2, Oxford, 309–329.

ROBERTS, J., CLAGUE-LONG, J.C. & JONES, P.J. 1991. Calibration of the Carboniferous and Early Permian of the southern New England orogen by SHRIMP ion microprobe zircon analyses. *Twenty Fifth Newcastle Symposium on Advances in the Study of the Sydney Basin, University of Newcastle, Newcastle, Australia*, 38–43.

SAUNDERS, A.D. & TARNEY, J. 1979. The geochemistry of basalts from a back-arc spreading centre in the East Scotia Sea. *Geochimica et Cosmochimica Acta*, **43**, 555–572.

SHERVAIS, J.W. 1982. Ti-V plots and the petrogenesis of modern ophiolitic lavas. *Earth and Planetary Science Letters*, **59**, 101–118.

SMELLIE, J.L. & STONE, P. 1992. Geochemical control on the evolutionary history of the Ballantrae Complex, SW Scotland, from comparisons with recent analogues. *In*: PARSON, L.M., MURTON, B.J. & BROWNING, P. (eds) *Ophiolites and their modern oceanic analogues*. Geological Society, London, Special Publications, **60**, 171–178.

STERN, R.J., LIN, P.-N., MORRIS, J.D., JACKSON, M.C., FRYER, P., BLOOMER, S.H. & ITO, E. 1990. Enriched back-arc basin basalts from the northern Mariana Trough: implications for the magmatic evolution of back-arc basins. *Earth and Planetary Science Letters*, **100**, 210–225.

TAYLOR, B., FUJIOKA, K., JANECEK, T. *et al.* 1990. *Proceedings of the Ocean Drilling Program, Initial Reports*, **126**, 1002 pp.

——, ——, ——, *et al.* 1992. *Proceedings of the Ocean Drilling Program, Scientific Results*, **126**, 709 pp.

TAYLOR, R.N., MURTON, B.J. & NESBITT, R.W. 1992. Chemical transects across intra-oceanic arcs: implications for the tectonic setting of ophiolites. *In*: PARSON, L.M., MURTON, B.J. & BROWNING, P. (eds) *Ophiolites and their modern oceanic analogues*. Geological Society, London, Special Publications, **60**, 117–132.

VALLANCE, T.G. 1969. Albitic basic rocks of the Tamworth Group in the Nundle district: *In*: PACKHAM, G.H. (ed.) The Geology of New South Wales. B. Albitic basic rocks of the Tamworth Group in the Nundle District. *Journal of the Geological Society of Australia*, **16**, 235–237.

VEEVERS, J.J. 1984. *Phanerozoic Earth History of Australia*. Oxford Geological Science Series 2, Oxford.

Cretaceous to Cenozoic volcanism in South Korea and in the Sea of Japan: magmatic constraints on the opening of the back-arc basin

ANDRÉ POUCLET[1], JIN-SOO LEE[2], PHILIPPE VIDAL[3], BRIAN COUSENS[4]
& HERVÉ BELLON[5]

[1] *Géotectonique et Pétrologie des volcanites, Faculté des Sciences and CNRS URA 1366, BP 6759, 45067 Orléans cedex 2, France*

[2] *Korea Institute of Geology, Mining and Materials, Daedok Science Town, Daejon, South Korea*

[3] *Origine, Evolution et Dynamique des Magmas, CNRS URA 10, Université, 5 rue Kessler, 63038 Clermont-Ferrand cedex, France*

[4] *Department of Earth Sciences, Carleton University, Ottawa, ON, K1S 5B6, Canada*

[5] *Genèse et Evolution des Domaines Océaniques, CNRS URA D1278, Université, 6 av. Le Gorgeu, 29287 Brest cedex, France*

Abstract: The major element, trace element, and radiogenic isotope compositions of volcanic rocks in the back-arc area of the eastern Eurasian continental margin provide insight into the nature of the mantle wedge and constrain the magmatic evolution of the Japan Sea back-arc basin linked to its tectonic history. Different phases of post-Early Cretaceous volcanic activity are identified along the Korean margin and in the Japan Sea. Volcanic rocks from Korea include (1) Cretaceous and early Cenozoic calc-alkaline lavas of a volcanic arc at an active margin, and (2) Pliocene and Quaternary intraplate flood basalts and volcanic islands of alkaline composition. Japan Sea volcanic rocks consist of (1) early Cenozoic andesite flows of a remnant arc in the Yamato Bank, (2) early Miocene basalts of the Japan Sea basin basement, which share compositional characteristics of island arc tholeiites, continental rift tholeiites and back-arc basin basalts, (3) late Miocene seamounts of tholeiitic and mildly alkaline compositions, and (4) Pliocene and Quaternary alkaline volcanic islands.

Geochemically, these rocks belong to three broad magmatic groups: (1) an arc-related, calc-alkaline group of a continental, Andean margin type, which prevailed prior to the opening of the Japan Sea between the Cretaceous and early Miocene, (2) continental rift tholeiites and back-arc basin basalts, formed during the rifting stage in the early Miocene, and (3) an intraplate alkaline group similar to OIB, erupted later during spreading, between late Miocene and Holocene times.

Trace element and Sr, Nd and Pb isotopic compositions of selected samples show that the sources of magma Group 1 calc-alkaline lavas and magma Group 2 tholeiitic lavas included varying contributions of two main mantle components: an Indian Ocean MORB-like depleted mantle source (DMM) and an enriched mantle component similar to EM II. The latter component could represent DMM contaminated by subducted oceanic sediments incorporated into the lower lithosphere during the long-lived subduction of west Pacific crust. During the opening of the Japan Sea back-arc basin, the relative proportion of the DMM component dramatically increased between the rifting and spreading stages. It is also necessary to postulate a third component present in the sources of the Group 3, post-opening alkaline lavas, perhaps enriched mantle of EM I composition, which may also have resided in the subcontinental lithospheric mantle.

Attempting to understand the geodynamic history and magmatic evolution of volcanic arc/ back-arc systems associated with subduction zones is a complicated problem. This paper aims to document the magmatic features of volcanic rocks in and around the Japan Sea back-arc area, including pre-, syn- and post-rifting lavas, in order to study the composition and dynamics of the mantle beneath this area.

Along the Japanese island arc, successive Cretaceous to Cenozoic volcanic products are more or less superimposed. However, the mid-Miocene opening of the Japan Sea rifted the active Japanese arc and terminated Korean

Table 1. *Summary of volcanic phases in Korea and the Japan Sea since the Early Cretaceous location and magmatic affinities of the selected samples are also indicated*

Korea margin					Japan Sea (Yamato Basin)				
Age (Ma)	Pulse	Volcanic context and location	Sample	Magmatic affinity	Age (Ma)	Pulse	Volcanic context and location	Sample	Magmatic affinity
115									
	K1	Volcanic arc Gyeongsang Basin	K-1	Calc-alkaline (andesitic)					
80									
	K2	Volcanic arc Gyeongsang Basin and Chugaryeong graben	K-2	Calc-alkaline (andesitic)					
65									
	K3	Volcanic arc Gyeongsang Basin	K-3	Calc-alkaline (andesitic)					
44									
					26				
23									
						J1	Volcanic arc Yamato Bank		Calc-alkaline (tholeiitic)
	K4	Volcanic arc Pohang Basin	K-4a,b	Calc-alkaline (tholeiitic)	22				
21						J2	Initial rifting basin spreading	J-2a,b,c	Continental tholeiite
					17	J3	Yamato Basin	J-3a,b,c	Back-arc basin basalt
					13	J4	Yamato Basin seamounts		Tholeiitic and/or alkaline
						J5	Yamato Bank	J-5	Calc-alkaline
6	K5	Dykes, Pohang Basin Fissural volcanism	K-5	Calc-alkaline		J6	Volcanic islands Ulleung Island	J-6	Alkaline (potassic)
	K6	Chugaryeong Graben	K-61a,b		0				
0		Cheju Island	K-62a,b	Alkaline (sodic)					

arc-related, calc-alkaline volcanic activity. Within the Japan Sea itself, drilling during ODP Legs 127 and 128 penetrated middle Miocene basaltic rocks which form the floor of the basin. After a short period of back-arc spreading, activity became focused in basinal seamounts and volcanic islands located on the continental shelves and ridges.

We have selected samples of volcanic rocks from South Korea and from the Yamato Basin (a sub-basin of the Japan Sea), for geochemical and isotopic investigations. We find that magma sources within the sub-Japan Sea mantle have varied with time and involved at least three isotopic components.

Geodynamical context and sample selection

Six pulses of volcanic activity (K1–6) can be distinguished in the Korean peninsula since the early Cretaceous. The Japan Sea also contains evidence for at least six volcanic pulses (J1–6) from early Miocene times, although there is limited overlap with those in Korea (Table 1). Sample locations are shown in Fig. 1.

Korean peninsula

Southeastern Korea was an active continental margin of Andean-type before the opening of the Sea of Japan. The volcanic arc of the combined Korean and Japan margin of the Eurasian plate was located in the Gyeongsang and Pohang basins in South Korea. The calc-alkaline magmatic activity is dated from the early Cretaceous to the early Miocene. We distinguish four main arc-related volcanic pulses, which can be explained by variable conditions of convergence and subduction of successive oceanic plates (Palaeo-Pacific (Kula)

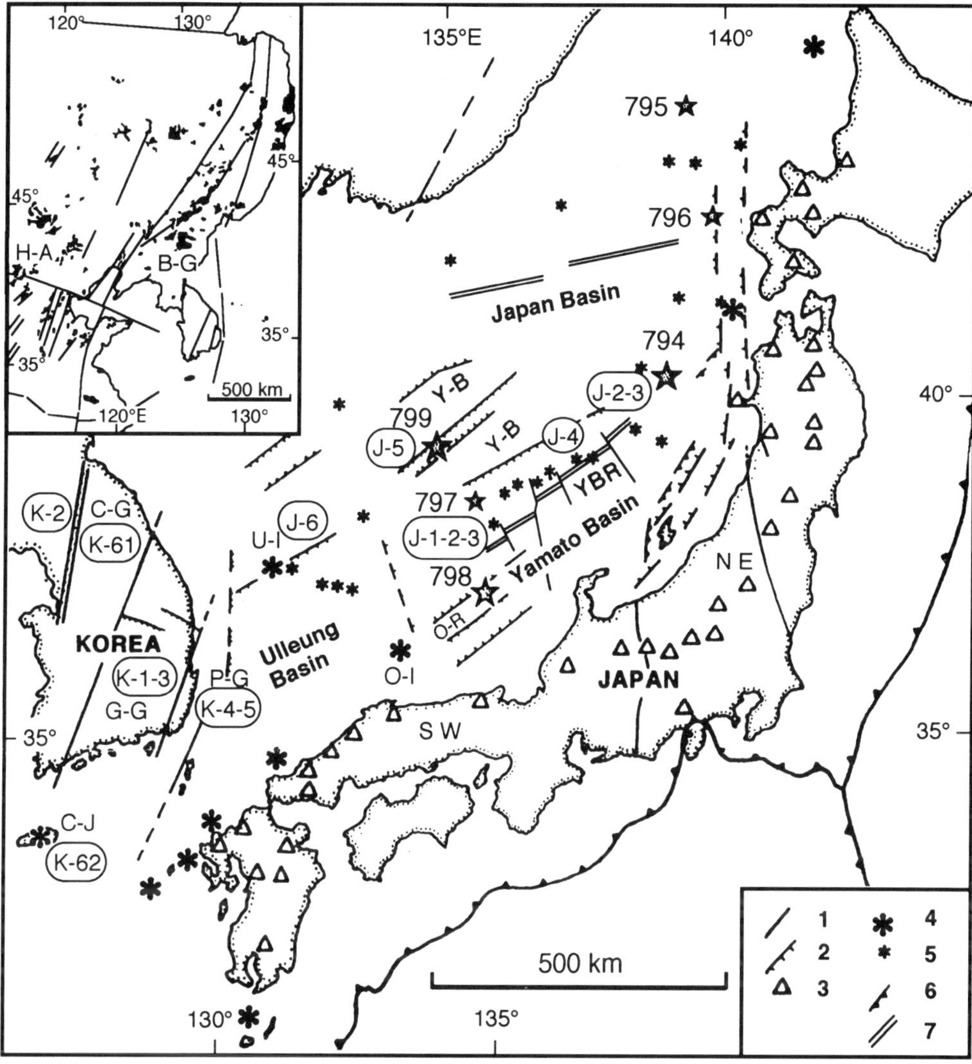

Fig. 1. Location and structural setting of the selected samples of volcanic rocks described in this paper. Symbols: small stars, Sites of ODP Leg 127; large stars, Sites of ODP Leg 128; 1, strike-slip faults; 2, normal faults; 3, main volcanoes; 4, volcanic islands; 5, seamounts; 6, thrusts and subduction zone; 7, spreading axis; C-G, Chugaryeong Graben; C-J, Cheju Island; G-G, Gyeongsang Basin; O-I, Oki Islands; O-R, Oki Ridge; P-G, Pohang Basin; U-I, Ulleung Island; Y-B, Yamato Bank; YBR, Yamato Basin Ridge spreading axis (according to Kimura *et al.* 1987). Inset: Distribution of Cenozoic basalts in eastern China (solid black ornament); B-G, Baegdusan; H-A, Hannuoba.

and Pacific plates), and variable tectonic stress conditions across the orogenic eastern Asian continental margin. After the opening of the Japan Sea back-arc area, two further volcanic pulses took place: in the Pohang Basin during the late Miocene, and in the hinterland and continental shelf region during the Plio-Quaternary.

Thus, the volcanic evolution of the Korean margin is schematically divided into six phases (K1–6 in Table 1).

K1. Between 115 and 80 Ma, volcano-tectonic activity was initiated by NNW convergence of the Izanagi and Eurasian plates and ended with a WNW reorientation of the oceanic plate movement at *c.* 80 Ma (Engebretson *et al.* 1985). The Gyeongsang Basin may have been created as an intra-arc basin. It shows a half-graben structure along NNE-trending normal faults, which are themselves cut by WNW transverse faults, and it is filled with fluvio-lacustrine sediments interbedded with volcanic products supplied from proximal arc volcanoes (Choi 1986). A complete magmatic sequence consists of (from base to top) basaltic andesite flows and sills intercalated with Barremian to Albian sedimentary rocks (Lee & Kim 1970), and thick andesitic pyroclastic flows followed by rhyodacite flows. The total thickness of the sequence reaches 2000 m (Kim & Lee 1981). The older lavas are generally severely altered and/or thermally metamorphosed by Late Cretaceous granitic intrusions. We sampled a relatively fresh flow interbedded with Albian strata (K–1). It is a plagioclase- and pyroxene- (diopside) phyric basaltic andesite with sparse microphenocrysts of pseudomorphed olivine.

K 2. Between 80 and 65 Ma, a distinct magmatic phase characteristic of a mature volcanic arc took place, including intrusion of the Bulgugsa granitic plutons, and the formation of subvolcanic ring complexes and volcanic calderas (Kim 1971; Jin *et al.* 1981, 1984). Northwest of the Gyeongsang area, transtensional rifting was responsible for the opening of the NNE-trending Chugaryeong Graben, which extends *c.* 200 km in length and 30 km in width. Moderate effusive and explosive volcanic activity formed calc-alkaline stratovolcanoes with basaltic to dacitic lava flows and pyroclastic deposits dated at 78–73 Ma (Lee 1989). A representative sample, K–2, an olivine-, plagioclase-, pyroxene- (augite) phyric basaltic andesite, was selected from the volcanic plug of Jeongog, the main basaltic vent in the Chugaryeong area.

K 3. In the early Cenozoic, renewed volcanic activity took place in the Gyeongsang Basin, especially during the Lutetian. It consists of basaltic flows and of N- to NE-trending dykes dated between 46 and 44 Ma (Pouclet & Lee 1988; Lee 1989; Jin *et al.* 1988). One of the dykes was sampled (K–3) and is a basaltic andesite with phenocrysts of plagioclase, pyroxene (calcic augite), and amphibole (pargasite).

Thermal effects of the K3 volcanic episode rejuvenated K–Ar ages in the Cretaceous lavas (Lee 1989) and fission track ages of apatite crystals in some Cretaceous granites (Jin *et al.* 1984, 1988). After the Lutetian, no significant volcanic activity occurred during the early Cenozoic, either in Korea or in the Japan arc (Isshiki 1977). This decrease in volcanic activity coincided with an anticlockwise rotation of the

Pacific plate at 42 Ma (Clague & Jarrard 1973).

K 4. A major change occurred in the early Miocene, marked by extensional tectonism in Korea and formation of the Pohang Basin, east of the Gyeongsang Basin. The new basin was filled initially with fluvio-lacustrine deposits and then with marine sediments, as a result of a migration of the marine Ulleung Basin. During the formation of the Pohang Basin, calc-alkaline volcanic activity was focussed along NE-trending fractures (Lee & Pouclet 1988). We distinguish the extrusion of andesitic to dacitic pyroclastic flows between 23 and 21 Ma, and an important outpouring of basaltic flows between 21 and 18 Ma. Two samples were selected from the younger episode (K–4a and b). The first sample is a plagioclase–diopside-phyric microlitic basaltic andesite, and the second is a diopside–olivine-phyric, doleritic-textured basaltic andesite.

The youngest lavas in the K4 period are dacitic and dated *c.* 17 Ma. They are overlain by marine sediments, which are devoid of volcanic products, reflecting the migration of the Japan arc away from Korea.

K 5. Weak magmatic activity has been recognized in the Pohang Basin (Lee 1989) and in northeastern North Korea (Polevaya *et al.* 1961) which produced basaltic to dacitic dykes. The age of this magmatic episode is not well determined but may be late Miocene (*c.* 6 Ma). A plagioclase-phyric basaltic andesite dyke (K–5) was sampled in the Pohang Basin for this study.

K 6. Late Korean volcanism extended from late Pliocene to recent times. It includes flood basalts in the Chugaryeong Graben (Won 1983), the Baegdusan volcano in North Korea and the volcanic island of Cheju (Won 1976). The lavas are alkaline and sodic in composition. Four samples were selected for study: K-61a and b (olivine- and Ti-diopside-phyric basanites from Pleistocene lava flows in the Chugaryeong Graben), and K-62a and b (olivine- and Ti-calcic augite-phyric basanites from Plio-Pleistocene lava flows of Cheju Island).

Japan Sea (Yamato Basin)

In the Yamato Basin, the cored basement samples from Legs 127 and 128 (Sites 794 and 797) consist of stacked basaltic and doleritic lava flows and sills, and rare interbedded sediments. Seven igneous units were recognized at Site 794 from 542 to 734 m below seafloor, and twenty one igneous units at Site 797 from 553 to 895 m (Ingle *et al.* 1990; Tamaki *et al.* 1990). Petrographical and geochemical data (Allan & Gorton 1992; Pouclet & Bellon 1992: Thy 1992 *a, b*) show that the units belong to superimposed volcanic complexes belonging to several magmatic series. At Site 794, the upper complex corresponds to mildly LREE-enriched tholeiites compositionally intermediate between continental basalts and arc tholeiites. The lower complex at Site 794 consists of tholeiites geochemically intermediate between island arc tholeiites and ocean floor basalts, which is typical of back-arc basin basalts. Two superimposed magmatic series are also present at Site 797. The upper units have compositional characteristics of back-arc basin basalts similar to those of the lower complex at Site 794. The lower units are incompatible- and light rare earth element-enriched tholeiites with Nb and Ta depletion and resemble calc-alkaline basalts. We tentatively correlate the back-arc basin basalts of Site 794 (lower complex) and Site 797 (upper complex). As the upper units at Site 797 are interpreted to be lava flows and the lower units at Site 794 consist of intruded sills, the back-arc basin basalt series represents the youngest volcanic activity of the Yamato Basin basement. The continental and/or calc-alkaline basalts must be slightly older. K–Ar and $^{40}Ar/^{39}Ar$ radiometric measurements have been performed on these samples (Kaneoka *et al.* 1992; Pouclet & Bellon 1992). Due to severe seawater and hydrothermal alteration, the calculated ages are poorly constrained but range between 22 and 17 Ma.

Various volcanic rocks have been dredged from the Yamato Basin and the Yamato Bank seafloor. They originated either from outcrops of the local basement or from the Yamato Seamount Chain. Their K–Ar and Ar–Ar radiometric ages are distributed in two age groups: 26–20 Ma, and 17–10 Ma (Kaneoka *et al.* 1988; Kaneoka & Yuasa 1988; Kaneoka 1990). In addition, rhyolitic explosive activity occurred in the Kita-Yamato Trough (Site 799 of Leg 128; Pouclet & Scott 1992) and was attributed to a proximal volcano on the Yamato Bank. The chronostratigraphical position of the volcanic products suggests an age between 13 and 12 Ma. At present, the volcanic activity in the Japan Sea is restricted to island volcanoes younger than 6.8 Ma (Iwata *et al.* 1988). Evolved products of these volcanoes are detected in the sedimentary pile of the basins from 4 Ma to the present time (Pouclet *et al.* 1992).

We have divided the history of the Yamato Basin in six phases (J1-J6).

J 1. Between 26 and 20 Ma, andesitic lava flows were erupted on the Yamato Bank, a continental fragment trapped in the Japan Sea by basin spreading. In the early Miocene, this fragment was close to the volcanic arc. At that time, volcanic activity was intense and resulted in the widespread deposition of the Green Tuff Formation. Volcanic rocks have been dredged and studied by Kaneoka & Yuasa (1988) and Kaneoka (1990). In this study, no sample is analyzed. However, we believe that the lower units at Site 797 (from the Yamato Basin basement) represent this arc-related volcanic phase (cf. analytical data of Allan & Gorton 1992). This magmatic pulse is essentially equivalent to the K 4 phase of the Pohang Basin.

J 2. The most recent and accurate age estimation of the formation of the Japan Sea is between 25 and 17 Ma (Isezaki 1986; Tamaki 1986; Kaneoka 1990). Rifting of the continental margin occurred in the early Miocene and was responsible for the opening of intra-arc and back-arc tectonic troughs. Multiple rifts, trending NNE–SSW, were created across the region from northern Japan to southeastern Korea (Iijima & Tada 1990): e.g. the Tohoku and Hokuriku–Saninoki rifts, and probably the Yamato initial rift and the Kita–Yamato rift. In the Yamato Basin, we think that the lavas of the upper complex at Site 794 (22–17 Ma) belong to this tectonic stage because they show some of the chemical characteristics of continental basalts. Three samples (J–2a, b and c) were selected from Hole 794D, from the most primitive and least-altered basaltic units.

J 3. Back-arc basin spreading followed the rifting stage. In the Yamato Basin, the spreading axis was located by Kimura *et al.* (1987). The back-arc basin basalts of Sites 794 (lower intrusive complex) and 797 (upper units) are linked to this spreading stage. They were formed following the 22–17 Ma calc-alkaline and continental lavas. Three samples (J–3a, b and c) were selected from less-altered units in Hole 794D.

J 4. After opening, the basins deepened by tectonic and thermal subsidence in the late early and middle Miocene (Tamaki *et al.* 1992). A seamount chain was built between 17 and 14 Ma in a near-ridge tectonic position in the central Yamato Basin (Kaneoka *et al.* 1990; Fig. 1). This volcanic activity is explained as remnant ridge axis magmatism erupted following the cessation of spreading. However, because of severe seawater alteration of the dredged samples, the

Table 2. *Major- and trace-element compositions of selected representative samples from Korea and the Japan Sea*

Sample	K-1 (1)	K-2 (1)	K-3 (1)	K-4a (1)	K-4b (1)	K-5 (1)	J-2a (2)	J-2b (2)	J-2c (2)	J-3a (2)	J-3b (2)	J-3c (2)	J-5 (3)	J-6 (4)	K-61a (1)	K-61b (1)	K-62a (1)	K-62b (1)
SiO_2	50.66	50.01	53.29	50.06	49.38	53.25	49.91	48.60	46.86	48.20	48.28	47.04	69.21	44.39	47.42	48.13	47.68	51.88
TiO_2	0.98	1.03	1.13	1.43	1.66	1.58	1.25	1.45	1.10	1.59	1.62	1.51	0.17	3.22	1.73	1.68	2.12	2.05
Al_2O_3	16.00	16.72	15.72	18.94	14.91	14.77	17.59	15.74	15.23	15.28	17.41	16.68	12.78	16.53	15.20	15.87	13.39	13.92
Fe_2O_3	0.93	1.20	0.97	1.13	1.45	1.34	1.18	1.07	1.15	1.39	1.03	1.25	0.36	1.45	1.37	1.33	1.45	1.49
FeO	6.21	7.98	6.44	7.54	9.66	8.92	7.84	7.11	7.66	9.30	6.86	8.32	1.50	9.67	9.12	8.88	9.66	9.92
MnO	0.12	0.15	0.14	0.15	0.18	0.31	0.12	0.13	0.15	0.18	0.24	0.17	0.03	0.20	0.18	0.18	0.17	0.15
MgO	7.03	6.50	6.44	4.00	4.75	3.01	6.09	9.93	10.70	8.13	6.71	8.58	2.29	3.52	9.83	9.36	9.98	6.12
CaO	6.57	4.94	5.35	9.54	9.13	7.62	9.38	6.80	7.65	10.66	10.52	10.16	0.59	9.71	8.47	8.75	9.22	8.16
Na_2O	3.91	4.11	3.89	3.47	3.13	3.60	3.51	2.89	2.65	3.20	3.40	2.99	1.37	3.58	3.36	3.29	3.06	3.25
K_2O	3.04	2.21	2.32	0.87	1.07	1.04	0.57	0.56	0.23	0.15	0.20	0.21	1.68	2.09	1.64	1.55	1.37	1.02
P_2O_5	0.47	0.18	0.36	0.31	0.34	0.26	0.21	0.22	0.12	0.13	0.23	0.18	0.11	0.79	0.34	0.31	0.43	0.36
LOI	3.34	3.55	3.79	1.71	3.16	3.62	2.51	5.34	6.41	1.52	3.52	2.48	9.93	3.40	−0.48	−0.34	−0.75	−0.30
Total	99.26	98.58	99.84	99.15	98.82	99.32	100.16	99.84	99.91	99.72	100.02	99.58	100.03	98.55	98.18	98.99	97.78	98.02
La	31.3	21.9	31.2	12.2	13.5	14.8	7.7	8.5	3.7	4.4	5.0	4.2	48.4	55.2	21.9	17.8	25.5	18.7
Ce	65.6	52.7	67.3	30.9	36.4	32.3	17.7	18.4	8.8	11.4	11.9	9.9	92.6	105.0	43.4	41.5	55.4	42.5
Nd	28.8	28.3	38.4	18.4	20.1	24.4	10.3	10.3	5.7	9.2	11.8	9.5	41.1	43.3	18.7	21.8	26.1	25.2
Sm	6.0	6.3	6.7	4.7	5.5	6.1	2.8	2.9	1.7	3.2	4.0	3.3	9.3	8.6	5.0	5.2	6.6	6.5
Eu	1.7	1.7	1.9	1.6	1.8	2.0	1.1	1.1	0.8	1.2	1.4	1.1	0.6	2.7	1.9	1.6	2.0	2.2
Gd	4.4	6.4	6.8	4.3	5.0	5.5	3.3	3.5	2.2	4.2	4.7	3.6	8.4	7.1	4.5	5.0	5.6	6.5
Dy	3.5	5.6	5.1	4.3	5.0	5.6	3.3	3.6	2.4	5.0	5.2	4.2	8.3	5.8	3.9	3.4	4.6	4.6
Er	2.0	2.1	2.0	2.4	2.7	2.9	1.9	2.0	1.4	2.9	2.9	2.4	4.8	2.5	2.0	1.4	2.1	1.8
Yb	2.0	2.0	2.0	2.5	2.9	3.1	1.6	1.7	1.2	2.5	2.8	2.2	4.9	1.6	2.1	1.2	1.8	1.6
Lu	0.3	0.3	0.3	0.3	0.4	0.5	0.2	0.2	0.2	0.3	0.3	0.3	0.8	0.2	0.3	0.2	0.3	0.2
Ba	597	825	677	228	265	343	154	121	57	21	20	53	520	805	297	282	401	243
Co	45	42		62	77									30	92	89	52	66
Cr	163	239		76	77		79	266	377	324	319	291	11	15	40	45	448	199
Cu	17	10	19	45	105		34	51	54	63	74	56	10	27	328	335	41	71
Nb	7.0	8.0	11.7	7.0	7.5	7.3	6.1	8.6	2.5	2.8	3.0	2.0	11.0	71.0	34.1	32.8	51.4	39.3
Ni	26	51	70	53	52		21	126	147	129	124	124	5	20	232	218	257	160
Rb	75.0	53.0	82.9	18.0	22.0	37.8	7.0	5.0	0.4	3.0	3.0	2.0	74.0	57.0	23.0	20.0	33.0	28.0
Sr	737	1092	938	529	394	464	370	277	275	194	200	173	178	907	509	529	471	324
Th	6.0	5.5	5.7	1.7	2.0	2.0	0.8	1.0	0.3	0.3	0.3	0.3	25.4	5.6	2.7	2.4	4.0	3.1
U	0.8	0.6		0.5	0.6		0.2	0.3	0.0	0.0					0.7	0.4	0.8	0.9
V	177	174		256	319		238	278	256	245	281	234	15	268	210	213	225	188
Y	23	25	27	29	35	34	20	20	16	31	33	28	50	26	27	22	27	25
Zn							49	49	47	85	83	53	45	102				
Zr	169	145	218	137	142	147	90	99	41	90	107	92	104	264	168	155	205	167

(1) Lee (1989); (2) Pouclet & Bellon (1992); (3) Pouclet & Scott (1992); (4) unpublished data. Ferric/ferrous ratio is 0.15 in mafic rocks and 0.24 in the acidic rock (J-5). LOI, loss on ignition.

Table 3. *Primitive mantle-normalized ratios of the incompatible and less mobile elements used to discriminate the magmatic groups and subgroups in Korea and the Japan Sea*

	Group 1						Group 2				Group 3		
	A K1-3		B K4-5		C J1		A J2		B J3		A K6		B J6
Volc. phases	n18	sd	n11	sd	n8	sd	n19	sd	n11	sd	n13	sd	n1
$(La/Yb)_N$	7.58	2.26	4.55	1.61	2.70	0.47	3.20	0.79	1.33	0.30	10.96	2.57	24.74
$(La/Sm)_N$	2.62	0.63	1.96	0.71	1.63	0.18	1.71	0.12	0.86	0.13	2.55	0.45	4.14
$(Sm/Yb)_N$	2.94	0.76	2.38	0.61	1.64	0.14	1.86	0.38	1.54	0.30	4.33	0.89	5.88
$(Th/Nb)_N$	4.64	2.07	3.15	1.81	2.29	0.47	1.10	0.37	0.93	0.19	1.08	0.42	0.66
$(Th/La)_N$	2.56	1.96	1.52	0.80	1.18	0.17	0.91	0.16	0.43	0.14	1.51	0.38	0.82
$(Nb/La)_N$	0.56	0.23	0.54	0.29	0.52	0.05	0.87	0.17	0.47	0.15	1.49	0.30	1.24

Values are means of n number of analyses s d, standard deviation. Normalizing values of Sun & McDonough (1989).

chemical composition of the volcanic rocks (tholeiitic and alkaline), is not well determined.

J 5. The rhyolitic explosive activity seen at Site 799 consists of a thick pumice flow dated between 13 and 12 Ma. This activity is linked to the faulting of the Kita–Yamato Trough developed on the Yamato Bank. It can be compared to the rhyodacitic volcanism of the Tohoku Trough (northern Japan), which occurred at the same time in a similar tectonic context and produced the acidic tuffs of the

Table 4. *New Sr, Nd, and Pb isotope data*

Sample	Rb	Sr	age (Ma)	$^{87}Sr/^{86}Sr_p$	$^{87}Sr/^{86}Sr_i$	$^{143}Nd/^{144}Nd$	ε_{Sr}	ε_{Nd}	$^{206}Pb/^{204}Pb$	$^{207}Pb/^{204}Pb$	$^{208}Pb/^{204}Pb$
K-1	75.0	737	(100)	0.709481 ± 0.000015	0.709062	0.512703 ± 0.000010	+64.75	+1.58			
K-2	53.0	1092	73.1	0.711722 ± 0.000025	0.711577	0.512158 ± 0.000010	+100.45	−9.05			
K-3	82.9	938	45.3	0.704551 ± 0.000017	0.704387	0.512716 ± 0.000011	−1.60	+1.83			
K-4a	18.0	529	19.5	0.704296 ± 0.000017	0.704269	0.512800 ± 0.000010	−3.28	+3.47			
K-4b	22.0	394	18.5	0.704359 ± 0.000016	0.704317		−2.60				
K-5	37.8	464	6.0	0.704505 ± 0.000014	0.704485		−0.21				
J-2a	7.0	370	(19)	0.704177 ± 0.000010	0.704162	0.512820 ± 0.000011	−4.80	+3.86			
J-2b	5.0	277	(19)	0.704356 ± 0.000012	0.704342	0.512789 ± 0.000012	−2.24	+3.26			
J-2c	0.4	275	(19)	0.704929 ± 0.000010	0.704928	0.512724 ± 0.000012	+6.08	+1.99	18.535	15.648	38.687
J-3a	3.0	194	(19)	0.704350 ± 0.000012	0.704338	0.512944 ± 0.000011	−2.29	+6.28	18.370	15.567	38.320
J-3b	3.0	200	(19)	0.704070 ± 0.000012	0.704058	0.512918 ± 0.000011	−6.27	+5.77	18.308	15.561	38.357
J-3c	2.0	173	(19)	0.704114 ± 0.000011	0.704103	0.512992 ± 0.000016	−5.64	+7.22	18.337	15.639	38.514
K-61a	23.0	509	0	0.705130 ± 0.000016	0.705130		+8.94				
K-61b	20.0	529	0	0.705197 ± 0.000017	0.705197		+9.90				
K-62a	33.0	471	0	0.704542 ± 0.000016	0.704542	0.512784 ± 0.000011	+0.60	+3.16			
K-62b	28.0	324	0	0.704583 ± 0.000016	0.704583		+1.18				

Subscript p, present-day measured ratios; subscript i, initial ratios. Ratios are normalized to $^{86}Sr/^{88}Sr = 0.1194$ and to $^{146}Nd/^{144}Nd = 0.7219$. Values for NBS standard: $^{87}Sr/^{86}Sr = 0.71025 \pm 4$ (2σ) and La Jolla standard $^{143}Nd/^{144}Nd = 0.51186 \pm 3$ (2σ). I Sr = 0.7045 and I Nd = 0.512638, present-day undifferentiated mantle reservoir (O'Nions *et al.* 1977). No corrections for Nd-decay.

Onnagawa Formation in northeastern Japan. One sample (J–5) was analysed and is a calc-alkaline rhyodacite.

J 6. From late Miocene to present times, localized volcanic activity built volcanic islands on the Korean plateau (Ulleung Island; Yoon 1987) and adjacent smaller volcanoes) and the Japan plateau (mainly Oki Islands; Iwata *et al.* 1988). The lavas are typically alkaline and potassic. A sample of alkaline basalt (J–6) was obtained from the Ulleung Island. Most of the volcanoes are basaltic, but phonolite and trachyte lavas are also numerous. In the proximal sedimentary sequences cored during Legs 127 and 128, these evolved products may account for half of the total volume of Quaternary tephra deposits, the remainder comprising calc-alkaline tephra of the Japan arc (Pouclet *et al.* 1992). The most important volume of tephra was erupted between 1.9 and 0.5 Ma (Pouclet & Scott 1992). This paroxysmal activity coincided with the development of a new tectonic stress pattern recorded by compressional deformation of the eastern edge of the Japan Sea (Jolivet 1987; Tamaki 1988) and uplift of the Okushiri and Oki Ridges (Ingle *et al.* 1990; Tamaki *et al.* 1990).

Geochemistry

Volcanic samples representing the different magmatic compositions were selected from the study area. They were analysed for major, minor and trace elements by ICP and ICP-MS methods at the University of Orléans (Tables 2 and 3). Sr, Nd and Pb isotopic measurements were performed at the Université de Clermont-Ferrand (France) and Université du Québec at Montréal (Canada) (Table 4).

Major and trace element geochemistry

Geochemical characteristics of the various volcanic rocks were delineated on a set of major and trace element analyses of mafic lavas: 41 lavas from the Korean margin (Lee 1989) and 38 from the Yamato Basin (Allan & Gorton 1992; Pouclet & Bellon 1992). Three major magmatic and chronological groups (Groups 1–3) are distinguished on the basis of trace element ratios and are farther subdivided into seven subgroups (Table 3). Figure 2 illustrates the distinction between the groups, using a covariation diagram adapted from Pearce (1982). Th/Yb and Nb/Yb ratios vary according to the degree of partial melting. Th/Nb ratios also clearly distinguish the field of arc-related basalts from those of variably enriched basalts from non-subduction settings.

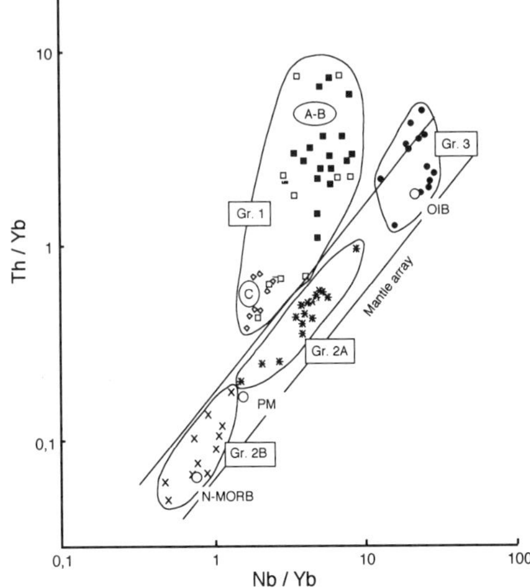

Fig. 2. Nb/Yb v. Th/Yb covariation diagram adapted from Pearce (1982 and 1983). Distinction of the magmatic groups; filled squares, Group 1A; open squares, Group 1B; diamonds, Group 1C; asterisk, Group 2A; crosses, Group 2B; filled circles, Group 3. PM, primordial mantle; N-MORB, normal MORB; OIB, Oceanic Island Basalt.

Distinctive compositional characteristics are also evident on primitive mantle-normalized multi-element diagrams (Fig. 3). Moreover, Rb and possibly Ba and Sr contents in the submarine basalts must also be considered with caution. Nonetheless, Th appears to be a reliable indicator of original LILE contents in the unaltered magmas (Saunders & Tarney 1984).

Group 1. The first group is characterized by an arc-related magmatic signature: Nb troughs, Th spikes, LILE-, light rare earth element (LREE)-, and Sr-enrichments, and moderate heavy rare earth element (HREE)-depletion. It includes the products from the volcanic phases K1, K2, K3, K4, and K5, the Cretaceous and Eocene to early Miocene lavas of the Korean margin, and the products from the early Miocene volcanic phase J1 of the Yamato Basin area. We are able to discriminate three sub-groups, based on decreasing LREE- and Th-enrichments, and HREE-depletion: Sub-Group 1A (K1 to K3 lavas), Sub-Group 1B (K4 and K5 lavas), and Sub-Group 1C (J1 lavas). Table 3 presents primitive mantle-normalized ratios for the mafic lavas of these sub-groups (La/Yb$_N$ = 7.58 to 2.70; Th/La$_N$ = 2.56 to 1.18; Sm/Yb$_N$ = 2.94 to 1.64), and their distinctive multi-element spidergrams are shown in Figure 3. Some Rb and

Sr anomalies of Sub-Group 1C may be due to seawater alteration.

Group 2. The second group is characterized by low LILE and LREE contents and the lack of a Th anomaly. It includes early Miocene tholeiitic lavas, J2 and J3 of the Yamato Basin basement. Two sub-groups are distinguished: Sub-Group 2A (J2 lavas) and Sub-Group 2B (J3 lavas). Sub-Group 2A exhibits slight LREE enrichment $(La/Yb_N = 3.20)$, low LILE abundances, and no Nb anomaly $(Nb/La_N = 0.87)$. Some Rb, Ba, and Sr anomalies in Fig. 3 are due to seawater alteration. The mineralogical and chemical compositions are similar to continental tholeiites (Pouclet & Bellon 1992), particularly tholeiites emplaced during initial rifting (cf. Holm 1985). Similar chemical compositions occur in Miocene basalts of the Akita-Yamagata oil field (Tsuchiya 1990), a tectonic trough of the Tohoku back-arc rift system. Spidergrams of Sub-Group 2B lavas show no enrichment or depletion of LILE $(Th/La_N = 0.43)$ and nearly flat REE patterns $(La/Yb_N = 1.33)$. Nb is depleted compared to La $(Nb/La_N = 0.47)$ but not to Th and the LILE $(Th/Nb_N = 0.93)$. The overall composition is typical of basaltic volcanism within back-arc basins and the low LIL/HFS element ratios are characteristics of basalts from basins associated with youthful subduction zone systems (Saunders & Tarney 1984).

Group 3. The third group contains LILE-, Nb-, and LREE-enriched lavas. It includes the Pliocene and Quaternary volcanic rocks from Korea and the Japan Sea islands (phases K 6 and J 6). All these lavas have compositions typical of alkaline intraplate basalts, with strong enrichment in LILE and LREE $(La/Sm_N = 10.96–24.74)$, depletion in HFSE and HREE $(Sm/Yb_N = 4.33–5.88)$, and moderate positive anomalies for Nb and Ti. LREE enrichment and HREE depletion require residual garnet in the source, which must be deeper than c. 60 to 80 km. Sodic (Chugaryeong Graben and Cheju Island) and potassic (Japan Sea islands) subgroups are distinguished. Greater element enrichment occurs in the potassic lavas of Ulleung Island. A few widely distributed lavas are slightly enriched in Th (Fig. 2); this may be due to minor crustal contamination (?).

To summarize, the geochemical patterns of the volcanic rocks in the back-arc area indicate the presence of at least three main magmatic groups, whose presence can be correlated with the evolution through time of the tectonic setting. (1) The Cretaceous and early Miocene lavas correspond to several, arc-related, calc-alkaline series (calc-alkaline basalts, CAB, and island arc tholeiites, IAT) at an active continental margin prior to back-arc basin opening (volcanic phases K1–K4 and J1). The subduction-related LILE (Th) enrichment characteristic of these lavas decreases between the Cretaceous and early Miocene, with eruption of the least- enriched lavas as tholeiites in the Yamato Basin basement (Site 797). (2) Successive, early Miocene, syn-opening lavas, in the Yamato Basin show the chemical signatures of continental initial rift tholeiitic series (IRT) and back-arc basin basalts (BABB), respectively. (3) Pliocene to Holocene, post-opening lavas have the composition of intra-plate alkaline basalts comparable to sodic and potassic oceanic island basalts (OIB).

Isotopic data

The succession of very different magmatic products erupted in the same geographical area must involve participation of several chemical reservoirs. Possible sources include (i) depleted asthenospheric upper mantle, (ii) enriched or undepleted primitive mantle, (iii) variably enriched sub-continental mantle, (iv) continental crust, and (v) the subducting slab. Isotopic and trace element ratios are particularly powerful in discerning the relative contributions of potential sources (Table 4). We integrate the measurements made by Cousens & Allan (1992) and by Nohda *et al.* (1992) on similar samples from Legs 127 and 128. Data are compared with previous isotopic studies of Japanese volcanic rocks: Nakamura *et al.* (1985, 1989, 1990), Kagami *et al.* (1986), Nohda & Wasserburg (1981, 1986), Nohda *et al.* (1988), Morris & Kagami (1989), Kaneoka (1990) and Tatsumoto & Nakamura (1991). Because of alteration and the inclusion of submarine samples, leaching procedures were undertaken prior to determining Sr–Rb concentrations. The leached samples may have lower $^{87}Sr/^{86}Sr$, but they show no significant modification of the Nd and Pb isotopic ratios.

Sr initial isotopic ratios range from 0.7041 to 0.7052. However, two Cretaceous lavas of the Korea Peninsula, K–1 and K–2, exhibit very high Sr ratios and low Nd ratios, suggesting that they have been strongly contaminated by a radiogenic component. This may be due either to contamination from the nearby Bulgugsa granites, or to a contribution from a primary, highly-enriched, sub-continental mantle. The high Sr content of these lavas is more consistent with the second hypothesis. Nd isotopic ratios range from 0.512703 to 0.512992, except for

Group 1

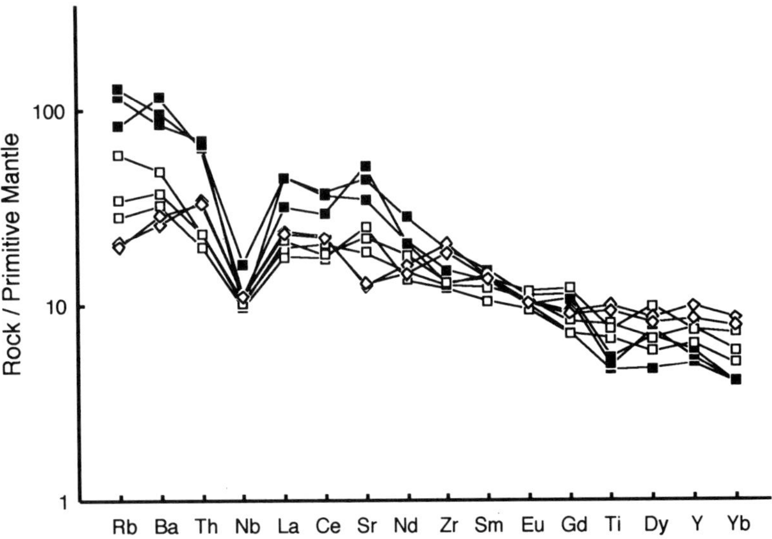

Group 2

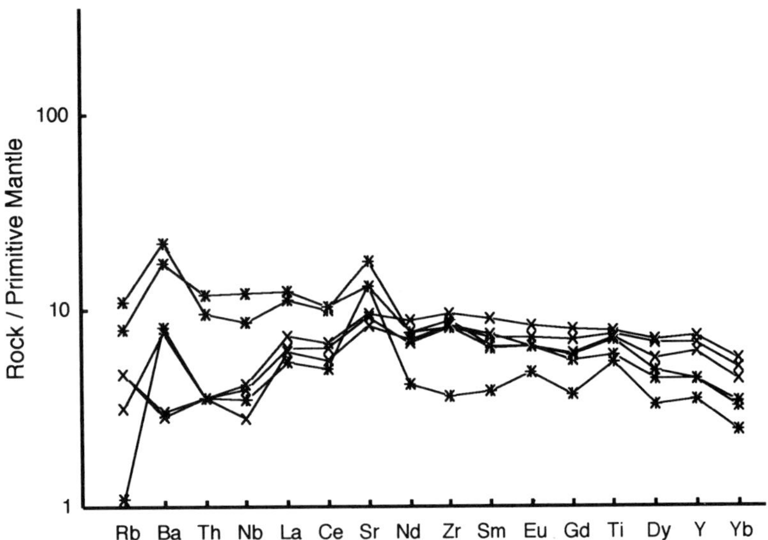

K–2. The new data are plotted on a Sr–Nd isotope variation diagram (Fig. 4) together with published data for the Japan Sea region adjusted to the present-day chondritic value (CHUR) of 0.512638 (O'Nions *et al.* 1977). The few analyses of the Korean Eocene and Miocene lavas plot in the middle part of the mantle array and above the bulk Earth value. The calc-alkaline tholeiites of the Yamato basement (J 1) fall in the same area, as do the tholeiites of initial rifting continental affinity (J 2). This area corresponds to the overlap for data for NE and SW Japan arc volcanic rocks (Nohda & Wasserburg 1981, 1986; Morris & Kagami 1989); it is also the compositional domain of the Miocene and Pliocene lavas of the back-arc side of the Japan arc (Nohda *et al.* 1988) and of Eastern China (Zhou & Armstrong 1982; Peng *et al.* 1986; Song *et al.* 1990; Basu *et al.* 1991). The back-arc basin basalts of the Yamato Basin basement and

Group 3

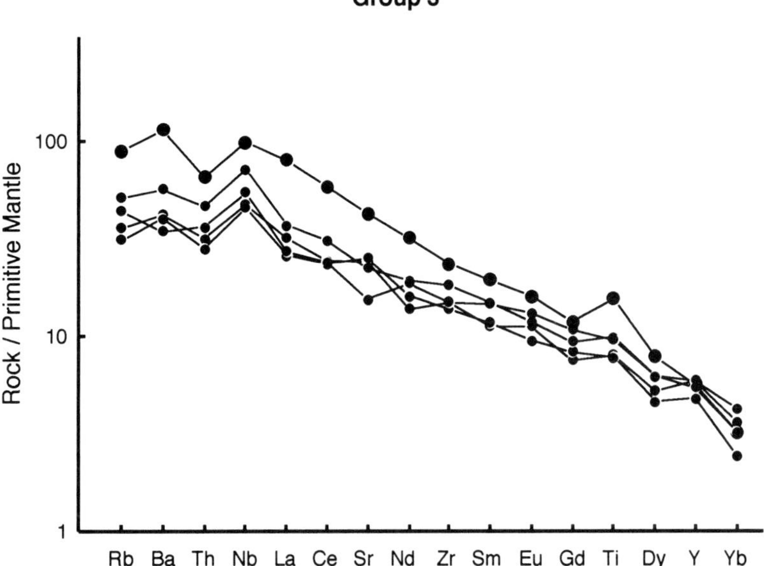

Fig. 3. Primitive mantle-normalized incompatible element diagrams for selected samples, to show the distinctive patterns of the three magmatic groups. Same symbols as for Fig. 2; Group 3: larger symbol is for the J–6 lava. Normalizing values from Sun & McDonough (1989).

seamounts (J3 and J4) have significantly lower $^{87}Sr/^{86}Sr$ and higher $^{143}Nd/^{144}Nd$ than the older lavas, but are similar isotopically to some NE Japan arc lavas. The alkaline lavas of the Japan Sea islands (J6) have lower $^{143}Nd/^{144}Nd$, except for Cheju Island basalts, which have similar isotope ratios to J1 and J2 lavas of the Yamato Basin. Within the J6 group, a slight distinction appears between the lavas of Oki Islands, which have relatively high $^{87}Sr/^{86}Sr$, and those of Ulleung and Dog islands with significantly lower $^{143}Nd/^{144}Nd$ and $^{87}Sr/^{86}Sr$.

As a whole, the data plot in a similar field to the Japan arc, and they define trends extending between depleted MORB-source mantle (DMM) and enriched mantle components EM I and/or EM II.

Our new Pb-isotope analyses, performed on samples from Site 794 are consistent with the data of Cousens & Allan (1992) from similar basalts from Sites 794 and 797. They are also comparable with the data of Tatsumoto & Nakamura (1991) for the Yamato seamounts and volcanic islands. In Pb–Pb isotope diagrams (Fig. 5), the volcanic rocks of the Miocene Yamato Basin basement plot above the Northern Hemisphere Reference Line (NHRL; Hart 1984) and define a linear array between depleted MORB mantle source (DMM) and the field for Pacific Ocean sediments or enriched component

EM II. Similar ranges in Pb isotope ratios are present in Japanese arc lavas (Hedge & Knight 1969; Tatsumoto 1969; Tatsumoto & Knight 1969; Tatsumoto & Nakamura 1991). Yamato Basin basalts also define linear array between DMM and Pacific sediments or EM II component in the Pb–Sr and Pb–Nd isotope ratio diagrams (Fig. 6). The more depleted back-arc basin basalts from Site 797 and the Yamato seamount basalts are the closest in composition to DMM and, more precisely, to Indian Ocean MORB, which have higher $^{87}Sr/^{86}Sr$, $^{207}Pb/^{204}Pb$ and $^{208}Pb/^{204}Pb$ than Pacific Ocean MORB (Dupré & Allègre 1983; Ito *et al.* 1987; Mahoney *et al.* 1989). In contrast, the Pliocene-Pleistocene alkaline lavas of the volcanic islands have distinctly higher $^{208}Pb/^{204}Pb$ and $^{87}Sr/^{86}Sr$ and lower $^{143}Nd/^{144}Nd$ than all the older volcanic rocks from the Japan Sea area (Figs 5 and 6). These isotopic characteristics imply a contribution from an EM I component, as has been demonstrated for Cenozoic lavas from eastern China (Tatsumoto *et al.* 1992), particularly the tholeiitic lavas of the Hannuoba area. This enriched component may have resided either in the asthenosphere or, more likely, in the subcontinental lithosphere (Song *et al.* 1990).

Variation diagrams of delta $^{208}Pb/^{204}Pb$ v. $^{87}Sr/^{86}Sr$, demonstrating Th and Rb enrichment or depletion in the mantle, and delta $^{208}Pb/^{204}Pb$

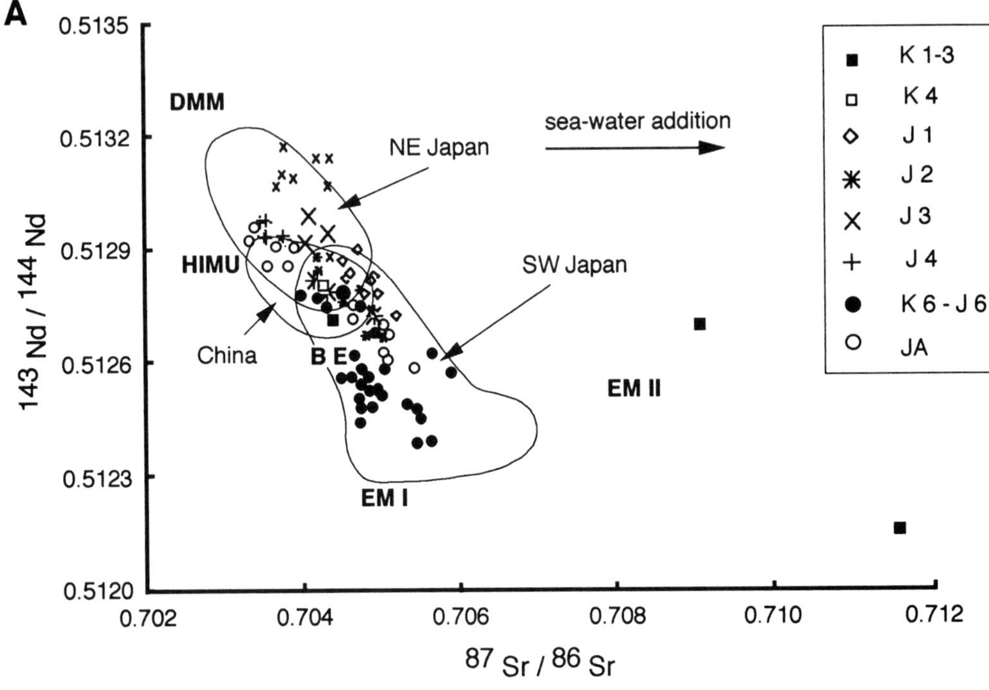

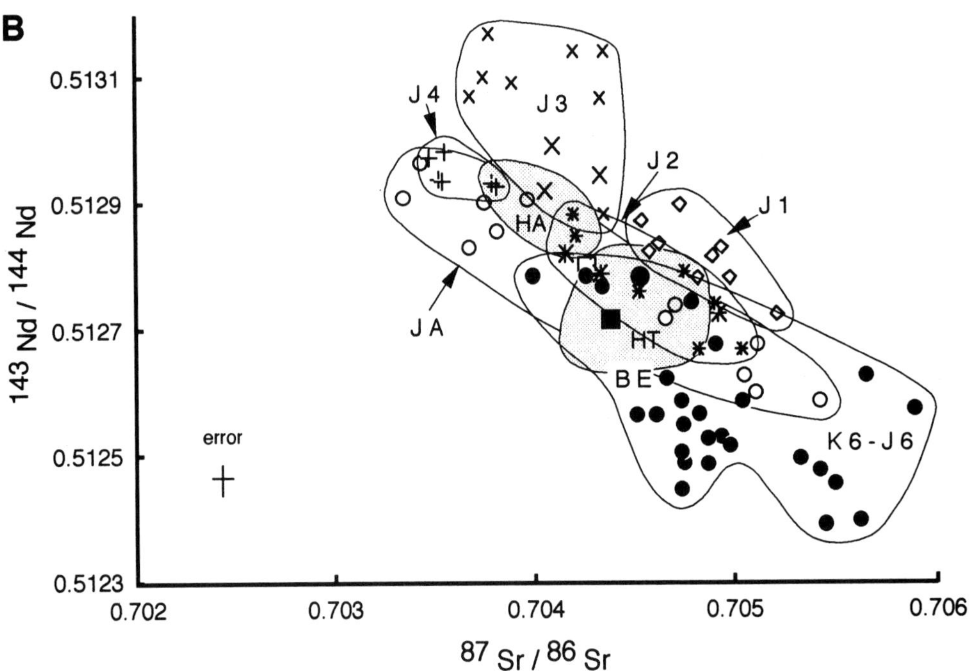

Fig. 4. Sr- and Nd-isotope variation diagram. B is an enlargement of A. Same symbols as for Fig. 2, plus: latin crosses, lavas of the Yamato Basin seamounts (J 4); open circles, Miocene–Pliocene lavas of the back-arc side of NE Japan arc (JA); larger symbols correspond to the samples analysed in this work; for additional data sources, see text. All Nd ratios are corrected for $^{143}Nd/^{144}Nd = 0.512638$ (CHUR). DMM, depleted MORB-type mantle; BE bulk Earth; HIMU, EM I and EM II, high- mantle and enriched mantle type 1 and type 2 (Zindler & Hart 1986). Composition of the mantle end-members after Hart (1988). HA, alkaline and transitional basalts of Hannuoba: HT, tholeiitic basalts of Hannuoba; China, NE Japan and SW Japan areas, see text.

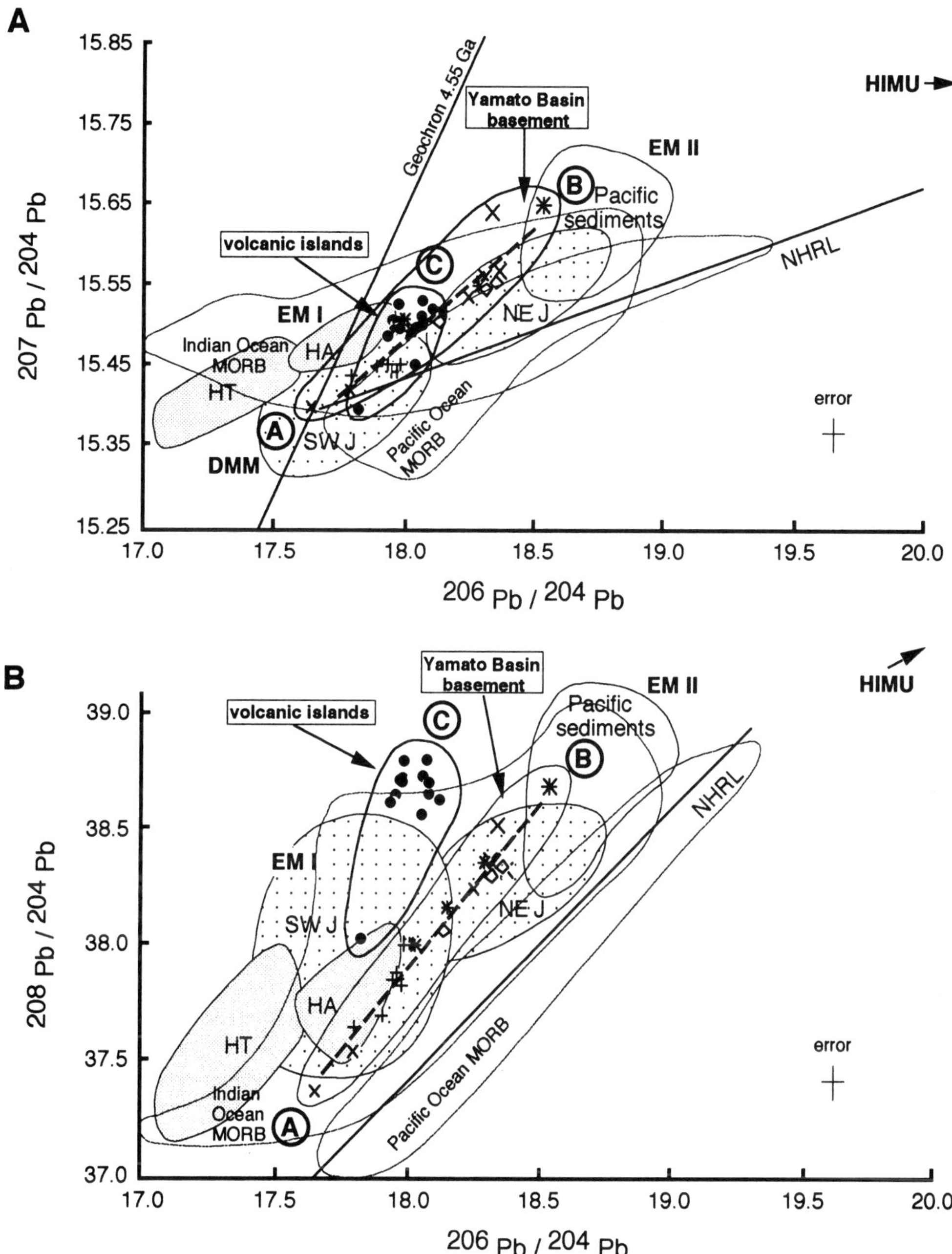

Fig. 5. Pb–Pb isotope plots of basalts from the Yamato Basin basement and volcanic islands. Note the good positive correlation of the Yamato Basin basement basalts that is explained by mixing of two end-members. Same abbreviations as for Figs 2 and 4. Circled A, B, and C, approximate compositions of sources. NE J, NE Japan; SW J, SW Japan; NHRL, Northern Hemisphere reference line (Hart 1984). Fields for Indian and Pacific Ocean MORB's after Hickey-Vargas (1991). Pacific subduction-related sediment composition according to Meijer (1976), Sun (1980), and Woodhead & Fraser (1985).

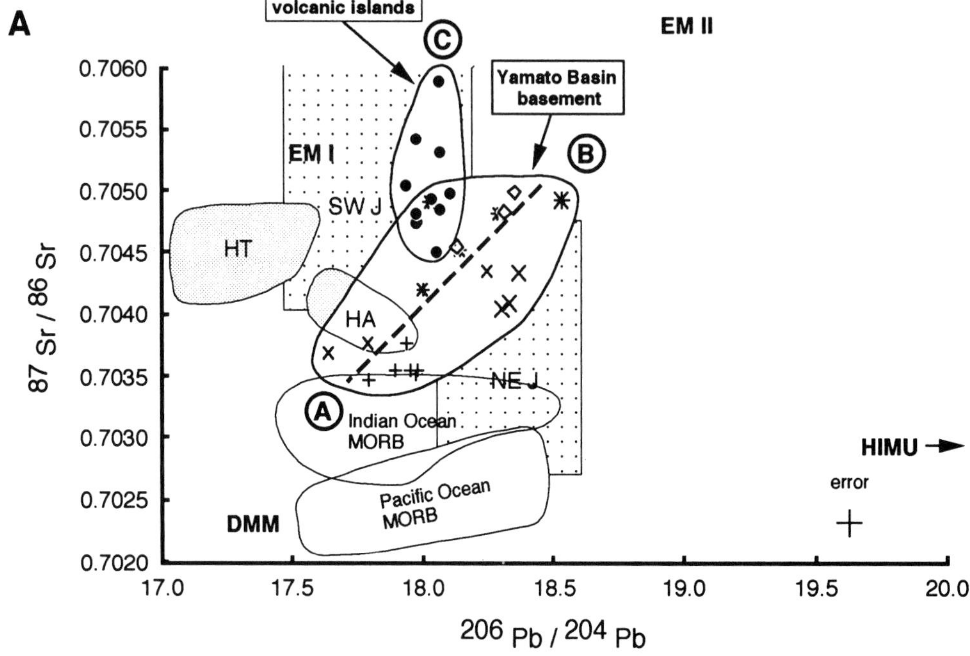

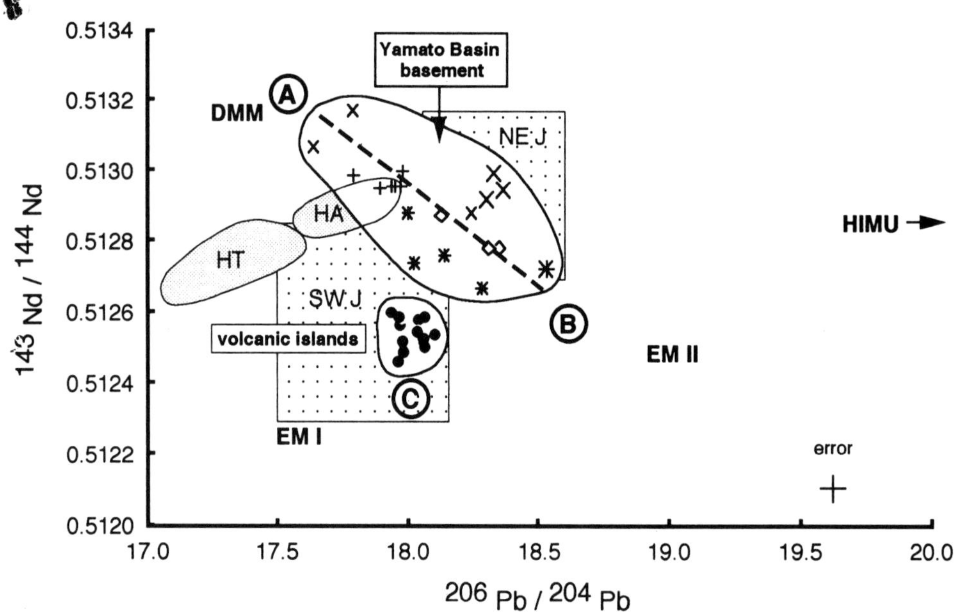

Fig. 6. Covariation diagrams of ^{206}Pb/^{204}Pb v. ^{87}Sr/^{86}Sr (**A**) and v. ^{143}Nd/^{144}Nd (**B**) in basalts from the Yamato Basin basement and volcanic islands. Same symbols as for Figures 2, 4 and 5. NE J and SW J areas according to Tatsumoto & Nakamura (1991). Fields for Indian and Pacific Ocean MORB's after Dupré & Allègre (1983) and Mahoney *et al.* (1989).

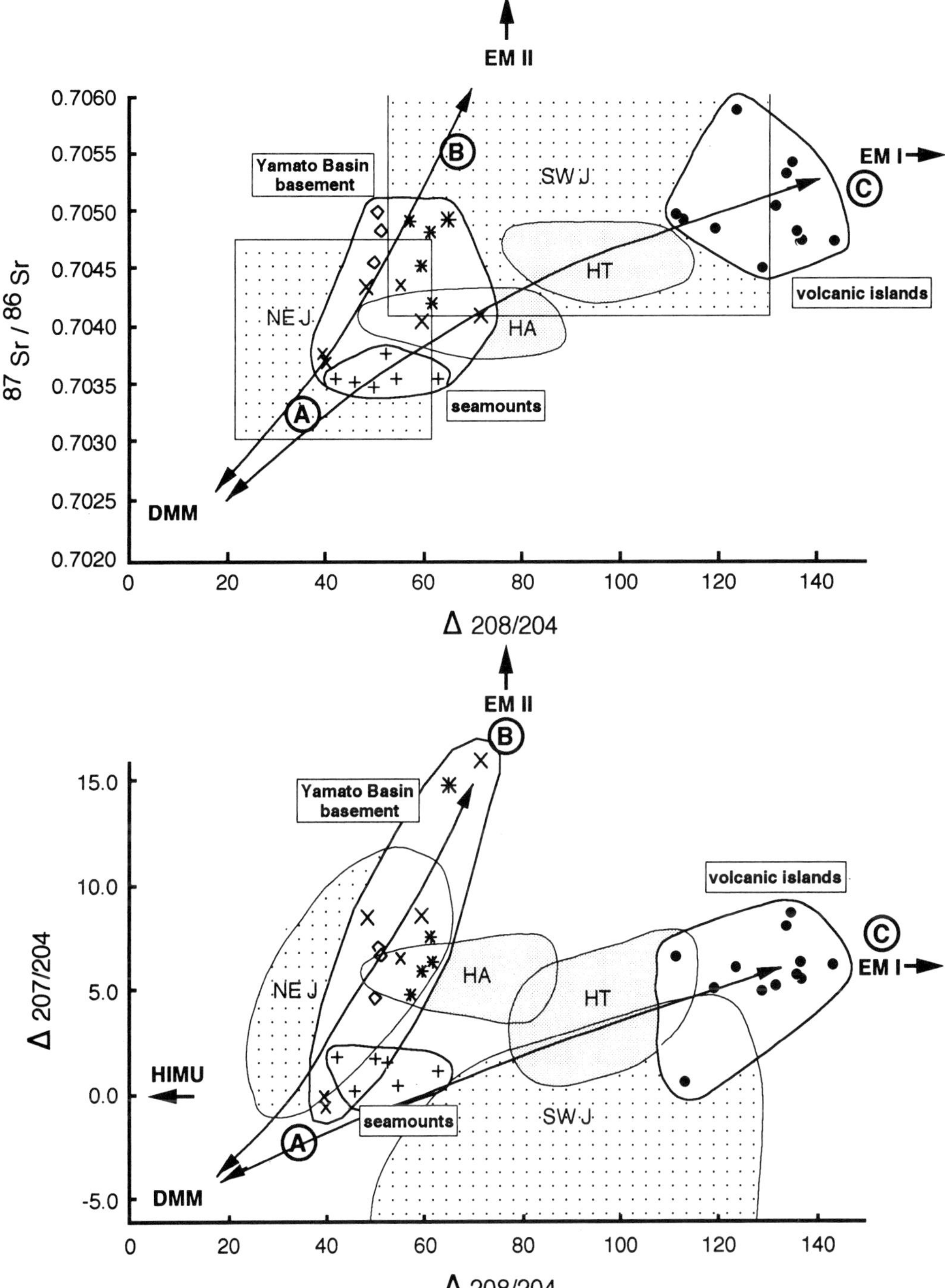

Fig 7. ^{87}Sr/^{86}Sr and $\Delta\,^{207}$Pb/^{204}Pb v. $\Delta\,^{208}$Pb/^{204}Pb in basalts from the Yamato Basin basement and volcanic islands. Same symbols as for Figures 2, 4 and 5. EM I and EM II 'branches' after Hart (1988).

v. delta $^{207}Pb/^{204}Pb$, showing relative Th and U enrichment or depletion (Hart 1988), distinguish two major trends in the Japan Sea area volcanic rocks (Fig. 7). The first trend, which links DMM to EM II, includes Yamato Basin basement basalts, lavas of the NE Japan volcanic arc, and some alkaline lavas of the Hannuoba area of China. The second trend goes from DMM to EM I and includes basalts from the Yamato Seamount chain, the alkaline lavas of the volcanic island, most of the SW Japan arc lavas, and the tholeiitic lavas of Hannuoba. These trends are similar to the EM II and EM I branches in correlation plots for OIB (Hart 1988).

Discussion

Determination of the mantle sources

At least three mantle sources are required to account for the isotopic variations among the lavas of the back-arc area.

In the Japan Sea, the pre-opening volcanic rocks (Magmatic Group 1) cover the same range of chemical (major and trace elements) and isotopic compositions than as in the NE Japan arc. The syn-opening tholeiite lavas of the Yamato Basin (Magmatic Group 2) are geochemically distinct from the arc lavas. However, their Sr, Nd and Pb isotopic signatures are similar to the NE Japan arc. Yamato Basin and NE Japan arc lavas may have been derived from similar mantle sources. The isotope ratio plots show mixing trends between two sources, 'A' and 'B', on the variation diagrams (Figs 5, 6 and 7). Source 'A' mainly involves a depleted MORB-type or DMM end-member having elevated $^{87}Sr/^{86}Sr$, $^{207}Pb/^{204}Pb$ and $^{208}Pb/^{204}Pb$, which are features of Indian Ocean MORB (Mahoney *et al.* 1989). This source 'A' is identical to source 1 of the Philippine Plate basin magmas determined by Hickey-Vargas (1991), a MORB mantle source perhaps slightly contaminated by EM I. Source 'B' includes a component with more-radiogenic Sr and Pb isotope ratios, which could represent subducted oceanic sediments, upper continental crust, and/or enriched mantle EM II. The Pb–Pb mixing relationship between the initial rifting tholeiites (Group 2A) and the back-arc basin basalts (Group 2B) of the Yamato Basin can be explained by contamination of a depleted MORB-type source by 0.5–2.5% of subducted sediments, where the sediment contribution decreases with time (Cousens & Allan 1992). The problem is to determine when and how sediments were incorporated into the mantle wedge. The trace element data indicate a rather weak participation of subduction-related components: there is no large enrichment in the LILE and LILE/HFSE ratios are also low. There is no evidence for crustal assimilation, since the lavas show neither strongly elevated $^{87}Sr/^{86}Sr$ at a given $^{143}Nd/^{144}Nd$ value, nor is there significant scatter on Pb–Pb isotope ratio plots.

Petrographical and geochemical features of the syn-opening lavas, including low pressure paragenesis and mineral compositions (Thy 1992*a*), and weak heavy REE depletion relative to middle REE (Allan & Gorton 1992; Pouclet & Bellon 1992), suggest a garnet-free mantle source of spinel or plagioclase lherzolite consistent with a depth of final melt equilibration shallower than 60 km. Melting was probably greatest in the subcontinental lithospheric mantle, or at the lithosphere-asthenosphere boundary subsequent to lithospheric thinning. Based on their Nd and Sr isotopic analyses of Tertiary lavas of the NE Japan arc and of Leg 127 basalts, Nohda *et al.* (1988, 1992) proposed that basalts from the back-arc area represent mixtures of melts from DMM and enriched subcontinental mantle (SCUM). The isotopic variation could be explained by a mechanism of replacement of SCUM by DMM as a result of asthenospheric injection during the process of back-arc rifting. While the depleted component, DMM (the asthenosphere), is well characterized geochemically, the composition of SCUM beneath Japan remains questionable. In contrast, the subcontinental lithosphere beneath eastern China appears to contain both an EM I (most important in basaltic lavas) and EM II (most important in mantle xenoliths and megacrysts) component (Song *et al.* 1990; Basu *et al.* 1991; Tatsumoto *et al.* 1992). Of course, the studied areas in eastern China are a few thousand kilometres west of the present-day subduction zone below Japan. In the Japan Sea region, two hypotheses for the origin of an EM II component are considered. First, the asthenospheric mantle wedge is contaminated by a sedimentary component (fluid or sediments themselves) and variable mixing in the source produces arc and back-arc magmas. Alternatively, the long-lived subduction processes, which have affected the East Eurasian continental margin since the early Mesozoic, may have imparted an EM II-like signature into the lower lithosphere. Primary melts from the asthenosphere may incorporate melts from this enriched lithosphere and thus acquire a variable EM II isotopic signature. It has been proposed that this process is important for numerous continental lithospheric mantles above subduction zones (Menzies 1990). The magma sources are derived from a mixing between this enriched lithosphere and depleted asthenosphere. For the case of the Japan Sea, the petrographical characteristics of the lavas, the inferred depth of magma generation, and the isotopic mixing trend of Magmatic Group 2 are consistent with the latter hypothesis. One may note that Pb isotope ratios of source 'B' of the Yamato Basin magmas are quite similar to those of source 2 of the Philippine Plate basin magmas, an OIB enriched-mantle source (Hickey-Vargas 1991). The enriched source for the NE Japan lavas is slightly different to source 'B', having higher $^{206}Pb/^{204}Pb$ and lower $^{87}Sr/^{86}Sr$, suggesting a weak contribution from the HIMU end-member or from a volcaniclastic sediment-derived component.

The post-opening lavas can be split into two sets. The Yamato seamount basalts (J4) have Sr and Nd isotopic compositions within the range of lavas from the NE Japan arc, close to Group 2B, and close to the DMM end-member. They may have been produced by a lesser degree of partial melting of the Group 2B source. In contrast, the alkaline basalts of Group 3 (K6 and J6) have some similarities with the SW Japan arc rocks. Their high $^{208}Pb/^{204}Pb$ ratios require a large contribution of an enriched source 'C' with a high Th/U ratio, such as EM I. This enriched component may reside in the asthenosphere (deep?) or in the subcontinental lithospheric mantle. The latter interpretation is supported by Tatsumoto & Nakamura (1991), referring to the presence of EM I in the lower lithosphere of eastern China. However, the REE patterns of the back-arc alkaline lavas suggest that they have a garnet-bearing mantle source, implying depths of melt equilibration deeper than 60 km. Thus, the back-arc alkaline lavas may have been generated by small degrees of partial melting at the base of the subcontinental lithosphere with an EM I composition, combined with contributions from depleted asthenosphere. Closer to the volcanic arc, the source of the alkaline lavas also contained an EM II-like component (Nakamura *et al.* 1985).

Dynamics of the mantle beneath eastern Eurasia

The back-arc region in the Japan Sea and a large area of eastern China are characterized by widespread tholeiitic to alkaline volcanism in the late Cenozoic (Fig. 1). As magma-genesis was due to partial melting of the subcontinental asthenosphere and of the lithosphere itself, a process of heat transfer is required. By what mechanism did upwelling hot mantle take place beneath eastern Eurasia and what is the relationship between this mechanism and back-arc basin opening? Is it an active mechanism, related to ascending of deep-seated mantle plumes, which are not dependent on the local geodynamics, or is it a passive event linked to convective circulation in the mantle as a result of subduction? The former process, was supported by Nakamura *et al.* (1990), who proposed the formation of numerous plumes beneath each volcanic area. A similar, but much larger, active plume system was proposed by Tatsumi *et al.* (1990) and Tatsumi & Kimura (1991). In their model, intense upwelling of asthenosphere has occurred beneath all of eastern China, thereby generating the widely distributed intraplate volcanism. Horizontal outwelling of the plume caused extension of continental rifts and back-arc basins along the eastern Eurasian margin. In the Japan area, alkaline magmatism was attributed to lateral injection of (enriched?) asthenosphere. Tatsumi *et al.* (1990) and Tatsumi & Kimura (1991) did not favour subduction processes to explain back-arc spreading and the generation of alkaline volcanoes. However, the location of uplifted regions in eastern China is inconsistent with the regional doming that would have resulted from a mega-plume (Lithospheric dynamics map of China, 1986). Moreover, these authors did not consider the global context of eastern Eurasian kinematics. Cenozoic tectonic and magmatic activities in eastern China are clearly distributed along the continental edge, orientated SSW–NNE, parallel to the active margin of the NW Pacific Ocean. The India–Eurasia collision and variations in the direction of convergence of the Pacific plate have created tectonic stresses producing important right-lateral movements (Tapponier *et al.* 1986; Jolivet *et al.* 1989, 1991). It is possible that hot mantle plumes may weaken pre-existing fracture zones and produce magmas by decompression melting in these weakened zones. Indeed, laboratory studies indicate that most of the mantle plumes are incapable of initiating continental break-up (Hill 1991). Thus, driving forces come from the plate-scale motions and an active plume hypothesis is not necessary to explain the tectonic and volcanic characteristics of the back-arc region.

Convective cell motion in the asthenosphere associated with the subducting slab is capable of generating magmas in the Japan Sea area, but may not explain the widespread magmatic activity in eastern China and also need not apply to the entire back-arc basin region. These large-scale processes require major mass-flux

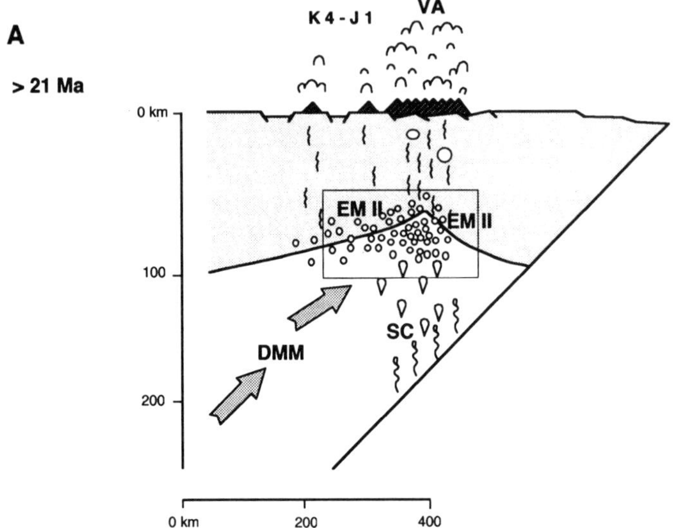

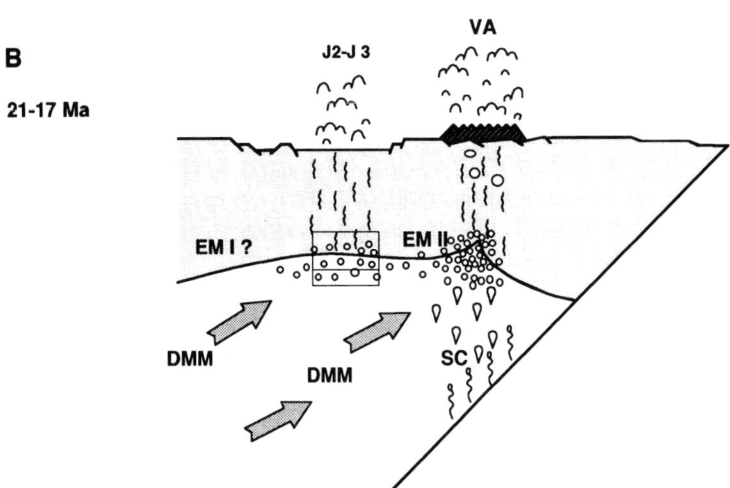

motion. In Mesozoic and Cenozoic times, along the western Pacific margin, continuous subduction of oceanic plates has supplied a considerable volume of megaliths at the thermal boundary layer between upper and lower mantle (650–700 km). The accumulation of cold material could have initiated convection, which resulted in emplacement of mantle 'blobs' beneath eastern China where heat transfer caused partial melting of the asthenosphere and of the lithosphere. In the Eurasian margin, this plate-scale, asthenospheric convective cell, combined with a steepening of the subducting older and colder Pacific plate could be responsible for the back-arc rifting and spreading of the Japan Sea.

Conclusions

In the back-arc basin area of the Japan Sea, a great diversity of volcanic rocks has been erupted since the late Mesozoic. We distinguish six main volcanic phases in the Korean margin from the early Cretaceous to the Quaternary and six volcanic stages in the Japan Sea from the early Miocene to the Quaternary. The volcanic rocks are classified into three compositional groups with distinctive magmatic affinities: (1) arc-related calc-alkaline series, which were erupted before the back-arc basin opened, (2) continental initial-rifting tholeiites closely associated with back-arc basin basalts during the opening of the Japan Sea, and (3) intra-plate

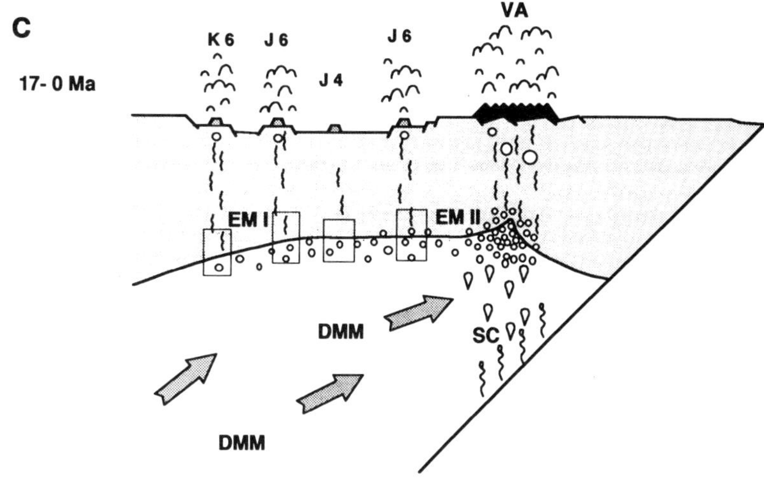

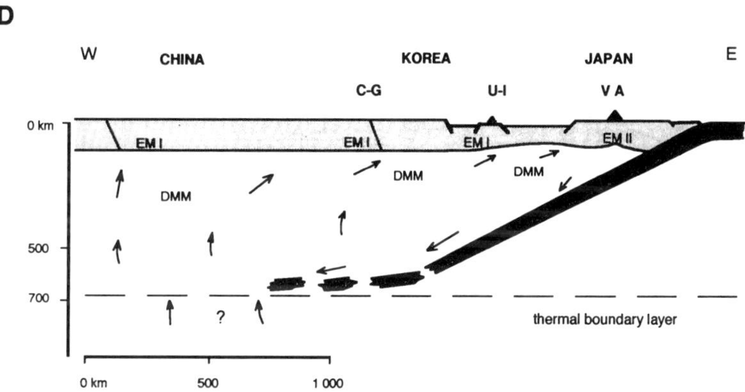

Fig. 8. Cartoon cross sections depicting the evolution of the east Eurasian margin in the Japan Sea area, from the early Miocene to the Present (approximate horizontal and vertical scales): (**A**) pre-opening stage of the Japan Sea; (**B**) opening stage in the Yamato Basin area between 21 and 17 Ma; (**C**) post-opening stage; (**D**) thermal convective cell induced by accumulation of cold megaliths of subducted slabs at the boundary layer between upper and lower mantle. K4–K6 and J1–J6, volcanic pulses, see text. Boxes correspond to the source material involved in the genesis of magmas. Abbreviations: VA, volcanic arc; DMM, depleted mantle; EM, enriched mantle; SC, subduction components; C-G, Chugaryeong Graben; U-I, Ulleung Island.

alkaline series in the post-opening stage. The occurrence of these different magmatic products is explained by conditions in the genesis of the magmas involving variable participation of three source components. Group 1 requires a contribution from depleted asthenospheric mantle (DMM) and mantle enriched by subduction components. The enriched component, EM II-like in composition, may have been incorporated in the lower subcontinental lithosphere and is a result of long-term subduction processes which occurred in the east Eurasian margin, at least since the early Mesozoic. However, involvement of recently subducted Pacific sediments is also possible. Lavas of Group 2 are also explained by mixing between DMM, which is probably an Indian Ocean MORB mantle source, and the former EM II-like component. A significant increase in the DMM contribution with time characterizes the back-arc basin basalts of the spreading stage. The alkaline lavas of Group 3 are produced from enriched mantle of EM I composition, which could be the lower continental lithosphere, and a minor contribution from depleted asthenosphere. An additional small contribution of an EM II-like component is suspected for some volcanic island lavas close to the active margin.

The conditions of magma genesis are correlated with the geodynamical evolution of the east Eurasian margin, which is summarized as follows: (i) thickening of an Andean-type active margin in response to the successive subduction of the palaeo-Pacific and Pacific plates during the Mesozoic and early Cenozoic, (ii) stretching of the margin and rifting of the back-arc area in the early Miocene, and (iii) thinning of the continental lithosphere and spreading in the back-arc basin. Successive cross sections in Figure 8 depict the magmatic evolution of the east Eurasian margin in the Japan Sea area. Before the opening of the back-arc basin (Fig. 8A), source regions involved in magma genesis are the asthenospheric depleted mantle (DMM) and the subcontinental lithosphere of EM II-like composition, strongly contaminated by the subduction-related components. These mixed magmas are responsible for the build-up of the volcanic arc. During opening of the Japan Sea (Fig. 8B), stretching of the continental plate in the back-arc area caused extensive partial melting of the subcontinental mantle heated by ascending asthenosphere. The final magmas resulted from mixing between DMM and EM II-like with decreasing participation of enriched component. In the spreading stage, the increasing rise of asthenosphere and thinning of the continental plate linked to the eastward migration of the subduction zone enhanced the contribution of the depleted mantle in the melt. At the end of opening or just after opening (Fig. 8C), residual melts of the mantle wedge were erupted and built several seamounts. Then, ascending 'blobs' of the asthenosphere heated the EM I lower lithosphere, which was partly melted beneath fractured and weakened zones, in the Japan Sea, Korea and eastern China. These magmatic activities and the back-arc kinematics could be explained by a general mass-flux motion of the asthenosphere beneath eastern Eurasia, which was initiated by accumulation of cold megaliths of subducted slabs at the thermal boundary layer between upper and lower mantle (Fig. 8D).

Financial support has been received from the CNRS-INSU (France) for chemical and Sr and Nd isotope analyses and from the NSERC (Canada) for Pb isotope analyses. Thanks to J. Carignan for running the samples on the mass spectrometer at the Université du Québec. Critical and constructive reviews by three anonymous referees were greatly appreciated and strongly improved the manuscript. We are grateful to J. Smellie for his comments on the paper and his help in improving the text.

References

ALLAN, J.F. & GORTON, M.P. 1992. Geochemistry of igneous rocks from Legs 127 and 128, Sea of Japan. *In*: Tamaki, K., Suyehiro, K., Allan, J., McWilliams, M. *et al.* (eds) *Proceedings of the Ocean Drilling Program, Scientific Results*, **127/128**, 905–929.

BASU, A.R., JIMWEM, W., WANKANG, H., GUANGHONG, X. & TATSUMOTO, M. 1991. Major element, REE, and Pb, Nd and Sr isotopic geochemistry of Cenozoic volcanic rocks of eastern China: implications for their origin from suboceanic-type mantle reservoirs. *Earth and Planetary Science Letters*, **105**, 149–169.

CHOI, H.I. 1986. Sedimentation and evolution of the Cretaceous Gyeongsang Basin, Southeastern Korea. *Journal of the Geological Society, London*, **143**, 29–40.

CLAGUE, D.A. & JARRARD, R.D. 1973. Tertiary Pacific plate motion deduced from the Hawaian Emperor Chain. *Geological Society of America Bulletin*, **84**, 1135–1154.

COUSENS, B.L. & ALLAN, J.F. 1992. A Pb, Sr, and Nd isotopic study of basaltic rocks from the Sea of Japan, ODP Legs 127/128. *In*: Tamaki, K., Suyehiro, K., Allan, J., McWilliams, M. *et al.* (eds) *Proceedings of the Ocean Drilling Project, Scientific Results*, **127/128**, 805–817.

DUPRE, B. & ALLEGRE, C.J. 1983. Pb-Sr isotopic variations in Indian Ocean basalts and mixing phenomena. *Nature*, **303**, 142–146.

ENGEBRETSON, D.C., COX, A. & GORDON, R.G. 1985. Relative motions between oceanic and continental plates in the Pacific Basin. *Geological Society of America, Special Paper*, **208**, 1–59.

HART, S.R. 1984. A large-scale isotope anomaly in the Southern Hemisphere mantle. *Nature*, **309**, 753–756.

—— 1988. Heterogeneous mantle domains, genesis and mixing chronologies. *Earth and Planetary Science Letters*, **90**, 273–296.

HEDGE, C.E. & KNIGHT, R.J. 1969. Lead and strontium isotopes in volcanic rocks from northern Honshu. *Geochemical Journal*, **3**, 15–24.

HICKEY-VARGAS, R. 1991. Isotope characteristics of submarine lavas from the Philippine Sea: implications for the origin of arc and basin magmas of the Philippine tectonic plate. *Earth and Planetary Science Letters*, **107**, 290–304.

HILL, R.I. 1991. Starting plumes and continental

break-up. *Earth and Planetary Science Letters*, **104**, 398–416.

HOLM, P. 1985. The geochemical fingerprints of different tectonomagmatic environments using hygromagmatophile element abundances of tholeiitic basalts and basaltic andesites. *Chemical Geology*, **51**, 303–323.

IIJIMA, A. & TADA, R. 1990. Evolution of Tertiary sedimentary basins of Japan in reference to opening of the Japan Sea. *Journal of the Faculty of Science, University of Tokyo, Sec. II*, **22**, 121–171.

INGLE, J.C. JR., SUYEHIRO, K., VON BREYMANN, M.T. et al. 1990. *Proceedings of the Ocean Drilling Program, Initial Reports*, **128**.

ISEZAKI, N. 1986. A magnetic anomaly map of the Japan Sea. *Journal of Geomagnetism and Geoelectricity*, **38**, 403–410.

ISSHIKI, N. 1977. Neogene and Quaternary. *In*: TANAKA, K. & NOZAWA, T. (eds) *Geology and mineral resources of Japan*. Geological Survey of Japan, 345–400.

ITO, E., WHITE, W.M. & GÔPEL, C. 1987. The O, Sr, Nd and Pb isotope geochemistry of MORB. *Chemical Geology*, **62**, 157–176.

IWATA, M., KAGAMI, H., TAKAHASHI, E. & KURASAWA, H. 1988. The Rb-Sr whole rock isochron age and genesis of Oki trachyte-rhyolite group in Oki Dogo Island. *Kazan, Ser. 2*, **33**, 79–86.

JIN, M.S., GLEADOW, A.J.W. & LOVERING, J.F. 1984. Fission track dating of apatite from the Jurassic and Cretaceous granites in South Korea. *Journal of the Geological Society, Korea*, **20**, 257–265.

——, KIM, S.Y. & LEE J.S. 1981. Granitic magmatism and associated mineralisation in the Gyeongsang Basin, Korea. *Mining Geology, Japan*, **31**, 245–260.

——, KIM, S.J. & SHIN, S.C. 1988. K/Ar and fission-track datings for granites and volcanic rocks in the southeastern part of the Korean Peninsula. *KIER Rep., Geol. Surv. South Korea*, KR–88–6D, 51–84.

JOLIVET, L. 1987. America-Eurasia plate boundary in eastern Asia and the opening of marginal basins. *Earth and Planetary Science Letters*, **81**, 282–288.

——, HUCHON, PH., BRUN, J.P., CHAMOT-ROOKE, N., LE PICHON, X. & THOMAS, J.C. 1991. Arc deformation and marginal basin opening, Japan Sea as a case study. *Journal of Geophysical Research*, **96**, 4367–4384.

——, HUCHON,, PH. & RANGIN, CL. 1989. Tectonic setting of Western Pacific marginal basins. *Tectonophysics*, **160**, 23–48.

KAGAMI, H., IWATA, M. & TAKAHASHI, E. 1986. Isotopic evidence for primitive mantle beneath the Sea of Japan, a young back arc basin. *Technical Report of the Institute for Study of the Earth's Interior, Okayama University, Japan, Series. A*, **7**, 1–13.

KANEOKA, I. 1990. Radiometric age and Sr isotope characteristics of volcanic rocks from the Japan Sea floor. *Geochemical Journal*, **24**, 7–19.

—— & YUASA, M. 1988. ^{40}Ar-^{39}Ar age studies on igneous rocks dredged from the central part of the Japan Sea. *Geochemical Journal*, **22**, 195–204.

——, HAYASHI, H., IWAGUCHI, T., YASUDA, A.,

TAKIGAMI, Y., FUJIOKA, K. & SAKAI, H. 1988. K-Ar dating of volcanic rocks dredged from the Yamato Seamount chain in the Japan Sea. *Bulletin of the Volcanological Society of Japan, Ser. 2*, **33**, 213–218.

——, NOTSU, K., TAKIGAMI, Y., FUJIOKA, K. & SAKAI, H. 1990. Constraints on the evolution of the Japan Sea based on ^{40}Ar-^{39}Ar ages and Sr isotopic ratios for volcanic rocks of the Yamato Seamount chain in the Japan Sea. *Earth and Planetary Science Letters*, **97**, 211–225.

——, TAKIGAMI, Y., TAKAOKA, N., YAMASHITA, S. & TAMAKI, K. 1992. ^{40}Ar-^{39}Ar analyses of volcanic rocks recovered from the Japan Sea floor by Leg 127/128: constraint on the formation age of the Japan Sea. *In*: TAMAKI, K., SUYEHIRO, K., ALLAN, J., McWILLIAMS, M. *et al.* (eds) *Proceedings of the Ocean Drilling Program, Scientific Results*, **127/128**, 819–836.

KIM, O.J. 1971. Study on the intrusion epochs of younger granites and their bearing to orogenies of South Korea. *Journal of the Korean Institute of Mining Geology*, **4**, 1–9.

KIM, S.W. & LEE, Y.G. 1981. Petrology and structural geology of the Late Cretaceous volcanic rocks in the Northern part of Yucheon Basin. *Journal of the Korean Institute of Mining Geology*, **14**, 35–49.

KIMURA, M., MATSUDA, T., SATO, H., KANEOKA, I., TOKUYAMA, H., KURAMOTO, S., OSHIDA, A., SHIMAMURA, K., TAMAKI, K., KINOSHITA, H. & UYEDA, S. 1987. Report on DELP 1985 cruises in the Japan Sea. Part VII: Topography and geology of the Yamato Basin and its vicinity. *Bulletin of the Earthquake Research Institute, University of Tokyo*, **62**, 447–483.

LEE, J.H. & KIM, S.Y. 1970. Mineralisation and ore deposits of native copper in Seachondong basalt flows in Yongyang Basin, Korea. *Journal of the Geological Society, Korea*, **6**, 230–244.

LEE, J.S. 1989. Pétrologie et relations structurales des volcanites crétacées à cénozoïques de la Corée du Sud: implications géodynamiques sur la marge est-eurasiatique. PhD. thesis, University of Orléans, France, 349p.

—— & POUCLET, A. 1988. Le volcanisme néogène de Pohang (SE Corée), nouvelles contraintes géochronologiques pour l'ouverture de la Mer du Japon. *Comptes Rendus de l'Académie des Sciences, Paris*, **307**, 1405–1411.

Lithospheric Dynamics Map of China and Adjacent Seas, 1: 4000000. 1986. State Seismological Bureau, China Geological Publishing House.

MAHONEY, J.J., NATLAND, J.H., WHITE, W.M., POREDA, R., BLOOMER, S.H. & FISHER, R.L. 1989. Isotopic and geochemical provinces of the western Indian Ocean spreading centers. *Journal of Geophysical Research*, **94**, 4033–4052.

MEIJER, A. 1976. Pb and Sr isotopic data bearing on the origin of volcanic rocks from the Mariana island-arc system. *Geological Society of America Bulletin*, **87**, 1358–1369.

MENZIES, M.A. 1990. Archean, Proterozoic, and Phanerozoic lithospheres. *In*: MENZIES, M.A. (ed.) *Continental Mantle*. Oxford Monographs on

Geology and Geophysics No. 16, Oxford Science Publications, 67–86.

MORRIS, P.A. & KAGAMI, H. 1989. Nd and Sr isotope systematics of Miocene to Holocene volcanic rocks from Southwest Japan: volcanism since the opening of the Japan Sea. *Earth and Planetary Science Letters*, **92**, 335–346.

NAKAMURA, E., CAMPBELL, I.H., McCULLOCH, M.T. & SUN. S.-S. 1989. Chemical geodynamics in a back arc region around the Sea of Japan: implications for the genesis of alkaline basalts in Japan, Korea, and China. *Journal of Geophysical Research*, **94**, 4634–4654.

——, CAMPBELL, I.H. & SUN, S.-S. 1985. The influence of subduction processes on geochemistry of Japanese alkaline basalts. *Nature*, **316**, 55–58.

——, McCULLOCH, M.T. & CAMPBELL, I.H. 1990. Chemical geodynamics in the back-arc region of Japan based on the trace element and Sr-Nd isotopic compositions. *Tectonophysics*, **174**, 207–233.

NOHDA, S., TATSUMI, Y., OTOFUJI, Y.-I., MATSUDA, T. & ISHIZAKA, K. 1988. Asthenospheric injection and back-arc opening: isotopic evidence from northeast Japan. *Chemical Geology*, **68**, 317–327.

——, TATSUMI, Y., YAMASHITA, S. & FUJII, T. 1992. Nd and Sr isotopic study of Leg 127 basalts: implications for the evolution of the Japan Sea back-arc basin. *In*: TAMAKI, K., SUYEHIRO, K., ALLAN, J., McWILLIAMS, M. *et al.* (eds) *Proceedings of the Ocean Drilling Program, Scientific Results*, **127/128**, 899–904.

—— & WASSERBURG, G.J. 1981. Nd and Sr isotopic study of volcanic rocks from Japan. *Earth and Planetary Science Letters*, **52**, 264–276.

—— & —— 1986. Trends of Sr and Nd isotopes through time near the Japan Sea in northeastern Japan. *Earth and Planetary Science Letters*, **78**, 157–167.

O'NIONS, R.K., HAMILTON, P.J. & EVENSEN, N.M. 1977. Variation in $^{143}Nd/^{144}Nd$ and $^{87}Sr/^{86}Sr$ ratios in oceanic basalts. *Earth and Planetary Science Letters*, **34**, 13–22.

PEARCE, J.A. 1982. Trace element characteristics of lavas from destructive plate boundaries. *In*: THORPE, R.S. (ed.) *Andesites*. John Wiley & Sons, New-York, 525–548.

—— 1983. Role of the sub-continental lithosphere in magma genesis at active continental margins. *In*: HAWKESWORTH, C.J. & NORRY, M.J. (eds) *Continental basalts and mantle xenoliths*. Shiva Geology Series, Shiva Publishing Limited, Nantwich, UK, 230–249.

—— & NORRY, M.J. 1979. Petrogenetic implications of Ti, Zr, Y and Nb variations in volcanic rocks. *Contribution to Mineralogy and Petrology*, **69**, 33–47.

PENG, Z.C., ZARTMAN, R.E., FUTA, K. & CHEN, D.G. 1986. Pb-, Sr-, and Nd-isotope systematics, and the chemical characteristics of Cenozoic basalts, eastern China. *Chemical Geology*, **59**, 3–33.

POLEVAYA, N.L., PUTINTZEN, V.K. & SPRINTZON, V.D. 1961. About the age of some magmatic and metamorphic rocks in north Korea. *Sovietskaya Geologiya* (English translation), **6**, 119–124.

POUCLET, A. & BELLON, H. 1992. Geochemistry and isotopic composition of volcanic rocks from the Yamato Basin: Hole 794D, Sea of Japan. *In*: TAMAKI, K., SUYEHIRO, K., ALLAN, J., McWILLIAMS, M. *et al.* (eds) *Proceedings of the Ocean Drilling Program, Scientific Results*, **127/128**, 779–789.

—— & LEE, J.S. 1988. Tectonic evolution of the Asian margin since the Cretaceous: implications from the geochronology of volcanism in the Korean peninsula. *In*: International Symposium '*Geodynamic Evolution of the Eastern Eurasian Margin*', Paris, Société Géologique de France, 90.

—— & SCOTT, S.D. 1992. Volcanic ash layers in the Japan Sea: tephrochronology of Sites 798 and 799. *In*: TAMAKI, K., SUYEHIRO, K., ALLAN, J., McWILLIAMS, M. *et al.* (eds) *Proceedings of the Ocean Drilling Program, Scientific Results*, **127/128**, 791–803.

——, FUJIOKA, K., FURUTA, T. & OGIHARA, S. 1992. Data Report: Geochemistry and mineralogy of ash layers from Legs 127 and 128 in the Japan Sea. *In*: TAMAKI, K., SUYEHIRO, K., ALLAN, J., McWILLIAMS, M. *et al.* (eds) *Proceedings of the Ocean Drilling Program, Scientific Results*, **127/128**, 1373–1393.

SAUNDERS, A.D. & TARNEY, J. 1984. Geochemical characteristics of basaltic volcanism within back-arc basins. *In*: KOKELAAR, B.P. & HOWELLS, M.F. (eds) *Marginal Basin Geology*. Geological Society, London, Special Publication, **16**, 59–76.

SONG, Y., FREY, F.A. & ZHI, X. 1990. Isotopic characteristics of Hannuoba basalts, eastern China: implications for their petrogenesis and the composition of subcontinental mantle. *Chemical Geology*, **85**, 35–52.

SUN, S.-S. 1980. Lead isotopic study of young volcanic rocks from mid-ocean ridges, ocean islands and island arcs. *Royal Society of London Philosophical Transactions*, **297**, 409–445.

—— & McDONOUGH, W.F. 1989. Chemical and isotopic systematics of oceanic basalts: implications for mantle composition and processes. *In*: SAUNDERS, A.D. & NORRY, M.J. (eds) *Magmatism in the Ocean Basins*. Geological Society, London, Special Publications, **42**, 313–345.

TAMAKI, K. 1986. Age estimation of the Japan Sea on the basis of stratigraphy, basement depth, and heat flow data. *Journal of Geomagnetism and Geoelectricity*, **38**, 427–446.

—— 1988. Geological structure of the Japan Sea and its tectonic implications. *Bulletin of the Geological Survey of Japan*, **39**, 269–365.

——, PISCIOTTO, K., ALLAN, J. *et al.* 1990. *Proceedings of the Ocean Drilling Program, Initial Reports*, **127**.

——, SUYEHIRO, K., ALLAN, J., INGLE, J.C. & PISCIOTTO, K.A. 1992. Tectonic synthesis and implications of Japan Sea ODP drilling. *In*: TAMAKI, K., SUYEHIRO, K., ALLAN, J. *et al.* (eds) *Proceedings of the Ocean Drilling Program, Scientific Results*, **127–128**, 1333–1348.

TAPPONIER, P., PELTZER, G. & ARMIJO, R. 1986 On the

mechanics of the collision between India and Asia. *In*: COWARD, M.P. & RIES, A.C. (eds) *Collision tectonics*. Geological Society, London, Special Publications, **19**, 127–137.

——, MARUYAMA, S. & NOHDA, S. 1990. Mechanism of backarc opening in the Japan Sea: role of asthenospheric injection. *Tectonophysics*, **181**, 299–306.

——, NOHDA, S. & ISHIZAKA, K. 1988. Secular variation of magma source compositions beneath the northeast Japan arc. *Chemical Geology*, **68**, 309–316.

TATSUMI, Y. & KIMURA, N. 1991. Backarc extension versus continental break-up: petrological aspects for active rifting. *Tectonophysics*, **197**, 127–137.

TATSUMOTO, M. 1969. Lead isotopes in volcanic rocks and possible ocean-floor thrusting beneath island arcs. *Earth and Planetary Science Letters*, **6**, 369–376.

—— & KNIGHT, J.R. 1969. Isotopic composition of lead in volcanic rocks from central Honshu with regard to basalt genesis. *Geochemical Journal*, **3**, 58–86.

—— & NAKAMURA, Y. 1991. DUPAL anomaly in the Sea of Japan: Pb, Nd, and Sr isotopic variations at the eastern Eurasian continental margin. *Geochemica et Cosmochimica Acta*, **55**, 3697–3708.

——, BASU, A.R., WANKANG, H., JUNWEN, W. & GUANGHONG, X. 1992. Sr, Nd, and Pb isotopes of ultramafic xenoliths in volcanic rocks of Eastern China: enriched components EM I and EM II in subcontinental lithosphere. *Earth and Planetary Science Letters*, **113**, 107–128.

THY, P. 1992*a*. Phenocryst and groundmass phase compositions of basaltic and andesitic sills and flows from the Japan Sea recovered during Legs 127 and 128. *In*: TAMAKI, K., SUYEHIRO, K., Allan, Z., McWilliams, M. *et al*. (eds) *Proceedings of the Ocean Drilling Program, Scientific Results*, **127/128**, 849–859.

—— 1992*b*. Liquid lines of descent of basalts drilled at Sites 794 and 797 in the Yamato Basin of the Japan Sea. *In*: TAMAKI, K., SUYEHIRO, K., ALLAN, J., McWILLIAMS, M. *et al*. (eds) *Proceedings of the Ocean Drilling Program, Scientific Results*, **127/128**, 869–882.

TSUCHIYA, N. 1990. Middle Miocene back-arc rift magmatism of basalt in the NE Japan arc. *Bulletin of the Geological Survey of Japan* **41**, 473–505.

WON, J.K. 1976. Study of petro-chemistry of volcanic rocks in Jeju Island. *Journal of the Geological Society of Korea*, **12**, 207–226.

—— 1983. A study on the Quaternary volcanism in the Korean Peninsula – in the Choogaryong rift valley. *Journal of the Geological Society of Korea*, **19**, 159–168.

WOODHEAD, J.D. & FRASER, D.G. 1985. Pb, Sr, and ^{10}Be isotopic studies of volcanic rocks from the Northern Mariana Islands. Implications for magma genesis and crustal recycling in the western Pacific. *Geochimica et Cosmochimica Acta*, **49**, 1925–1930.

YOON, H.D. 1987. *The geochemical characteristics and origin of alkaline magmas in the Ulleung Island, Korea*. PhD thesis, Seoul Univ., South Korea.

ZINDLER, A. & HART, S.R. 1986. Chemical geodynamics. *Annual Reviews of Earth and Planetary Science*, **14**, 493–571.

ZHOU, X. & ARMSTRONG, R.L. 1982. Cenozoic volcanic rocks of eastern China -secular and geographic trends in chemistry and strontium isotopic composition. *Earth and Planetary Science Letters*, **58**, 301–329.

Arc and back-arc geochemistry in the southern Kermadec arc–Ngatoro Basin and offshore Taupo Volcanic Zone, SW Pacific

J.A. GAMBLE[1], I.C. WRIGHT[2], J.D. WOODHEAD[3] & M.T. McCULLOCH[3]

[1] *Department of Geology, Victoria University of Wellington, PO Box 600, Wellington, New Zealand*

[2] *New Zealand Oceanographic Institute, National Institute of Water and Atmospheric Research, PO Box 14–901, Wellington, New Zealand*

[3] *Research School of Earth Sciences, Australian National University, GPO Box 4, Canberra, ACT 2601, Australia.*

Abstract: Back-arc basin basalts from the Ngatoro Basin (the southern end of the Havre Trough) are similar geochemically to, yet subtly distinct from, basalts of the Havre Trough to the north. Whole rock and glass chemistry are consistent with derivation from a fertile mantle source with subsequent evolution by fractionation of olivine (+ Cr-spinel) + plagioclase, and then clinopyroxene. Basalts from the vicinity of Rumble IV seamount at the southern end of the Kermadec island arc, and the eastern Ngatoro rift escarpment, are strongly porphyritic relative to the back-arc basin basalts and show trace element (high LIL abundances and highly depleted HFS abundances) and isotopic signatures of subduction zone basalts.

At its southern end, the Ngatoro Basin penetrates the continental crust of New Zealand creating a major, 3000 m deep bathymetric re-entrant in the slope-break; the slope-break marks the transition from oceanic to continental crust. Basalts from the floor of the Ngatoro Basin re-entrant are isotopically distinct from the basalts of the oceanic sector in that they have higher Sr and correspondingly lower Nd isotope ratios and are comparable to basalts of the Taupo Volcanic Zone (TVZ) to the south.

In contrast to the basalt-dominated oceanic sector, basalts from the offshore TVZ, a 100 km long area extending roughly NNE from White Island to the submarine Whakatane arc volcano at the edge of the continental slope-break, occur in association with andesites, dacites and rhyolites. These basalts are generally strongly porphyritic (olivine + plagioclase + clinopyroxene) and show trace element abundances typical of suprasubduction zone rocks. However, offshore TVZ basalts show subtle distinctions from onshore TVZ basalts to the south; the former have more radiogenic Sr isotopes. Furthermore, their high field strength element and transition element systematics appear to overlap with those of basalts from the Kermadec arc to the north.

The authors attribute these lateral (along arc) and transverse (across arc) variations to source heterogeneity and variable fertility in the sources of the arc and back-arc basin magmas. Sources of the arc-front magmas are more refractory and also more susceptible to contamination by slab-derived fluids than sources for back-arc basin magmatism, reflecting the dynamic nature of flow from the back-arc into the mantle wedge beneath the volcanic front.

At its southern end, the 2000 km long NE–SW orientated oceanic Tonga–Kermadec island arc and Lau Basin–Havre Trough back-arc basin system impinges on the continental edge of New Zealand. Changes in volcanism associated with this transition from oceanic to continental crust are spectacular, with the dominantly mafic volcanism of the oceanic arc–back-arc system giving way to explosive rhyolitic volcanism in the continental TVZ of New Zealand. Basaltic magmas, formed by partial melting of peridotite mantle in the wedge between the subducting Pacific ocean plate and sub-arc lithosphere are common to both areas, although they are minor in volume in comparison to rhyolite in the TVZ (Wilson *et al.* 1984; Gamble *et al.* 1990).

In this paper the authors report new major element, trace element and Nd and Sr isotope geochemical data on rocks dredged from the arc volcanoes of the southern Kermadec arc (KA), the Ngatoro Basin, which forms the southern extremity of the Havre Trough back-arc basin axial rift, and the offshore TVZ (Fig. 1). These samples, which were recovered by rock dredging

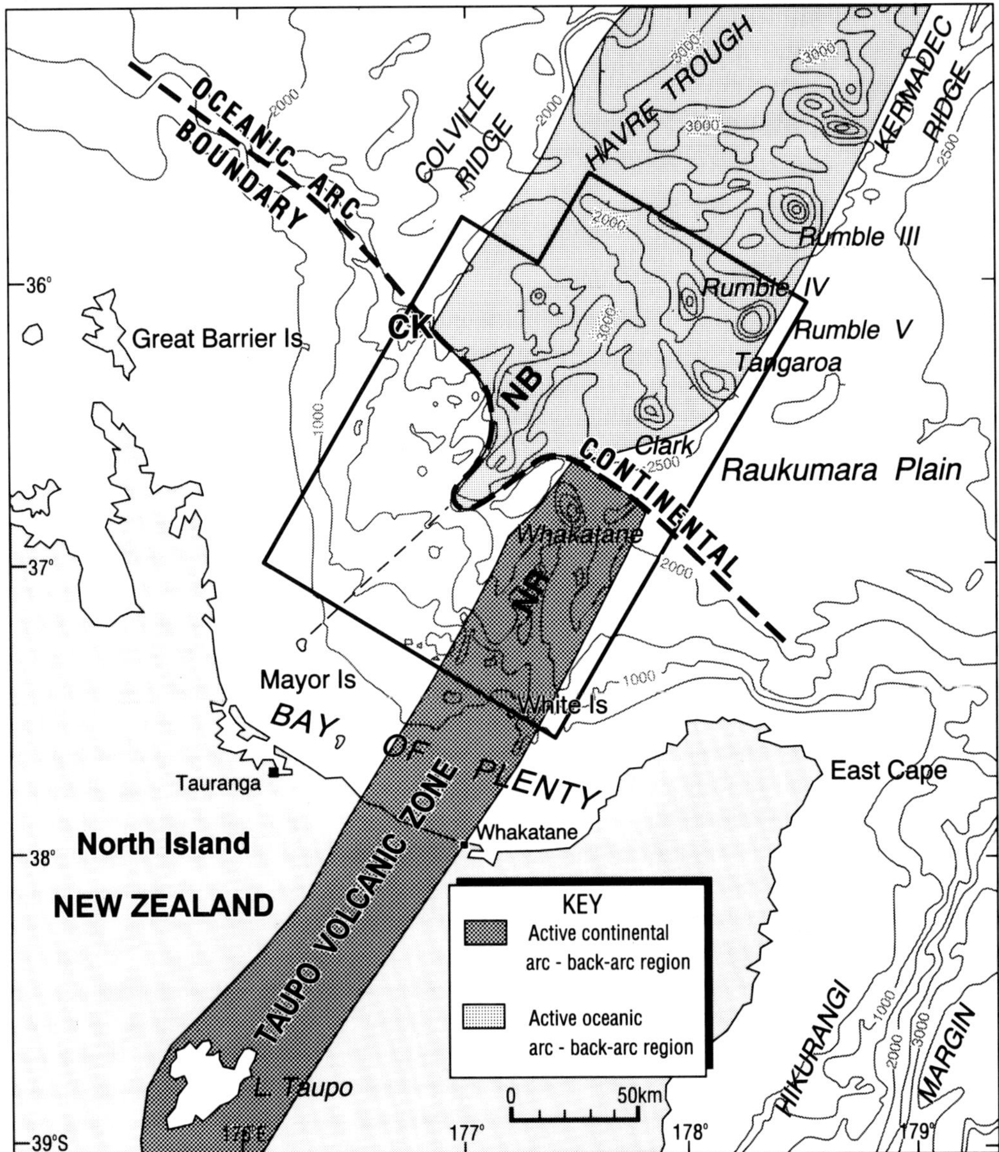

Fig. 1. Location map showing area studied (in rectangle) in relation to the Taupo Volcanic Zone, Kermadec arc and Havre Trough. The inferred transition zone between continental and oceanic settings is marked by a dashed line. Note the steep bathymetry associated with the re-entrant Ngatoro Basin (NB). Colville Knolls (CK) are basement New Zealand continental crust greywacke. The major volcanic features of Rumble IV, V, Tangaroa and Clark, Whakatane and Ngatoro Ridge (NR) are also shown.

during cruises of the R.V. *Rapuhia* between 1988 and 1990, complimented by recent tectonic studies (Wright *et al.* 1990; Wright 1990, 1992, 1993*a,b*), provide an important insight into changes in magmatism and tectonism along an active contiguous oceanic–continental arcback–arc system. Our major aim here is to integrate these new geochemical data into our existing database from the TVZ, KA and Havre Trough (Gamble *et al.* 1990, 1993*a*) and to document fine detail in trace element and isotopic systematics in arc–back-arc magmatism within the transition between oceanic and continental lithospheric settings.

Geological setting

The KA–Havre Trough and TVZ have a complex volcano-tectonic structure. The region investigated here encompasses a generally rectangular area whose eastern boundary extends 200 km north-east from White Island in the south to the Rumble V volcano at the southern end of the KA, and whose western boundary is defined by the Colville Knolls and Colville Ridge (Fig. 1).

To the north, the Havre Trough (HT) is a typical actively widening back-arc basin characterized by shallow seismicity, high heat flow, complex and rugged seafloor topography, and relatively minimal or absent sediment cover (Pelletier & Loutat 1989; Wright *et al.* 1990; Caress 1991; Wright 1992). At its southern extremity, the HT comprises a complex axial rift-graben system (the >3000 m deep Ngatoro Basin) flanked to the east and west by a heterogeneous terrane of axis-parallel ridges and basins and isolated knolls. The axial rift system consists of three partially sediment-infilled, contiguous, segmented, *en-echelon* sub-grabens, each some 10–12 km wide and 24–26 km long. The bounding rift escarpments have relief of 500–1000 m. Constructional volcanic forms are, in general, absent within the graben, and volcanism associated with rifting is apparently restricted to the graben escarpments (Wright 1993*b*).

East of the axial rift, the morphology of the back-arc region is complicated by the spatial position of the major Rumble IV, V, Tangaroa and Clark stratovolcanoes. All four of these volcanoes have basal diameters exceeding 12 km and relief exceeding 1000 m. Farther east, a series of major normal faults downthrown to the west, marks the eastern margin of the back-arc region (Wright 1990).

At the landward extremity of the Ngatoro Basin rift system, the southernmost sub-graben penetrates New Zealand continental crust, forming a major bathymetric and tectonic re-entrant along the continental–oceanic crustal boundary (Fig. 1). Although lacking relevant deep-crustal geophysical data, elsewhere the northeastern continental edge of New Zealand appears to form a distinct, near-linear boundary.

At the continental–oceanic arc boundary, the axis of the presently active TVZ is offset sinistrally from the Ngatoro Basin rift by some 45–50 km (Lewis & Pantin 1984; Wright *et al.* 1990). A series of oblique synthetic shears accommodates the motion between the oceanic and continental segments (Wright 1992).

The northern limit of the offshore TVZ is marked by the 1000 m high submarine Whakatane stratovolcano (Fig. 1). South of this, three essentially parallel volcano-tectonic units comprise the Ngatoro Ridge. This ridge is flanked by graben structures. The western graben is studded with numerous circular, elliptical, crescentic and irregular knolls (Wright 1989), which are known or presumed to be volcanic structures (Lewis & Pantin 1984; Wright 1992, 1993*a*).

Analytical methods

All chemical analyses were undertaken on fresh chips of rock or pillow glass. Samples were initially washed and rinsed with distilled water and an aliquot of oven dried chips reduced to powder by grinding in a Tema tungsten carbide disc mill for 1 minute.

Major and trace elements were determined by XRF using routine techniques of the Analytical Facility of Victoria University (Palmer 1990). Electron microprobe analyses were determined on a JEOL – 733 Superprobe in the Analytical Facility of Victoria University. Glass analyses were determined using a defocused electron beam and reduced beam current. Full details of the methodology are contained in Watanabe *et al.* (1981) and Gamble & Kyle (1987).

Sr and Nd isotope analyses were undertaken in the Research School of Earth Sciences, Australian National University. All samples for isotopic analysis were selected from a split of the rock chips by careful hand-picking. These chips were washed in ultra-pure water and then leached with strong hot acid (6N HCL) in an ultrasonic bath for 1 hour. The residual chips were then rinsed and dried prior to dissolution in HF and HNO_3. Sr and Nd were separated by standard ion exchange methods and loaded on single Ta (Sr) and Re – Ta double (Nd) filament assemblies. All isotope measurements were made on a Finnigan MAT 261 multiple collector mass spectrometer and normalized to $^{86}Sr/^{88}Sr = 0.1194$ and $^{146}Nd/^{144}Nd = 0.7219$. Blank measurements were negligible. During our analytical runs $^{87}Sr/^{86}Sr = 0.710209 \pm 13$ 1 s ($n = 79$) and $^{143}Nd/^{144}Nd = 0.511872 \pm 7$, 1 s ($n = 85$) on NBS–987 and La Jolla respectively.

Petrology

In this section, the authors concentrate on summarizing the major petrographical features of newly analysed rocks. Full details of the petrology and mineral chemistry of these rocks are contained in Gamble *et al.* 1993*b*. For ease of description, we subdivide the area into 3 sections:

(1) southern Kermadec arc seamounts and eastern Ngatoro Basin;

(2) western Ngatoro Basin, both of which are located on oceanic crust; and

(3) offshore TVZ, an extension of the TVZ located on continental crust.

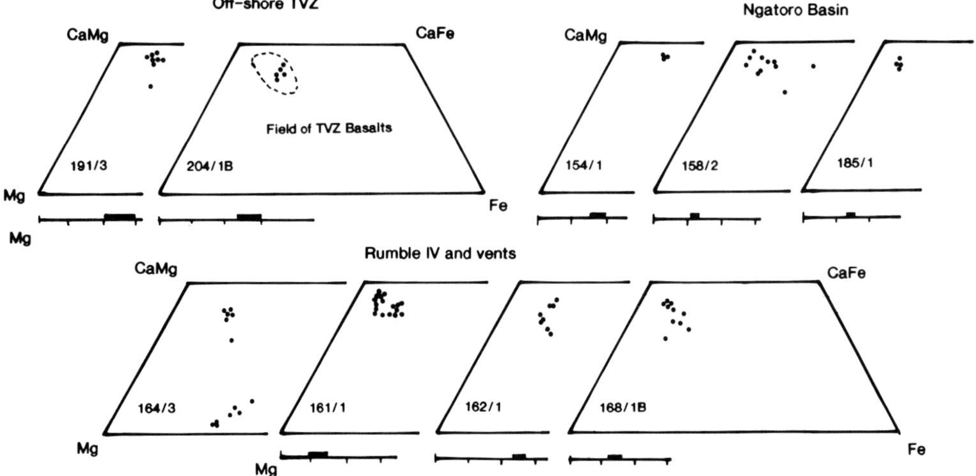

Fig. 2. Electron microprobe analyses of pyroxenes (quadrilateral) and olivine from selected basalts and an andesite (164/1) from Ngatoro Basin, Rumble IV and Tangaroa volcanoes and Ngatoro Ridge (offshore TVZ). The field of TVZ basalts, outlined by dashed line (Gamble *et al.* 1990) is shown for comparison. Ticks on olivine composition axes are at intervals of 10% Fo.

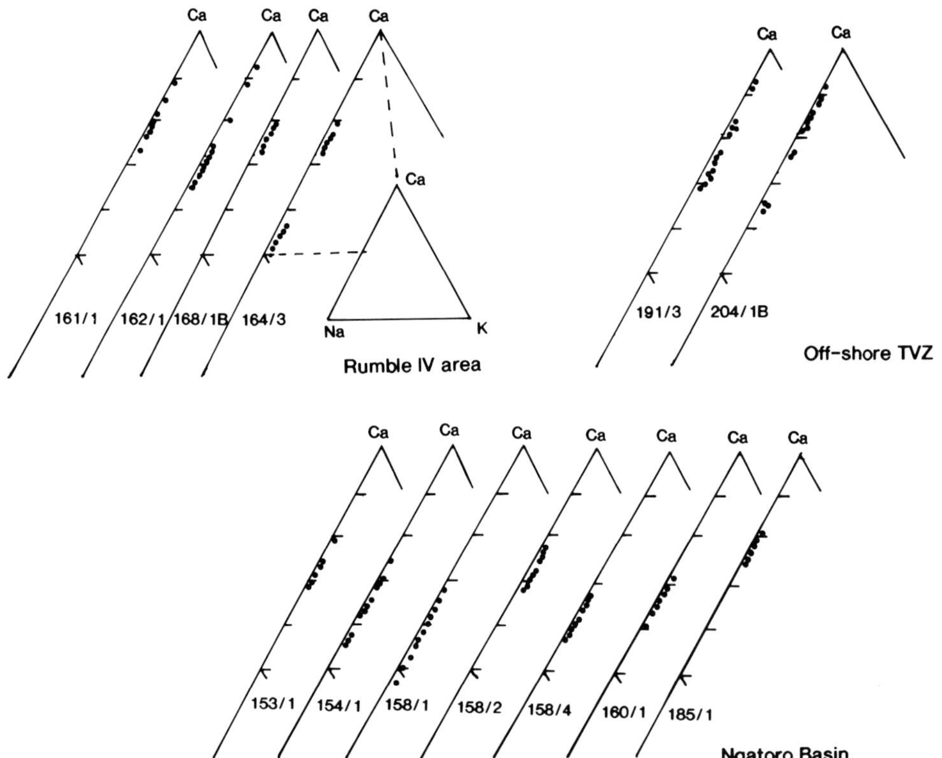

Fig. 3. Electron microprobe compositions (Ca-Na-K) for plagioclase from Ngatoro Basin, Rumble IV, Tangaroa and offshore TVZ (Ngatoro Ridge) samples.

Oceanic section

Southern Kermadec arc seamounts and eastern Ngatoro Basin. Rumble IV seamount at the southern end of the Kermadec island arc (Fig. 1) is a large (*c.* 1000 m relief) basalt–andesite stratovolcano. Two basaltic blocks dredged previously from the Rumble IV volcano are vesicular and show glassy outer surfaces with variably porphyritic textures dominated by plagioclase (Smith & Brothers 1988). The samples reported here were dredged from the flanks of Rumble IV and surrounding environs, and from the flanks of the newly discovered Tangaroa volcano (Fig. 1). These samples range from basalts (X161, X162) to andesite (X164). Basalts (X168, X169) were sampled from the eastern flank of the Ngatoro rift graben. All samples are vesicular and variably porphyritic with phenocrysts which include olivine ($100.Mg/Mg + Fe^{2+} = 84$–90), plagioclase (An_{64-93}) and clinopyroxene ($100.Mg/Mg + Fe^{2+} = 90$) in the basalts, plus orthopyroxene in the andesite which contains no olivine. Cr-spinel ($100.Cr/Cr + Al \approx 60$) occurs as inclusions in the olivine phenocrysts. Temperatures based on the olivine–spinel geothermobarometer (Ballhaus *et al.* 1991) yielded values between 975 and 1050°C and f_{O_2} 1–2 log units above QFM. Compositions and composition fields are shown in Figs 2 to 4 and complete mineral analyses are contained in Gamble *et al.* 1993*b*)

Western Ngatoro Basin. These samples were recovered from the western escarpment of the axial Ngatoro Basin rift (Fig. 1). The samples (X153, 154, 158 and 160) are all basalts with glassy pillow rinds in which fresh brown glass encloses sparse phenocryst assemblages dominated by olivine ($100.Mg/Mg + Fe^{2+} = 75$–88) and plagioclase (An_{47-78}). Cr spinel ($100.Cr/Cr + Al = 44$–70) occurs as tiny red-brown octahedral inclusions in the olivines. These spinels are intermediate in composition between those in TVZ lavas and typical MORB (Fig. 4). Temperatures based upon the olivine-spinel geothermobarometer (Ballhaus *et al.* 1991) range from 950 to 1075°C, with f_{O_2} ranging from QFM values to 2 log units above QFM. Clinopyroxene phenocrysts occur only in more evolved samples (e.g. X185, higher SiO_2, lower $Mg^{\#}$) from the extreme south end of the basin (Fig. 2). The crystallization sequence of western Ngatoro Basin samples is consistently (Cr-spinel) – olivine and then plagioclase, with clinopyroxene appearing somewhat later. This compares favourably with the sequence identified for primitive TVZ basalts to the south

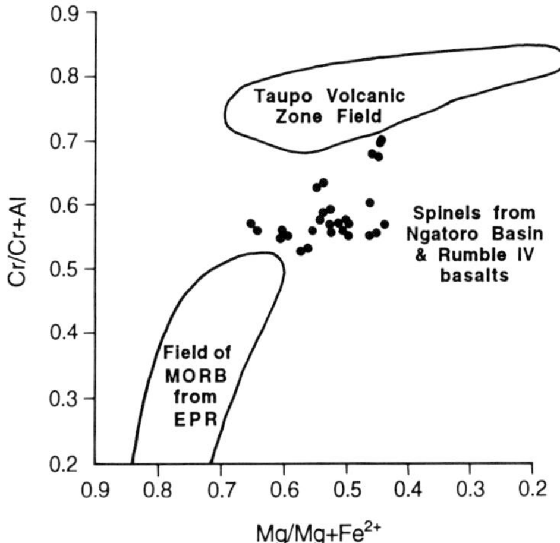

Fig. 4. Compositions of Cr-spinels from Ngatoro Basin and Rumble IV seamount compared to Cr-spinels from basalts from TVZ (Gamble, unpublished data) and typical MORB (Natland 1989).

(Gamble *et al.* 1990) and contrasts with basalts from eastern escarpment of the Ngatoro Basin and Rumble IV at the southern end of the KA, which show more complicated sequences of crystallization (e.g. early clinopyroxene), in part deriving from mixing processes, and more abundant phenocrysts. Compositions of olivine, clinopyroxene and plagioclase are shown in Figs 2 and 3.

Continental section

Offshore Taupo Volcanic Zone. The main distinguishing feature of the sample suite from the continental sector compared with the oceanic section is the diversity of rock types. Although basalts are relatively common, andesites, dacites and rhyolites, similar to those of the onshore TVZ region, occur in greater abundance in the continental sector, evidently reflecting the continuity of settings between onshore TVZ and the offshore region.

Whakatane seamount is a major, basalt–andesite, arc stratovolcano at the northern extremity of the offshore TVZ. The andesites are porphyritic with phenocryst assemblages dominated by plagioclase + orthopyroxene + clinopyroxene and minor olivine, similar petrographically to many Ruapehu Type 1 andesites (Graham & Hackett 1987). The Ngatoro Ridge (Fig. 1) extends northward from the White Island massif and appears in the main to comprise basalts (X190, 196, 201 and 204), andesites (X201 and 204) and rhyolites (X204).

Table 1. *Major oxide (wt%), trace element (ppm) and Sr and Nd isotope data for western Ngatoro Basin and Havre Trough basalts*

Sample no. Field no. Locality	VUW TVZ-19 Havre Trough	VUW 185-1 Ngatoro Basin	VUW 158-2 Ngatoro Basin	VUW 158-1 Ngatoro Basin	VUW 158-3 Ngatoro Basin	VUW 160 Ngatoro Basin	VUW 154-1 Ngatoro Basin	VUW 154-5 Ngatoro Basin	VUW 153-1 Ngatoro Basin	VUW 158-4 Ngatoro Basin
SiO_2	50.66	51.51	50.19	51.05	51.70	50.09	50.47	50.43	49.99	50.74
TiO_2	1.49	0.92	0.91	1.43	1.45	1.45	1.66	1.62	1.14	1.45
Al_2O_3	16.09	17.68	16.86	16.10	16.05	15.92	16.27	16.18	16.52	16.15
Fe_2O_3	1.11	1.18	0.93	1.14	1.12	1.21	1.19	1.19	1.02	1.14
FeO	7.37	7.87	6.23	7.58	7.45	8.07	7.90	7.95	6.76	7.59
MnO	0.16	0.16	0.13	0.16	0.16	0.17	0.18	0.18	0.15	0.16
MgO	7.49	4.78	7.66	6.92	7.29	7.04	6.89	7.06	8.06	7.03
CaO	10.26	10.83	12.61	9.45	9.69	9.95	9.31	9.36	10.60	9.42
Na_2O	3.95	2.81	2.93	3.73	3.76	3.35	4.39	4.25	3.33	3.75
K_2O	0.29	0.54	0.24	0.51	0.41	0.56	0.52	0.47	0.40	0.42
P_2O_5	0.18	0.12	0.07	0.15	0.15	0.16	0.18	0.18	0.13	0.15
LOI	0.54	0.71	0.58	1.13	0.16	1.35	0.28	0.50	1.19	1.35
Total	100.36	99.97	100.04	100.19	100.22	100.22	100.11	100.25	100.05	100.20
MgNo.	66.7	52.0	68.7	61.9	63.6	60.9	60.9	61.3	68.0	62.3
Sc	34	32	34	30	28	32	29	31	33	30
V	246	209	209	257	248	311	286	282	234	255
Cr	280	222	280	293	280	310	209	217	363	300
Ni	97	65	86	100	103	120	90	96	128	104
Cu	58	64	74	49	40	60	43	45	52	48
Zn	78	113	62	81	77	79	79	78	67	80
Ga	16	17	17	19	18	18	20	20	17	19
Rb	3	10	4	7	5	10	7	6	7	6
Sr	174	263	190	173	168	209	213	210	217	172
Y	35	22	19	32	30	30	33	34	25	31
Zr	127	113	60	110	106	116	119	119	97	107
Nb	2	4	1	4	3	4	3	3	3	2
Ba	23	196	58	63	64	75	95	88	82	66
La	6	11	4	5	5	5	6	5	4	5
Ce	21	25	10	17	17	22	23	28	16	17
Pb	3	8	6	5	5	5	4	5	5	6
Th	1	2	2	–	–	2	–	2	–	2
U	–	–	–	–	–	–	–	1	1	–
$^{87}Sr/^{86}Sr$	0.702556±30	0.704267±14		0.702917±10		0.703052±10		0.703103±14	0.703090±14	0.702922±15
$^{143}Nd/^{144}Nd$	0.513129±8	0.512845±2		0.513105±7		0.513097±6		0.513086±8	0.513076±7	0.513084±26
ϵNd	+9.3	+4.0		+9.1		+9.0		+8.7	+8.5	+8.7

All analyses normalized to $Fe_2O_3/FeO = 0.15$, MgNo. $= 100.$ $Mg/(Mg + Fe^{2+})$.

Table 2. *Major oxide (wt%), trace element (ppm) and Sr and Nd isotope data for basalts from Rumble IV (161/1, 162/1, 162/2) and the eastern escarpment of Ngatoro Basin (168/1A, 168/1B, 169/1)*

Sample No.	VUW	VUW	VUW	VUW	VUW	VUW
Field no.	168-1A	168-1B	162-2	162-1	161	169/1
Locality	RIV	RIV	RIV	RIV	RIV	RIV
SiO_2	52.03	50.25	50.38	50.88	48.60	49.64
TiO_2	0.82	0.65	0.84	0.87	0.82	0.82
Al_2O_3	16.56	15.40	18.40	18.30	15.54	16.96
Fe_2O_3	0.97	1.11	1.14	1.15	1.04	0.95
FeO	6.46	7.37	8.71	7.68	6.95	6.34
MnO	0.16	0.14	0.17	0.17	0.16	0.14
MgO	7.27	8.31	5.16	5.08	10.01	9.10
CaO	10.55	12.52	11.22	11.12	12.96	12.22
Na_2O	2.96	2.11	2.64	2.62	2.27	2.78
K_2O	0.59	0.45	0.47	0.45	0.27	0.25
P_2O_5	0.20	0.06	0.11	0.11	0.08	0.09
LOI	1.00	1.14	0.94	0.77	0.51	0.00
Total	100.28	100.34	99.97	100.05	99.97	99.99
MgNo.	66.70	66.90	51.40	54.10	72.00	71.80
Sc	29	38	32	30	35	30
V	290	278	301	300	245	196
Cr	24	355	72	64	418	345
Ni	23	90	33	32	147	125
Cu	118	95	89	90	57	66
Zn	96	68	76	76	55	55
Ga	17	14	17	19	14	15
Rb	9	8	7	8	3	7
Sr	263	239	289	288	203	222
Y	20	14	19	20	17	17
Zr	62	40	58	57	54	59
Nb	2	2	3	–	2	2
Ba	206	235	190	187	110	123
La	4	3	4	5	4	5
Ce	18	12	14	16	10	15
Pb	6	5	5	7	5	5
Th	3	1	5	1	–	1
U	1	1	–	–	–	–
$^{87}Sr/^{86}Sr$	0.704251±13	0.704270±13		0.703976±12	0.703616±12	0.703348±11
$^{143}Nd/^{144}Nd$	0.513014±3	0.512989±12		0.513000±6	0.513030±14	0.513050±7
ϵNd	+7.3	+6.8		+7.1	+7.6	+8.0

All data normalized to $Fe_2O_3/FeO = 0.15$, MgNo. = 100. $Mg/Mg + Fe^{2+}$.

All the continental sector basalts are relatively porphyritic in comparison to the Ngatoro Basin samples. Phenocryst assemblages are dominated by plagioclase (An_{94} – An_{66}) and clinopyroxene (augite) with lesser amounts of olivine ($Mg/Mg + Fe^{2+} = 68$–79). In the andesites, olivine is replaced by orthopyroxene and, in X164, hypersthene phenocryst cores are rimmed by pigeonite (Fig. 2) and plagioclase compositions differentiate into two distinctive fields (An_{80-73} and An_{57-50}; Fig. 3) suggestive of magma mixing.

Geochemistry

Major oxide, trace element and Sr and Nd isotope analyses for the samples are given in Tables 1 to 4. In Fig. 5, all the data are plotted in terms of the total alkali versus SiO_2 (TAS) diagram (Le Bas *et al.* 1986). This diagram serves to highlight the dominantly basaltic nature of the oceanic suite in contrast to the variable (basalt-andesite-dacite-rhyolite) nature of the continental suite. For comparison, we have added some of our published data on basalts from TVZ, the Kermadec arc and Havre Trough (Gamble *et al.* 1990) and Ruapehu volcano (Graham & Hackett 1987). The basalts from the HT and western Ngatoro Basin define a near vertical array of increasing alkalis whereas the offshore basalts occupy a field in the middle of the TVZ and KA arrays.

Electron microprobe analyses of groundmass glasses, together with host whole rock analyses, are given in Table 5. Consistent with the olivine-bearing phenocryst assemblages, whole rock analyses show higher Mg-numbers than

Table 3. *Major oxide (wt%), trace element (ppm) and Sr and Nd isotope data for basalts from offshore Bay of Plenty*

Sample no. Field no. Locality	VUW 191-3 W. Alderman T	VUW 204/1B Ngatoro R.	VUW 190/1 Ngatoro R.	VUW 201/2 Ngatoro R.	VUW 196/1A Ngatoro R.	VUW 196/1B Ngatoro R.	VUW 174/1 Alderman T.	VUW 146/1 Alderman T.	VUW A-002 Volcanolog	VUW B-002 Volcanolog	VUW C-002 Volcanolog
SiO_2	49.14	48.41	50.55	48.41	49.57	49.96	48.29	47.09	51.33	51.05	49.21
TiO_2	1.25	0.87	0.80	0.93	0.99	0.89	0.94	1.05	0.81	0.81	0.89
Al_2O_3	18.18	19.23	17.77	20.01	19.15	19.10	16.54	17.89	17.90	18.05	19.96
Fe_2O_3	1.46	1.33	1.30	1.20	1.28	1.31	1.72	1.53	1.10	1.13	1.20
FeO	9.74	8.86	8.67	8.00	8.51	8.73	8.76	7.78	7.34	7.54	7.98
MnO	0.21	0.17	0.19	0.15	0.18	0.18	0.44	0.20	0.15	0.15	0.18
MgO	4.83	5.44	5.75	5.17	3.94	3.91	6.37	7.51	4.97	5.02	4.52
CaO	10.70	12.34	11.34	12.49	12.05	11.88	11.51	11.66	10.35	10.39	12.75
Na_2O	2.78	2.13	2.33	2.26	2.48	2.39	2.62	2.71	2.28	2.30	1.93
K_2O	0.39	0.27	0.45	0.37	0.41	0.37	0.29	0.31	0.82	0.80	0.38
P_2O_5	0.15	0.07	0.07	0.10	0.08	0.08	0.07	0.18	0.25	0.26	0.10
LOI	0.35	0.19	0.09	0.39	0.19	0.32	1.30	1.61	1.34	1.48	0.00
Total	100.26	99.31	99.31	99.48	98.83	99.12	98.85	99.52	99.47	99.83	99.90
MgNo.	46.90	52.20	54.20	53.50	45.20	44.40	68.20	63.20	55.7	54.3	50.5
Sc	38	41	38	35	47	43	43	36	35	37	42
V	429	334	315	320	425	357	333	292	257	265	334
Cr	25	17	35	58	15	11	58	173	41	41	16
Ni	12	9	20	22	5	4	83	69	11	10	10
Cu	43	50	102	72	40	35	139	46	26	18	46
Zn	104	87	85	83	108	73	102	99	85	83	86
Ga	21	19	17	20	20	19	18	17	20	19	21
Rb	8	5	11	7	5	5	6	3	27	26	8
Sr	320	276	257	350	320	315	259	299	305	300	284
Y	22	17	20	19	19	18	14	23	23	22	18
Zr	57	40	52	47	51	50	40	75	94	93	42
Nb	1	2	1	3	2	2	1	4	4	4	–
Ba	140	168	200	165	229	221	126	95	312	293	181
La	6	3	4	7	6	5	2	10	10	11	2
Ce	22	17	18	19	20	18	13	35	25	24	10
Pb	5	6	7	6	5	6	7	8	9	7	4
Th	1	–	1	–	1	–	1	2	–	–	–
U	0.4	0.4	0.4	0.4	0.4	0.4	0.3	0.3	0.2	0.2	0.4
$^{87}Sr/^{86}Sr$		0.705297±14	0.705128±12		0.705128±14	0.705132±10					
$^{143}Nd/^{144}Nd$		0.512196±7	0.512908±9		0.512843±7	0.512824±3					
ϵNd		+3.7	+5.3		+4.0	+3.6					

Samples labelled 'Volcanolog' were dredged north of White Island during the *Volcanolog* cruise of April 1979. All data normalized to $Fe_2O_3/FeO = 0.15$, MgNo. = 100. $Mg/Mg + Fe^{2+}$.

Table 4. *Major oxide (wt%), trace element (ppm) and Sr and Nd isotope data for andesites, dacites and rhyolites from offshore Bay of Plenty*

Sample no. Field no. Location	VUW S800 Ngatoro Ridge	VUW 139/1 NW of White Isl	VUW 200/1 E of Ngatoro R	VUW 207/1 due W of White	VUW 221/1 S Ngatoro R.	VUW 201/1 Ngatoro R.	VUW 201/4B Ngatoro R.	VUW 184/1 Whakatane SM	VUW 184/3 Whakatane SM	VUW 204/1A Ngatoro Ridge	VUW 132/1 E of White	VUW 132/2 E of White	VUW 120/1 NW of White	VUW 140/1 NW of White	VUW 140/2 NW of White	VUW 127/1 NW of White
SiO_2	57.84	64.13	67.06	59.88	67.38	59.75	60.37	57.15	58.41	69.38	68.81	68.21	68.42	66.64	67.94	58.02
TiO_2	0.76	0.60	0.52	0.63	0.48	0.65	0.66	0.68	0.70	0.32	0.40	0.29	0.31	0.49	0.29	0.87
Al_2O_3	16.86	15.73	14.92	16.68	15.54	16.20	16.40	15.95	16.05	13.46	14.41	13.65	13.63	14.72	13.45	17.49
Fe_2O_3	8.04	5.90	4.85	7.58	4.32	7.00	6.41	8.61	7.90	2.68	4.01	2.63	2.83	4.61	2.67	7.94
MnO	0.14	0.11	0.09	0.13	0.10	0.12	0.12	0.15	0.16	0.10	0.14	0.06	0.07	0.12	0.07	0.20
MgO	3.61	2.33	1.43	3.19	1.34	3.30	3.31	4.82	4.34	0.68	1.07	0.84	0.86	0.98	0.84	2.99
CaO	7.70	5.36	3.88	6.95	4.51	6.88	6.95	8.66	8.03	2.52	3.30	2.90	2.99	3.32	2.81	7.33
Na_2O	2.81	3.42	3.66	2.93	3.59	3.00	3.10	2.86	2.89	3.89	3.49	3.93	3.90	3.81	3.87	3.33
K_2O	1.30	1.94	2.65	1.60	1.82	1.95	1.87	0.90	1.24	2.69	2.62	2.49	2.49	2.39	2.52	0.51
P_2O_5	0.12	0.10	0.10	0.08	0.10	0.10	0.10	0.08	0.11	0.07	0.08	0.06	0.06	0.12	0.06	0.20
LOI	0.60	0.45	0.48	0.41	1.14	1.13	0.82	0.40	0.36	4.30	1.04	4.36	4.21	2.18	5.36	1.06
Total	99.10	100.08	99.64	100.06	100.32	100.08	100.12	100.27	100.19	100.08	99.37	99.40	99.76	99.38	99.88	99.93
MgNo.	46	35	26	42	29	42	42	55	50	19	23	21	22	23	21	42
Sc	27	15	11	23	13	22	23	30	24	8	12	6	8	11	6	25
V	204	108	78	181	59	167	167	246	219	22	53	32	40	30	35	122
Cr	8	13	1	14	3	23	22	53	70	–	1	–	1	–	–	3
Ni	7	3	2	5	1	7	6	17	23	2	2	0	1	1	1	2
Cu	35	15	17	56	6	15	14	80	58	3	4	5	3	70	44	12
Zn	79	55	55	61	51	67	69	73	70	46	50	42	43	44	99	99
Ga	17	18	17	19	17	16	18	17	16	14	16	14	13	17	14	20
Rb	40	68	55	55	60	64	63	23	32	87	99	88	87	81	87	10
Sr	254	209	242	242	215	229	234	213	273	176	187	171	171	222	167	282
Y	24	29	30	17	28	25	23	21	23	22	29	20	20	30	20	31
Zr	104	146	182	92	150	141	141	77	94	168	172	147	144	160	148	78
Nb	3	6	4	3	5	5	4	2	3	6	7	5	5	7	6	3
Ba	515	582	704	610	623	508	505	406	533	698	789	762	758	647	755	282
La	26	17	16	13	15	16	16	7	12	17	21	17	18	20	18	10
Ce	24	34	33	25	34	36	35	21	21	37	44	34	32	39	31	22
Pb	7	10	13	11	10	11	12	7	7	13	16	13	13	13	12	6
Th	5	7	7	5	7	6	6	2	2	9	11	9	9	9	8	2
U	1	1	1	1	2	2	1	–	–	3	3	2	2	2	2	1
$^{87}Sr/^{86}Sr$								0.704776±11								
$^{143}Nd/^{144}Nd$								0.512886±8								
ϵNd								+4.8								

All data normalized to $Fe_2O_3/FeO = 0.15$, MgNo. = 100. $Mg/Mg + Fe^{2+}$.

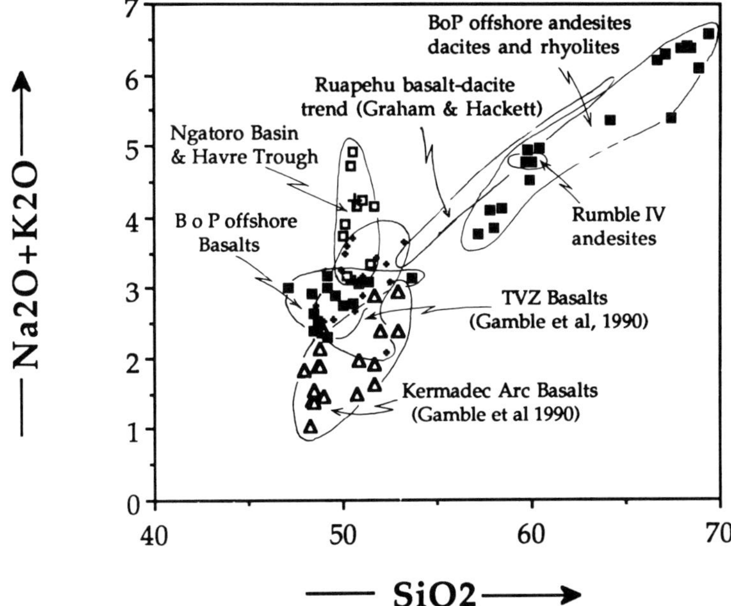

Fig. 5. Total alkali – silica diagram (Le Bas *et al.* 1986) for rocks from the Ngatoro Basin, Rumble IV and offshore Bay of Plenty. Ruapehu basalt – dacite field from Graham and Hackett (1987). Fields of Kermadec arc basalts (open triangles) and TVZ basalts (small crosses) and the Havre Trough basalt (large cross) are from Gamble *et al.* (1990) and are shown for comparison.

coexisting glasses. CaO/Al_2O_3 ratios of the basaltic glasses are generally slightly higher than the whole rocks consistent with the later appearance of clinopyroxene in most rocks. Calculated CIPW norms of glasses and coexisting whole rocks show that compositions extend from weakly nepheline normative ($\sim 1\%$) to quartz normative (*c.* 12%). In Fig. 6, chemical analyses for western Ngatoro Basin basalts are plotted in the plagioclase projection of the CMAS system (Walker *et al.* 1979). Data are shown for glass (gl) and coexisting whole rocks, the two being joined by tie-lines. Fields for basalts from TVZ and KA (Gamble *et al.* 1990), Lau Basin (Sunkel 1990) and Egmont (Price *et al.* 1992) are added for comparison as are the Havre Trough Basalt (designated by PPTUW/5) and a primitive TVZ basalt (TVZ–15, from Gamble *et al.* 1990). The data are consistent with an initial stage of evolution by fractionation of olivine + plagioclase and later olivine + plagioclase + clinopyroxene towards and along the 1 atmosphere cotectic. In Fig. 7, basalts from the Ngatoro Basin (subdivided into groups > 7% MgO and <7% MgO), Rumble IV seamount and nearby vents and offshore TVZ are shown in a series of MORB normalized (Pearce 1983) multi-element plots. For comparative purposes basalt data from TVZ, KA and the Havre Trough (Gamble

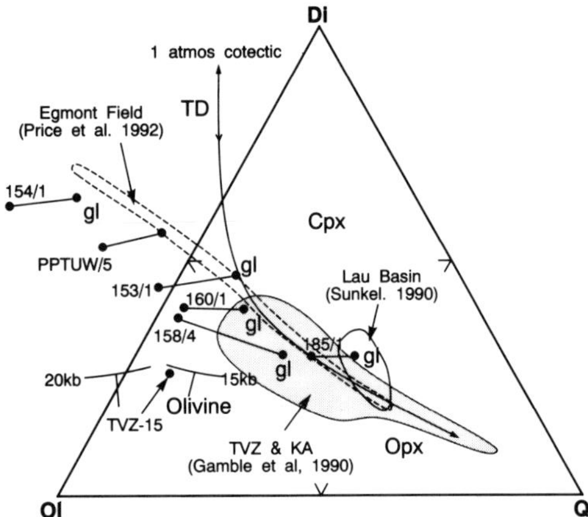

Fig. 6. CMAS pseudo-ternary plagioclase projection (Walker *et al.* 1979) for basatic glasses (gl) and whole rocks from the Ngatoro Basin. The primitive TVZ sample (TVZ–15, from Gamble *et al.* 1990) is shown as is the field occupied by most TVZ and KA basalts (Gamble *et al.* 1990) and the field of young Egmont basalts (Price *et al.* 1992). Note the evolved nature of the Lau basin basalts (Sunkel 1990). The 1 atmosphere cotectic is from Walker *et al.* 1979 and the approximate positions of the 20 kb and 15 kb cotectics are indicated. TD = Thermal Divide.

Table 5. *Representative glass (electron microprobe) and whole rock (XRF) analyses of basalts from Ngatoro Basin, Havre Trough, offshore Bay of Plenty and TVZ (sample TVZ-15, Gamble et al. 1990, 1993b)*

Sample no. Location	153/1(wr) Ngatoro Basin	153/1(gl)	154/1(wr) Ngatoro Basin	154/1(gl)	158/4(wr) Ngatoro Basin	158/4(gl)	160/1(wr) Ngatoro Basin	160/1(gl)	185/1(wr) Ngatoro Basin	185/1(gl)	TVZ-19(wr) Havre Trough
SiO_2	51.03	52.73	51.00	52.05	51.78	53.56	51.13	52.41	52.35	54.85	51.15
TiO_2	1.16	1.20	1.68	1.95	1.48	1.62	1.48	1.54	0.93	1.04	1.50
Al_2O_3	16.86	17.09	16.44	16.55	16.48	16.87	16.25	16.22	17.97	16.19	16.24
Fe_2O_3	1.04	0.91	1.20	1.20	1.16	1.09	1.24	1.26	1.20	0.97	1.12
FeO	6.90	6.07	7.98	7.98	7.74	7.27	8.24	8.37	8.00	6.44	7.44
MnO	0.15	0.21	0.18	0.19	0.16	0.18	0.17	0.22	0.16	0.14	0.16
MgO	8.23	6.62	6.96	5.37	7.17	5.79	7.19	5.77	4.86	6.09	7.56
CaO	10.82	11.26	9.41	9.65	9.61	9.49	10.16	10.20	11.01	10.88	10.35
Na_2O	3.40	3.71	4.44	4.62	3.83	3.96	3.42	3.79	2.86	3.05	3.99
K_2O	0.41	0.20	0.53	0.43	0.43	0.16	0.57	0.22	0.55	0.36	0.29
P_2O_5	0.13	–	0.18	–	0.15	–	0.16	–	0.12	–	0.18
Total	100.05		100.11		100.20		100.22		99.97		100.36
MgNo.	68.00	66.00	60.90	54.50	62.30	58.70	62.30	55.20	52.00	62.70	66.00
CaO/Al_2O_3	0.64	0.66	0.57	0.58	0.58	0.56	0.58	0.63	0.61	0.67	0.64

All data normalized to $Fe_2O_3/FeO = 0.15$, MgNo. $= 100$. $Mg/Mg + Fe^{2+}$.

Table 5. *continued*

	TVZ-19(gl)	162/1(wr) Rumble IV	162/1(gl)	164/3(wr) E. of RIV	164/3(gl)	204/1B(wr) Ngatoro Ridge	204/1B(gl)	201/2(wr) Ngatoro Ridge	201/2(gl)	TVZ-15(wr) TVZ	TVZ-15(gl)
SiO_2	52.33	51.69	54.67	61.19	66.29	48.83	51.13	48.84	51.07	48.61	51.38
TiO_2	1.65	0.88	1.45	0.82	1.01	0.88	1.43	0.94	1.63	0.89	2.24
Al_2O_3	16.55	18.59	14.53	15.72	12.81	19.39	12.73	20.19	13.44	18.12	14.58
Fe_2O_3	1.30	1.17	1.54	1.10	1.16	1.33	2.62	1.52	2.28	1.05	1.91
FeO	6.63	7.80	10.27	7.35	7.73	8.86	13.35	7.79	11.65	7.00	9.71
MnO	0.20	0.17	0.21	0.20	0.20	0.17	0.26	0.15	0.24	0.13	0.31
MgO	6.00	5.16	4.48	2.32	1.21	5.49	5.03	5.22	5.70	9.14	5.61
CaO	10.20	11.30	8.99	6.29	4.74	13.45	10.29	12.60	11.00	11.18	10.50
Na_2O	4.20	2.66	3.27	4.11	4.04	2.15	2.68	2.28	2.45	2.49	3.16
K_2O	0.10	0.46	0.57	0.75	0.81	0.27	0.57	0.37	0.72	0.27	0.61
P_2O_5	–	0.11	–	0.14	–	0.07	–	0.10	–	0.14	–
Total		100.05		98.64		100.29		100.08		99.82	
MgNo.	61.70	54.10	43.70	36.00	21.80	52.20	40.20	53.50	46.60	72.00	50.70
CaO/Al_2O_3	0.64	0.61	0.62	0.40	0.37	0.69	0.81	0.62	0.82	0.62	0.72

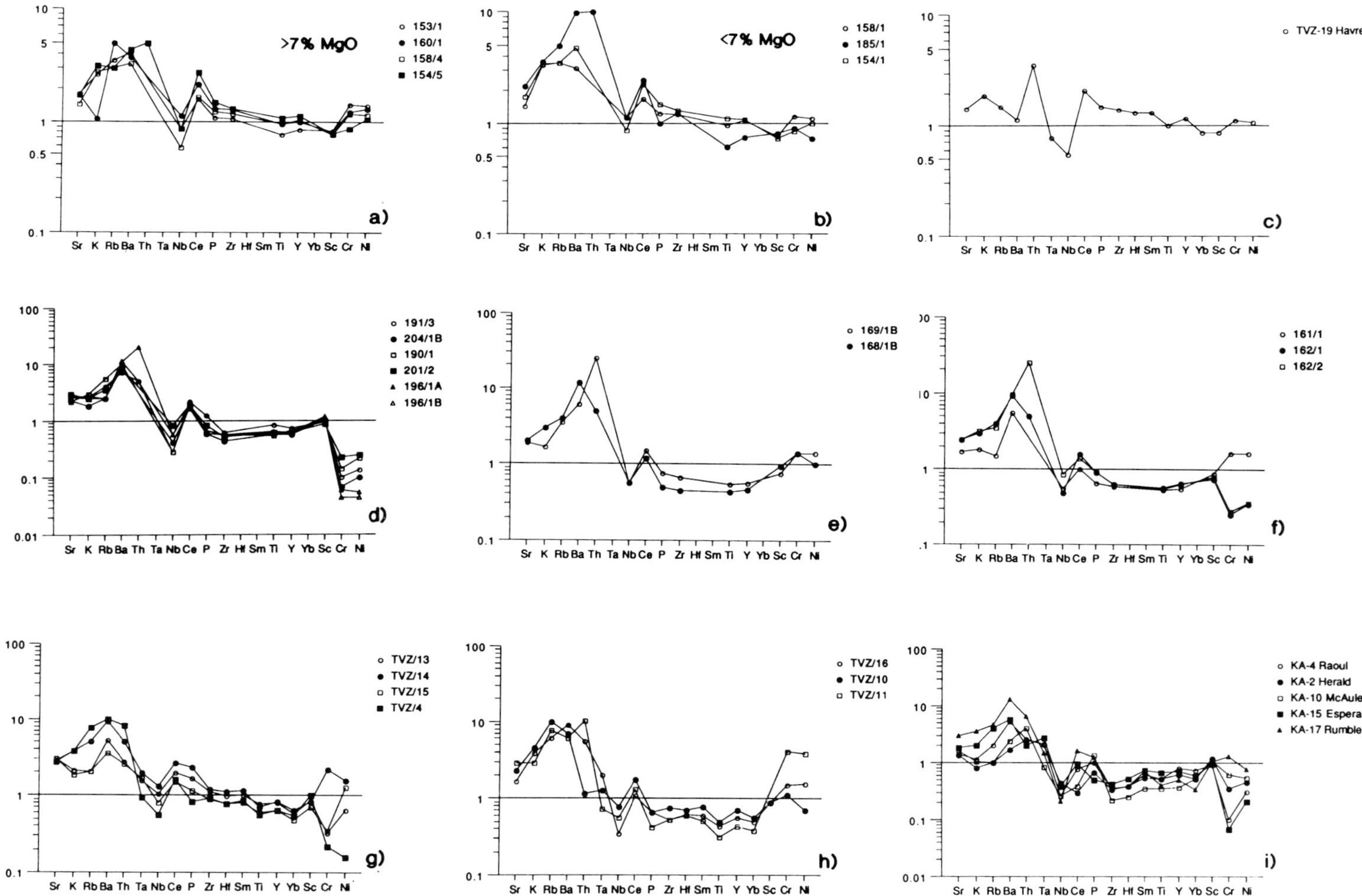

>7% MgO
153/1
160/1
158/4
154/5
a)
Sr K Rb Ba Th Ta Nb Ce P Zr Hf Sm Ti Y Yb Sc Cr Ni

<7% MgO
158/1
185/1
154/1
b)
Sr K Rb Ba Th Ta Nb Ce P Zr Hf Sm Ti Y Yb Sc Cr Ni

TVZ-19 Havre
c)
Sr K Rb Ba Th Ta Nb Ce P Zr Hf Sm Ti Y Yb Sc Cr Ni

191/3
204/1B
190/1
201/2
196/1A
196/1B
d)
Sr K Rb Ba Th Ta Nb Ce P Zr Hf Sm Ti Y Yb Sc Cr Ni

169/1B
168/1B
e)
Sr K Rb Ba Th Ta Nb Ce P Zr Hf Sm Ti Y Yb Sc Cr Ni

161/1
162/1
162/2
f)
Sr K Rb Ba Th Ta Nb Ce P Zr Hf Sm Ti Y Yb Sc Cr Ni

TVZ/13
TVZ/14
TVZ/15
TVZ/4
g)
Sr K Rb Ba Th Ta Nb Ce P Zr Hf Sm Ti Y Yb Sc Cr Ni

TVZ/16
TVZ/10
TVZ/11
h)
Sr K Rb Ba Th Ta Nb Ce P Zr Hf Sm Ti Y Yb Sc Cr Ni

KA-4 Raoul
KA-2 Herald
KA-10 McAuley
KA-15 Esperanc
KA-17 Rumble II
i)
Sr K Rb Ba Th Ta Nb Ce P Zr Hf Sm Ti Y Yb Sc Cr Ni

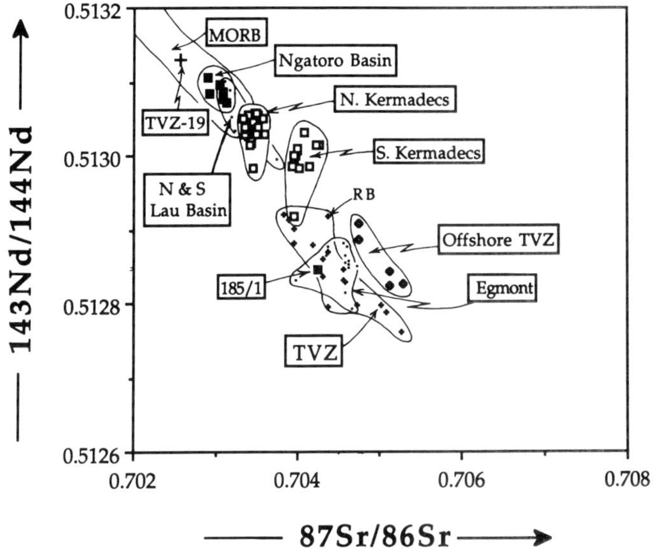

Fig. 8. Sr and Nd isotope covariation diagram for basalts from Ngatoro Basin (closed squares), Havre Trough (TVZ–19), (Gamble *et al.* 1993*a*), Rumble IV and offshore TVZ (open diamonds). Other sources of data are Kermadec arc (open squares): Ewart & Hawkesworth (1987), Gamble *et al.* (1993*a*), Woodhead (unpublished data). TVZ (small crosses): Gamble *et al.* 1993*b*, McCulloch & Corlette (unpublished data). Egmont (dots): Price *et al.* (1992). North and South Lau Basin (dots): Loock *et al.* (1990) and MORB field: Ito *et al.* (1987). RB: basalt from Ruapehu Volcano (Gamble *et al.* 1993*a*). Note how the basalt (185/1) from the floor of the Ngatoro Basin re-entrant plots in the field of TVZ basalts.

et al. 1990, 1993*a*) have been included. The main feature of the western Ngatoro Basin suite (Fig. 7a) is its LIL- and LREE-enriched character relative to N-MORB and the MORB-like HFSE abundances. In comparison, the Havre Trough basalt (Fig. 7c) shows less marked LIL-enrichment and broadly similar HFSE to the normalizing MORB composition. Contrasting with this, the southern KA volcanoes and eastern Ngatoro Basin rocks (Fig. 7e & f) display the unmistakable patterns of subduction-related basalts, with strongly depleted HFSE and enriched LIL and LREE abundances relative to MORB. These latter features (high Ba, Th and Ce) generate the apparent Nb troughs which show depletions of a similar or greater magnitude to the other HFSE such as Zr, (Hf) and Ti. Comparison of the western Ngatoro Basin basalts with basalts from TVZ and KA (Fig. 7g, h & i) indicates that they compare more favourably with TVZ basalts from the axial (back-arc) part of the TVZ. Basalts from the volcanic front volcanoes of TVZ (e.g. TVZ–16, Ruapehu and TVZ–10,

Tongariro) display patterns with greater depletions in HFS elements, more closely resembling patterns from the arc front lavas of the Kermadec Ridge (cf. Rumble IV).

Sr and Nd isotope compositions are shown on a conventional Sr–Nd covariation diagram (Fig. 8). For comparison, we have added data from other published work (Ewart & Hawkesworth 1987; Loock *et al.* 1990; Price *et al.* 1992; Gamble *et al.* 1993*a*) and some previously unpublished data. The analyses plot in distinctive fields with the western Ngatoro Basin rocks (apart from sample 185/1) showing lower Nd and higher Sr isotope ratios than the Havre Trough basalt (TVZ–19 of Gamble *et al.* 1993*a*) but less radiogenic Sr and correspondingly higher Nd than the KA basalts. The Rumble IV samples plot in *both* the north Kermadec arc and south Kermadec arc fields (as defined by Gamble *et al.* 1993*a*) and the offshore basalts correspond closely to basalts from the TVZ and Mt Egmont (Gamble *et al.* 1993*a*; Price *et al.* 1992) but have somewhat higher Sr isotope ratios. The basalt

---

Fig. 7. Multi-element MORB normalised (after Pearce 1983) abundance diagrams for western Ngatoro Basin (back-arc) basalts (**a & b**), Havre Trough (**c**), offshore TVZ (**d**), Eastern escarpment of Ngatoro Basin (**e**), Rumble IV (**f**), TVZ axial (**g**), TVZ volcanic front (**h**), and Kermadec arc (**i**).

(185/1) from the floor of the Ngatoro Basin in the crustal indentor shows appreciably higher Sr and lower Nd isotope ratios than the other Ngatoro Basin rocks, plotting in the field of TVZ basalts defined by Gamble *et al.* (1993*a*), commensurate with contamination by New Zealand continental crust.

Discussion

A major debate on the petrogenesis of volcanic arc and back-arc basin magmas concerns the mechanisms and extent of transfer of material from the descending slab of oceanic lithosphere to the overlying mantle wedge and the eventual tapping of this slab-contaminated source by arc volcanoes (Saunders & Tarney 1984; Ellam & Hawkesworth 1988; Davies & Bickle 1991; Hawkesworth *et al.* 1991; McCulloch & Gamble 1991). In this regard an important observation, summarized in Plank & Langmuir (1988), concerns the relationship between the position of the volcanic arc and the depth to the Wadati-Benioff zone. Such scale parameters, combined with the length scale between volcanic front and back-arc axis and the thickness of the subarc lithosphere, serve to delimit the volume of mantle wedge available for melt generation above the descending slab. Gamble *et al.* (1993*a*) calculated that the volume of convecting mantle per unit km of arc reduced by a factor of around two from the KA and Havre Trough in the north toward the TVZ in the south. Assuming a constant slab flux into the mantle wedge they suggested that this could in part explain the isotopic and trace element systematics of basalts along the plate boundary. A similar situation can be envisaged to the north in the Lau Basin. There, basalts from the northern Lau Basin show MORB-like geochemistry whereas those from the southern Lau Basin are more frac-tionated, notably those from the Valu Fa Ridge which carry strong subduction signatures (Loock *et al.* 1990; Jenner *et al.* 1987; Vallier *et al.* 1991).

For basalts from the TVZ and KA, Gamble *et al.* (1993*a*) noted that ratio plots of HFSE and transition elements v. a HFSE, such as Zr, could be usefully employed to distinguish arc front basalts from back-arc basin basalts and more-over, to identify source features such as relative fertility. Subsequently, this approach was ex-tended and applied to arc and back-arc basin basalts on a world scale (Woodhead *et al.* in press). In Fig. 9, basalts from Ngatoro Basin, Rumble IV seamount and offshore TVZ are plotted in terms of Ti/Zr, Ti/Sc and Ti/V v. Zr. Other TVZ and KA data from the literature (Ewart *et al.* 1977; Smith & Brothers 1988;

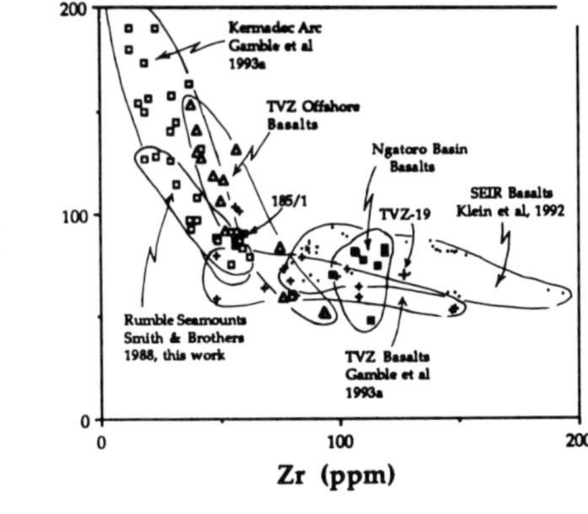

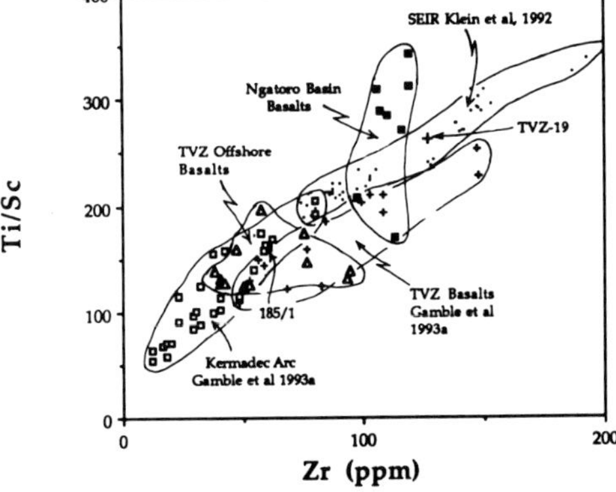

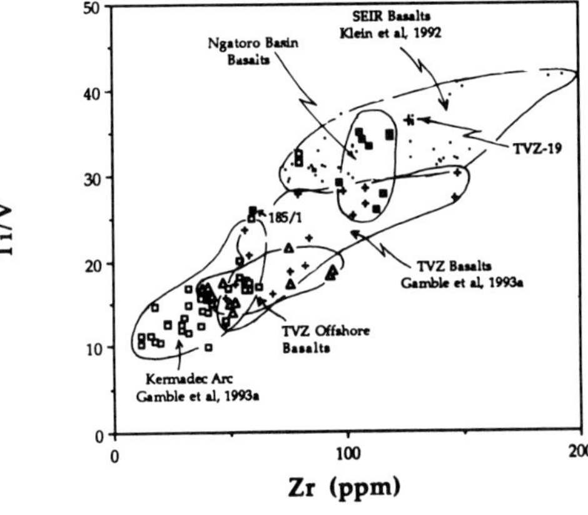

Fig. 9. High field strength versus transition element ratio plots for basalts from Ngatoro Basin (solid squares), Rumble IV (solid diamonds) and Offshore TVZ (open squares). Data are added for TVZ basalts (small crosses) (Gamble *et al.* 1993), Kermadec arc basalts (solid triangles) (Ewart & Hawkesworth 1987; Smith & Brothers 1988; and Gamble *et al.* 1993*a*). The SE Indian Ridge MORB field (dots) is from Klein *et al.* (1991).

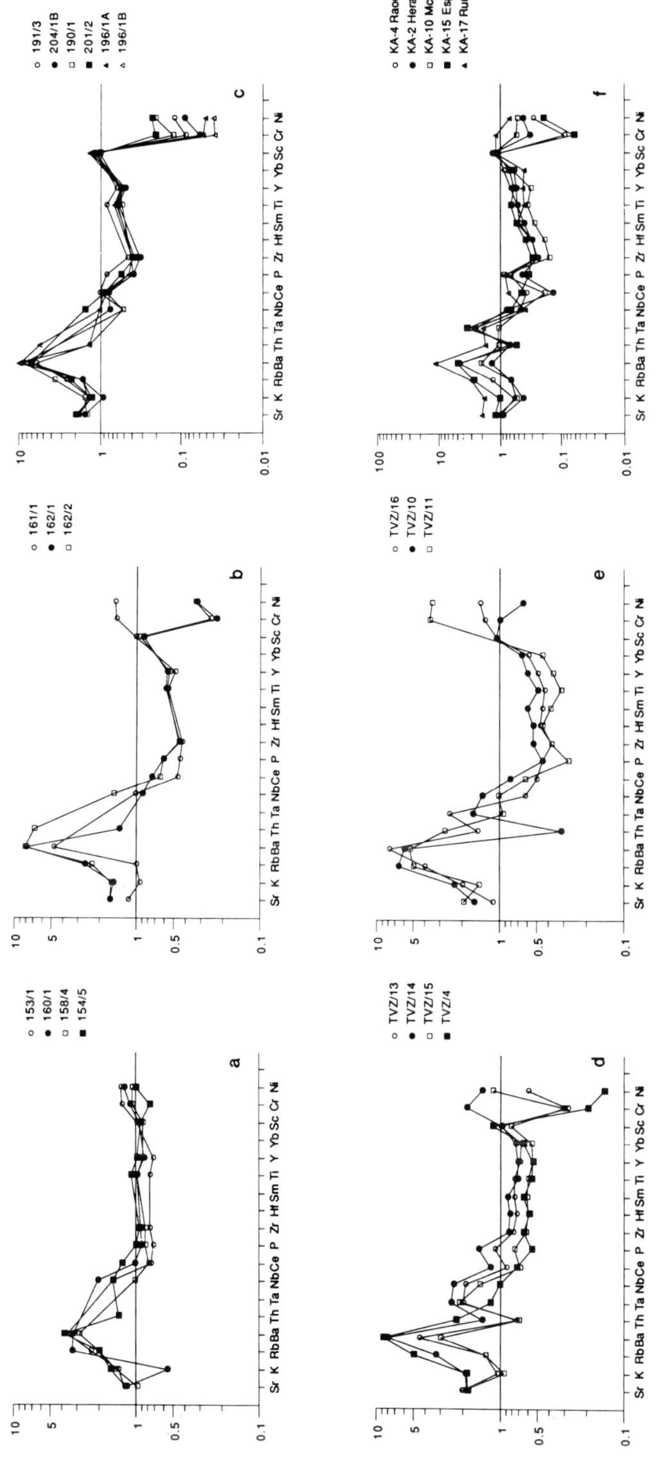

Fig. 10. Multi-element plots of basalts from Ngatoro Basin (**a**), Rumble IV (**b**), Offshore TVZ (**c**), Axial TVZ (**d**), volcanic front TVZ (**e**) and Kermadec arc (**f**) normalized to the Havre Trough basalt (TVZ–19) composition.

Gamble *et al.* 1993*a*), have been added as have data from the SE Indian Ridge (Klein *et al.* 1991). The western Ngatoro Basin basalts compare favourably with the Havre Trough sample (TVZ–19), with basalts from TVZ, the Lau Basin (Sunkel 1990) and MORB as exemplified by the SE Indian Ridge (Klein *et al.* 1991). In contrast, the Rumble IV data and the TVZ offshore basalts define steep arrays on the Ti/Zr v. Zr plot, lying towards the low Ti/Zr end of the KA array. These positions are mirrored in the Ti/Sc and Ti/V plots. Woodhead *et al.* (1993) have argued that the steep array of the Ti/Zr v. Zr plots for many arc basalt suites can be explained by prior extraction of melt from their mantle sources implying that their sources had become 'ultra-depleted' in incompatible elements. Furthermore, this process could make the source even more susceptible to contamination by fluids originating in the slab, leading to the prediction that for a given slab flux, Sr isotope ratios will show the greatest shift for the more depleted sources.

A means of addressing this question is to follow the strategy advocated by Hansen (1989), and normalizing the multi-element plots to a 'primitive' composition within the suite studied. This procedure has the advantage of normalizing to a composition which is internally consistent with the other data in terms of the analytical methods used and with the processes experienced. Furthermore, it permits closer monitoring of local influences and here we use it to examine relative LIL and HFSE enrichments and depletions in the basalts.

In Fig. 10, we have normalized our data to the Havre Trough basalt (TVZ–19) which was previously shown to have trace element and isotope abundances similar to MORB (Gamble *et al.* 1993*a*), thereby reflecting a source region relatively uncontaminated by fluids derived from the subducting slab. Note that all the multi-element plots in Fig. 10 show enrichment of LIL elements, which we infer are derived from slab-mantle interaction. Measured as a 'Ba-spike' (Ba* = Ba measured in sample/Ba in normalization sample, cf. Arculus & Johnson 1981; Arculus 1985) this increases from a factor of around 2× in the western Ngatoro Basin basalts to between 5× and 10× in the other suites. At the same time, the HFS elements (Ta–Yb) show variable degrees of depletion across (normal to) the arc. For example, the western Ngatoro Basin basalts (> 7% MgO) are only slightly depleted (Fig. 10a) in HFS elements whereas the basalts from Rumble IV (Fig. 10b) and the other volcanoes of the KA (Fig. 10f) are more strongly depleted. This situation extends to the TVZ where HFS elements in basalts from the TVZ front (Fig. 10e) are more depleted than in basalts from the central part of the TVZ (Fig. 10d). We suggest that this observation adds weight to the argument for differential source depletion across arcs.

Returning to the Sr and Nd isotope variations displayed in Fig. 8, Gamble *et al.* (1993*a*) interpreted the array of TVZ, KA and Havre Trough Sr and Nd isotope data to result from a combination of lateral source heterogeneity and variable slab-mantle interaction to which, in the south, was added the ingredient of contamination by the continental crust of New Zealand. Our new isotopic data permit closer examination of several features in Fig. 8 and raise several questions:

(1) How can one account for the large isotopic shifts encountered in the Rumble seamount data, where samples plot in both north and south Kermadec arc fields?

(2) How can one account for the relatively large isotopic shifts recorded between basalts from the western flanks of the Ngatoro Basin, to the eastern flanks of the basin – over a distance of <15 km.

(3) Why are the offshore TVZ arc front (Ngatoro Ridge) basalts shifted to more radiogenic Sr isotope ratios than the TVZ basalts further to the south?

The authors believe that explanations for these features may lie in the interplay between relative fertility of the basalt magma sources across any segment of arc and the extent of slab fluxing. Ewart & Hawkesworth (1987) and Gamble *et al.* (1993*a*) documented source variability along the KA towards New Zealand and the latter concluded that basalts from TVZ appear to have been derived from a more fertile source than those of the northern KA. Furthermore, Gamble *et al.* (1993*a*) and Woodhead *et al.* (1993) have developed a theme of variable depletion and/or fertility between arc and coupled back-arc basin magma sources. The HFSE data in the multi-element diagrams (Figs 7 & 10) suggest that the sources of the arc front Rumble IV and offshore TVZ (Ngatoro Ridge) basalts are depleted relative to the western Ngatoro Basin basalts. The elevated LILE abundances recorded in all these basalts (Figs 7 & 10), but to a lesser extent in the Ngatoro Basin basalts, are a direct indication of slab involvement. The accompanying isotopic shifts for Sr at relatively constant Nd compositions suggest that depleted sources will be very susceptible, if not swamped by slab-derived Sr (cf. Hawkesworth *et al.* 1991). This being the case, small variations in the slab-derived flux can be expected to produce significantly different isotope ratios in temporally discrete batches of melt. An alternative model is that the basaltic magmas are interacting with variable composition sub-arc crust. For example, this would require that the crust beneath the offshore TVZ arc front (i.e. the Ngatoro Ridge) had more radiogenic Sr than crust subjacent to the TVZ itself. This may be indicative of mixing processes involving more radiogenic 'Torlesse-like' crust to the east and 'Waipapa-like' crust to the west of the central part of the TVZ (e.g. Price *et al.* 1992; Graham *et al.* 1992; Gamble *et al.* 1993*a*). We are presently undertaking Pb isotope studies on these samples and based on combined Sr–Nd–Pb isotope measurements intend to attempt to quantify the slab flux process and assess the extent of crust-mantle interaction.

Conclusions

Dredge samples from the sea floor to the north of New Zealand document the transition from the oceanic Kermadec arc – Havre Trough – Ngatoro Basin southward across the continental edge of New Zealand. The following conclusions emerge from our study.

(1) Basalts from the western Ngatoro rift escarpment contain uncomplicated (MORB-like) phenocryst assemblages of olivine (+ Cr-spinel) and plagioclase, with later clinopyroxene appearing in more evolved samples.

(2) Isotopically the western Ngatoro rift escarpment basalts are more radiogenic in Sr isotopes and less radiogenic in Nd isotopes than the Havre Trough basalt (Gamble *et al.* 1993a) to the north. Basalts from the southern end of the Ngatoro rift graben, where it forms a re-entrant at the continental edge of New Zealand, have elevated Sr isotopes and low Nd isotopes suggestive of contamination by continental lithosphere.

(3) HFSE and transition element trace element abundances and ratios of the Ngatoro Basin basalts are similar in many ways to those of basalts from the HT and TVZ to the north and south respectively. LIL abundances of the Ngatoro Basin rocks hint at the presence of a subduction component but it is subdued in comparison with the KA and TVZ rocks.

(4) Rumble IV seamount basalts and andesites and the eastern Ngatoro rift escarpment basalts are unmistakably suprasubduction zone rocks with porphyritic textures, strong compositional zoning in the phenocryst assemblages and multi-element incompatible element plots showing enriched LIL and depleted HFS elements. Isotopically these rocks are quite heterogeneous and the authors ascribe this to variable fluxing of the magma source by slab fluids.

(5) The rocks dredged from the offshore extension of the TVZ range from basalt to andesite, dacite and rhyolite similar to the onshore TVZ rocks. Trace elements and isotopes in basalts are subtly distinctive from those of the on-shore rocks described in Gamble *et al.* (1993*a*), in that their HFSE and transition element abundances overlap those of the southern KA and their Sr isotope ratios show consistently higher ratios than those of TVZ.

(6) It is proposed that many of the local heterogeneities identified in the isotopes result from variable fluxing of variably depleted mantle wedge sources. Furthermore, the authors confirm the concept of variable lateral source fertility, raised by Ewart & Hawkesworth (1987), for the Tonga–Kermadec–New Zealand subduction system.

J.A.G. thanks the Royal Society and the Trans-Antarctic Association for financial assistance towards attending the meeting 'Volcanism Associated with Extension at Consuming Plate Margins'. Thanks to John Smellie and the organizers of the parallel 'Western Pacific Back-arc Basins' meeting for a great job. I.C.W. thanks the officers and crew of the R.V. *Rapuhia* and R.R.S. *Charles Darwin*. J.D.W. acknowledges a QE II post-doctoral fellowship at RSES ANU. Also, thanks to J. Baker and K. Palmer in the Analytical Facility of Victoria University and to many individuals, not least G. Mortimer, J. Hergt, S. Eggins and M. Fanning in the isotope geochemistry group, RSES, ANU. Finally, we thank J. Smellie, S. Weaver and an anonymous reviewer for valued comments on a draft manuscript. This work was in part funded by the New Zealand Foundation for Research Science and Technology geological oceanography programme and by VUW internal grants committee.

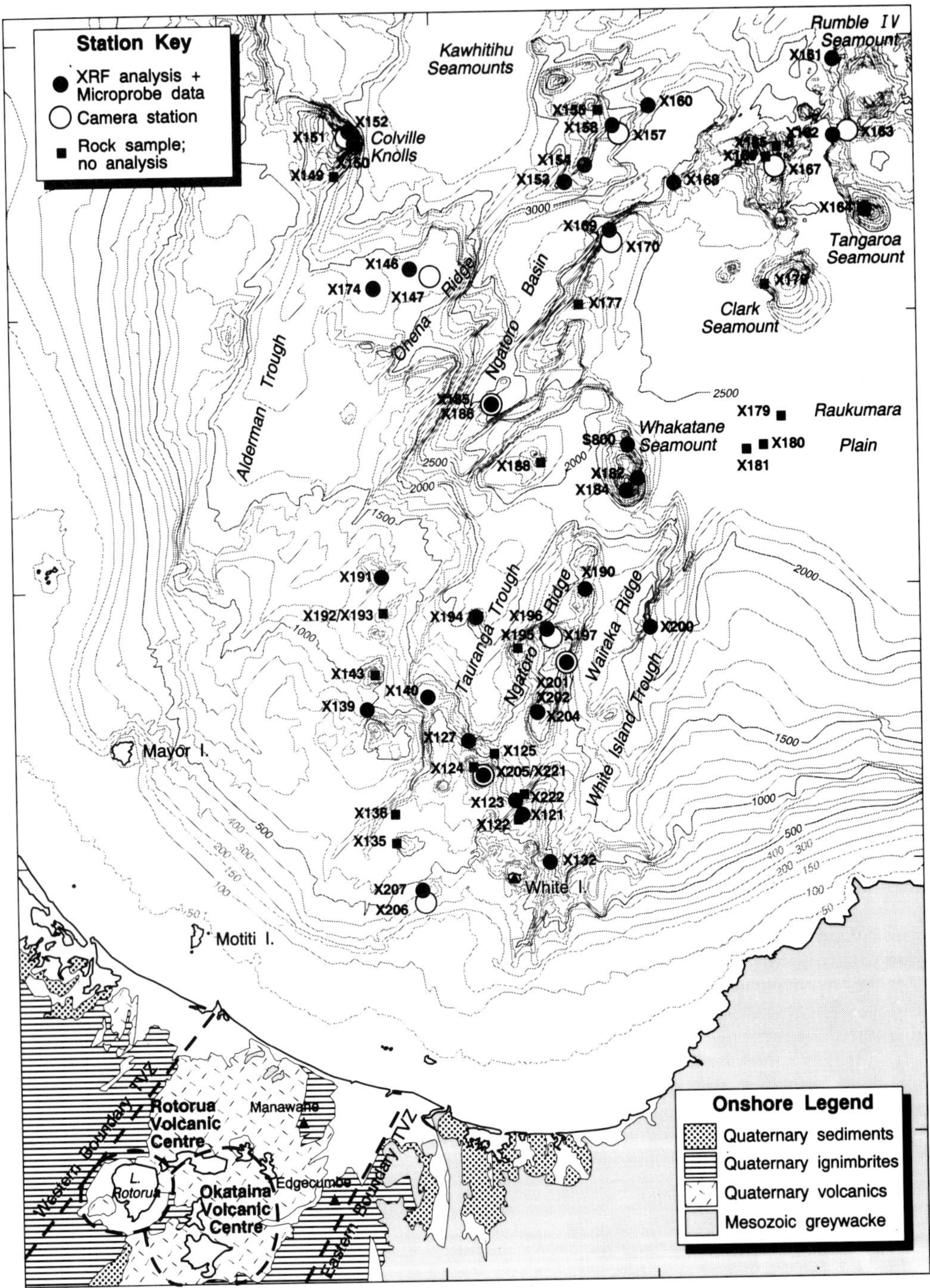

Appendix Map showing location of dredge samples analysed during this study.

References

ARCULUS, R.J. 1985. Arc Magmatism – an unresolved problem of sources, material fluxes, tectonic evolution and thermochemical regions of subduction zones. *In*: NASU, N. *et al*. (eds) *Formation of active ocean margins*. Terra Scientific Publishing Company, Tokyo, 367–397.

—— & JOHNSON, R.W. 1981. Island-arc magma sources: a geochemical assessment of the roles of slab-derived components and crustal contamination. *Geochemical Journal*, **15**, 109–133.

BALLHAUS, C., BERRY, R.F. & GREEN, D.H. 1991. High pressure experimental calibration of the olivine-orthopyroxene-spinel oxygen geobarometer: implications for the oxidation state of the upper mantle. *Contributions to Mineralogy and Petrology*, **107**, 27–40.

CARESS, D.W. 1991. Structural trends and back-arc extension in the Havre Trough. *Geophysical Research Letters*, **18**, 853–856.

DAVIES, J.H. & BICKLE, M.J. 1991. A physical model for the volume and composition of melt produced by hydrous fluxing above subduction zones. *Philosophical Transactions of the Royal Society, London*, **A335**, 355–364.

ELLAM, R.M. & HAWKESWORTH, C.J. 1988. Elemental and isotope variations in subduction related basalts: evidence for a three component model. *Contributions to Mineralogy and Petrology*, **98**, 72–80.

EWART, A. & HAWKESWORTH, C.J. 1987. The Pleistocene to Recent Tonga-Kermadec arc lavas: interpretation of new isotope and rare earth data in terms of a depleted mantle source model. *Journal of Petrology*, **28**, 495–530.

——, BROTHERS, R.N. & MATEEN, A. 1977. An outline of the geology and geochemistry, and the possible petrogenetic evolution of the volcanic rocks of the Tonga–Kermadec–New Zealand island arc. *Journal of Volcanology and Geothermal Research*, **2**, 205–250.

GAMBLE, J.A. & KYLE, P.R. 1987. The origins of glass and amphibole in spinel-wehrlite xenoliths from Foster Crater, McMurdo Volcanic Group, Antarctica. *Journal of Petrology*, **28**, 755–779.

——, SMITH I.E.M., McCULLOCH, M.T., GRAHAM, I.J. & KOKELAAR, B.P. 1993*a*. The geochemistry and petrogenesis of basalts from the Taupo volcanic Zone and Kermadec island arc, SW Pacific. *Journal of Volcanology and Geothermal Research*, **54**, 265–290.

——, SMITH, I.E.M., GRAHAM, I.J., KOKELAAR, B.P., COLE, J.W., HOUGHTON, B.F. & WILSON, C.J.N. 1990. The petrology, phase relations and tectonic setting of basalts from the Taupo Volcanic Zone, New Zealand and the Kermadec island arc-Havre Trough, S.W. Pacific. *Journal of Volcanology and Geothermal Research*, **43**, 253–270.

——, WRIGHT, I.C. & BAKER, J.A. 1993*b*: Seafloor geology and petrology in the oceanic to continental transition zone of the Kermadec-Havre-Taupo Volcanic Zone arc system, New Zealand. *New Zealand Journal of Geology and Geophysics*, 36, 417–435.

GRAHAM, I.J. & HACKETT, W.R. 1987. Petrology of calc-alkaline lavas from Ruapehu Volcano and related vents, Taupo Volcanic Zone, New Zealand. *Journal of Petrology*, **28**, 531–567.

——, GULSON, B., HEDENQUIST, J.W. & MIZON, K. 1992. Petrogenesis of late Cenozoic volcanic rocks from the Taupo Volcanic Zone, New Zealand; In the light of new lead isotope data. *Geochimica et Cosmochimica Acta*, **56**, 2797 – 2819.

HANSEN, G.N. 1989. An approach to trace element modelling using a simple igneous system as an example. *In*: LIPIN, B.R. & McKAY, G.A. (eds) *Geochemistry and mineralogy of rare earth elements*. Reviews in Mineralogy, **21**, 79–97.

HAWKESWORTH, C.J., HERGT, J.M., ELLAM, R.M. & McDERMOTT, F. 1991. Element fluxes associated with subduction related magmatism. *Philosophical Transactions of the Royal Society, London*, **A** **335**, 393–405.

ITO, E., WHITE, W.M. & GOPEL, C. 1987. The O, Sr, Nd and Pb isotope composition of MORB. *Chemical Geology*, **62**, 157–176.

JENNER, G.A., CAWOOD, P.A., RAUTENSCHLEIN, M. & WHITE, W.M. 1987. Composition of back-arc basin volcanics, Valu Fa Ridge, Lau Basin: Evidence for a slab-derived component in their mantle source. *Journal of Volcanology and Geothermal Research*, **32**, 209–222.

KLEIN, E.M., LANGMUIR, C.H. & STAUDIGEL, H. 1991. Geochemistry of basalts from the Southeast Indian Ridge, 115°E–138°. *Journal of Geophysical Research*, **92**, 8089 – 8115.

LE BAS, M.J., LE MAITRE, R.W., STRECKEISEN, A. & ZANETTIN, B. 1986. A chemical classification of igneous rocks based on the total alkali – silica diagram. *Journal of Petrology*, **27**, 745–750.

LEWIS, K.B. & PANTIN, H.W. 1984. Intersection of a marginal basin with a continent: structure and sediments of the Bay of Plenty, New Zealand. *In*: KOKELAAR, B.P. & HOWELLS, M.F. (eds) *Marginal basin geology: volcanic and associated sedimentary and tectonic processes in modern and ancient marginal basins*. Geological Society, London, Special Publication, **16**, 121–135.

LOOCK, G., McDONOUGH, W.F., GOLDSTEIN, S.L. & HOFMANN, A.W. 1990. Isotopic compositions of volcanic glasses from the Lau Basin. *Marine Mining*, 9, 235–245.

McCULLOCH, M.T. & GAMBLE, J.A. 1991. Geochemical and geodynamical constraints on subduction zone magmatism. *Earth and Planetary Science Letters*, **102**, 358–374.

NATLAND, J.H. 1989. Partial melting of a lithologically heterogeneous mantle: inferences from crystallisation histories of magnesian abyssal tholeiites from the Siqueiros Fracture Zone. *In*: SAUNDERS, A.D. & NORRY, M.J. (eds) *Magmatism in the oceanic basins*. Geological Society, London, Special Publication, **42**, 41–70.

PALMER, K. 1990. *XRF Analyses of granitoids and associated rocks, St. Johns Range, south Victoria Land, Antarctica*. Research School of earth Sciences, Geology Board of Studies Publications **5**, Victoria University of Wellington.

PEARCE, J.A. 1983. The role of sub-continental lithosphere in magma genesis at active continental margins. *In*: HAWKESWORTH, C.J. & NORRY, M.J. (eds) *Continental basalts and mantle xenoliths*. Shiva, Nantwich, U.K., 230–249.

PELLETIER, B. & LOUTAT, R. 1989. Seismotectonics and present-day relative plate motions in the Tonga-Lau and Kermadec – Havre Trough Region. *Tectonophysics*, **165**, 237–250.

PLANK, T. & LANGMUIR, C.H. 1988. An evaluation of the global variations in the major element chemistry of arc basalts. *Earth and Planetary Science Letters*, **90**, 349–370.

PRICE, R.C., McCULLOCH, M.T., SMITH, I.E.M. & STEWART, R.B. 1992. Pb-Nd-Sr isotopic compositions and trace element characteristics of young volcanic rocks from Egmont Volcano, New Zealand and comparisons with basalts and andesites from the Taupo Volcanic Zone. *Geochimica et Cosmochimica Acta*, **56**, 941–953.

SAUNDERS A.D. & TARNEY, J. 1984. Geochemical characteristics of basaltic volcanism within back-arc basins. *In*: KOKELAAR, B.P. & HOWELLS, M.F. (eds) *Marginal basin geology: volcanic and associated sedimentary and tectonic processes in modern and ancient marginal basins*. Geological Society, London, Special Publication, **16**, 59–76.

SMITH, I.E.M. & BROTHERS, R.N. 1988. Petrology of Rumble Sea Mounts, southern Kermadec Ridge, South West Pacific. *Bulletin of Volcanology*, **50**, 139–147.

SUNKEL, G. 1990. Origin of petrological and geochemical variations of Lau Basin lavas (SW Pacific). *Marine Mining*, **9**, 205–234.

VALLIER, T.L., JENNER, G.A., FREY, F.A. *et al*. 1991. Subalkaline andesite from Valu Fa Ridge, a back-arc spreading centre in southern Lau Basin: Petrogenesis, comparative chemistry, and tectonic implications. *Chemical Geology*, **91**, 227–256.

WALKER, D., SHIBATA, T. & DE LONG, S.E. 1979. Abyssal tholeiites from the Oceanographer Fracture Zone, II, phase equilibria and mixing. *Contributions to Mineralogy and Petrology*, **70**, 111–125.

WATANABE, T., GRAPES, R.H. & PALMER, K. 1981. Quantitative analysis of rock forming minerals by JXA–733 electron probe microanalyser. *JEOL News*, **19E**, 15–19.

WILSON, C.J.N., ROGAN, A.M., SMITH, I.E.M., NORTHEY, D.J., NAIRN, I.A. & HOUGHTON, B.F. 1984. Caldera volcanoes of the Taupo Volcanic Zon New Zealand. *Journal of Geophysical Research*, **98**, 8463–8484.

WOODHEAD, J.D., EGGINS, S. & GAMBLE, J.A. 1993. High field strength and transition element systematics in island arc and back-arc basin basalts: evidence for multi-stage melt extraction and an ultra-depleted mantle wedge. *Earth and Planetary Science Letters*, **114**, 491–504.

WRIGHT, I.C. 1989. *Bay of Plenty Bathymetry (2nd Edition) 1:200,000*. NZ Oceanographic Institute, Coastal Series, New Zealand Department of Science and Industrial Research.

—— 1990. *Bay of Plenty – southern Havre Trough Physiography, 1:400,000 NZ*. Oceanographic Institute Miscellaneous Series, No.68, New Zealand Department of Science and Industrial Research.

—— 1992. Shallow structure and active tectonism of an offshore continental spreading system: The Taupo Volcanic Zone, New Zealand. *Marine Geology*, **103**, 287–309.

—— 1993a. Southern Havre Trough – Bay of Plenty (New Zealand): Structure and seismic stratigraphy of an active back-arc basin complex. *In*: BALLANCE, P.F. (ed) *South Pacific sedimentary basins. Sedimentary basins of the world*. Elsevier, Amsterdam, 195–208.

—— 1993b. Pre-spread rifting and heterogeneous volcanism in the southern Havre Trough back-arc basin. *Marine Geology*, **113**, 179–200.

——, CARTER, L. & LEWIS, K.B. 1990. GLORIA survey of the oceanic – continent transition of the Havre – Taupo back-arc basin. *Geo-Marine Letters*, **10**, 59–67.

Neogene volcanoes of Chios, Greece: the relative importance of subduction and back-arc extension

G. PE-PIPER[1], D.J.W. PIPER[2], C.N. KOTOPOULI[3] & A.G. PANAGOS[4]

[1] *Department of Geology, Saint Mary's University, Halifax, NS Canada B3H 3C3*
[2] *Atlantic Geoscience Centre, Geological Survey of Canada, Bedford Institute of Oceanography, P.O. Box 1006, Dartmouth, NS, Canada B2Y 4A2*
[3] *Department of Geology, University of Patras, Patras 26110, Greece*
[4] *Department of Applied Geology, Technical University of Athens, Athens, Greece*

Abstract: The Aegean area has experienced both extension and subduction throughout the Neogene and Quaternary. Small volcanic centres with a wide range of rock types occupy a 'back-arc' position with respect to the subduction zone, but their geochemistry and chronological correlation with local basin subsidence indicates that their origin is primarily the result of extension. The position of voluminous andesite–dacite volcanism in the modern south Aegean arc relative to the seismically defined Benioff zone is similar to that of other active subduction zones, although the timing of vigorous volcanic activity correlates with periods of most rapid basin subsidence. The voluminous early Miocene intermediate igneous rocks of the northeast Aegean also appear closely related to subduction. Variations in subduction-related products through time are correlated with changes in the rate and direction of subduction inferred from regional tectonic data.

The island of Chios, in the eastern Aegean Sea, has several small, mid-Miocene (14–17 Ma) volcanic centres containing high-Mg calc-alkaline (adakitic) andesite, ne-normative basalt, and alkaline and calc-alkaline rhyolite. Both the ne-normative basalt and calc-alkaline andesite have Pb isotopic compositions similar to other Neogene volcanic rocks in the central Aegean region. The small size of the calc-alkaline centres ($<10^{-1}$ km^3) distinguishes them from the nearby voluminous (1000 km^3), 17–22 Ma, calc-alkaline to shoshonitic volcanic centres of Lesbos and Karaburum.

The minor volcanism on Chios occurred near the site of earlier voluminous arc volcanism. The rapid decrease in the rate of subduction in the mid-Miocene reduced both the production of slab-derived volatiles and silicic melts, as well as the volume of mantle wedge from which arc-type magma was extracted. At the same time, extension resulted in decompression of the lithosphere and the generation of small-volume partial melts. Volcanic products are not found near the major extensional centres: rather, they are associated with minor basin subsidence, which provided brittle pathways for magma flow.

The Aegean Sea is an area of widespread Cenozoic igneous activity. It is a region of thinned continental crust situated north of the compressional margin between the African plate and the Aegean microplate on the southern edge of Eurasia. The modern south Aegean volcanic arc is located about 130 km above the seismically defined Benioff zone (Makropoulos & Burton 1984). The subducted slab has been imaged by seismic tomography to a depth of 600 km beneath the northern Aegean Sea (Spakman *et al.* 1988), implying at least 26 Ma of subduction of the African plate beneath Eurasia (Meulenkamp *et al.* 1988). The locus of voluminous igneous activity appears to have migrated southward through the Cenozoic: from north Greece and Bulgaria (Eocene–Oligocene volcanism and plutonism), through the north Aegean islands (voluminous early Miocene volcanism) and the Cyclades (Miocene I-type plutonism) to the modern south Aegean arc (Pliocene–Quaternary). Many authors have interpreted these rocks as representing a volcanic arc that migrated southwards as a result of the rapid post-middle Miocene extension of the Aegean Sea (e.g. Fytikas *et al.* 1984). Not all of the volcanism follows this general pattern: minor late Miocene to Quaternary volcanic rocks occur throughout the Aegean area in a back-arc position (Pe-Piper & Piper 1989*a*). The rapid extension in the Aegean has led some authors to interpret the Cenozoic igneous activity primarily in terms of extension (e.g. Jones *et al.* 1992; Seytoğlu *et al.* 1992).

The term back-arc is used here to refer to an extensional region above a subduction zone. In

From Smellie, J.L. (ed.), 1995, *Volcanism Associated with Extension at Consuming Plate Margins*, Geological Society Special Publication No. 81, 213–231.

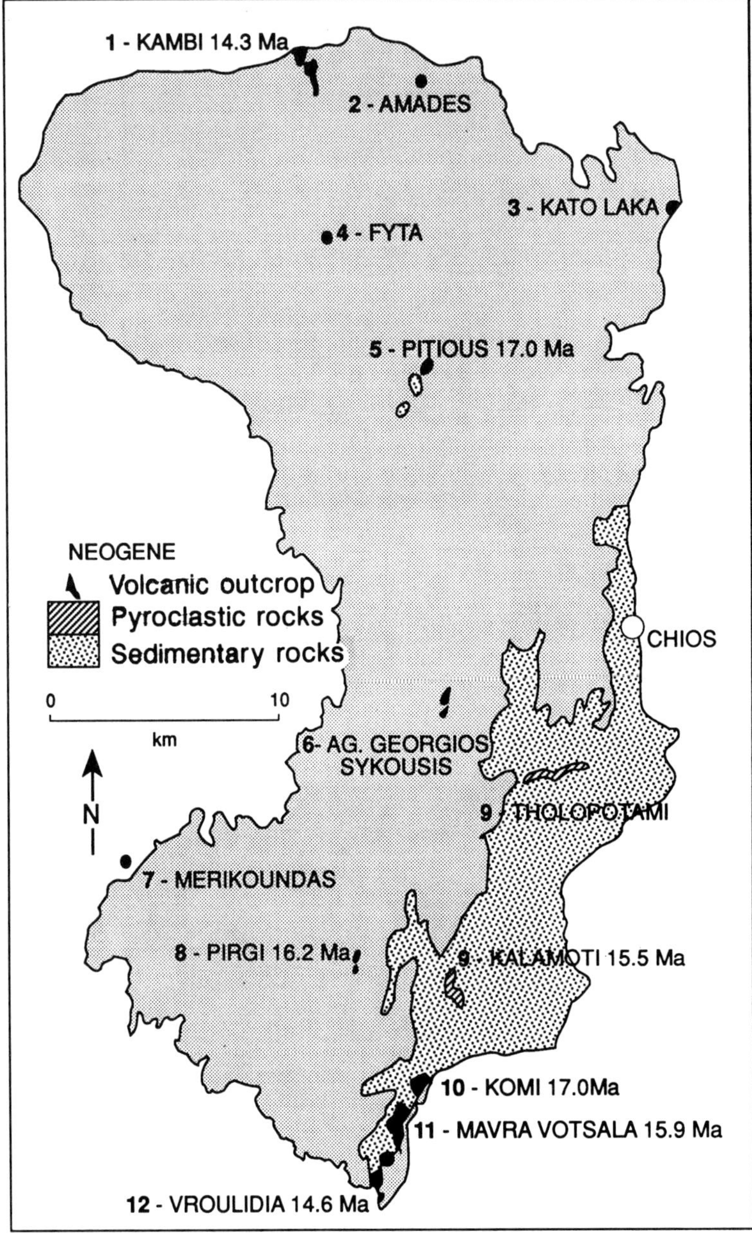

Fig .1. Generalized geological map showing distribution and radiometric age of the Neogene volcanic rocks of Chios. Numbering scheme for localities after Besenecker 1973. Radiometric ages from Bellon *et al.* (1979) (with standard error of ± 0.8 Ma), except for new date of 16.2 ± 0.6 from Pirgi reported here. (For location of Chios in the Aegean Sea, see Fig. 7)

such a setting, subduction processes may influence the petrogenesis of the back-arc volcanic rocks. A causal mechanism for the extension is not implied. In the Aegean area, various authors have related extension primarily to back-arc processes (e.g. Le Pichon & Angelier 1981; Angelier *et al.* 1982; Kastens 1991), strike-slip tectonics (e.g. Dewey & Sengör 1979; Mercier *et al.* 1989) or to orogenic collapse (e.g. Seytoğlu & Scott 1992; Sokoutis *et al.* 1993).

Table 1. *Neogene volcanic localities in Chios*

Locality no.	Locality name	Rock type	Age	Notes
1	Between Agii Pantes and Kambi	Rhyolite	14.3±0.7	
2	Amades (6 km E of Kambi)	Rhyolite	nd	Not examined
3	*Kato Laka* (Rhodosi)	Basaltic andesite flows, intermediate agglomerate and tuffs		
4	Fyta	Andesite	nd	Not found (see Herget 1968)
5	2 km SW of *Pitious*	Altered rhyolitic tuff	17.0±0.8	Ignimbrite (Besenecker 1973)
6	2 km NW of *Agios Georgios Sykousis*	Hematitized rhyolitic tuff & andesite	nd	
7	Merikoundas island, off SW coast	'Rhyodacite'	nd	Not examined (See Lüdtke 1969)
8	1 km NE of *Pirgi*	Alkaline basalt & andesitic vitric tuff	16.2±0.6	New date
9	near *Kalamoti* and NE of *Tholopotami*	Bedded felsic tuffs	15.5±0.8	Within Neogene section
10	Profitis Ilias near *Komi*	Rhyolite plug	17.0±0.8	
11	*Mavra Votsala* (Emborios, Psaronas)	Several columnar jointed andesite flows	15.9±0.8	
12	*Vroulidia* Bay	Flow banded rhyolite dome, cut by pipe breccias with hematitization & silicification	14.6±0.8	

Numbering scheme is that of Besenecker (1973). Radiometric dates from Bellon *et al.* (1979) except as noted.

The island of Chios (Fig. 1), in the eastern Aegean Sea, contains a series of small volcanic centres that developed during mid-Miocene basin formation (Besenecker & Pichler 1974). To the north and east are voluminous early Miocene (22–17 Ma) stratovolcanoes in Lesbos and the Izmir area of Anatolia; to the south and west are the mid- to late Miocene plutons of the Cyclades and the 11 to 5.6 Ma volcanoes of the southeastern Aegean islands. Chios thus lies in a pivotal position to examine the relative importance of subduction and extension in the Cenozoic igneous activity of the Aegean area.

Geological setting of the Chios volcanics

Small Neogene volcanic centres occur throughout Chios (Besenecker *et al.* 1968), both within Neogene sedimentary basins and on alpine basement. The locality numbering scheme of Besenecker (1973) is used in this paper (Fig. 1; Table 1). Most of the localities of Besenecker (1973) were re-examined and sampled for geochemical analysis. Rock nomenclature follows Le Bas *et al.* (1986). Small areas of rhyolite, andesitic flows and pyroclastic rocks occur throughout the island, with volumes of volcanic products individually <0.1 km^3. Ne-normative basalt crops out near Pirgi (locality 8, Fig. 1) and float of andesitic vitric tuff has been found in the same area. Felsic pyroclastic rocks up to 60 m thick (locality 9) are conformably interbedded with the base of the Sarmatian (middle Miocene) Keramaria Formation in the middle of the Miocene basin sequence, and a 40 cm-thick felsic tuff occurs within the overlying Sarmatian–Pannonian Nenita Formation. The rhyolites of southern Chios (localities 10, 12) and the Mavra Votsala andesite (locality 11) deform pre-Sarmatian (middle Miocene) sedimentary rocks (Böger 1983) and have yielded radiometric dates of 14.3–15.5 Ma (Bellon *et al.* 1979). A new K/Ar whole-rock radiometric age of 16.2 ± 0.6 Ma was obtained by the authors on the Pirgi basalt, indicating synchroneity with other igneous activity on Chios and contradicting the suggestion of Bellon *et al.* (1979), that this *ne*-normative basalt might correlate with the 7 Ma *ne*-normative basalts on Samos.

Petrography

The main petrographical characteristics of the analyzed rocks are summarized in Table 2. The

Table 2. *Petrographical characteristics of the studied volcanic rocks*

Sample	Locality	Macroscopic character*	Phenocrysts	Groundmass	Remarks
CH13	8	Basalt	cpx(Ti-augite); ol	Hyalopilitic. cpx; plag; op	Alt: cc; op; clays; idd.; serp. Corroded qtz xenocrysts and quartzite xenoliths
CH21, 27	8	Andesitic vitric lapilli tuff	qtz; feld; alt crystals (cc and/or clays)	Colourless to brownish vesicular glass and pumice fragments	Some clasts of fine-grained felsic rocks
CH29, 30, 34	12	Rhyolite	qtz; K-feld; plag	Microcrystalline	
CH37	12	Rhyolite	idiom K-feld; plag; qtz	Microcrystalline	Rare zircon, some alt red-brown crystals
P3	12	Rhyolite	idiom K-feld; plag; qtz	Microcystalline	Rare mu
CH40	11	Andesite	ol resorbed	Hyalopilitic. cpx; plag; glass; op	Plag laths show flow texture
CH41	11	Andesite	ol resorbed	Hyalopilitic. cpx; plag; glass; op	Amygdaloidal Amygdules filled with cc.
P4	11	Andesite	ol resorbed often with cpx	Hyalopilithic. cpx; plag; glass; op	Clots of cpx+glass
CH47, 51, 48, 49	1	Dacitic crystal tuff	qtz; feld	Crystal dust; mu	Veined and spotted with qtz+hem+sulf
P1, CH69	1	Rhyolites	K-feld; plag; qtz; bi	Microcystralline. qtz; feld; op; bi	Highly porphyritic. Alt: mu; hem. Rare gt, zircon
CH80, 81	6	Andesite	ol	Hyalopilitic. cpx; plag; glass; op	Amygdaloidal. Alt: op, chl, idd; serp. Amygdules filled with qtz+chl
CH83	10	Rhyolite	K-feld; qtz; plag	Microcystalline. qtz; feld; op; altered crystals	
CH84	9	Rhyolitic crystal tuff	idiom plag; K-feld; qtz; rare bi	Dusty glass	Undistorted pumice fragments. Glass about 50%
CH108, 109, 117	3	Andesite	opx; cpx; plag (glass-charged zones)	Hyalopilitic. cpx; plag; op; glass	Alt: op, cc, mu. Some amygdules filled with cc.
P2	10	Rhyolite	K-feld; qtz; plag	Microcrystalline qtz; feld; bi	
P6	9	Rhyolitic crystal tuff	K-feld; qtz; plag	Dusty glassy microcrystalline	Undistorted glass fragments. Glass about 50%. Similar to CH84.

* Rock nomenclature according to IUGS classification system (LeBas *et al.* 1986). alt, altered/alteration; bi, biotite; cc, calcite; chl, chlorite; cpx, clinopyroxene; gt, garnet; hem, hematite; idd, iddingsite; idiom, idiomorphic; K-feld, K-feldspar; mu, muscovite; ol, olivine; op, opaque oxide minerals; qtz, quartz; plag, plagioclase; serp, serpentine; sulf, sulphide minerals.

lavas are basalts, andesites and rhyolites, whereas the pyroclastic samples are andesitic, dacitic and rhyolitic tuffs.

The basalt from Pirgi has a porphyritic texture with phenocrysts of Ti-augite and olivine set in a hyalopilitic groundmass of clinopyroxene, plagioclase and opaque oxides. The basalt contains basement xenoliths and quartz xenocrysts rimmed with clinopyroxene. Carbonate alteration is common. In the field these rocks are associated with minor andesitic vitric lapilli-tuffs.

The andesites (localities 3, 6, 11) are also porphyritic, with phenocrysts of plagioclase, olivine, orthopyroxene and clinopyroxene. Olivine is commonly partly resorbed and has iddingsite rims. The orthopyroxene is very Mg-rich, up to En_{87}; some phenocrysts have

Table 3. *Representative electron microprobe analyses of olivine and pyroxenes in andesites from Chios*

	Olivine		Orthopyroxene						Clinopyroxene				
	CH40		CH108					CH80	CH108		CH80		CH40
	loc. 11 Ph		Ph			Ph(?x)		loc. 6	loc. 3 Ph		loc. 6 Ph		loc. 11
					loc. 3								
	c	r	c	r	gr	c	r	gr	r	c	r	c	mph
SiO_2	38.83	56.52	56.44	50.89	50.64	56.15	53.96	53.15	53.65	52.37	54.21	54.33	
TiO_2	0.00	0.00	0.12	0.11	0.27	0.39	0.00	0.33	0.25	0.28	0.45	0.16	0.11
Al_2O_3	0.00	0.00	2.08	2.18	2.38	5.59	2.96	1.09	3.67	3.21	2.37	1.66	1.25
FeO_t	18.76	21.63	7.00	7.61	25.43	18.08	7.73	17.30	4.91	5.49	9.22	4.46	5.01
MnO	0.47	0.34	0.00	0.22	0.65	0.28	0.14	0.55	0.09	0.12	0.30	0.19	0.37
MgO	41.76	39.17	33.05	32.31	18.68	23.71	31.40	24.84	17.43	17.51	15.72	17.52	17.90
CaO	0.13	0.17	1.26	1.58	0.90	1.53	1.59	1.96	20.28	20.00	19.04	21.16	20.27
Na_2O	0.27	0.23	0.19	0.14	0.40	0.37	0.17	0.34	0.48	0.46	0.44	0.39	0.35
NiO	0.19	nd	0.14	nd	nd	nd	nd	nd	nd	nd	nd	nd	nd
Cr_2O_3	nd	nd	0.47	0.44	nd	0.11	0.66	nd	0.32	0.09	nd	0.60	0.43
Total	100.95	100.37	101.30	101.47	99.60	100.81	101.46	100.37	100.90	100.90	99.91	100.95	100.45
Structural formulae	[4]	[4]	[6]	[6]	[6]	[6]	[6]	[6]	[6]	[6]	[6]	[6]	[6]
Si	1.000	1.003	1.948	1.947	1.940	1.846	1.941	1.965	1.922	1.937	1.942	1.967	1.978
Ti	0.000	0.000	0.003	0.003	0.008	0.011	0.000	0.009	0.007	0.008	0.013	0.004	0.003
Al	0.000	0.000	0.085	0.089	0.107	0.240	0.121	0.047	0.156	0.137	0.104	0.071	0.054
Cr	0.000	0.000	0.013	0.012	0.000	0.003	0.018	0.000	0.009	0.003	0.000	0.017	0.012
Fe	0.399	0.467	0.202	0.220	0.811	0.551	0.223	0.527	0.149	0.166	0.286	0.135	0.153
Mn	0.010	0.007	0.000	0.006	0.021	0.009	0.004	0.017	0.003	0.004	0.009	0.006	0.011
Mg	1.581	1.508	1.697	1.661	1.061	1.288	1.618	1.348	0.939	0.942	0.869	0.947	0.971
Ca	0.004	0.005	0.047	0.058	0.037	0.060	0.059	0.076	0.786	0.774	0.756	0.823	0.791
Na	0.013	0.012	0.013	0.009	0.030	0.026	0.011	0.024	0.034	0.032	0.032	0.027	0.025

Ph, phenocryst; c, core; r, rim; gr, groundmass; mph, microphenocryst; x, xenocryst. nd, not detected.

Mg-rich rims, which may indicate magma mixing (Table 3). The groundmass has a hyalopilitic texture and consists of plagioclase laths, clinopyroxene, brown glass and opaque oxides.

The rhyolites (localities 1, 10, 12) have phenocrysts of sanidine, quartz, plagioclase and biotite. The groundmass is microcrystalline and may show flow lamination. The dacitic (locality 1) and rhyolitic (locality 9) crystal tuffs have phenocrysts of quartz, K-feldspar, plagioclase and subordinate biotite.

Geochemistry

A total of about 45 samples from all the localities were selected for chemical analysis: representative analyses are given in Table 4. Analyses were made by X-ray fluorescence, with REE determined by neutron activation analysis: analytical methods are described by Pe-Piper & Piper (1989b). All pyroclastic analyses reported were based on bulk samples and must therefore be treated with caution (Wolff 1985). The whole-rock geochemistry shows the presence of basalt,

andesite, dacite and rhyolite in the IUGS nomenclature of Le Bas *et al.* (1986; Fig. 2). In the nomenclature of Ewart (1982), the andesites, dacites and rhyolites are all high-K varieties. The distribution of lithologies is as follows:

(1) *ne*-normative basalt and andesitic tuff are found only at Pirgi.
(2) andesite with lower TiO_2 than at Pirgi occurs at Mavra Votsala, Agios Georgios Sykousis and Kato Laka;
(3) dacitic crystal tuffs crop out only at Kambi;
(4) rhyolite crops out at Kambi, Vroulidia and Komi. Bedded tuffs at Kalamoti and Tholopotami are also rhyolitic.

The Pirgi ne-normative basalt has high Na+K; a spidergram of trace element distribution (Fig. 3) does not show the low values of Nb_N, Ta_N or Ti_N typical of subduction-related rocks. However, the enrichment in LILE compared with HFSE is greater than in OIB-type alkaline basalts (Table 5). The Pirgi basalt is similar to late Miocene basalts from Samos and Patmos, but is not as alkaline as Quaternary

Table 4. *Representative chemical analyses of igneous lithologies from Chios*

	8-Pirgi			12-Vroulidia					11-Mavra Votsala			1-Kambi						6-Ag. Georgios Syk.			10-Komi		9-Kalam.-Thol.		3-Kato Laka		
	CH13	CH21	CH27	CH34	CH30	CH29	CH37	P3	CH40	CH41	P4	CH47	CH51	CH48	CH49	P1	CH69	CH82	CH80	CH81	CH83	P2	P6	CH84	CH109	CH117	CH108
Major elements (wt%)																											
SiO_2	46.00	60.95	61.37	69.80	72.25	73.45	74.28	73.90	57.72	56.30	59.21	61.58	64.15	66.97	68.06	76.19	74.90	56.83	57.68	59.00	73.86	73.55	71.52	69.77	55.00	55.63	56.31
TiO_2	1.68	0.85	0.87	0.02	0.02	0.02	0.02	0.02	0.59	0.57	0.58	0.66	0.30	0.38	0.36	0.05	0.06	0.58	0.58	0.60	0.01	0.01	0.03	0.05	0.56	0.55	0.56
Al_2O_3	15.14	18.70	18.91	14.23	15.66	14.50	14.55	14.20	16.34	16.06	16.25	18.54	16.85	17.42	17.14	13.40	13.70	15.54	15.75	16.15	15.11	15.01	13.95	13.69	16.38	16.42	16.76
Fe_2O_{3t}	8.79	7.27	7.38	0.73	0.17	0.34	0.43	0.49	5.52	5.39	5.10	2.30	3.30	3.30	2.04	1.00	1.16	5.29	5.10	4.78	0.59	0.58	0.70	1.09	5.71	5.59	5.73
MnO	0.13	0.08	0.08	0.12	0.01	0.04	0.04	0.05	0.10	0.12	0.10	0.08	0.06	0.06	0.04	0.07	0.08	0.09	0.10	0.08	0.13	0.13	0.07	0.05	0.10	0.10	0.10
MgO	6.18	2.63	2.89	0.86	0.79	0.28	0.13	0.16	5.01	3.53	4.53	1.24	1.84	0.68	0.83	0.02	0.05	5.26	5.12	4.10	0.02	0.00	0.22	0.37	6.08	5.28	5.90
CaO	12.02	2.26	1.64	1.42	0.38	0.56	0.52	0.64	6.37	6.60	6.12	1.72	1.44	0.35	0.40	0.50	0.42	8.53	8.27	7.83	0.53	0.54	0.87	1.15	6.64	6.54	6.51
Na_2O	2.55	1.15	1.31	1.69	1.79	2.45	2.12	2.47	2.90	2.96	2.86	4.44	3.17	2.82	2.21	3.50	3.31	2.52	2.47	2.66	4.12	4.01	2.54	2.14	2.51	2.55	2.63
K_2O	1.94	3.64	3.83	4.36	4.66	4.44	4.33	4.19	2.34	2.49	2.45	2.79	3.04	3.65	3.97	4.66	4.37	2.08	2.12	2.21	4.03	4.06	3.67	3.74	2.08	2.21	2.15
P_2O_5	0.42	0.14	0.17	0.05	0.04	0.05	0.04	0.04	0.19	0.19	0.18	0.12	0.09	0.15	0.14	0.00	0.03	0.18	0.18	0.19	0.02	0.01	0.04	0.03	0.12	0.12	0.13
L.O.I.	4.90	1.30	0.30	6.40	4.70	2.20	2.40	2.50	1.50	5.50	1.70	5.30	5.80	3.70	4.00	0.90	0.80	2.70	2.20	2.10	0.90	1.10	5.40	7.50	4.10	4.40	2.60
Total	99.75	98.97	98.75	99.68	100.47	98.33	98.86	98.66	98.58	99.71	99.38	98.77	100.04	99.48	99.19	100.29	98.88	99.60	99.57	99.70	99.32	99.00	99.01	99.58	99.28	99.39	99.38
CIPW Norms* (wt%)																											
Qz	0.00	27.84	27.34	41.24	43.83	42.07	45.44	43.26	8.99	9.26	11.98	19.85	27.77	34.49	38.86	36.37	37.89	8.72	10.19	12.38	33.36	33.74	42.80	42.63	6.90	8.37	7.50
Or	12.20	22.19	23.16	27.64	28.76	27.31	26.54	25.76	14.33	15.71	14.90	17.68	19.13	22.60	24.70	27.74	26.36	12.76	12.93	13.45	24.21	24.52	23.19	24.03	12.99	13.83	13.21
Ab	11.82	10.04	11.34	15.34	15.82	21.57	18.61	21.75	25.42	26.74	24.91	40.29	28.56	25.00	19.69	29.83	28.59	22.13	21.58	23.18	35.44	34.68	22.98	19.69	22.45	22.85	23.13
An	25.69	10.78	7.34	7.27	1.76	2.61	2.46	3.10	25.55	24.75	24.97	8.48	7.23	0.99	1.35	2.52	1.95	25.89	26.46	26.36	2.55	2.68	4.38	6.10	28.84	28.41	28.67
Ne	6.04	0.00	0.00	0.00	0.00	0.00	0.00	0.00	0.00	0.00	0.00	0.00	0.00	0.00	0.00	0.00	0.00	0.00	0.00	0.00	0.00	0.00	0.00	0.00	0.00	0.00	0.00
Ol	10.88	0.00	0.00	0.00	0.00	0.00	0.00	0.00	0.00	0.00	0.00	0.00	0.00	0.00	0.00	0.00	0.00	0.00	0.00	0.00	0.00	0.00	0.00	0.00	0.00	0.00	0.00
Hy	0.00	17.84	18.51	3.80	2.33	1.35	1.12	1.32	19.04	14.33	17.31	6.36	10.27	6.94	5.17	1.76	2.13	15.01	15.08	12.66	1.27	1.21	1.90	2.97	23.02	20.76	22.61
Di	29.03	0.00	0.00	0.00	0.00	0.00	0.00	0.00	5.11	7.65	4.42	0.00	0.00	0.00	0.00	0.00	0.00	13.97	12.24	10.40	0.00	0.00	0.00	0.00	4.41	4.40	3.50
Ilm	3.40	1.67	1.69	0.04	0.04	0.04	0.04	0.04	1.16	1.16	1.13	1.34	0.61	0.76	0.72	0.10	0.12	1.14	1.14	1.17	0.02	0.02	0.06	0.10	1.12	1.11	1.11
Ap	1.04	0.34	0.40	0.12	0.10	0.12	0.10	0.10	0.46	0.47	0.43	0.30	0.22	0.37	0.34	0.00	0.07	0.43	0.43	0.45	0.05	0.02	0.10	0.08	0.29	0.30	0.31
C	0.00	9.32	10.21	4.56	7.37	4.94	5.71	4.69	0.00	0.00	0.00	5.71	6.24	8.89	9.20	1.69	2.88	0.00	0.00	0.00	3.10	3.13	4.60	4.42	0.00	0.00	0.00
Trace elements (ppm) by XRF																											
Ba	1114	566	492	185	192	196	180	189	742	788	737	307	671	353	403	68	75	749	747	803	20	26	111	397	477	503	497
Rb	49	173	172	178	189	185	181	176	120	113	127	122	143	166	179	484	587	84	102	80	1086	1094	176	160	92	104	94
Sr	848	121	124	71	83	65	62	73	607	655	580	295	338	352	424	15	18	589	598	615	<5	<5	49	87	532	517	528
Y	32	37	38	15	14	12	15	11	19	18	18	11	27	33	33	73	117	20	19	29	20	23	16	11	16	15	16
Zr	234	184	189	36	40	36	35	35	150	148	152	162	190	220	222	67	62	136	135	142	47	48	35	40	171	172	170
Nb	42	18	17	16	21	20	18	18	11	12	13	9	11	11	11	36	43	10	10	11	149	155	19	17	12	10	9
Th	<10	16	15	<10	<10	11	<10	<10	<10	<10	10	<10	23	20	22	29	23	<10	<10	<10	46	23	<10	<10	<10	<10	<10
Pb	21	26	26	41	29	42	34	34	38	40	38	17	30	23	38	53	65	34	36	34	92	95	36	39	31	31	26
Ga	15	20	22	18	19	16	16	15	19	19	15	15	17	18	13	16	20	15	17	18	33	33	15	13	17	17	15
Zn	80	108	112	36	13	26	17	25	60	59	53	30	50	69	55	41	41	59	59	61	117	115	41	38	58	62	59
Cu	55	48	43	<5	<5	<5	<5	<5	35	31	31	21	9	13	10	<5	<5	36	38	38	<5	<5	<5	8	31	26	28
Ni	86	78	74	<5	<5	<5	<5	<5	53	44	47	16	6	7	7	<5	<5	54	55	47	<5	<5	24	21	148	132	144
V	190	160	163	<5	<5	<5	<5	<5	106	103	98	129	49	63	54	<5	<5	112	115	116	<5	<5	7	7	88	82	88
Cr	162	139	141	6	5	<5	<5	5	166	162	165	174	37	27	18	5	12	217	206	228	13	5	31	46	292	261	284

Table 4. *continued*

	8-Pirgi			12-Vroulidia					11-Mavra Votsala			1-Kambi					6-Ag. Georgios Syk.				10-Komi		9-Kalam.-Thol.		3-Kato Laka		
	CH13	CH21	CH27	CH34	CH30	CH29	CH37	P3	CH40	CH41	P4	CH47	CH51	CH48	CH49	P1	CH69	CH82	CH80	CH81	CH83	P2	P6	CH84	CH109	CH117	CH108
REE and other trace elements (ppm) by INAA																											
La	38.2	42.2	nd	nd	nd	10.6	nd	nd	28.6	nd	nd	nd	nd	nd	58.0	nd	8.2	28.4	nd	27.9	4.14	5.37	nd	nd	nd	nd	24.9
Ce	79.3	89.5	nd	nd	nd	21.9	nd	nd	56.8	nd	nd	nd	nd	nd	120	nd	20.6	48.9	nd	50.7	10.2	21.4	nd	nd	nd	nd	49.5
Nd	36.5	38.4	nd	nd	nd	8.6	nd	nd	20.4	nd	nd	nd	nd	nd	50.6	nd	11.5	22.0	nd	22.8	11.5	16.3	nd	nd	nd	nd	20.3
Sm	7.5	7.5	nd	nd	nd	2.55	nd	nd	4.60	nd	nd	nd	nd	nd	9.8	nd	6.1	4.6	nd	4.8	5.9	8.4	nd	nd	nd	nd	3.8
Eu	2.07	1.47	nd	nd	nd	0.23	nd	nd	1.08	nd	nd	nd	nd	nd	1.57	nd	bd	1.13	nd	1.23	bd	bd	nd	nd	nd	nd	0.97
Tb	0.88	1.05	nd	nd	nd	0.46	nd	nd	0.63	nd	nd	nd	nd	nd	1.16	nd	2.34	0.67	nd	0.67	1.24	1.54	nd	nd	nd	nd	0.49
Yb	1.92	3.03	nd	nd	nd	0.62	nd	nd	1.54	nd	nd	nd	nd	nd	2.88	nd	12.93	1.67	nd	2.26	0.51	0.55	nd	nd	nd	nd	1.40
Lu	0.30	0.43	nd	nd	nd	bd	nd	nd	0.26	nd	nd	nd	nd	nd	0.40	nd	1.81	0.28	nd	0.36	bd	bd	nd	nd	nd	nd	0.21
Hf	5.02	4.89	nd	nd	nd	2.08	nd	nd	3.57	nd	nd	nd	nd	nd	6.62	nd	3.98	3.46	nd	3.56	7.16	7.32	nd	nd	nd	nd	4.17
Sc	26.4	19.0	nd	nd	nd	1.73	nd	nd	16.3	nd	nd	nd	nd	nd	11.8	nd	7.7	19.8	nd	20.6	13.6	13.7	nd	nd	nd	nd	14.9
Ta	2.36	1.80	nd	nd	nd	2.03	nd	nd	1.56	nd	nd	nd	nd	nd	1.43	nd	10.2	1.10	nd	1.27	»15[†]	»15[†]	nd	nd	nd	nd	1.48
Th	10.1	16.5	nd	nd	nd	9.0	nd	nd	12.7	nd	nd	nd	nd	nd	25.3	nd	23.7	10.9	nd	11.3	20.7	21.0	nd	nd	nd	nd	12.1
U	2.5	3.8	nd	nd	nd	5.5	nd	nd	5.1	nd	nd	nd	nd	nd	6.8	nd	9.7	4.6	nd	4.5	9.8	32.2	nd	nd	nd	nd	3.7

* Norms were calculated using the method of Irvine and Baragar (1971).

[†] Ta abundance substantially greater than that in standard: determination approximate

Pb isotope determinations

Sample	Rock type	Locality	$^{208}Pb/^{204}Pb$	$^{207}Pb/^{204}Pb$	$^{206}Pb/^{204}Pb$
CH13	basalt	8	18.872	15.689	39.097
CH40	andesite	11	18.796	15.706	39.055

basalts from Psathoura and Kula (Table 5). The associated andesitic tuffs have lower Nb_N and Ta_N and strong negative anomalies for Sr, P and Ti, suggesting fractional crystallization of plagioclase, apatite and opaque oxides. They have rather higher Y, Zr, Nb, Ti, Th and K than other andesites. REE spectra for the basalt and andesite (Fig. 4) are very similar, except that the andesite is slightly more enriched in REE and has a Eu anomaly.

The andesite flows show similar trace element patterns to the andesitic pyroclastic rocks, but La_N, Ta_N and Ti_N troughs on the spidergrams are more pronounced (Fig. 3). The Kato Laka andesites are distinctive in having rather lower Y and Ba and higher Zr and MgO than other andesites (Fig. 5), but are indistinguishable in their REE spectra (Fig. 4). In their high MgO and low Y contents they resemble adakites (Defant & Drummond 1990). Their MgO contents and Sr/Y ratios are intermediate between those of the calc-alkaline rocks of Lesbos and the mid-Miocene adakites of Skyros (Pe-Piper 1991) and Evia (Pe-Piper & Piper 1994).

Rhyolites at Komi and Kambi have higher total alkalies, Nb, Y, Th and Na_2O and lower Ba than the dacitic tuffs at Kambi and rhyolite at Vroulidia (Fig. 5). The group with higher alkalis shows a small increase in normalized REE abundance from La to Tb, with a large Eu anomaly (Fig. 4). Using criteria from Pearce *et al.* (1984) for 'within-plate granites' (Y + Nb >50 ppm, Nb > 35 ppm, Yb + Ta >6 ppm), these rocks have an alkaline character. The role of minor mineral phases is unclear: there is no visible mineralogical host for the high Ta in samples CH83 and P2 (Table 4) (cf. Wolff 1984) and no allanite, which might account for the light REE depletion. The lower-alkalis group at Kambi and Vroulidia shows a gradual decrease in normalized REE abundance from LREE to HREE, and a small Eu anomaly. Using criteria of Pearce *et al.* (1984), these rocks resemble volcanic arc granites.

Lead isotope ratios for the Pirgi basalt and Mavra Votsala andesite (Table 4) fall within the narrow range determined for Aegean Neogene rocks (Pe-Piper & Piper 1992 and unpublished data). The similarity between the two rocks suggests a similar mantle source.

The regional setting: Cenozoic igneous activity in the Aegean area

The south Aegean arc

The volcanism of the south Aegean arc provides a reference point for interpreting older volcanic

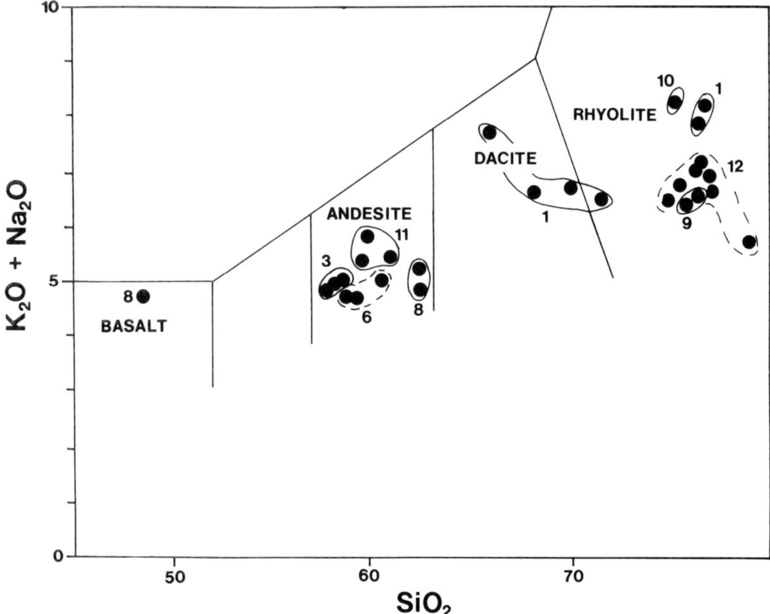

Fig. 2. Plot of K$_2$O+Na$_2$O v. SiO$_2$ showing nomenclature of the volcanic rocks of Chios. Numbers indicate localities in Fig. 1.

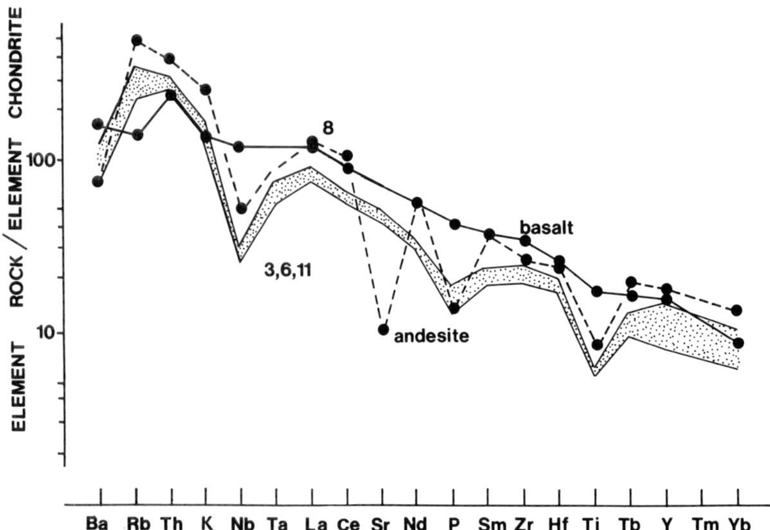

Fig. 3. Spidergrams of normalized trace element abundances for basalt, andesite and dacite from Chios (normalizing values from Thompson *et al.* 1984). Lines are for andesite and basalt from Pirgi (locality 8); stippled area is abundance range for andesites from localities 3, 6 and 11.

rocks of the Aegean region. The deformation of the Mediterranean Ridge accretionary wedge (Kastens 1991) and the seismically-defined Benioff zone (Makropoulos & Burton 1984) are evidence of late Cenozoic subduction. The correspondence of the active volcanic arc with a Benioff zone depth of about 130 km is evidence for a role for subduction in the genesis of the arc (Gill 1981; Spiegelman & McKenzie 1987). The volcanism is voluminous, with individual centres

Table 5. *Selected geochemical parameters for Pyrgi basalt and andesite compared with other rocks*

	Pirgi bas.	Pirgi and.	Samos	Patmos	Psath- oura	Kula	Ezine	Urla	IAT	OIT	e-MORB
SiO_2*	48.5	62.4	52.1	50.3	50.2	46.3	46.4	49.0			
Na_2O/K_2O	1.31	0.33	1.40	1.14	1.60	1.51	1.81	2.19			
K/Ba	18.4	59.8	21.8	12.5	33.4	26.1	nd	nd	60	35	37
Ba/La	29.1	12.5	29.4	27.3	17.1	13.8	nd	nd	25–40	8–12	6–10
Nb/Th	4.2	1.1	1.8	1.8	6.6	12.3	nd	nd	4.6	12	14
Nb/Ta	17.8	10.0	19.2	18.6	15.4	16.5	nd	nd	17	17	17
Ti/Y	554	231	452	nd	382	378	nd	nd	296	590	270
Ce/Yb	41.3	29.5	36.1	54.6	30.2	37.6	45.5	26.3	3.5	37	6
Hf/Yb	2.1	1.6	1.9	2.0	3.1	2.4	nd	nd	0.6	3.5	0.9
Tb/Yb	0.46	0.35	0.38	nd	0.33	0.42	0.55	0.30		0.5	0.2

(Aegean data from Pe-Piper & Piper (1989a); OIB and E-MORB from Sun & McDonough (1989); IAT from Pearce (1982), with selected ratios from Dupuy *et al.* (1988).)
* Volatile free.
nd, no data available.

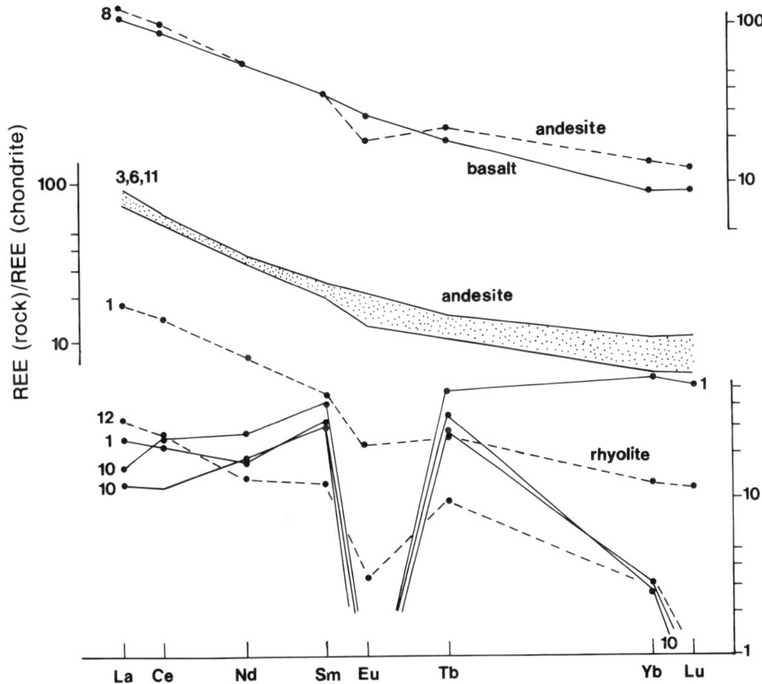

Fig. 4. Chondrite-normalized REE of selected rocks from Chios. Numbers indicate localities in Fig. 1. Two groups of rhyolites, distinguished on the basis of trace and REE abundance, shown by solid and dashed lines respectively (for explanation, see text).

having volumes above sea level of the order of 10–30 km³.

Some centres in the modern arc (Aegina, Crommyonia, Milos) began their activity in the Pliocene. There was a general cessation of activity in the late Pliocene and then activity renewed in the Quaternary (early Pleistocene in Aegina and Santorini, mid Pleistocene in Methana and Nisyros). This pattern of volcanism corresponds in timing with two phases of basin subsidence in the arc (Doutsos & Piper 1990; Perissoratis *et al.* in press). Thus, in addition to more enriched in TiO_2 and Nb (Table 4) than nepheline trachybasalt from Patmos (Wyers & Barton 1986, 1987) and ne-normative basalts from Samos (Pe-Piper, unpublished data), but total alkalis and HFSE abundance are not unusually high: for example total alkalis are higher in the Patmos basalts. Although there is no evidence for a subduction-related geochemical influence in the ne-normative basalt, it has Pb-isotopic values typical of K-rich subduction-related rocks and is similar to many Aegean subduction-related rocks (Pe-Piper & Piper 1992, unpublished data).

The Pb-isotope evidence suggests that the alkaline character of the basalt results from partial melting conditions rather than a different source. Reagan & Gill (1989) ascribed the difference between Nb-enriched and Nb-depleted lavas from Costa Rica to small degree partial melts produced under respectively reduced and oxidizing conditions. Falloon & Crawford (1991) similarly argued for a role for CO_2 in the volatile component of the source mantle.

Discussion: the role of subduction and extension

The Aegean area has experienced both subduction and extension at least throughout the Neogene and Quaternary (Fig. 8). Although possibly not genetically related to subduction, the extension has been in a back-arc position with respect to the subduction zone. Igneous activity appears to be a result of both subduction- and extension-related magma-generation processes.

The sedimentary record shows that extension has been more active in different places in the Aegean region at different times, but there is no strong regional evidence for any particular time interval showing a much greater rate of extension (Fig. 6). Why, then, are the most voluminous volcanic rocks found either in the Pliocene–Quaternary south Aegean arc or in the early Miocene northeast Aegean centres? Not only are these rocks voluminous, but they are also dominated by andesites and dacites, in contrast to the basalts, trachyandesites and rhyolites of many smaller 'back-arc' centres (Fig. 8).

If the geological evidence for the role of subduction in the modern south Aegean arc is accepted, then subduction was also a significant factor contributing to the origin of early Miocene volcanic rocks of the northeast Aegean. Variation in the importance of subduction processes may be related to rate of

subduction. Changes in subduction rate will produce changes in the abundance, distribution and type of igneous products. Although precise data are lacking (Crisp 1984), several authors have argued that the volume of igneous products is directly related to the rate of subduction (Reymer & Schubert 1984; Page & Engebretson 1984; McCulloch & Gamble 1991). The modelling of Spiegelman & McKenzie (1987) showed that, for any constant value of mantle porosity, the volume of the melt extraction zone that feeds arc volcanoes decreases as subduction rate decreases and the distance of the arc volcanism behind the trench decreases. Thus, an area receiving strongly slab-influenced melts under conditions of rapid subduction will receive less slab-influenced back-arc magmas if subduction rate decreases. Oblique subduction resulting in strike-slip motion may change the proportion of plutonic to volcanic rocks (Glazner 1991).

Independent evidence for the rate of subduction may be provided by geological evidence of major compression near the margins of the present Aegean microplate (Fig. 8). On Crete and the southern Peloponnese, nappes were emplaced over autochthon between early Oligocene and middle Miocene (Seidel & Wachendorf 1986). Radiometric dates suggest a late Oligocene–early Miocene high-pressure/low-temperature metamorphism (Seidel *et al.* 1982). The last phase of nappe emplacement in the external Hellenides of the northern Peloponnese and continental Greece was Burdigalian to middle Miocene (Jacobshagen 1986), representing the convergence of Apulia and the main Aegean plate.

The net relative motion of Africa and Eurasia showed an abrupt change in the middle Miocene, through about 90° (Savostin *et al.* 1986). At the same time, the collision of the Arabian and Anatolian plates (Dewey *et al.* 1986) set up the late Neogene westward motion of the Anatolian plate (Sengör & Yilmaz 1981). Thus, a change in the relative motion of Africa and the Aegean plate probably occurred in the middle Miocene. In western Greece, minor compressive deformation has occurred from late Miocene to recent times in the Ionian and Pre-apulian zones (Doutsos *et al.* 1987; Underhill 1989). This suggests that middle Miocene rates of convergence of the African and Aegean plates in the south Aegean decreased to about 1.0–1.5 cm/yr (Savostin *et al.* 1986), which may have been accompanied by a change in subduction vector (Fig. 8). There may have been a small change in rate or convergence direction in the late Miocene, corresponding to the onset of late Miocene compressive deformation in the Ionian

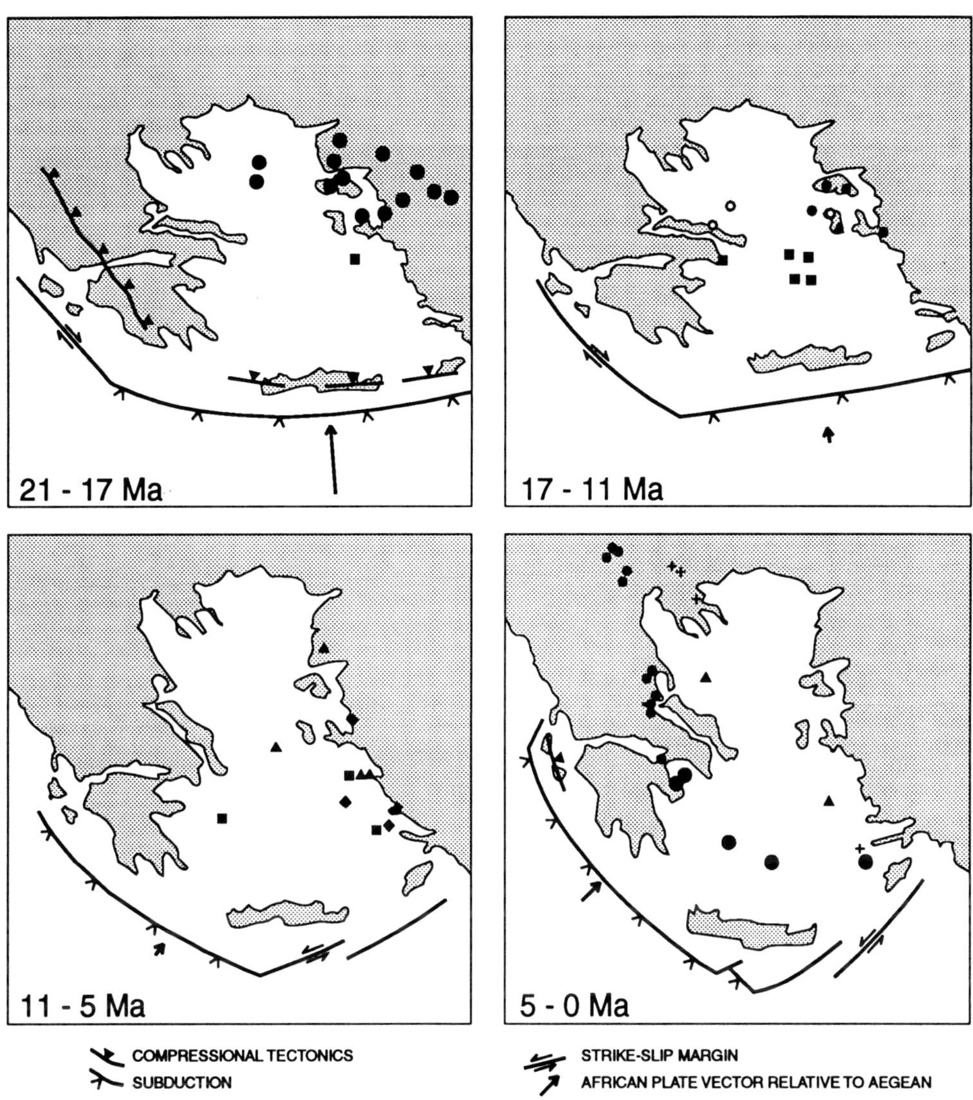

Fig. 8. Schematic maps showing speculative Neogene evolution of the Aegean region, taking into account crustal extension. Modern coastline shown for reference only: reconstructions based on Le Pichon & Angelier (1981), with modifications. Location of igneous rocks from Fig. 7; symbols for volcanic products as in Fig. 6. Plate boundaries and kinematics speculative, based on evidence discussed in text.

and pre-Apulian zones and the marked change in rotation of Ionian islands at about 56 Ma, as implied by palaeomagnetic data (Kissel *et al.* 1984). Kastens (1991) has shown that the Mediterranean Ridge accretionary prism has been continuously accreted since before the middle Miocene, but that the rate of accretion increased in the Pliocene, a phenomenon that she ascribed to the mechanical role of the Messinian evaporites (Kastens *et al.* 1992).

Crustal extension also played an important role in determining the location and character of the volcanic rocks of the Aegean. The volcanism of at least parts of the modern arc can be correlated in both time and space with extensional faulting. The back-arc volcanic rocks are closely associated with the development of small sedimentary basins, whose scale is insufficient to produce significant decompression melting. Conversely, volcanic rocks are rare around the major extensional basins (Fig. 7). Such a situation is similar to that observed on some rifted continental margins, where magmatism occurs not at the incipient oceanic rift but in more distant half grabens, and was ascribed by Sawyer & Harry (1991) to lateral migration of extension-related melts along brittle crustal faults. Such a pattern implies a complex melt extraction history for the back-arc volcanic rocks. The change through time, from calc-alkaline or shoshonitic magmas to ne-normative basalts in several back-arc volcanic centres (Pe-Piper & Piper 1989a), may reflect progressive dehydration of subduction-influenced lithospheric mantle as volcanism continued.

The development of the major extensional basins, such as the north Aegean trough–Orfanou basin and the Cretan Sea, is associated with asthenospheric upwelling. The greater contribution of asthenospheric melting to volcanic products is inferred from Sr and Nd isotopes (Güleç 1991), Pb isotopes (Pe-Piper & Piper 1992, unpublished data) and rock and mineral chemistry (Mitropoulos & Tarney 1992). In the Aegean, the volcanic rocks of Psathoura and Santorini most reflect this asthenospheric upwelling.

The scattered volcanic rocks of Chios resemble other 'back-arc' volcanism of the Aegean in their association with minor basin-forming faults. By the time that they were erupted, significant extension of the Aegean (inferred from mylonitic detachments and early Miocene basin sedimentation) had been under way for a few million years. Decompression remelting of older subduction-related magma batches may have produced the adakitic volcanic rocks of Chios and the similar minor volcanism on Lesbos. The ne-normative basalts of Pirgi and the alkaline rhyolites may represent melts derived from the major extensional areas that acquired some subduction-related trace element characteristics from either the source lithospheric mantle or from assimilation during melt migration.

Conclusions

Major changes in the volume of volcanic products through the later Cenozoic in the Aegean are probably a consequence of changes in the rate of subduction. Back-arc extension has played an important role in melt generation and in providing pathways for melt migration, but the effect of asthenospheric upwelling is seen only locally. The pattern of back-arc volcanism suggests significant lateral migration of melts, which may also be responsible for much of their geochemical variability. Both extension and subduction have played an important role in the petrogenesis of the back-arc rocks. The gross pattern of Cenozoic volcanism in the Aegean was ultimately a consequence of the collision of the Arabian promontory with Eurasia. Analogous decreases in arc-related volcanism associated with changes in subduction rate can be expected in other orogens following collision of promontories.

This work was supported partly by an NSERC operating grant to G. Pe-Piper. The geochemical work was carried out at the Saint Mary's University regional geochemical centre. Geological Survey of Canada contribution number 18692.

References

ANGELIER, J., LYBERIS, N., LE PICHON, X., BARRIER, E. & HUCHON, P. 1982. The tectonic development of the Hellenic arc and the Sea of Crete: a synthesis. *Tectonophysics*, **86**, 159–196.

ARMIJO, R., LYON-CAEN, H. & PAPANASTASSIOU, D. 1992. East–west extension and Holocene normal-fault scarps in the Hellenic arc. *Geology*, **20**, 491–494.

BELLON, H., JARRIGE, J.-J. & SOREL, D. 1979. Les activités magmatiques égéennes de l'Oligocène à nos jours et leurs cadres géodynamiques. Données nouvelles et synthèse. *Revue du géologie dynamique et géographie physique*, **21**, 41–55.

BESENECKER, H. 1973. *Neogen und Quartär der Insel Chios (Ägäis)*. Doctoral dissertation, Freien Universität, Berlin.

—— & PICHLER, H. 1974. Die jungen Vulkanite der Insel Chios (östliche Agäis, Griechenland). *Geologische Jahrbuch*, **D9**, 41–65.

——, DÜRR, ST., HERGET, G. *et al.* 1968. Geologie von Chios (Agäis). *Geologica et Palaontologica*, **2**, 121–150.

BÖGER, H. 1983. Stratigraphische und tektonische Verknüpfungen kontinentale Sedimente des Neogens in Agäis-Raum. *Geologische Rundschau*, **72**, 771–814.

BUICK, I.S. 1991. The late Alpine evolution of an extensional shear zone, Naxos, Greece. *Journal of the Geological Society, London*, **148**, 93–103.

BÜTTNER, D. & KOWALCZYK, G. 1978. Late Cenozoic stratigraphy and paleogeography of Greece: a review. *In:* CLOSS, H., ROEDER, D. & SCHMIDT, K. (eds) *Alps, Apennines, Hellenides*. Schweizerbart, Stuttgart, 494–501.

CHORIANOPOULOU, P., GALEOS, A. & IOAKIM, C. 1984. Pliocene lacustrine sediments in the volcanic succession of Almopias, Macedonia, Greece. *In:* DIXON, J.E. & ROBERTSON, A.H.F. (eds) *The geological evolution of the eastern Mediterranean*. Geological Society, London, Special Publications, **17**, 95–806.

COLLIER, R.E.LL. & DART, C.J. 1991. Neogene to Quaternary rifting, sedimentation and uplift in the Corinth Basin, Greece. *Journal of the Geological Society, London*, **148**, 1049–1065.

CRISP, J.A. 1984. Rates of magma emplacement and

volcanic output. *Journal of Volcanology and Geothermal Research*, **20**, 177–211.

DEFANT, M.J.& DRUMMOND, M.S. 1990. Derivation of some modern arc magmas by melting of young subducted lithosphere. *Nature*, **347**, 662–665.

DEWEY, J.F. & SENGÖR, A.M.C. 1979. Aegean and surrounding regions: complex multiplate and continuum tectonics in a convergent zone. *Geological Society of America Bulletin*, **90**, 84–92.

——, HEMPTON, M.R., KIDD, W.S.F., SAROGLU, F. & SENGÖR, A.M.C. 1986. Shortening of continental lithosphere: the neotectonics of eastern Anatolia – a young collision zone. *In:* COWARD, M.P. & RIES, A.C. (eds) *Collision Tectonics*. Geological Society, London, Special Publications, **19**, 3–36.

DOUTSOS, T. & PIPER, D.J.W. 1990. Listric faulting, sedimentation and morphological evolution of the Quaternary eastern Corinth rift, Greece: first stages in continental rifting. *Geological Society of America Bulletin*, **102**, 812–829.

——, KONTOPOULOS, N. & FRYDAS, D. 1987. Neotectonic evolution of northwestern continental Greece. *Geologische Rundschau*, **76**, 433–450.

DUPUY, C., MARSH, J., DOSTAL. J., MICHARD, A. & TESTA, S. 1988. Asthenospheric and lithospheric sources for Mesozoic dolerites from Liberia (Africa): trace element and isotopic evidence. *Earth and Planetary Science Letters*, **87**, 100–110.

EWART, A. 1982. The mineralogy and petrology of Tertiary–Recent orogenic volcanic rocks, with special reference to the andesitic-basaltic compositional range. *In:* THORPE, R.S. (ed.) *Andesites*. Wiley, Chichester, U.K., 26–87.

FALLOON, T.J. & CRAWFORD, A.J. 1991. The petrogenesis of high-calcium boninite lavas dredged from the northern Tonga ridge. *Earth and Planetary Science Letters*, **102**, 375–394.

FYTIKAS, M., INNOCENTI, F., MANETTI, P., MAZZUOLI, R., PECCERILLO, A. & VILLARI, L. 1984. Tertiary to Quaternary evolution of volcanism in the Aegean region. *In:* DIXON, J.E. & ROBERTSON, A.H.F. (eds.) *The geological evolution of the eastern Mediterranean*. Geological Society, London, Special Publication, **17**, 687–699.

GILL, J.B. 1981. *Orogenic andesites and plate tectonics*. Springer, New York.

GLAZNER, A.F. 1991. Plutonism, oblique subduction, and continental growth: an example from the Mesozoic of California. *Geology*, **19**, 784–786.

GÜLEÇ, N. 1991. Crust-mantle interaction in western Turkey; implications from Sr and Nd isotope geochemistry of Tertiary and Quaternary volcanics. *Geological Magazine*, **128**, 417–435.

GÜLEN, L. 1989. Isotopic characterization of Aegean magmatism and geodynamic evolution of Aegean subduction. *In:* HART, S.R. *et al.* (eds) *Crust/ Mantle recycling at convergence zones*. NATO Advanced Studies Institute Series, C258, 143–166.

HERGET, G. 1968. *Die Geologie von Nord-Chios (Ägäis)*. Dissertation, University of Marburg, 206p.

IRVINE, T.N. & BARAGAR, W.R.A. 1971. A guide to the chemical classification of the common volcanic rocks. *Canadian Journal of Earth Sciences*, **8**, 523–548.

JACOBSHAGEN, V. 1986. *Geologie von Griechenland*. Borntrager, Berlin.

JONES, C.E., BAKER, J.H., TARNEY, J. & GEROUKI, F. 1992. Tertiary granitoids of Rhodope, N. Greece: magmatism related to extensional collapse of the Hellenic Orogen. *Tectonophysics*, **210**, 295–314.

KASTENS, K.A. 1991. Rate of outward growth of the Mediterranean Ridge accretionary complex. *Tectonophysics*, **199**, 25–50.

——, BREEN, N.A. & CITA, M.B. 1992. Progressive deformation of an evaporite-bearing accretionary complex: SeaMARC I, SeaBeam and piston core observations from the Mediterranean Ridge. *Marine Geophysical Researches*, **14**, 249–298.

KATSIKATSOS, G., DE BRUIJN, H. & VAN DER MUELEN, A.J. 1981. The Neogene of the island of Euboea (Evia), a review. *Geologie en Mijnbouw*, **60**, 509–516.

KISSEL, C., JAMET, M. & LAJ, C. 1984. Paleomagnetic evidence of Miocene and Pliocene rotational deformations of the Aegean area. *In:* DIXON, J.E. & ROBERTSON, A.H.F. (eds) *The geological evolution of the eastern Mediterranean*. Geological Society, London, Special Publications, **17**, 669–679.

KOUKOUVELAS, I. & DOUTSOS, T. 1990. Tectonic stages along a traverse cross-cutting the Rhodopian zone (Greece). *Geologische Rundschau*, **79**, 753–776.

KOUSPARIS, D. 1979. *Seismic stratigraphy and basin development: Nestos Delta area, northeastern Greece*. PhD thesis, University of Tulsa, Oklahoma, USA.

LALECHOS, N. & SAVOYAT, E. 1979. La sédimentation néogène dans le fosse Nord Égéen. *6th Colloquium on the Geology of the Aegean region*, **2**, 591-603.

LE BAS, M.J., LE MAITRE, R.W., STRECKEISEN, A. & ZANETTIN, B. 1986. A chemical classification of volcanic rocks based on the total alkali – silica diagram. *Journal of Petrology*, **27**, 745–750.

LE PICHON, & ANGELIER, J. 1981. The Aegean Sea. *Philosophical Transactions of the Royal Society, London*, **A300**, 357–372.

LISTER, G.S., BANGA, G. & FEENSTRA, A. 1984. Metamorphic core complexes of the Cordilleran type in the Cyclades, Aegean Sea, Greece. *Geology*, **12**, 221–225.

LÜDTKE, G. 1969. *Die Geologie von Südwest-Chios (Ägäis)*. Dissertation, University of Marburg.

McCULLOCH, M.T. & GAMBLE, J.A. 1991. Geochemical and geodynamical constraints on subduction zone magmatism. *Earth and Planetary Science Letters*, **102**, 358–374.

MAKROPOULOS, K.C. & BURTON, P.W. 1984. Greek tectonics and seismicity. *Tectonophysics*, **106**, 275–304.

MASCLE, J. & MARTIN, L. 1989. Shallow structure and recent evolution of the Aegean Sea: a synthesis based on continuous reflection profiles. *Marine Geology*, **94**, 271–299.

MERCIER, J.L., SOREL, D., VERGELY, P. & SIMEAKIS, K. 1989. Extensional tectonic regimes in the Aegean

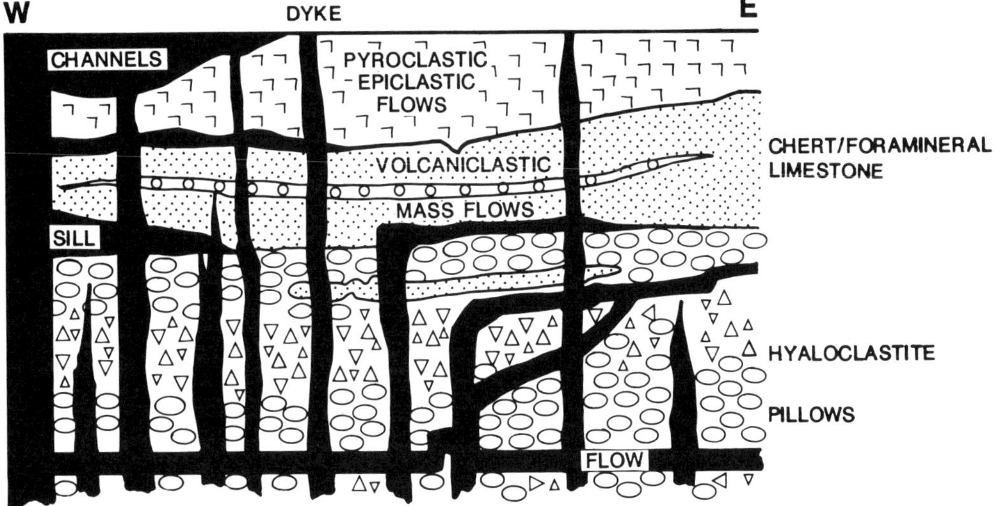

Fig. 3. Generalized facies model for the Casma volcanic and volcaniclastic rocks of the Huarmey basin. The facies association is similar to those that characterize slow spreading MOR systems (after Atherton & Webb 1989).

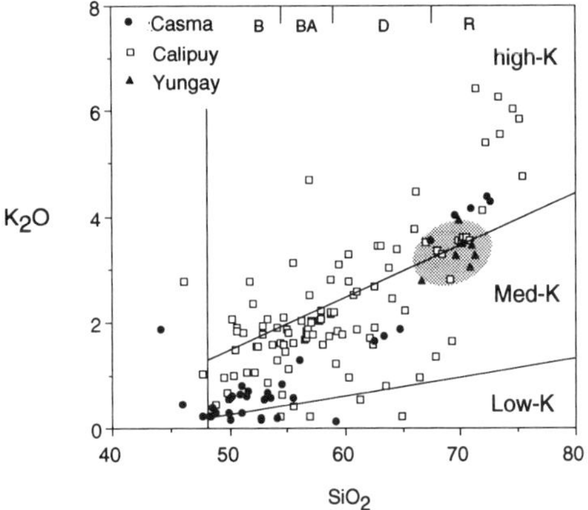

Fig. 4. Plot of K_2O v. SiO_2 for the volcanic rocks of the WPT. Note that the generally high-K trend seen in the data represents nearly 100 Ma of volcanism erupted during three periods of discrete crustal extension. B (basalt), BA (basaltic andesite) D (dacite) R (rhyolite). Shaded area shows field of Yungay volcanic rocks.

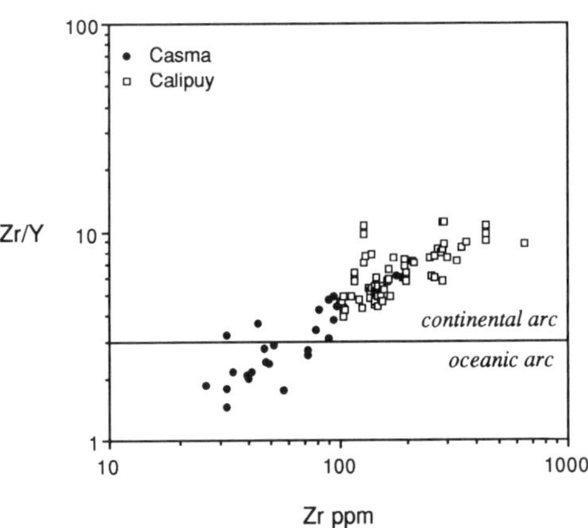

Fig. 5. Zr/Y v. Zr tectonic discrimination diagram for the volcanic rocks of the WPT (after Pearce 1982). Note that the Casma volcanic rocks of the Huarmey basin have elements characteristic of oceanic and continental arc magmas, even though they were generated during a single basin-forming event. In contrast, the Calipuy volcanics are confined entirely to the continental arc field.

have, on the basis of metamorphism, geophysical evidence and facies analysis, recently interpreted the Casma marginal basin as a mini-ocean, developed in response to extensive crustal rifting along the western margin of South America during the Cretaceous. Rifting began with the splitting of continental crust parallel to the coast, and the basin ultimately developed an ocean floor. The basin fill is ophiolitic in character (the term 'aborted ophiolite' has been used by Aguirre & Offler (1985) to describe the basinal Casma sequence) and is in many ways similar to, though less complete than, ophiolitic sequences of the same age from Sarmiento and Tortuga in southern Chile (Stern 1980). Metamorphism of the Casma rocks is non-deformational and ranges from zeolite through

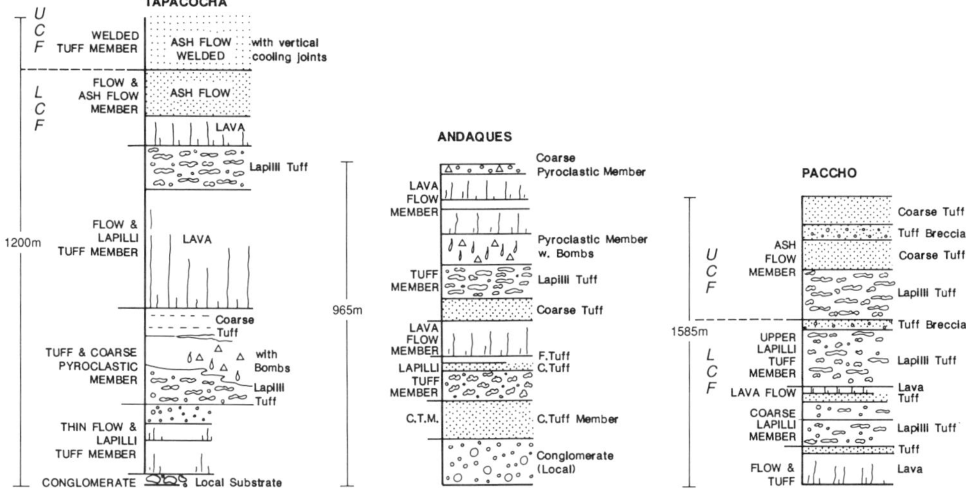

Fig. 6. North-south section (*c.* 100 km) through the Calipuy Group from Tapacocha to Paccho along strike of the Tapacocha axis (after Webb 1976). Lavas make up 31%, 52.3% and 9.8% of the section at Andaques, Tapacocha and Paccho, respectively. 60% of the lithologies are primary deposits. UCF: upper Calipuy Formation; LCF: lower Calipuy Formation.

prehnite-pumpellite to amphibolite facies. Aguirre & Offler (1985) have suggested local thermal gradients in excess of 300°C/km, similar to those seen in ocean floor environments and in rifting/hydrothermal systems such as Iceland.

In summary, the geophysical, geological, geochemical and metamorphic facies analysis all indicate a major spreading centre in an ensialic marginal basin with no continental input (Atherton 1990). Comparisons were made with the present east Scotia Sea and Bransfield Strait, as well as the Gulf of California. The absence of an arc to the west made the Gulf of California an attractive model (Atherton 1990).

The Calipuy Group: plateau-type volcanic rocks

The Calipuy Group (Fig. 6) forms an areally extensive, thick volcanic succession (generally in excess of 2000 m and about 40 km wide) lying inboard of the Casma Group rocks of the Huarmey basin (Fig. 2). It consists of a lower sequence of basic to intermediate lavas, coarse- to fine-grained lapilli-tuffs and agglomerates, often with bombs, and an upper sequence of mainly acid pyroclastics rocks (Webb 1976). The rocks are terrestrial in origin, and rest unconformably upon an erosion surface cut in the main intrusions of the batholith and the volcanic rocks of the Casma Group (Cobbing & Pitcher 1972). K–Ar ages range from 53 to *c.* 15 Ma (Wilson

1975, Farrar & Noble 1976). The group as a whole is calc-alkaline, and belongs to the high alumina basalt–andesite–dacite–rhyolite association that characterizes much of the present-day Andes. Detailed chemical variation within the Calipuy Group is discussed elsewhere (Atherton *et al.* 1985*b*), but it is likely that the acid parts of the Calipuy were derived from the more basic facies by fractional crystallization. Comparison of the basic rocks of the Calipuy with similar rocks from the Casma Group show that the former have higher K_2O (generally 1–3 wt%; Fig. 4), higher incompatible element contents (Ba, Rb, Y, Zr) but lower FeO and MgO than their Casma equivalents, indicating their more evolved nature. Zr/Y values are greater than 3 indicating a continental plate enrichment distinguishing them clearly from the less enriched Casma rocks (Fig. 5). The low Cr and Ni (<20 ppm) contents of the Calipuy basalts exclude these rocks from being primary mantle-derived melts (Atherton *et al.* 1985*b*).

Tectonic setting

The Calipuy Group lie above and to the east of the Casma volcanic rocks and are separated by a major unconformity marking Albian marginal basin closure and inversion. The time elapsed between the inversion of the marginal basin and the earliest Calipuy basalts is about 48 Ma, during which magmatism switched from volcanic

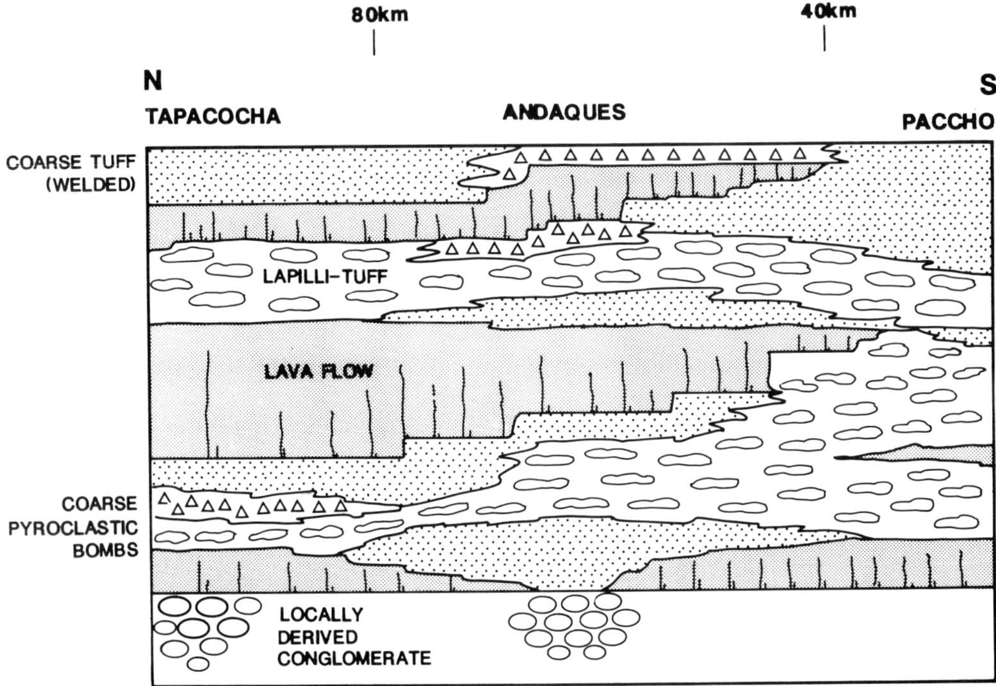

Fig. 7. Schematic facies model for the Calipuy Group based on the sections shown in Fig. 6. The facies relationships are generally dissimilar to those normally associated with active stratovolcanoes as seen in southern Peru and northern Chile (see text for discussion).

to entirely plutonic, culminating with the intrusion of the main part of the Coastal Batholith axially within the marginal basin over a period of 100 to about 50 Ma (Fig. 15 and Atherton 1990).

Myers (1980) considered NNW–SSE dykes, which cut the the lower Calipuy Group, to be feeders for the upper part of the sequence and that lava extrusion occurred through fissure vents and calderas. In his detailed mapping he found no sequence evidence for the stratovolcanoes that characterize the present-day Andean volcanic zones (Thorpe *et al.* 1982). Furthermore, no evidence of major vent systems or magma chambers near the surface have been found and he concluded that extrusion took place on a surface of low relief. Consideration of the facies associations within the Calipuy Group (Fig. 7) supports this model, where epiclastic deposits characteristic of large stratovolcanoes, are generally lacking (Cas & Wright 1987). The Calipuy group is instead characterized by tuffs, agglomerates and breccias, all of which are primary deposits from relatively small scale eruptions.

The setting of the Calipuy volcanic rocks is markedly different from that of the Casma Group. Thus, the latter formed in a deep marine marginal basin apparently isolated from the continent, while the Calipuy volcanic rocks were erupted subaerially through extensional fissure-type vents on a low-relief surface located inland, away from the trench and orientated parallel to the main Andean trend in Peru (Bussell 1975; Atherton *et al.* 1985b). The volcanic magmas passed through the newly cratonized and uplifted continental edge to the east of marginal basin and have a clear continental arc character (Fig. 5). Plumbing systems thus moved east and formed in response to extension in the continental lip.

Incaic orogeny. The age of the oldest basal units of the Calipuy Group (*c.* 53 Ma), which effectively dates the onset of this second major phase of volcanism in central-northern Peru, coincides almost exactly with the start of the Incaic orogeny at about 50 Ma (Megard 1984). The Incaic orogeny has been tentatively related to a major increase in Eocene convergence rate between the Nazca and South American plates, from *c.* 100 to 150 mm a^{-1} (Pardo-Casas & Molnar 1987). It is thus interesting to speculate that the Calipuy Group volcanism is also related, through its initiation at the beginning of the

Table 1. *Representative major (wt%) and trace (ppm) element abundances in the volcanic rocks*

Sample	Casma				Calipuy				Yungay			
	V37	A189	7.65	A201	P5	7.13	V174	V183b	70	89	A142	CB11
SiO_2	48.19	53.06	67.39	72.24	46.20	56.44	66.12	75.04	71.07	71.33	68.88	69.91
TiO_2	0.48	0.94	0.43	0.23	1.08	0.84	0.64	0.19	0.32	0.26	0.29	0.23
Al_2O_3	14.64	18.92	15.23	13.90	20.10	19.69	15.68	13.43	15.81	15.59	16.67	15.73
Fe_2O_3	1.04	10.65	1.93	0.87	7.00	3.50	1.82	1.57	0.89	1.20	1.04	0.84
FeO	7.44	nd	1.97	1.36	2.76	3.53	2.12	0.16	0.52	0.25	0.70	0.51
MnO	0.13	0.17	0.70	0.06	0.14	0.13	0.11	0.30	0.03	0.05	0.30	0.03
MgO	10.33	3.36	1.20	0.40	2.49	2.48	1.06	0.17	0.53	0.21	0.33	0.32
CaO	12.10	8.07	2.94	1.73	9.70	6.83	1.66	0.49	2.27	1.51	2.48	2.36
Na_2O	1.45	2.72	3.72	4.00	4.46	3.19	4.19	2.90	3.44	5.24	4.46	3.33
K_2O	0.22	1.20	3.52	4.37	2.76	1.68	4.46	5.84	3.39	3.24	2.71	4.01
P_2O_5	0.05	0.05	0.07	0.04	0.33	0.19	0.17	0.04	0.08	0.06	0.12	0.10
LOI	3.00	0.46	0.60	0.80	1.61	0.52	1.76	nd	1.40	1.00	1.13	2.09
Total	99.1	99.6	99.06	100	98.63	99.02	99.88	99.86	99.75	99.94	99.11	99.36
Ba	269	575	882	736	406	562	896	942	836	892	796	794
Ce	6	15	nd	53	40	29	71	71	40	39	50	48
Co	43	31	7	10	23	15	9	1	46	42	nd	nd
Cr	373	7	8	146	nd	12	8	3	9	8	nd	nd
Hf*	0.7	nd	6.1	6	7.2	4.1	2	3	4.16	3.95	4.2	4.1
La	1	10	nd	34	31	17	29	29	26	24	26	23
Nd	8	12	nd	32	26	23	17	nd	19	23	19	19
Ni	96	3	nd	12	nd	nd	nd	nd	5.3	5.01	nd	nd
Pb	8	18	13	22	11	13	11	18	18	17	nd	nd
Rb	5	37	111	137	92	40	148	213	156	94	84	146
Sc	40	34	14	2	21	24	nd	nd	2.2	2	nd	nd
Sr	222	401	219	66	151	434	354	71	632	548	662	476
Ta*	0.1	nd	0.8	1	1	0.55	2	1	1.32	1.29	0.62	0.63
Th*	0.6	1.6	11	16	10	6	11	21	12	10.6	8.4	11
U*	0.2	1.1	3.3	4.6	3.3	1.8	2	3	4.83	2.33	2.5	2.3
V	235	274	51	11	130	176	36	9	23	15	nd	nd
Y	14	25	34	48	40	26	41	44	5.1	7	7	7
Zn	79	93	71	46	88	91	101	35	56	48	nd	nd
Zr	26	48	217	215	301	139	261	270	148	137	135	133

All analyses by XRF (Liverpool University except *INAA (Risley).
V37: Basalt, Western Huarmey basin. A189: Basaltic andesite, Western Huarmey basin. 7.65: Dacite, Eastern Huarmey basin. A201: Rhyolite, Eastern Huarmey basin. P5: Basalt, Paccho. 7.13: Basaltic andesite, Tapacocha. V174: Dacite, Fortaleza. V183b: Rhyolite, Fortaleza. 70: Rhyolitic ignimbrite, Callejon de Huaylas. 89: Rhyolitic ignimbrite, Callejon de Huaylas. A142: Dacitic ignimbrite, Callejon de Huaylas. CB11: Dacitic ignimbrite, Callejon de Huaylas.

Incaic orogeny, to a period of crustal extension in the overriding South American plate triggered by a sudden increase in relative plate convergence. Note that shortening during the Incaic orogeny was taken up to the east by the Maranon and sub-Andean thrust and fold belts bounding the eastern margin of the WPT (Fig. 1, Megard 1984), while the region to the west was buttressed and behaved passively (Cobbing 1985). It was this block in which brittle, extensional fissures and dykes acted as conduits for the Calipuy volcanic rocks. During this time there was a decrease in the dip of the slab (Soler & Bonhomme 1990), which may well have caused arching, uplift and extension and volcanism in the overriding plate.

Yungay volcanic rocks: strike-slip basinal fill extrusive rocks

The Yungay volcanic rocks lie above the deep crustal keel of the Andes. They have not been mapped or sampled to the same degree as the Casma and Calipuy volcanic rocks, and for this reason are the most poorly understood in terms of both field geology and chemistry of the three volcanic groups described here. However, they

Table 2. *Selected rare earth element abundances (ppm) of Yungay volcanic rocks*

Sample	70	89	899	CB10	CB11
La	26	24	18	21	23
Ce	40	39	36	33	48
Nd	19	23	24	19	19
Sm	4.3	3.4	nd	2.9	2.7
Eu	0.97	0.87	0.83	0.81	0.73
Gd	nd	nd	nd	2.1	1.4
Tb	0.28	0.29	0.21	0.24	0.21
Yb	0.95	1.04	0.64	0.67	0.69

nd, not determined

are important in that, along with the Cordillera Blanca batholith (Fig. 2), they represent the final magmatic episode in the Andean cycle in central northern Peru. As a result, new data on these rocks are presented here (Tables 1 and 2). They are medium to high-K dacitic to rhyolitic ignimbrites and non-welded tuffs with a relatively narrow SiO_2 range of 68–72 wt%. Characteristic features of these rocks include high Al_2O_3, high Sr (generally >500 ppm) and significantly lower Y contents (by up to a factor of 8) compared with dacites and rhyolites of the Casma and Calipuy volcanic groups (Table 1). The high Sr/Y values are characteristic of high-Al trondhjemite suites (Atherton & Petford 1993). K-Ar and $^{40}Ar/^{39}Ar$ mica cooling ages of 7.0 to 4.65 Ma have been reported for these rocks by Wilson (1975) and Bonnot (1984), while recent total gas ages of K-feldspars give dates of 4.5 to 4.1 Ma (Petford unpublished data).

Tectonic setting

The Yungay ignimbrite is confined entirely to the NW–SE-trending Pliocene Callejon de Huaylas strike-slip basin. The basin lies axially above the thickened crustal keel (50–60 km) of the Andes, inland from the rocks of the Casma and Calipuy. The basin forms a half graben, and is effectively the hanging wall to the Cordillera Blanca fault, with subsidence in the basin related to periods of NE–SW then E–W directed extension (Bonnot *et al.* 1988). The ignimbrite which is at the bottom of the basin is most extensive in the north where it reaches a maximum thickness of 800 m compared to *c.* 150 m in the south (Fig 8; Wilson & Garayar 1967). Facies analysis of the basin fill shows that the ignimbrites are closely associated with fluvial, glacial and lake deposits (Fig. 8), with much of the coarse material comprising the basin fill in the upper part of the sequence derived

from the Cordillera Blanca batholith. The ignimbrite and batholith, and the Callejon de Huaylas basin in which the ignimbrites ponded are all related to transtensional (strike-slip) movements along the Cordillera Blanca fault system, with the fault acting as a conduit for magma ascent and creating space for magma emplacement (Petford & Atherton 1992). Furthermore, the timing of volcanism (and plutonism) can be related directly to extensional pulses occurring between the Quechua 2 and 3 compressional phases of the Late Miocene (Megard 1984; Petford & Atherton 1992). Thus, the Yungay volcanic rocks represent extrusion into an actively subsiding (and extending) intermontane strike-slip basin. The probable driving mechanism behind basin formation was gravity collapse in thickened continental crust (England & Houseman 1988; Deverchere *et al.* 1989; Mercier *et al.* 1991).

Discussion

The plutonic–volcanic association

We consider there to be an intimate relationship between the volcanic and plutonic rocks in northern Peru, but not in the sense of Hamilton & Myers (1967), namely volcanic rocks representing ejecta vented from an associated batholith. Clearly, the pillow basalts and hyaloclastites of the Casma group were not vented from the Coastal Batholith, as it is younger and demonstrably intrudes and metamorphoses them (Atherton & Brenchley 1972). The batholith is related to the basin fill volcanic rocks by partial melting (Atherton 1990).

More problematical has been the relationship between the Calipuy volcanic rocks and the high level ring complexes exposed in the upper units of the batholith, where many workers have proposed that the Calipuy rocks originated from the ring complexes (Knox 1971; Bussell 1975; Pitcher 1978). The reasons why this association is unlikely have been set out by Atherton *el al.* (1979) and Atherton *et al.* (1985a) but are mainly based on age considerations, since it is now known the whole of the Calipuy Group is simply too young to have vented from the ring complexes. Thus, the youngest pluton of the Huaura centered complex (indeed in all the complexes) is Canas with a K–Ar age of 58 Ma, while the maximum age for a basalt flow from the Calipuy Group at Tapacocha is 53 Ma.

The relationship between the Yungay volcanic rocks and the Cordillera Blanca batholith may appear more consistent with a venting-type

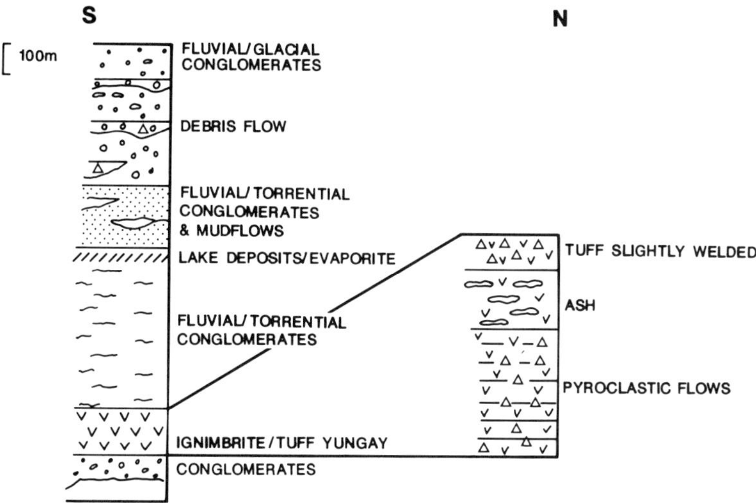

Fig. 8. Basin fill facies of the Callejon de Huaylas strike-slip basin. The maximum thickness of the Yungay ignimbrite occurs in the north of the basin, with the sequence thinning southwards. Preliminary K–Ar age dating suggests that the volcanic rocks are also oldest in the north (*c.* 7 Ma) but young southwards, where $^{40}Ar/^{39}Ar$ ages of 5.4 to 4.1 Ma have been reported.

origin, with the ages of both rock types overlapping. K/Ar and $^{40}Ar/^{39}Ar$ mineral ages for the ignimbrites vary from 7.6 to 4.1 Ma (Cobbing *et al.* 1981; Bonnot 1984; Petford unpublished data). K/Ar $^{40}Ar/^{39}A$ mica and U/Pb zircon dates on the batholith vary from 13.6–3 Ma (Cobbing *et al.* 1981; Mukasa 1984; Petford & Atherton 1992). However, estimated emplacement pressures of *c.* 3.5 kbar for the batholith magmas (Petford & Atherton unpublished data) are inconsistent with subvolcanic intrusion. Furthermore, rare earth modelling of the Yungay volcanic rocks (see later section) suggests that they are derived from a source different than the Cordillera Blanca magmas. Clearly, in this case the association between the volcanic and plutonic rocks is a structural one in the sense that magmatism has been controlled fundamentally by the location of pre-existing crustal discontinuities determining the plumbing systems (Cobbing *et al.* 1981). In both instances the magmas appear to have originated from a deep crustal source, the difference being that, whereas the Cordillera Blanca magmas were emplaced into an active shear zone and crystallized as a batholith, the Yungay magmas were able to vent at the surface.

Chemical variation between the volcanic rocks

Although the volcanic rocks of the west Peruvian trough are calc-alkaline in the sense of

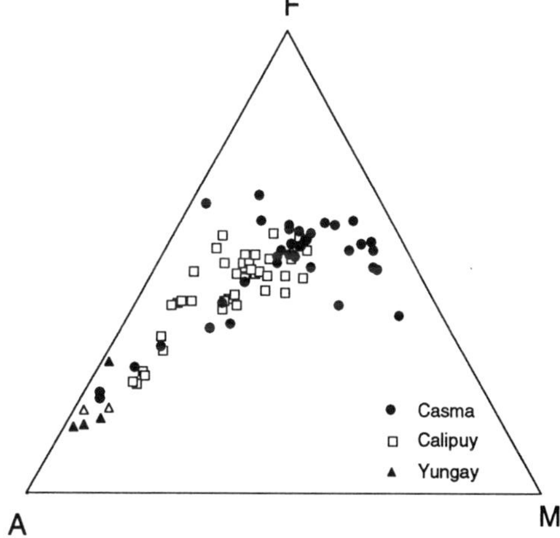

Fig. 9. AFM plot for the Casma, Calipuy and Yungay volcanic rocks showing a calc-alkaline trend typical of arc rocks.

Jakes & White (1972) and Dostal *et al.* (1977; see also Fig. 9), the preceding tectonic synthesis has revealed a pattern in which volcanism is closely associated with discrete (but episodic) periods and styles of extension. This being so, it is useful to compare the general chemistry of the volcanic rocks in relation to their age and position within the Andean belt.

A much-discussed tectonomagmatic relationship in active margin volcanism is the apparent

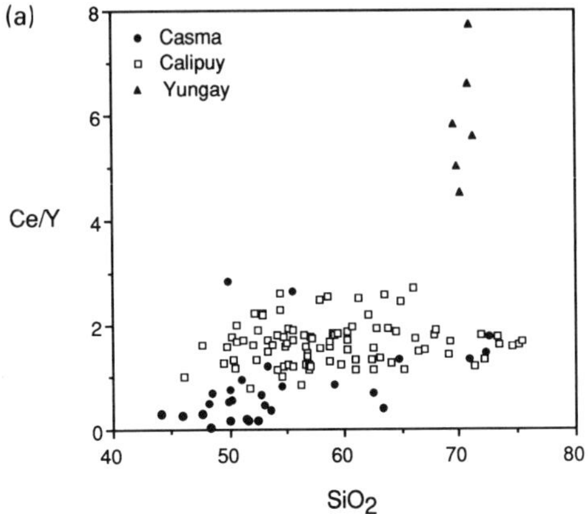

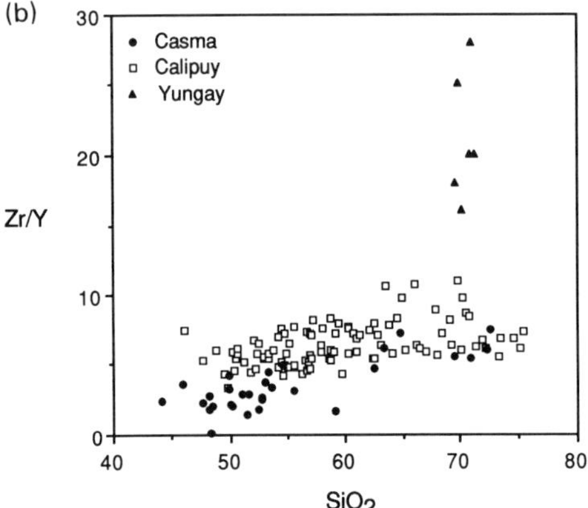

Fig. 10. (**a**) Plot of Ce/Y v. SiO$_2$ for the volcanic rocks of the WPT showing a systematic increase in Ce/Y with decreasing age and distance from the present-day trench. The low Y (and Yb) contents of the Yungay volcanic rocks separate them clearly from the rocks of the Casma and Calipuy (see also Table 1). (**b**) Zr/Y v. SiO$_2$ plot for the same rocks shown in (**a**). Note again the distinctive grouping of the Yungay volcanic rocks.

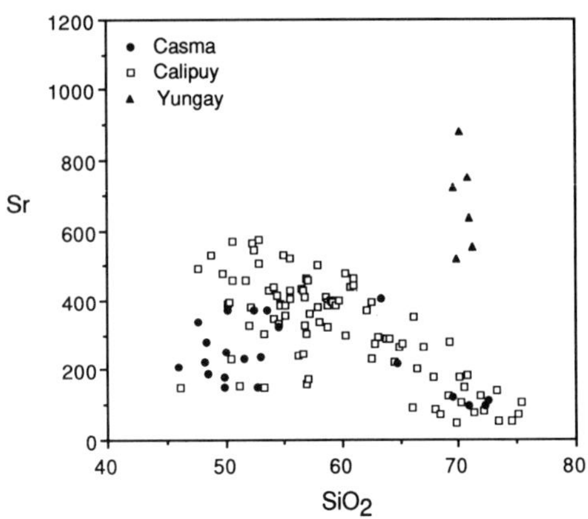

Fig. 11. Plot of Sr v. SiO$_2$ for the volcanic rocks of the WPT. Sr contents for each group increase generally in the sequence Casma-Calipuy-Yungay (i.e. with increasing distance and decreasing age from the present-day trench).

systematic variation between chemistry, isotopes and distance from the trench (Dostal *et al.* 1977; Rogers & Hawkesworth 1989). Given that the position of the Peru–Chile trench has remained more or less static relative to the Peruvian continental margin since the Cretaceous period (James 1978), the volcanic rocks of the WPT may be expected so show such relationships. In the Andes between 26°S and 29°S, Dostal *et al.* (1977) found an eastward increase in K$_2$O, Rb, Ba, Sr and Zr at a given SiO$_2$ value. Here we see similar relations in rocks <55 wt% SiO$_2$ of the Casma and Calipuy groups, i.e. all four elements generally are higher in the more easterly Calipuy group (e.g. Fig. 4, K$_2$O v. SiO$_2$). However, in rocks with SiO$_2$ > 55% SiO$_2$, there is no corresponding increase in these elements in outcrops further east. In fact the Zr content in the Yungay volcanic rocks is lower and Ba and Rb values lie within the fields of the more westerly Casma and Calipuy groups.

In contrast, a plot of Ce/Y against SiO$_2$ (Fig. 10a) reveals clear differences between each of the three groups, with the highest Ce/Y ratios found in the Yungay volcanic rocks. There is also a clear increase in Ce/Y in both space and time. The average Ce/Y of the Casma volcanic rocks is 0.57. This ratio increases inland (and with decreasing age), away from the trench, from 1.61 in the Calipuy rocks to a maximum of 5.47 in the Yungay rocks. This is also seen in the Zr/Y ratios, where the low Y contents of the Yungay rocks account for their distinctive grouping (Fig. 10b). As Zr and Y are apparently not fractionated by subduction processes this distinctive character must relate to different source compositions or melting conditions. We suggest that melting conditions were probably more important and that the presence of garnet in the mantle source and small-degree melting could have produced the Zr/Y ratios characteristic of the Yungay magmas (see also Nicholson & Latin 1992).

A similar trend is seen in strontium contents. Thus, the Casma Group have the lowest mean Sr value at a given SiO$_2$ whereas mean Sr values in the Calipuy and Yungay rocks increase respectively (Fig. 11). The lack of a clear trend in the

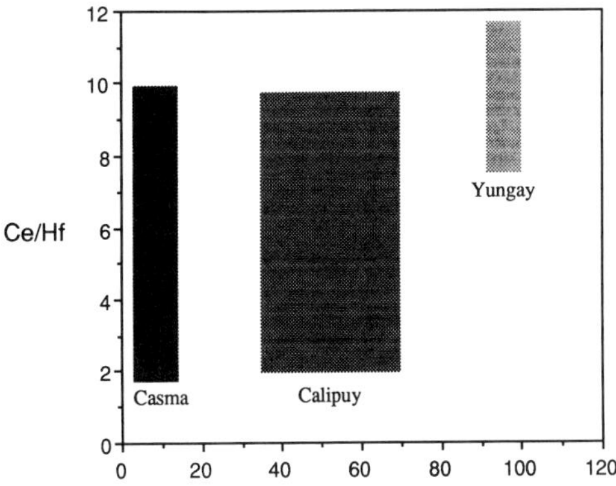

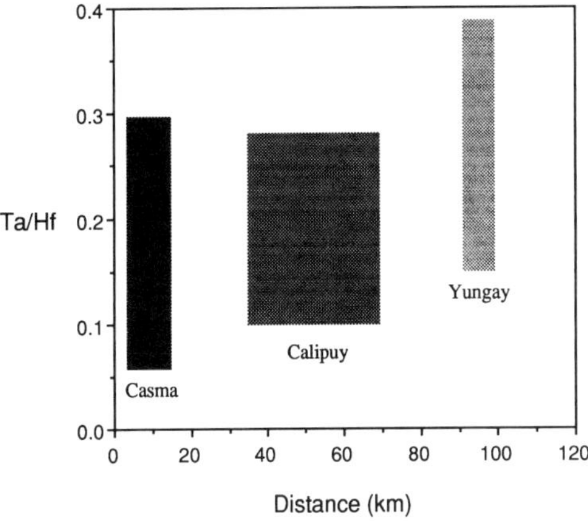

Fig. 12. Plot showing the range in (**a**) Ce/Hf and (**b**) Ta/Hf for the Casma-Calipuy-Yungay volcanic rocks in relation to distance from the present-day continental margin. Both ratios show a crude increase in localities further inland, with highest values in the Yungay rocks. (cf. Ce/Y and Zr/Y; Fig. 10).

Casma rocks relates to the fact that only the more acid rocks ($SiO_2 > 62$ wt%) are related by fractionation (Atherton *et al.* 1985*b*; Atherton & Webb 1989). Similar arguments have been put forward for the Calipuy trend (see Fig. 11). In contrast the Yungay rocks were derived from individual small partial melts and show no evidence of large-scale high-level fractionation. This is supported by the wide spread in Sr (and Ce/Y, Zr/Y, Ce/Hf and Ta/Hf) in the range SiO_2 range of 70–73 wt%. The easterly increasing, relatively high Ce/Hf and Ta/Hf ratios (Fig. 12) are consistent with an increasing contribution from enriched subcontinental mantle (Pearce 1982). Similar trends in Ce, Sr and Ta in magmatic rocks from from a traverse in Chile at 22°S were noted by Rogers and Hawkesworth

(1989), which they attributed to lateral changes in the composition of mantle lithosphere.

The generally increasing Ce/Y, Zr/Y, Ce/Hf, Ta/Hf and Sr contents inland must reflect changes either in source composition, or variations in melting conditions. Fractionation is not a viable mechanism for the reasons given above. The idea of changing source characteristics with time is discussed in relation to the rare earth elements in the following section.

Rare earth elements

The REE abundances in samples from all three volcanic groups are summarized in Fig. 13. The lateral variation of the Casma rocks across the marginal basin are clearly defined, with the western Casma basalts showing generally gently sloping chondrite normalized patterns ($Ce_N/Yb_N = 2$–3) while the acid rocks of similar age in the east are highly fractionated, with $Ce_N/Yb_N \approx 6$ and large negative Eu anomalies ($Eu/Eu^* \approx 0.5$, Atherton *et al.* 1985*a*). In contrast, the basic Calipuy samples have much higher total REE than their Casma counterparts, and show a marked LREE enrichment ($Ce_N/Yb_N \approx 6$). The acid Calipuy rocks are highly fractionated, with very similar REE characteristics to the acid

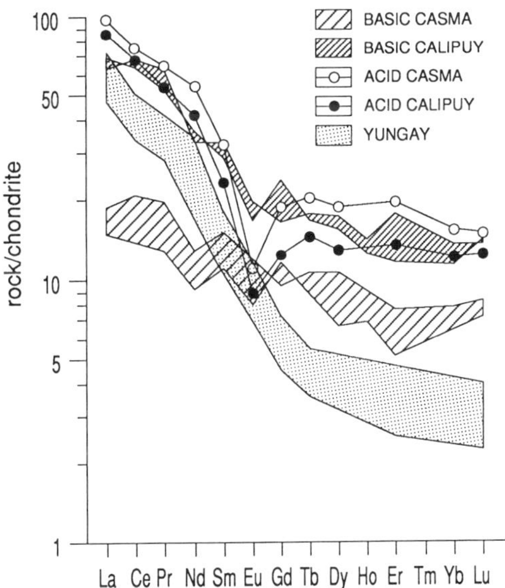

Fig. 13. REE abundances in samples of the Casma, Calipuy and Yungay volcanic rocks. Note the similarity between the acid Casma and Calipuy. Although similar in LREE abundances, the Yungay volcanics show a depletion in HREE. The generally low La/Yb$_{(N)}$ and Y contents in the Yungay volcanics implies a garnetiferous source for these magmas.

eastern Casma rocks, although REE is lower in the latter. Arguably the most striking comparative feature of Fig. 13 is the large HREE depletion seen in the Yungay volcanic rocks ($Ce_N/Yb_N = 10.3$–21.2). Furthermore, although covering a range in SiO_2 similar to the acid facies of the Casma and Calipuy groups, these rocks are less enriched in the LREE and do not show a corresponding negative Eu anomaly.

In terms of magma generation, Atherton & Webb (1989) and Atherton (1990) concluded that the western Casma basalts formed as separate magma batches by c. 5–36% shallow melting of hydrated spinel-plagioclase lherzolite mantle, with the eastern rocks modified by extensive plagioclase and clinopyroxene fractionation. The basic Calipuy rocks may be generated from c. 3% melting of a spinel-clinopyroxene-orthopyroxene-olivine lherzolite source, with subsequent high-level plagioclase, clinopyroxene and Ti-oxide phase fractionation to produce the more evolved dacitic and rhyolitic compositions (Atherton *et al.* 1985*b*, p. 281).

In contrast, the fractionated HREE pattern seen in the Yungay volcanic rocks implies a garnetiferous source. Steep REE patterns ($La_N/Yb_N \approx 30$–70) also characterize the leucocratic rocks of the Cordillera Blanca batholith, and while we do not believe that the volcanic rocks are the vented products of the batholith, the close temporal and spatial relationship between both rock types suggests that these lineament-controlled magmas share a broadly similar deep source (Atherton & Petford 1993). However, although the Yungay magmas can be modelled as 15–18% (batch) partial melts of underplated basaltic protolith similar to that used to model the generation of the Cordillera Blanca leucogranites, their REE patterns are less depleted in the HREE. As a consequence, the Yungay source residue requires much less garnet (c. 11–12%) compared to 25–35% for the leucogranites (Petford & Atherton unpublished data). The results of the REE modelling are shown in Fig. 14; k_ds used are listed in Table 3.

To summarize, the variation in Ce/Y, Zr/Y, Ce/Hf, Ta/Hf, REE and Sr concentrations seen in the Casma–Calipuy–Yungay volcanic rocks suggests that eastwardly migrating Cretaceous–Tertiary magmatism within the WPT was accompanied by a change in the mineralogy of the source region, from essentially garnet-free spinel facies mantle (Casma and Calipuy) to a garnet-bearing, plagioclase-poor source for the Yungay magmas. Such a source for the Yungay volcanic rocks is consistent with REE modelling of the Cordillera Blanca Batholith source involving partial melting of a basaltic garnet-

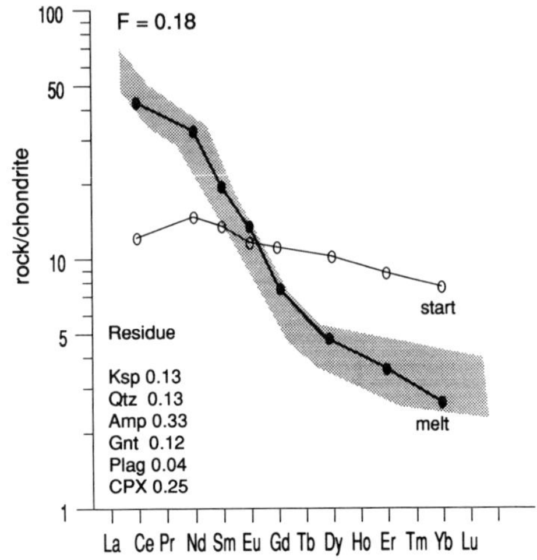

Fig. 14. Summary of REE modelling of the Yungay rocks. 18% melting of a basaltic lower crustal protolith, produces a melt composition that lies inside the envelope (shaded) defined by the Yungay rocks, with a residue composed of amphibole, clinopyroxene, garnet, K-feldspar, quartz and minor plagioclase. Kd values used are shown in Table 3.

amphibolite-bearing crustal underplate (Atherton & Petford 1993). Rapid underplating and crustal thickening beneath the western Cordillera of central Peru during the Miocene (Kono *et al.* 1989) provided both a source for the volcanic rocks and batholith, and a mechanism for crustal extension through uplift and subsequent isostatic adjustment.

Back arc basins of the western Pacific – an Andean analogue?

At this stage it may be worth considering if the Huarmey basin, and indeed other Albian deposystems such as the Canete basin and associated volcanic rocks of the Quilmana Group (Atherton & Aguirre 1992) that characterize much of the Cretaceous western margin of Peru, were ever true back-arc basins in the sense of Karig (1971). With the exception of Bransfield Strait (Weaver *et al.* 1979), the majority of present-day back-arc basins are located in the western Pacific. In terms of gross tectonic setting, the Huarmey basin and Casma volcanic rocks may be compared to ophiolitic-type rocks from the Jurassic–Cretaceous Rocas Verdes basin in southern Chile (Saunders *et al.* 1979) and the Larsen Harbour complex of South Georgia (Alabaster & Storey 1990). Indeed, the Rocas Verdes basin has been compared to the

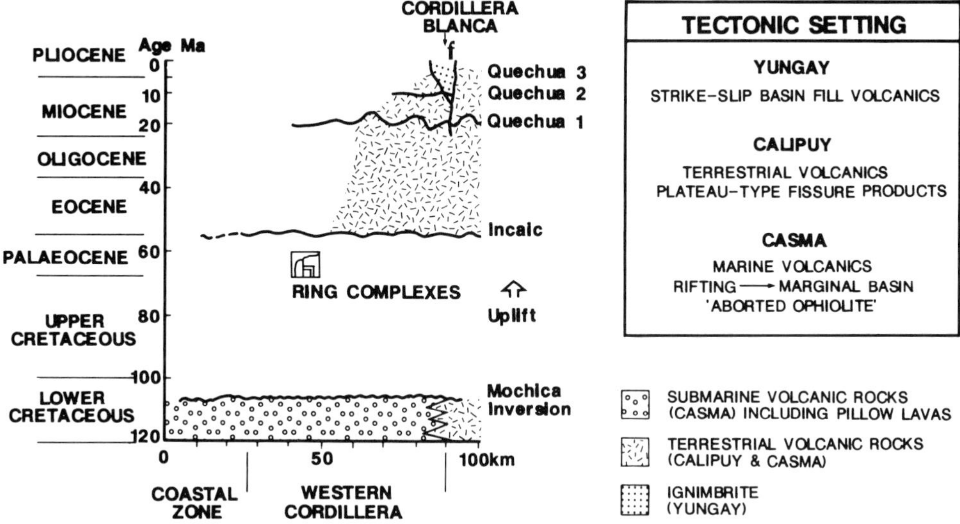

Fig. 15. Summary of the space–time relationships of volcanism, plutonism and tectonism in the Western Cordillera north of Lima. Note the eastwards migration of magmatism with time, into the thickened Cordilleran crust.

Gulf of California (Dalziel 1981) and such an interpretation may hold for many of the elongate Jurassic–Cretaceous marine basins along the length of the Andes (Alabaster & Storey 1990; Atherton & Aguirre 1992). If correct, then back-arc basins of modern western Pacific type may never have existed along the Andean continental margin of South America.

Summary

Three discrete phases of volcanism related to crustal extension characterize the western Cordillera of central northern Peru and are summarised in Fig. 15. These are: (1) The Casma marine volcanic rocks of the Huarmey marginal basin, formed during a major Albian rifting episode; (2) the terrestrial, plateau-type volcanic rocks of the Calipuy Group, associated with the Eocene Incaic orogeny; and (3) the acid volcanic rocks of the Miocene Yungay Formation, extruded into an extensional strike-slip basin situated in the high Andean Cordillera. Migration of magmatism inland was accompanied by changes in rock chemistry, with Ce/Y, Zr/Y, Sr, and La_N/Yb_N (and to a lesser extent Ce/Hf and Ta/Hf) increasing in the volcanic rocks with both distance from the trench and decreasing age.

The Casma basin-fill volcanic rocks show a clear compositional asymmetry, with a western facies characterized by basalts and andesites, and a more evolved eastern facies composed mainly of dacites and rhyolites. Facies relationships are reminiscent of slow spreading MOR systems. Similar spatial variations are absent in Calipuy and Yungay rocks.

The Calipuy Group belongs to the classic Andean calc-alkaline basalt–andesite–dacite–rhyolite association. These magmas were extruded through extensional fissures orientated parallel to the main Andean trend, but located inland from the older Casma volcanic rocks. Consequently, these rocks passed through the continental edge to the east of the basin which might explain their medium to high-K character. Facies analysis suggests extrusion onto a surface of low topographical relief and not from strato-volcanoes that characterize much of the present-day Andean volcanic zones.

The Yungay volcanic rocks were erupted within an actively subsiding extensional strike-slip basin situated above magmatically thickened continental crust. Extrusion was controlled by a deeply-penetrating crustal lineament that tapped a source region in newly underplated basaltic (garnetiferous) lower crust. Here, upper

Table 3. *Partition coefficients (K_d values) used in trace element modelling calculations for Yungay magmas (from Martin 1987)*

	CPX	Garnet	Amphibole	Plagioclase	Quartz	K-feldspar	Magnetite	Ilmenite
Ce	0.2	0.08	0.3	0.11	0.014	0.04	0.26	0.006
Nd	0.4	0.2	0.8	0.07	0.01	0.03	0.3	0.008
Sm	0.6	1	1.1	0.05	0.01	0.02	0.35	0.01
Eu	0.6	0.98	1.3	1.3	0.06	0.5	0.26	0.007
Gd	0.7	3.8	1.8	0.04	0.01	0.01	0.32	0.01
Dy	0.7	11	2	0.03	0.01	0.006	0.28	0.02
Yb	0.6	21	1.7	0.02	0.01	0.01	0.18	0.07

crustal extension and basin formation may be due to compensation effects at high topography, itself a likely consequence of magmatic underplating.

The variations in magma chemistry inboard relate to changes in the mantle source mineralogy from garnet-free, plagioclase-bearing for the Casma Group, spinel-bearing for the Calipuy rocks, to garnetiferous, essentially plagioclase-free mantle-derived underplate for the Yungay rocks. Whereas some high-level fractionation appears to have been involved in further modifying magma compositions of the Casma and Calipuy rocks, chemical variation in the Yungay rocks is due to different amounts of partial melting.

In summary, the volcanic and tectonic evoluton of the Andean margin of northern Peru over the last 200 Ma records a complex interplay between magmatism and continental extension. The punctuated nature of the volcanism related to specific periods of extension raises the important question of how much calc-alkaline magmatism at continental arcs is extensional, and challenges the precise role played by subduction in controlling the spatial and temporal occurrence of volcanic rocks at continental arcs.

We thank J. Smellie and two anonymous referees for helpful comments.

References

AGUIRRE, L. & OFFLER, R. 1985. Burial metamorphism in the Western Peruvian trough: its relation to Andean magmatism and tectonics. *In*: PITCHER, W.S., ATHERTON, M.P., COBBING, E.J. & BECKINSALE, R.D. (eds) *Magmatism at a plate edge: the Peruvian Andes*. Blackie Halsted, Glasgow, 59–71.

ALABASTER, T. & STOREY, B.C. 1990. Modified Gulf of California model for South Georgia, north Scotia Ridge, and implications for the Rocas Verdes back-arc basin, southern Andes. *Geology*, **18**, 497–500.

ATHERTON, M.P. 1990. The Coastal Batholith of Peru:the product of rapid recycling of new crust formed within rifted continental margin. *Geological Journal*, **25**, 337–349.

—— & AGUIRRE, L. 1992. Thermal and geotectonic setting of Cretaceous volcanic rocks near Ica, Peru, in relation to Andean crustal thinning. *Journal of South American Earth Sciences*, **5**, 47–69.

—— & BRENCHLEY, P.J. 1972. A preliminary study of the structure, stratigraphy and metamorphism of some contact rocks of the western Andes, near the Quebrada Venado Muerto, Peru. *Geological Journal*, **8**, 161–178.

—— & PETFORD. N. 1993. Generation of sodium-rich magmas from newly underplated basaltic crust. *Nature*, **362**, 144–146.

—— & WEBB, S. 1989. Volcanic facies, structure and geochemistry of the marginal basin rocks of central Peru. *Journal of South American Earth Sciences*, **2**, 241–261.

——, PITCHER, W.S. & WARDEN, V. 1983. The Mesozoic marginal basin of Central Peru. *Nature*, **305**, 303–306.

——, WARDEN, V. & SANDSERSON, L.M. 1985a. The Mesozoic marginal basin of central Peru: a geochemical study of within-plate edge volcanism. *In*: PITCHER, W.S., ATHERTON, M.P., COBBING, E.J. & BECKINSALE, R.D. (eds) *Magmatism at a plate edge: the Peruvian Andes*. Blackie Halsted, Glasgow, 47–58.

——, McCOURT, W.J., SANDERSON, L.M. & TAYLOR, W.P. 1979. The geochemical character of the segmented Peruvian Coastal Batholith and associated volcanics. *In*. ATHERTON, M.P. & TARNEY, J. (eds) *Origin of granite batholiths: geochemical evidence*. Shiva, Cheshire, 45–64.

——, SANDERSON, L.M., WARDEN, V. & McCOURT, W.J. 1985b. The volcanic cover: chemical composition and the origin of the magmas of the Calipuy Group. *In*: PITCHER, W.S., ATHERTON, M.P., COBBING, E.J. & BECKINSALE, R.D. (eds) *Magmatism at a plate edge: the Peruvian Andes*. Blackie Halsted, Glasgow, 273–284.

AUBOUIN, J., BORELLO, A.V., CECIONI, G. *et al*. 1973. Esquisse paleogeographique et structurale des Andes. *Revue Geographie Physique et Geologie Dynamique, Paris*, **15**, 11–72.

BARAZANGI, M. & ISACKS, B.L. 1976. Spatial distribution of earthquakes and subduction of the Nazca plate beneath South America. *Geology*, **4**, 686–692.

—— & ISACKS, B.L. 1979. Subduction of the Nazca plate beneath Peru: evidence from spatial distribution of earthquakes. *Geophysical Journal of the Royal Astronomical Society*, **57**, 537–555.

BONNOT, D. 1984. *Neotectonique et tectonique active de la Cordillera Blanca et Callejon de Huaylas (Andes nord-Peruviennes)*. PhD thesis, University Paris-Sud, France.

——, SEBRIER, M. & MERCIER, J. 1988. Evolution geodynamique Plioquaternaire du bassin intra-Cordilleran du Callejon de Huaylas et de la Cordillera Blanca, Perou. *Geodynamique*, **3**, 57–83.

BUSSELL, M.A. 1975. *The structural evolution of the Coastal Batholith in the Provinces of Ancash and Lima, central Peru*. PhD thesis, University of Liverpool, England.

CAS, R.A.F & WRIGHT, J.V. 1987. *Volcanic successions: modern and ancient*. Allen and Unwin, London.

COBBING, E.J. 1978. The Andean geosyncline in Peru and its distinction from Alpine geosynclines. *Journal of the Geological Society, London*, **135**, 207–218.

COBBING, E.J. 1985. The tectonic setting of the Peruvian Andes. *In*: PITCHER, W.S., ATHERTON, M.P., COBBING, E.J. & BECKINSALE, R.D. (eds) *Magmatism at a plate edge: the Peruvian Andes*. Blackie Halsted, Glasgow, 167–176.

—— & PITCHER, W.S. 1972. The Coastal Batholith of central Peru. *Journal of the Geological Society, London*, **128**, 421–460.

——, ——, WILSON, J.J. *et al.* 1981. The geology of the western Cordillera of Northern Peru. Overseas *Memior, Institute of Geological Sciences, London*.

DEVERCHERE, J., DORBATH, C. & DORBATH, L. 1989. Extension related to a high topography: results from a microseismic survey in the Andes of Peru and tectonic implications. *Geophysical Journal International*, **98**, 281–292.

DICKINSON, W.R. 1975. Potash-depth (K–h) relations in continental margin and intra-ocean magmatic arcs. *Geology*, **3**, 53–56.

DALZIEL, I.W.D. 1981. Back-arc extension in the southern Andes, a review and critical reappraisal. *Philosophical Transactions of the Royal Society, London*, **A300**, 319–335.

DOSTAL, J., ZENTILLI, M., CAELLES, J.C & CLARKE, A.H. 1977. Geochemistry and origin of volcanic rocks of the Andes (26°–28°S). *Contributions to Mineralogy and Petrology*, **63**, 113–128.

ENGLAND, P.C. & HOUSEMAN, G.A. 1988. Extension during continental convergence with application to the Tibetan Plateau. *Journal of Geophysical Research*, **94**, 17561–17579.

FARRAR, E. & NOBLE, D.C. 1976. Timing of Late Tertiary deformation in the Andes of Peru. Bull. *Geological Society of America Bulletin*, **87**, 1247–1250.

HAMILTON, W. & MYERS, W.B. 1967. The nature of batholiths. *Professional Papers of the US Geological Survey*, **554C**, 30 pp.

JAMES, D.E. 1971. Andean crustal and upper mantle structure. *Journal of Geophysical Research*, **6**, 3246–3271.

—— 1978. Subduction of the Nazca plate beneath central Peru. *Geology*, **6**, 174–178.

JAKES, P & WHITE, A.J.R. 1972. Major and trace element abundances in volcanic rocks of orogenic areas. *Geological Society of America Bulletin*, **83**, 29–40.

KARIG, D.E. 1971. Origin and development of marginal basins in the western Pacific. *Journal of Geophysical Research*, **76**, 2542–2561.

KONO, M., FAUKO, Y. & YAMAMOTO, A. 1989. Mountain building in the central Andes. *Journal of Geophysical Research*, **94**, 3891–3905.

KNOX, G.J. 1971. *The structure and emplacement of the Rio Fortaleza centred acid ring complex, Ancash, Peru*. PhD thesis, University of Liverpool, UK.

MEGARD, F. 1984. The Andean orogenic period and its major structures in central northern Peru. *Journal of the Geological Society, London*, **141**, 893–900.

MARTIN, H. 1987. Petrogenesis of Archean trondhjemites, tonalites and granodiorites from eastern Finland: major and trace element geochemistry. *Journal of Petrology*. **28**, 921–953.

MERCIER, J.L., CAREY-GAILHARDIS, E. & SEBRIER, M. 1991. Palaeostress determinations from fault kinematics: application to neotectonics of the Himalayas-Tibet and the Central Andes. *In*: WHITMARSH, R.B., BOTT, M.H.P., FAIRHEAD, J.D. & KUSZNIR, N.J. (eds) Tectonic Stress in the Lithosphere, *The Royal Society, London*, 41–52.

MUKASA, B.S. 1984. *Compatative Pb isotope systematics in batholithic rocks from the Coastal, San Nicholas and Cordillera Blanca Batholiths, Peru*. PhD thesis, University of California, USA.

MYERS, J.S. 1975. Vertical crustal movements of the Andes in Peru. *Nature*, **254**, 672–674.

MYERS, J.S. 1980. *Geologica de los cuadrangulos de Huarmey y Huayllapampa*. Boletin Carta Geologica Nacional Instituto Geologico Minero y Metalurgico, Energia y Minas, Peru. **33**.

NICHOLSON, H. & LATIN, D. 1992. Olivine tholeiites from Krafla, Iceland: evidence for variations in melt fraction within a plume. *Journal of Petrology*, **33**, 1105–1124.

PARDO-CASAS & MOLNAR, P. 1987. Relative motion of the Nazca (Farallon) and South American plates since Late Cretaceous time. *Tectonics*, **6**, 233–248.

PEARCE, J.A. 1982. Trace element characteristics of lavas from destructive plate boundaries. *In*: THORPE, R.S. (ed.), *Andesites, orogenic andesites and related rocks*, Wiley, New York, 525–548.

PETFORD, N. & ATHERTON, M.P. 1992. Granitoid emplacement and deformation along a major crustal lineament: the Cordillera Blanca, Peru. *Tectonophysics*, **205**, 171–185.

PITCHER, W.S. 1978. The anatomy of a batholith. *Journal of the Geological Society, London*, **135**, 157–182.

ROGERS, G. & HAWKESWORTH, C.J. 1989. A geochemical traverse across the north Chilean Andes: evidence for crust generation from the mantle wedge. *Earth and Planetary Science Letters*. **91**, 271–285.

SAUNDERS, A.D., TARNEY, J., STERN, C.R. &

DALZIEL, I.W.D. 1979. Geochemistry of Mesozoic marginal basin floor igneous rocks from southern Chile. *Geological Society of America Bulletin*, **90**, 237–258.

SOLER, P. & BONHOMME, M.G. 1990. Relation of magmatic activity to plate dynamics in central Peru from the Late Cretaceous to Present. *In*: KAY, S.M. & RAPELA, C.W. (eds) *Plutonism from Antarctica to Alaska*. Geological Society of America Special Papers, **241**, 173–191.

STERN, C.R. 1980. Geochemistry of Chilean ophiolites: evidence for the compositional evolution of the mantle source of back-arc basin basalts. *Journal of Geophysical Research*, **85**, 955–966.

THORPE, R.S. & FRANCIS, P.W. 1979. Variations in Andean andesite compositions and their petrogenetic significance. *Tectonophysics*, **75**, 53–70.

THORPE, R.S., FRANCIS, P.W., HAMILL, M. & BAKER, M.C.W. 1982. The An'des. *In*: THORPE, R.S. (ed.) *Andesites: orogenic andesites and related rocks*. Wiley, New York, 187–205.

WEAVER, S.D., SAUNDERS, A.D., PANKHURST, R.J. & TARNEY, J. 1979. A geochemical study of magmatism associated with the initial stages of back-arc spreading: the Quarternary volcanics of Bransfield Strait, from South Shetland Islands. *Contributions to Mineralogy and Petrology*, **68**, 151–169.

WEBB, S.E. 1976. The volcanic envelope of the Coastal Batholith in Lima and Ancash, Peru. PhD thesis, University of Liverpool, England.

WILSON, J.J.1963. Cretaceous stratigraphy of the Central Andes of Peru. *Bulletin of the American Association of Petroleum Geologists*, **47**, 1–34.

WILSON, J.J. & GARAYAR, J. 1967. Geologica de los cuadranglos de Mollebamba, Tayabamba, Huaylas, Pomabamba, Carhuaz Y Huari. *Boletin Servicio de Geologia y Mineria, Peru*, **16**, 95.

WILSON. P.A. 1975. *K–Ar age studies in Peru with special reference to the emplacement of the Coastal Batholith*. PhD thesis, University of Liverpool, England.

Diverse shoshonite magma series in the Kamchatka Arc: relationships between intra-arc extension and composition of alkaline magmas

PAVEL KEPEZHINSKAS

Department of Oceanic Lithosphere, Institute of Lithosphere, Russian Academy of Sciences, Moscow 109180, Russia
Present address: Department of Geology, University of South Florida, Tampa, Florida 33620–5200, USA

Abstract: High-potassium magmas were erupted in the western and eastern volcanic zones of the northern segment of the Kamchatka arc (Russia) during late Eocene to Pliocene time. Shoshonite suites of similar age show across-arc trends in major and trace element chemistry suggesting their derivation from different mantle sources. Low-Ti shoshonites from the western (rear-arc) volcanic zone are thought to have been generated in a refractory mantle wedge affected by the addition of an LILE- and LREE-enriched component possibly identified as the slab-derived hydrous fluid. High-Ti shoshonites from the eastern (frontal arc) volcanic zone were probably derived through partial melting of a more fertile source modified by the addition of a high-Na component as suggested by major and trace element systematics and radiogenic isotope relationships. This component is compositionally similar to a felsic trondhjemitic melt generated during partial melting of the hot downgoing slab in the amphibolite facies. The volcanism in the northeastern zone of the Kamchatka arc is associated with the subduction of hot oceanic lithosphere of the Komandor Basin which was generated during two stages of extension (late Eocene–Oligocene and late Miocene–Pliocene). Slab-derived melts (adakites) were erupted along with Neogene high-Ti shoshonites in the eastern volcanic zone of the northern Kamchatka arc.

Two types of shoshonites can be found in volcanic arcs. Low-Ti shoshonites are related to the temporal evolution of the subduction zone and occur at the latest stages of its development following predicted depth–K relationships. High-Ti shoshonites are associated with melting of a slab-modified mantle during the earliest stages of intra-arc rifting followed by the eruption of high-Na, Nb-enriched arc basalts. Nb-enriched arc basalts and high-Ti shoshonites are spatially and temporally associated with slab melts (adakites) in the northern segment of the Kamchatka arc.

The close genetic link between the potassium content of arc lavas, distance from the trench and maturity of the volcanic arc is a widely recognized phenomenon of subduction zone systems (Gill 1981). Traditionally, shoshonites are regarded as late-stage magmatic products of mature volcanic arcs (Morrison 1980). Their occurrence is commonly explained by the potential increase in crustal contamination as well as polybaric fractionation effects resulting from progressive thickening of the crust in continental volcanic arcs (Gill 1981; Meen 1989).

However, several exceptions have been recently reported from the Mariana (Stern *et al.* 1988) and the Kamchatka (Kepezhinskas *et al.* 1990) arcs. Derivation of these high-K volcanic suites cannot be easily explained by simple models of crustal contamination or greater depths of primary melt fractionation. An extensional setting has been suggested for the shoshonite suites from the northeastern volcanic zone of the Kamchatka arc on the basis of field mapping and major and trace element geochemistry (Kepezhinskas *et al.* 1988). These lavas are chemically distinct from the high-K lavas erupted within northwestern, rear-arc volcanic centres of the same age. High-K volcanic rocks in the northern Kamchatka arc were erupted through the same crustal terranes, which limits the possible extent of crustal input affecting the composition of the arc shoshonite magmas. Therefore, the northern segment of the Kamchatka arc provides an opportunity to investigate the additional influence of extensional processes at consuming plate margins on the chemistry of erupted magmatic rocks.

This paper presents a comparative geochemical study of late Eocene to Pliocene shoshonites from the western and eastern volcanic zones of the northern Kamchatka arc. The aims of the

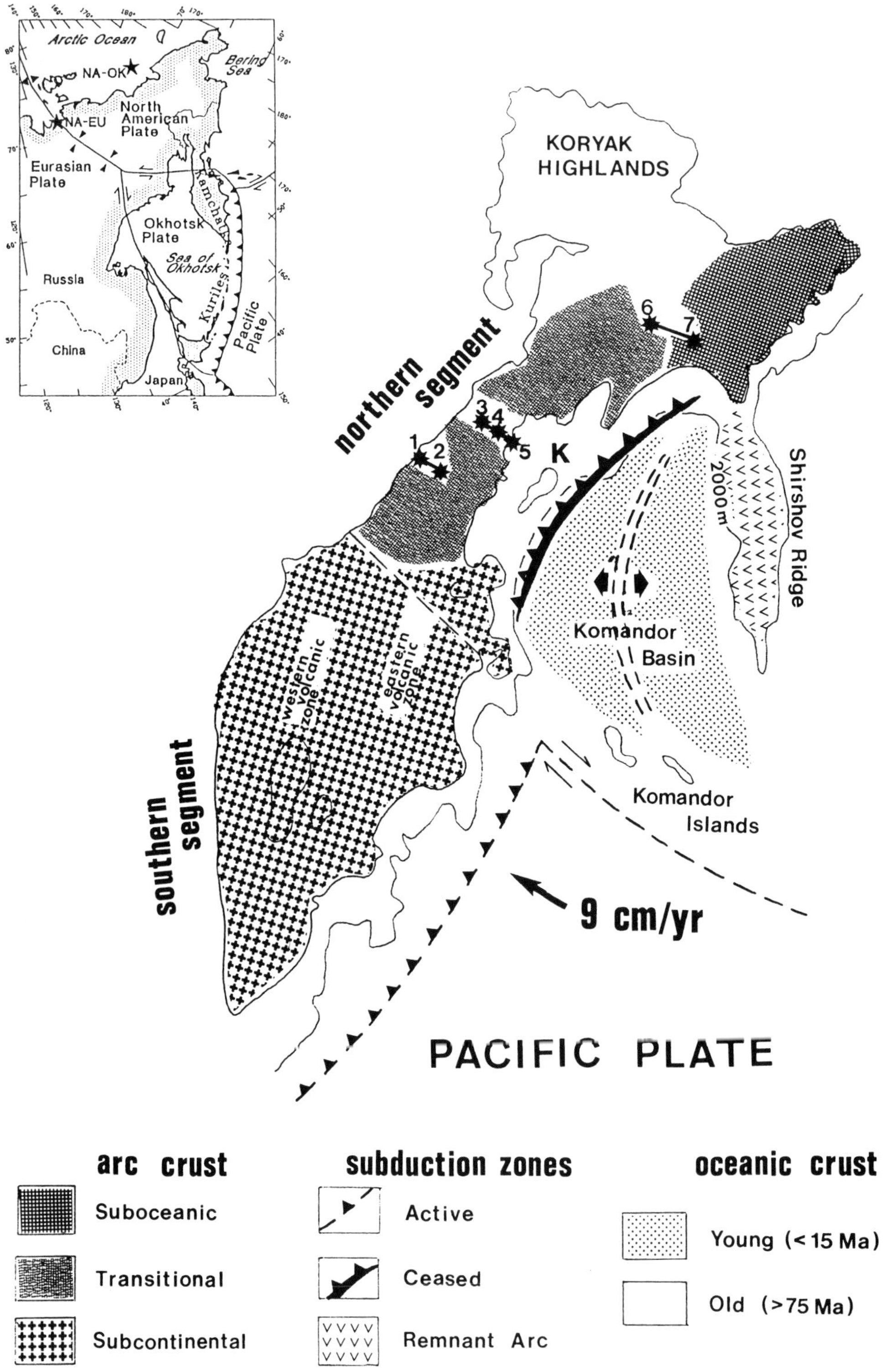

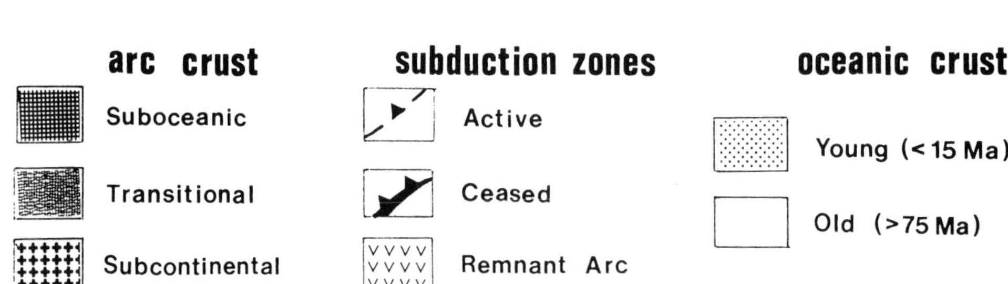
Arctic Ocean
NA-OK
North American Plate
NA-EU
Eurasian Plate
Okhotsk Plate
Russia
Sea of Okhotsk
Kamchatka
Kuriles
Bering Sea
Pacific Plate
China
Japan
KORYAK HIGHLANDS
northern segment
southern segment
6
7
3 4
1 2
5
K
Shirshov Ridge
2000 m
Komandor Basin
Komandor Islands
western volcanic zone
eastern volcanic zone
9 cm/yr
PACIFIC PLATE
arc crust
Suboceanic
Transitional
Subcontinental
subduction zones
Active
Ceased
Remnant Arc
oceanic crust
Young (< 15 Ma)
Old (> 75 Ma)

paper are: (1) to compare geochemical characteristics of high-K volcanic rocks erupted since late Eocene times across the Kamchatka arc; (2) to evaluate potential mantle sources and processes resulting in the production of different shoshonite magma series; and (3) to relate the chemistry of magmatic series to extension in magmatic arcs.

Tectonic setting of the Kamchatka arc

The Kamchatka magmatic arc extends from the northernmost Kurile islands to the southern part of the Koryak Highlands (Fig. 1). Two segments are commonly recognized within the arc and are separated by an on-shore continuation of the Bering Fracture Zone in the Komandor Basin (Kepezhinskas *et al.* 1990; Hochstaedter *et al.* in press). The southern segment extends south of the junction with the westernmost Aleutian arc (Komandor Islands) and is associated with the northwestward subduction of old Pacific oceanic lithosphere (Scholl *et al.* 1975). The northern segment comprises the northern portion of the Sredinny Range and the southern part of the Koryak Highlands (Fig. 1). It includes volcanic suites of late Eocene to Pliocene age containing arc tholeiite, calc-alkaline and shoshonite lava series (Kepezhinskas *et al.* 1990; Bogdanov & Kepezhinskas in press). These rocks were derived during two major arc-related magmatic episodes. The late Eocene–early Oligocene episode includes eruption of large volumes of calc-alkaline, high-K calc-alkaline and shoshonite lavas and pyroclastic rocks. Palaeomagnetic evidence from Eocene–Oligocene arc-derived pyroclastic rocks suggests that they were formed at 61°N, which is similar to their present geographical position (Kovalenko 1990). This implies that pre-Miocene shoshonite-bearing volcanic units were formed *in-situ* in the Tertiary magmatic arc (Stavsky *et al.* 1990). The formation of this arc is thought to represent either the northwestward subduction of a trapped fragment of the Kula plate or subduction of the young oceanic crust of an expanding 'proto-Komandor Basin' (Bogdanov 1988; Stavsky *et al.* 1990).

Late Miocene to Pliocene shoshonitic rocks were erupted in the northern segment of the Kamchatka arc after the Miocene emplacement of oceanic and arc-derived accreted terranes. The formation of the northern Kamchatka volcanic province is related to the westward subduction of young oceanic crust of the Komandor Basin (Bogdanov 1988; Kepezhinskas *et al.* 1990).

The Kamchatka volcanic arc exhibits lateral variability in thickness and composition of the crust as suggested by geophysical data and studies of crustal xenolith suites (Bogdanov & Kepezhinskas in press). The northernmost termination of the arc (east of the Pakhachinsky Range in Fig. 1) is underlain by thickened (up to 20 km) suboceanic crust which gradually changes into transitional crust (attenuated subcontinental crust with a thickness of 22 to 27 km) in a southwest direction (Fig. 1). A further increase in crustal thickness and development of granulite-facies lower crust has resulted in the occurrence of typical subcontinental crust of variable thickness (30 to 45 km) beneath the southern segment of the Kamchatka arc (Fig. 1). Shoshonite rocks were emplaced above the transitional crust, except for the Pakhachinsky Range shoshonite suite which appears to be emplaced along the boundary between suboceanic and attenuated subcontinental (transitional) blocks in the northern part of the Kamchatka arc. Geophysical data available for the Pakhachinsky Range and interpretation of crustal xenolith assemblages (amphibolites, orthogneisses, granites and metasediments) suggest that crustal structure and composition below this region are similar to areas located further to the southwest (Stavsky *et al.* 1990; Bogdanov & Kepezhinskas in press). These data suggest that the Pakhachinsky Range high-K suite represents the same igneous province, formed during extension of subcontinental (transitional) crust. Both northern and southern segments of the Kamchatka arc are built upon accreted Mesozoic to Tertiary oceanic, arc- and back-arc basin-derived terranes. The arc consists of two sub-parallel, linear volcanic belts, which are referred to here as the western

---

Fig. 1. Sketch map showing the regional tectonic setting of the Kamchatka arc. The plate boundaries, relative plate motions and major tectonic structures are from Scholl *et al.* (1975) and Bogdanov (1988). The northern trench is sediment-filled and is identified by a gravity anomaly after Shapiro *et al.* (1987). Kamchatka arc crustal types are from Bogdanov & Kepezhinskas (in press). K represents the location of the Karaginsky Island accretionary complex (Kravchenko-Berezhnoy *et al.* 1990). Shoshonite localities (Volynets *et al.* 1986; Kepezhinskas *et al.* 1988): 1, coastal part of the Kinkil Range; 2, Kinkil volcanic rocks in the western part of the Sredinny Range; 3, crest of the Sredinny Range; 4, eastern part of the Sredinny Range; 5, Belaya River area; 6, Malinovsky Range; 7, Pakhachinsky Range (PR).

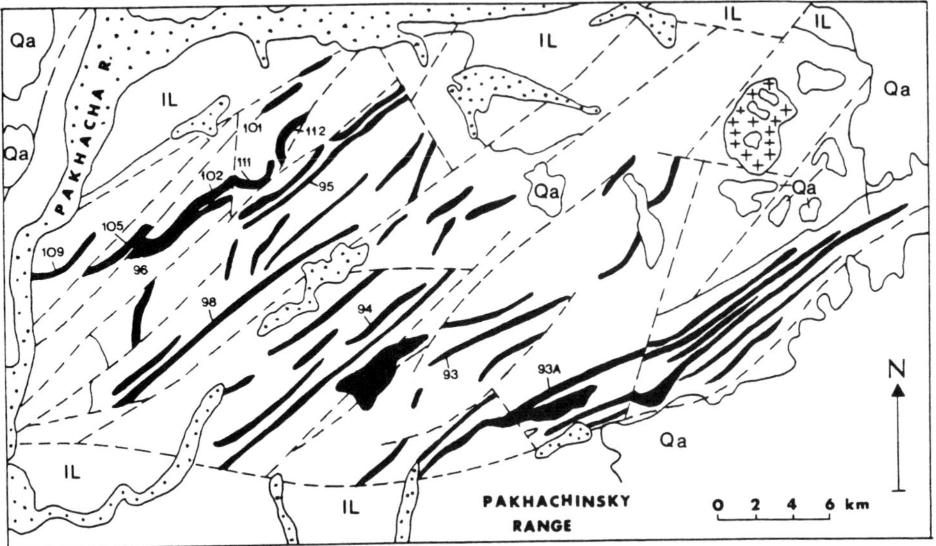

Fig. 2. Geological map of the central portion of the Pakhachinsky Range, after Kepezhinskas (1990). Map symbols: IL, Ilpinsky Series (Oligocene conglomerates, sandstones, mudstones); Qa, Quaternary (Early Pleistocene) andesites and basaltic andesites; crosses, Late Miocene (11 Ma) shallow-level granitic plutons; dots, Quaternary alluvium deposits. Oligocene shoshonite dykes and sills are shown in black; numbers refer to samples included in Table 2. Extension is indicated by abundant shoshonite dyke swarms and sill-sediment complexes.

(rear-arc) and eastern (frontal arc) volcanic zones (Fig. 1). These zones have slightly different histories of crustal accretion, with a northwestern zone including Late Jurassic–Early Cretaceous terranes and a northeastern zone containing mainly Late Cretaceous–Early Eocene accreted complexes (Stavsky *et al.* 1990; Fedorchuk & Izvekov 1992). They are separated by the Vivenka Fault which is a major terrane boundary in this region (Fig. 1). The Vivenka Fault was also tectonically active during Miocene to Quaternary time (Ufimtsev 1975).

Geological setting of the shoshonite magmatism in the Kamchatka arc and evidence for extension

Late Eocene to Pliocene shoshonite magmatism is documented in both western and eastern volcanic zones of the northern Kamchatka arc (Volynets *et al.* 1986; Firsov 1987; Kepezhinskas *et al.* 1988; Kepezhinskas *et al.* 1993). The thickness of volcanic sequence in the western zone is approximately 1500 m. Shoshonites form flows and domes and are associated with several large calc-alkaline volcanoes. Dyke swarms are absent and ring dykes are commonly related to caldera formation within eroded central-type volcanic complexes. Shoshonite lava flows are

commonly accompanied by abundant fine-grained pyroclastic deposits and volcanic breccias.

Shoshonites from the eastern volcanic zone are represented by sills or thick lava flows intercalated with coarse- to medium-grained sedimentary rocks (Figs 2 & 3). Abundant high-K dyke swarms and sill-sediment complexes are common, indicating extensional conditions during shoshonite emplacement (Kepezhinskas *et al.* 1990). Shoshonites in the Pakhachinsky Range area were emplaced mainly as dyke swarms and sill-sediment complexes and are commonly associated with listric faults (Fig. 2). The area shown in Fig. 2 structurally represents a half-graben separated from Cretaceous rocks farther east by a major fault which was active in the Pliocene period (Stavsky *et al.* 1990). The Belaya River graben represents a northwest–southeast-orientated structure filled with late Eocene to Pleistocene coarse-grained sedimentary rocks intruded by abundant shoshonite dykes and sills (Fig. 3). The extension was terminated by massive eruption of plateau-type shoshonite lavas (also including feeder dykes) which are morphologically distinct from central type volcanic centres of the same age formed within the western (rear-arc) volcanic zone. The shoshonitic plateau-type lavas are concentrated within

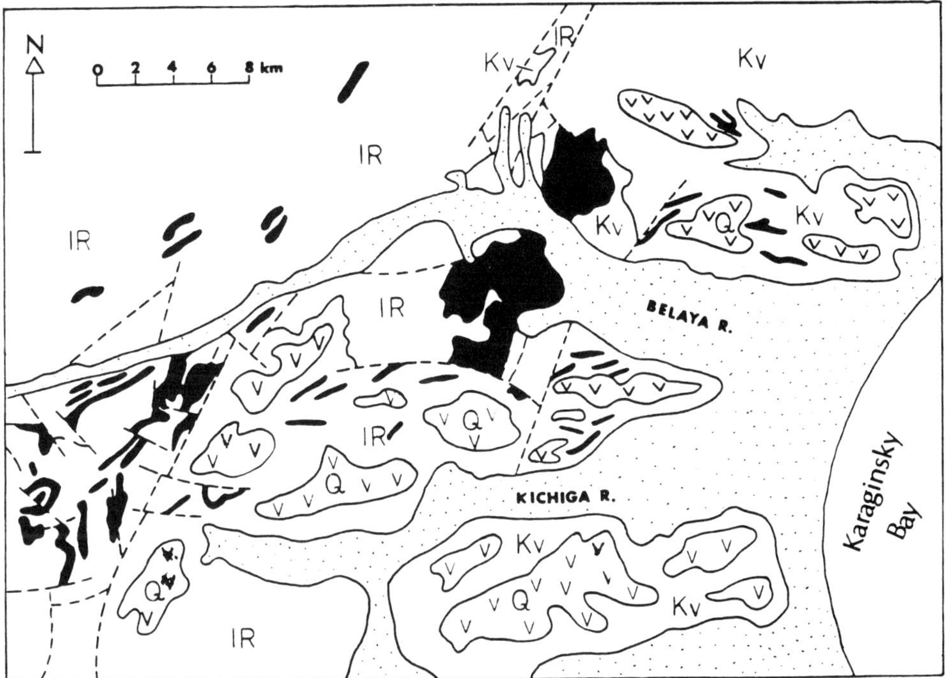

Fig. 3. Geological map of the Belaya River area, after Kepezhinskas (1990). Map symbols: IR, Iruneisky Series (Late Cretaceous arc-derived lavas, pyroclastic and siliceous sedimentary rocks); Kv, Kavachinsky Series (Oligocene conglomerates, sandstones, mudstones), Q, Pliocene to Quaternary (Early Pleistocene) calc-alkaline, high-K calc-alkaline and shoshonite lavas. Older (late Eocene to Oligocene) shoshonite dykes and sills are shown in black, and Quaternary alluvium by dotted ornament. The Belaya River area is a west–east orientated graben filled with Late Eocene to Pliocene coarse-grained clastic sediments and intruded by shoshonite dyke swarms and sill-sediment complexes.

the Belaya River graben and do not extend beyond its listric fault boundaries. Late Miocene to Pliocene rocks to the south and north of the Belaya River area include adakites, high-Na, Nb-enriched basalts and calc-alkaline lavas (Kepezhinskas 1989; Hochstaedter *et al.* in press).

Petrography of the shoshonites

Representative phenocryst compositions from north Kamchatka shoshonites are listed in Table 1. The Tertiary shoshonites commonly have Cpx–Pl–Kfsp–Amph phenocryst assemblages, whereas younger shoshonites contain Cpx–Opx–Pl–Kfsp and Cpx–Opx–Amph–Pl assemblages. Clinopyroxene is Fe-rich in older shoshonites, and tends to have higher MgNo. (Mg/Mg+Fe) in younger shoshonite dykes and flows (Table 1). Plagioclase commonly has An content less than 50, and oligoclase (An_{10-13}) is typical of shoshonites in the Pakhachinsky Range. This feature is potentially an effect of alteration in the older shoshonite suites (Kepez-

hinskas *et al.* 1988) but, in general, alteration in these relatively fresh rocks appear to have a negligible effect on original compositions. Amphibole is Fe-rich edenite-ferroedenite and has been described from several shoshonite suites (Morrison 1980; Arculus *et al.* 1983). Sphene and Ti-magnetite (with 7 to 22 wt.% TiO_2) are abundant in both late Eocene–Oligocene and Pliocene shoshonites from the eastern volcanic zone indicating a high-Ti composition of the parental shoshonite melt. High-Ti magnetites are also common inclusions in plagioclase and clinopyroxene phenocrysts from northeastern Kamchatka shoshonites which is consistent with a general enrichment of these rocks in high-field strength elements.

Geochemistry of the shoshonites

High-K volcanic rocks from the northern Kamchatka arc range in composition from absarokites to potassic andesites and dacites and plot in the field of the shoshonite magma series (Fig. 4).

Table 1. *Representative mineral compositions from North Kamchatka shoshonites*

Locality	5				5				7			7				
Age	T				P				P			T				
Sample	1900				5486				4386			9585				
Phase	CPx	Pl	K-fsp	Sphene	CPx	OPx	Pl	Ti-Mt	CPx	OPx	Amph	Pl	CPx	Amph	Ti-Mt	Sphene
SiO_2	51.29	55.56	65.08	31.59	51.92	53.38	54.58	0.07	51.75	54.49	42.99	65.04	50.59	47.31	0.13	31.93
TiO_2	0.47	0.06	0.02	27.08	0.71	0.29	0.09	18.95	0.51	0.07	3.41	0.02	0.27	0.82	17.63	23.1
Al_2O_3	1.86	26.27	18.62	6.32	2.4	1.21	27.73	3.2	2.8	0.48	10.74	21.88	0.86	5.3	0.53	7.81
Cr_2O_3	0.03	0	0.06	0.06	0	0	0	0.01	0	0	0	0.07	0.07	0.05	0.09	0.1
FeO	10.31	0.64	0.06	1.85	8.35	15.43	0.56	71.31	8.06	16.51	11.21	0.04	16.5	21.87	80.15	9.51
MnO	0.48	0.01	0.01	0.04	0.42	0.7	0	0.74	0.38	1.29	0.26	0.04	0.86	0.56	1.68	0.2
MgO	13.97	0.08	0.01	0.32	15.02	26.53	0.08	4.35	14.79	25.95	14.21	0	10.46	11.08	0.15	1.32
CaO	19.8	9.18	0	28.84	19.87	1.57	10.58	0.01	20.39	1.03	10.94	2.49	20.07	10.01	0.04	23.51
Na_2O	0.29	5.67	0.18	0	0.42	0.04	5.09	0.03	0.42	0.05	2.65	8.98	0.48	2.03	0.04	0.04
K_2O	0	0.59	16.59	0.06	0.01	0.01	0.73	0	0	0	0.88	0.2	0.01	0.66	0	0
Total	98.49	98.06	100.96	96.16	99.12	99.16	99.44	98.68	99.1	99.87	99.28	98.74	100.17	99.69	100.45	97.51
MgNo.	70.1	–	–	–	76	75.4	–	–	76.5	73.6	69.3	–	51.9	47.3	–	–
An	–	46	–	–	–	–	52	–	–	–	–	13	–	–	–	–

Mineral compositions were determined using CAMEBAX microprobe at the Institute of Geology and Geophysics, Siberian Branch of Russian Academy of Sciences. Analytical conditions were: a 20 kV accelerating potential, a probe current of 14 nA, a counting time of 10 s, and a beam diameter of approximately 10 microns. A set of natural and synthetic mineral standards was used. T, Tertiary (late Eocene to Oligocene) and P, Pliocene shoshonites. MgNo. = Mg/(Mg+Fe). Numbers of individual shoshonite localities refer to numbers used in Figs 1 & 6.

Table 2. *Representative major (wt%) and trace (ppm) element analyses of Kamchatka arc shoshonites*

Locality	1	2	3	3	4	4	5	5	5	5	6	7	7	7
Age	T	T	T	P	T	P	T	P	P	P	T	T	T	T
Sample	2286	3686	2736	3186	3286	3886	1900	5486	4486	4386	67683	9685	10985	9585
SiO_2	51.13	53.17	53.86	52.21	50.88	53.14	56.14	58.2	48.0	48.03	52.7	48.63	52.7	54.4
TiO_2	0.63	0.71	0.66	0.73	0.61	0.81	0.82	1.16	1.14	1.12	0.91	1.03	1.19	1.16
Al_2O_3	12.21	11.03	13.01	13.16	13.76	13.57	15.81	17.59	17.74	17.88	15.05	14.03	14.81	14.1
Fe_2O_3	7.68	9.16	8.17	9.03	9.64	10.13	2.87	7.92	6.51	5.84	9.78	4.92	5.91	6.23
FeO							4.31		3.75	4.39		5.11	4.75	4.56
MnO	0.31	0.37	0.33	0.4	0.28	0.29	0.22	0.15	0.19	0.18	0.14	0.28	0.27	0.28
MgO	8.09	6.14	6.05	5.66	5.61	5.17	5.07	1.92	5.02	5.45	5.62	11.12	5.31	3.4
CaO	9.15	7.63	8.84	8.36	8.16	7.66	5.14	3.59	9.43	9.11	5.71	8.1	6.21	4.97
Na_2O	1.96	2.68	1.76	3.15	2.66	3.43	4.16	3.68	2.97	3.08	4.42	2.72	3.59	3.86
K_2O	4.88	5.21	4.1	5.28	4.31	4.48	3.82	4.37	2.16	2.08	3.25	1.61	2.53	2.9
P_2O_5	0.86	0.63	0.66	0.58	0.39	0.41	0.26	0.47	0.52	0.51	0.36	0.32	0.76	0.58
LOI	2.11	2.36	1.75	1.25	3.17	1.08	2.31	0.5	1.66	1.83	1.56	2.03	2.76	2.8
Total	99.01	100.09	99.19	99.81	99.47	100.17	100.93	99.5	99.17	99.5	99.5	99.7	100.79	99.24
Cr	226	56	50	26	31	21	23	12	5	6	53	226	66	40
Ni	88	38	29	35	20	26	16	10	20	19	46	68	28	32
Ba	1992	1805	1816	1670	1371	1552	986	1870	1555	1560	1080	860	689	762
Rb	33	56	61	65	41	48	39	84	45	43	35	25	60	58
Sr	585	604	660	543	580	636	835	578	970	990	466	215	324	246
Zr	158	125	133	117	131	115	154	363	180	180	182	195	210	204
Y	29	27	29	27	30	28	34	39	20	22	23	34	38	34
La	91	50	35	58	26	33	20	39	46	44	48	42	60	48
Ce	192	148	94	150	72	86	50	90	108	105	112	100	140	115
Nd	94	58	50	60	33	47	32	56	66	65	52	61	84	69
Sm	18	9.1	8.2	9.4	5.4	6.4	6.2	15	14	14	15	13	19	14
Eu	3.5	1.7	1.6	1.8	1.2	1.3	1.3	2	2.6	2.6	2.5	2.4	3.6	2.6
Gd	11	8.1	7.9	8.3	6	6.2	5.6	9.4	9.8	9.6	12	8	12	9.8
Er	4	3	2.8	3.1	2.2	2.3	2.6	4	3.4	3.2	3.5	3.3	4.2	3
Yb	3.9	2.6	2.5	2.8	2	2.1	2.2	4	2.4	2.4	3.1	2.6	2.8	2.6
La/Sm N	3.14	3.38	2.65	3.83	3.04	3.18	1.99	1.6	2.03	1.93	1.97	1.99	2.13	2.12
Na_2O/K_2O	0.4	0.46	0.43	0.6	0.62	0.77	1.1	0.84	1.38	1.48	1.36	1.69	1.42	1.33

Numbers of individual shoshonite localities refer to numbers used in Figs 1 & 6. Major and trace (Cr to Y) elements by XRF at the Institute of Lithosphere (Moscow, Russia).
REE by ICP-AES (Hilger Analytical MONOSPEC1000 spectrometer) with preliminary concentration of rare earth elements on ion-exchange resins (DOWEX-HCR-W2) following the procedure of Kepezhinskas *et al.* (1990).
International standard samples AGV-1, W-2, STM-1 and RGM-1 were used to estimate accuracy of the method.
Average ordinary chondrite values of Nakamura (1974) were used for normalizing La/Sm ratios in shoshonites.
T, Tertiary (late Eocene to Oligocene); P, Pliocene.

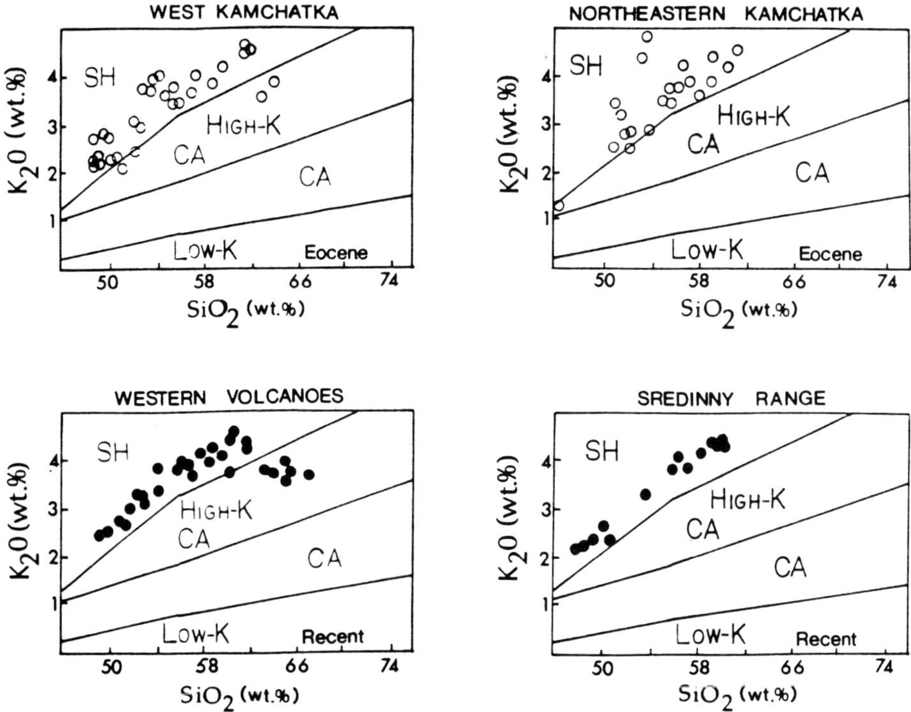

Fig. 4. K_2O vs. SiO_2 plot for Eocene to Pliocene shoshonites from the Kamchatka arc. Solid lines separate the fields of low-K, calc-alkaline (CA), high-K calc-alkaline (high-K CA) and shoshonite (SH) volcanic series, adapted from Peccerillo & Taylor (1976). Data for West Kamchatka (Eocene) and Western Volcanoes (Pliocene) rocks include shoshonites from localities 1–3 (Fig. 1). Eocene to Oligocene shoshonites from northeastern Kamchatka are from localities 6 and 7 (Fig. 1). Recent lavas from the Sredinny Range represent Late Cenozoic shoshonites from localities 4 and 5 (Fig. 1).

However, several compositional differences exist between the shoshonite sub-groups. Representative compositions of less evolved shoshonites from both northwestern (rear-arc) and northeastern (frontal arc) suites are listed in Table 2.

Shoshonites from the western volcanic zone display a nearly continuous compositional array on the K_2O vs. SiO_2 diagram with a few dacite compositions showing lower potassium contents similar to those of high-K calc-alkaline dacites. They commonly have low TiO_2 contents and fall below the line discriminating between the high-Ti and low-Ti shoshonites on the TiO_2–K_2O and K_2O–SiO_2 plots (Fig. 5). They are similar to typical subduction-related potassic lavas from continental volcanic arcs, such as the Andes (Dostal *et al.* 1977; Kontak *et al.* 1986), Aeolian arc (Keller 1974) and Greece (Pe-Piper 1980). Contemporaneous shoshonites from the eastern volcanic zone are predominantly basaltic andesites and andesites but also include some absarokites (Table 2; Fig. 4). They show higher

TiO_2 concentrations compared to the rear-arc shoshonites at similar contents of MgO and SiO_2 (see also Table 2). This possibly implies that fractionation is not solely responsible for the distinctive HFSE signatures in the two potassic suites, and Kepezhinskas *et al.* (1990) demonstrated that the shoshonite magma series in Kamchatka display across-arc compositional variations which are generally not controlled by crustal contamination or polybaric fractionation.

Across-arc variations in the shoshonite magma chemistry

The compatibility of the structures, terrane relationships and general volcanic stratigraphy along the northern Kamchatka arc as well as lack of evidence for significant compressional tectonics implies a relatively stable configuration of the arc-trench system since latest Eocene times (Shapiro *et al.* 1987; Stavsky *et al.* 1990). Late

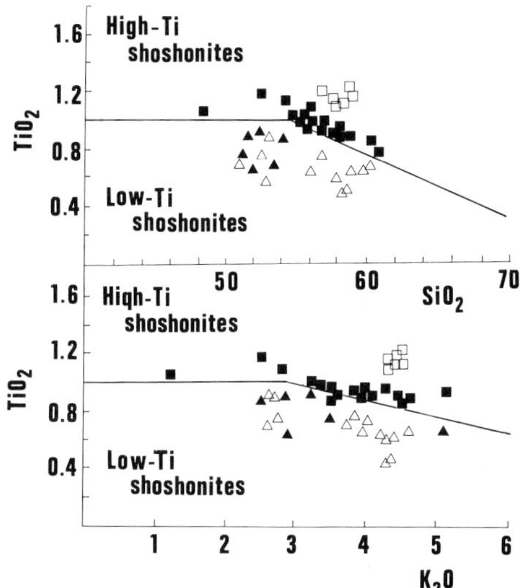

Fig. 5. TiO_2–SiO_2–K_2O relationships for Kamchatka high-K volcanic rocks. Fields for high-Ti (extension-related) and low-Ti (late-stage or rear-arc) shoshonites are from Kepezhinskas *et al.* (1988). The high-Ti shoshonites may appear at any stage of arc development and are commonly found in frontal arc and fore-arc settings. The data used to distinguish the high-Ti shoshonite field include suites from Papua New Guinea (Arculus *et al.* 1983), Puerto Rico (Jolly 1971), Lesbos, Greece (Pe-Piper 1980), Eastern Srednegorie, Bulgaria (Manetti *et al.* 1979) and Fiji (Gill & Whelan 1989). The low-Ti shoshonites display K-depth relationships characteristic of subduction zones and are commonly emplaced in the rear-arc environment. The low-Ti shoshonite field includes data from the Andes (Dostal *et al.* 1977; Kontak *et al.* 1986), Indonesia (Foden, 1986) and the Aeolian arc (Keller 1974). Western volcanic zone (triangles) and eastern volcanic zone (squares) shoshonites of late Eocene–Oligocene (closed symbols) and late Miocene–Pliocene (open symbols) age show similar compositional relationships suggesting similar conditions for their origin. Western zone, rear-arc shoshonites from Kamchatka belong to the low-Ti type, whereas eastern zone, frontal arc shoshonitic rocks plot in the high-Ti shoshonite field.

Eocene–Oligocene and Pliocene shoshonites within the same volcanic zone (western or eastern) share major geological, mineralogical and chemical features (e.g. Tables 1 and 2, Fig. 5). Therefore the eruption of shoshonite magmas within both western and eastern volcanic zones provides a basis for across-arc comparison of the shoshonite magma chemistry in the northern Kamchatka arc.

Chemical data for the Kamchatka arc shoshonites are presented as a transect from the western rear-arc volcanic zone eastwards to the trench. A few assumptions have been made: (1) the sedimented trench associated with a large negative gravity anomaly in the northern part of the Kamchatka arc-trench system was a locus of active subduction of Komandor Basin oceanic lithosphere during Late Eocene to Pliocene time. Subduction probably ceased around 3 Ma due to plate re-arrangement in the northern Pacific (Hochstaedter *et al.* in press). (2) The northernmost portion of the Kamchatka arc was built on relatively thin mafic crust, as suggested by the xenolith assemblages found in shoshonite and associated calc-alkaline lavas (Bogdanov & Kepezhinskas in press). Therefore, the role of crustal contamination was probably negligible. (3) Terrane accretion was completed before the late Eocene and none of the post-middle Eocene magmatic suites is accreted (Fedorchuk & Izvekov 1992). The last assumption is supported by a wealth of geological evidence including 45 Ma granites intruding the Vivenka Fault associated with the terrane accretion and a 44 Ma age for the high-K volcanic sequences from the western volcanic zone, which unconformably overlie the thrust zone within the northern segment of the arc (Fedorchuk & Izvekov 1992). Palaeomagnetic evidence also suggests that the Eocene–Oligocene arc units were formed at 61°N which is similar to their present geographical position (Kovalenko 1990). However, a comparatively complicated subduction history is suggested by the presence and development of an accretionary complex along the eastern coast of Karaginsky Island (shown as K in Fig. 1). The accretionary prism is late Eocene–Oligocene in age and demonstrates that considerable accretion took place along the trench at that time (Kravchenko-Berezhnoy *et al.* 1990). The shortening in the forearc area due to this process was estimated from the thickness of accreted crustal units of various age. The calculated distance from the trench (corrected for the forearc accretion) was then used in geochemical considerations. The calculated distances used in this study are believed to represent displacement of shoshonite units relative to each other rather than their absolute position in space during the evolution of the convergent margin. In order to avoid the influence of crystal fractionation processes involved in the genesis of differentiated shoshonite series (Morrison 1980; Pe-Piper 1980), only mafic high-K lavas with MgO contents greater than 5 wt% (absarokites and shoshonitic basalts) were used to construct the transects shown in Fig. 6. Each data set represents the individual shoshonite volcanic unit along the transect and

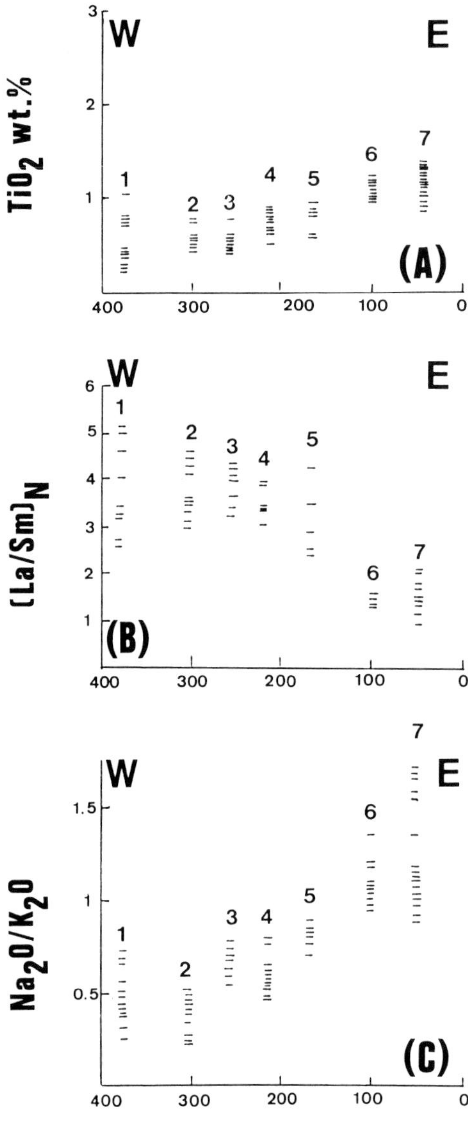

Fig. 6. Geochemical west–east transect across the Kamchatka arc based on samples from the shoshonite magma series. Samples of both Eocene–Oligocene and Mio-Pliocene age groups share chemical characteristics within any geographical locality and, therefore, were used together to construct the individual data columns. For selection procedure and trench distance estimates, see text. Data sources: Volynets *et al.* (1986), Kepezhinskas *et al.* (1988), Kepezhinskas *et al.* (1990), P.K.Kepezhinskas and A.V.Koloskov (unpublished data). (**a**) TiO₂; (**b**) (La/Sm)ₙ; (**c**) Na₂O/K₂O. Despite overlapping compositions, there are clear, gradual changes in shoshonite magma chemistry with respect to geographical position.

the numbers in Fig. 6 refer to the numbers in Fig. 1 and Tables 1 & 2.

Several petrogenetically important incompatible elements were used to investigate across-arc systematics and variations in source compositions during the generation of these arc-related potassic magmas. High field strength (HFS) element concentrations and chondrite-normalized light rare earth element (LREE)/middle rare earth element (MREE) ratios were utilized as possible source composition indicators (Pearce 1982). Na/K ratio is an important parameter during arc petrogenesis and is commonly used for the tectonomagmatic classification of island-arc magma series (Gill 1981). In general, the technique used to process the Kamchatka data is that applied by Taylor *et al.* (1992) for the intra-oceanic Izu-Bonin and Mariana arc-trench systems.

Data are presented in Fig. 6 in which the *x*-axis represents the actual distance or estimated and corrected palaeo-distance from the trench, and the *y*-axis the concentration of the element or inter-element ratio. Only data for the least fractionated shoshonite lavas from the Kamchatka arc are shown in Fig. 6. TiO₂ contents increase gradually towards the frontal arc high-K suites although there is a large overlap between the individual geographical localities. This may be due in part to fractionation since Fe–Ti oxide is an early phase in oxidized shoshonite magmas (Morrison 1980). It is important to note that this effect is not due to increasing modal proportions of phenocryst and groundmass Fe–Ti oxide and titanite phases although this process could have a limited influence on the variations among the HFSE group. This is illustrated by the general positive correlation between TiO₂ and Zr in Kamchatka shoshonite suites (Fig. 7) suggesting that Ti was buffered or has a bulk K_D less than 1. The across-arc variations in HFSE probably reflect differences in magma sources beneath the western (rear-arc) and eastern (frontal arc) volcanic zones. This suggestion is further supported by regional variations in chondrite-normalized La/Sm ratios (Fig. 6b), which decrease markedly towards the trench suggesting that eastern volcanic zone shoshonites were derived from a less LREE-enriched source. Shoshonites of the same age from the western rear-arc volcanic zone are strongly enriched in light rare earth elements (Kepezhinskas *et al.* 1990). The changes in the Na₂O/K₂O ratio parallel the changes in HFSE contents within the Kamchatka shoshonite suites (Fig. 6c). While western rear-arc low-Ti magmas are always enriched in K, even approaching ultrapotassic

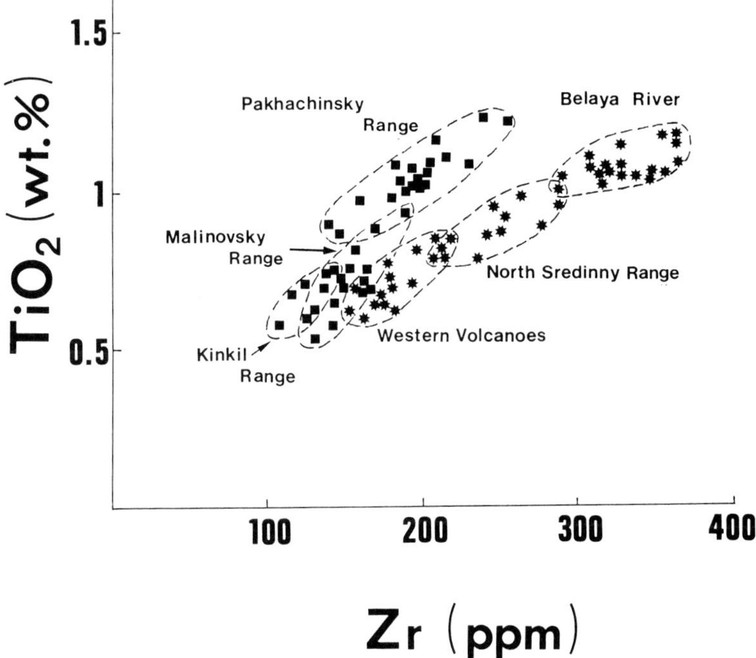

Fig. 7. TiO$_2$–Zr variations in the Kamchatka arc shoshonite suites. Individual localities are identified in Fig. 1. Late Eocene–Oligocene shoshonite suites are shown by filled squares and late Miocene–Pliocene shoshonites are plotted as stars. Shoshonite rocks of the same age plot along linear trends on this diagram, suggesting that no significant Fe-Ti oxide fractionation was involved in their evolution. Western zone, rear-arc shoshonites show low TiO$_2$ and Zr concentrations, while eastern zone, frontal arc shoshonites display elevated TiO$_2$ and Zr contents, probably reflecting differences in magma source compositions rather than fractionation processes. Only mafic compositions with MgO>5% were used to construct this diagram (see text for details).

compositions (up to 9 wt% of K$_2$O at 7 wt% of MgO; Volynets *et al.* 1986), the high-Ti frontal arc shoshonites show a gradual increase in Na concentrations and Na$_2$O/K$_2$O ratios towards the trench. However, they still fall in the shoshonite compositional range on the K$_2$O–SiO$_2$ classification diagram (Fig. 4). This trend is probably not related to differentiation since no specific K-rich phase (phlogopite or leucite) is involved and the chosen data are restricted to the least fractionated compositions. One explanation requires addition of a Na-rich component to the mantle source of the eastern zone high-Ti shoshonite magmas. Alternatively, this trend can represent a lower degree of melting and a greater input of K beneath the western rear-arc volcanic zone. The role of lateral crustal variability cannot be completely eliminated, although it seems unlikely to play an important role since all the shoshonite suites used in this study were erupted through similar crust. The crustal thickness in this part of the arc varies only

between 22 and 27 km, and even the Pakhachinsky Range shoshonites, which were emplaced at the boundary between transitional (attenuated subcontinental) and suboceanic crust, share chemical and petrological characteristics of the high-K rocks erupted elsewhere in the northeastern volcanic zone.

Possible nature of source components

High-Ti eastern zone shoshonites display relatively low LREE/MREE and LREE/HREE ratios suggesting derivation from only slightly enriched sources, while low-Ti western zone shoshonites of similar age have high (La/Sm)$_N$ and (La/Yb)$_N$ ratios requiring the involvement of a LREE-enriched component. Further consideration of the Sr–Nd isotopic data confirms this observation. Preliminary whole-rock data from the Kamchatka shoshonite series are plotted on a ^{143}Nd/^{144}Nd vs. ^{87}Sr/^{86}Sr diagram together with data for MORB and other island

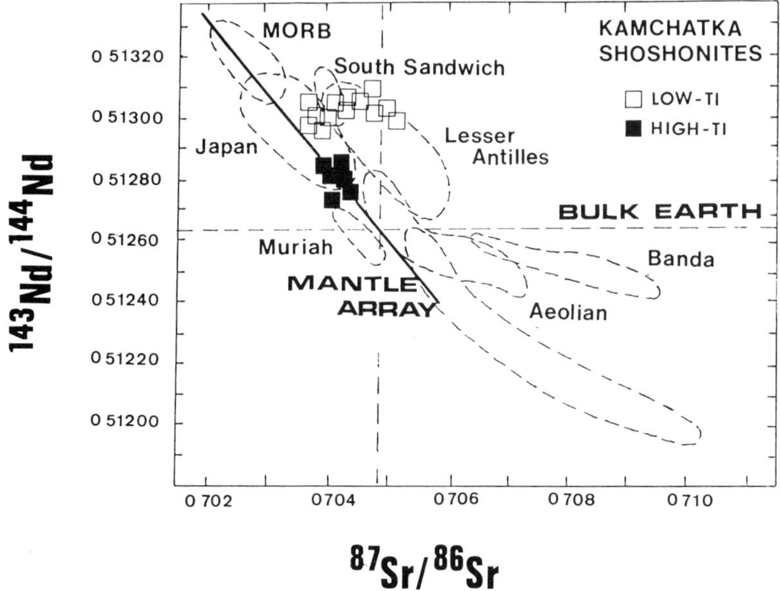

Fig. 8. ^{143}Nd/^{144}Nd vs. ^{87}Sr/^{86}Sr isotope diagram showing Kamchatka shoshonite data relative to MORB and selected island arcs. The data are from: Kamchatka (unpublished information of the author, F. McDermott, C. Hawkesworth and A. Koloskov); MORB (White & Hofmann 1982); South Sandwich (Hawkesworth *et al.* 1977); Japan (Nohda & Wasserburg 1981); Lesser Antilles (Hawkesworth *et al.* 1979; Davidson 1985); Muriah complex, Indonesia (Edwards *et al.* 1991); Banda arc, Indonesia (Whitford *et al.* 1981); Aeolian arc (Ellam *et al.* 1988). Low-Ti shoshonites include both late Eocene–Oligocene and Miocene–Pliocene shoshonites from localities 1–4. High-Ti shoshonites encompass samples of both age groups from localities 5–7.

arc volcanic suites (Fig. 8). The high-Ti shoshonites have low ^{143}Nd/^{144}Nd ratios and plot close to undepleted arc lava compositions. On the contrary, the low-Ti shoshonites are relatively depleted in terms of Nd isotopes and plot closer to the MORB field. They display variable but generally high ^{87}Sr/^{86}Sr ratios plotting away from the mantle array (Fig. 8). Leaching experiments on the high ^{87}Sr/^{86}Sr samples confirmed the primary nature of these characteristics suggesting the possible involvement of a component with a radiogenic Sr isotopic signature.

The diverse shoshonite magma types in the Kamchatka arc were probably produced through partial melting of a variably depleted sub-arc mantle to which additional components have been introduced. The nature of the added component is different in the western (rear-arc) and eastern (frontal arc) volcanic zones. In the eastern zone it is a high-Na component as suggested by a Na/K ratio vs. Na$_2$O plot (Fig. 9). Various mantle-derived magmas define a mantle enrichment trend on this diagram suggesting that there is a nearly continuous array of mantle sources from K-rich, sub-continental mantle reservoirs to K-depleted, MORB-type oceanic

sources. Experimentally produced slab melts show Na/K ratios close to continental flood basalts (CFB) and oceanic island basalts (OIB) but with distinctly different Na$_2$O concentrations (Rapp *et al.* 1991; Rapp & Watson in press; Fig. 9). Natural slab-derived melts, such as adakites from eastern Panama (Drummond & Defant 1990), plot in the same field showing strong enrichment in Na$_2$O at low K$_2$O concentrations. Western zone shoshonites plot in the field of continental high-K series (Montana potassic basalts), showing a slight increase in Na$_2$O concentrations potentially due to crystal fractionation. Eastern zone shoshonites exhibit increase in both Na/K ratio and Na$_2$O content suggesting that a Na-rich component was probably added to their source (Fig. 9). In terms of major element chemistry, the latter is similar to the high-Na and Al component involved in the petrogenesis of some boninites in the Mariana arc (Pearce *et al.* 1992) and Nb-enriched basalts in the northern Kamchatka arc (Defant *et al.* 1993). High-Na, Nb-enriched basalts which are spatially and temporally associated with high-Ti shoshonites in the frontal arc, eastern volcanic zone display high Na/K ratios at high MgO

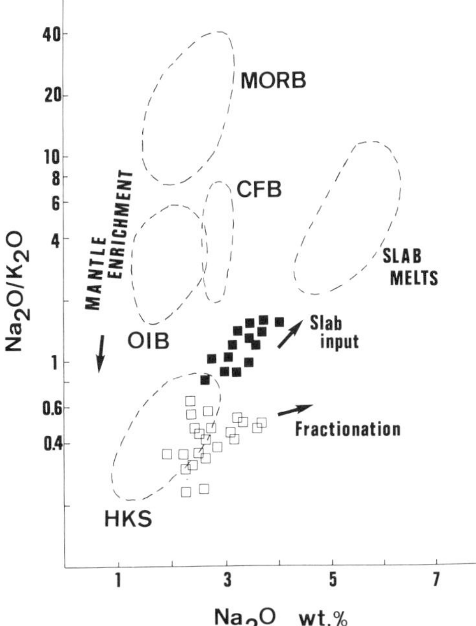

Fig. 9. Na₂O/K₂O vs. Na₂O plot showing low-Ti (open squares) and high-Ti (filled squares) shoshonites from the Kamchatka arc compared with mantle- and slab-derived melts. Data sources: MORB, Sun *et al.* (1979); continental flood basalts (CFB), Hooper (1988); oceanic island basalts (OIB), Nicholls & Stout (1988) and Albarede & Tamagnan (1988); high-K series mafic lavas (HKS), Macdonald *et al.* (1992); experimental slab melts from amphibolite source (Slab Melts), Rapp *et al.* (1991). Mafic basaltic compositions with MgO>7 wt.% were used to construct this diagram. Slab melts are low-Mg melts derived from a metamorphosed basaltic source. Consequently, all experimental data from Rapp *et al.* (1991) were used to define the field of partial melts from the downgoing slab (see text for details).

contents suggesting a relatively Na-enriched source (Defant *et al.* 1993). The young and hot oceanic lithosphere of Komandor Basin, which has been subducting below the northern Kamchatka arc during the Cenozoic, was potentially capable of generating amphibolite-facies slab melts (Kepezhinskas 1989). These melts would have been rich in Na and may have had low ^{143}Nd/^{144}Nd ratios resembling some compositional features observed in the frontal arc shoshonites (Bloomer *et al.* 1989; Kepezhinskas *et al.* 1990). Interaction of these high-Na melts with the sub-arc mantle will result in production of mantle metasomatic mineral assemblages which include metasomatic high-Ti spinels. Ti-spinels are abundant in the metasomatized mantle xenoliths found in Nb-enriched arc

basalts spatially and temporally associated with high-Ti shoshonites in the eastern volcanic zone of the northern Kamchatka arc (Defant *et al.* 1993). Melting of a slab-metasomatized mantle source can potentially account for the chemical characteristics (including relative Ti-enrichment) of the high-Ti shoshonite magmas. A similar high-Na slab-melt component was recorded in high-Mg dacite lavas from the submarine Piip Volcano behind the Komandor Islands eastward from the northern Kamchatka arc. The geochemical characteristics of Piip Volcano lavas account for slab melt-mantle interaction through addition of 3–5% of Na-trondhjemite (adakite) melt to the sub-arc mantle source (Yogodzhinski *et al.* 1993).

Western zone shoshonite magmas were probably derived through the melting of a depleted source affected by a component with high LILE/HFSE and LREE/HFSE ratios. This component is commonly identified as the slab-derived fluid enriched in LREE and LILE (Gill 1981; Bloomer *et al.* 1989). In the case of low-Ti shoshonite magmas from the western Kamchatka arc, high (Ba/Ti)$_N$ and (Ba/Zr)$_N$ ratios, enrichment in LREE compared to MREE and HREE and radiogenic Sr isotope composition (Volynets *et al.* 1986; Kepezhinskas *et al.* 1990) also imply the possible involvement of pelagic sediment or a sub-continental mantle source since these shoshonites were erupted above attenuated sub-continental crust at the rear of the northern segment of the Kamchatka arc.

Relation to extensional tectonics and geodynamic implications

An extensional setting has been proved for many subduction-related alkaline magmas on the basis of field mapping and geochronological studies (Smellie 1987; Gill & Whelan 1989). The high-Ti shoshonite suites in northern Kamchatka are also clearly associated with a period of extension terminating or following periods of subduction (Kepezhinskas *et al.* 1988; Kepezhinskas *et al.* 1990; Fedorchuk & Izvekov 1992). K–Ar dating suggests that the first rifting episode was early Oligocene (32–34 Ma), whereas the second extensional phase occurred during Late Miocene to Pliocene times (11–5 Ma) (Kepezhinskas *et al.* 1993). The volcanic sequences commonly fill graben-like structures orientated perpendicular to the trench. Dyke swarms and sills are abundant while lavas were erupted from central vents within the central fissure zone. Conversely, low-Ti shoshonite lavas were erupted within central-type volcanoes lacking dykes except for ring dyke complexes associated with calderas.

Temporally, high-Ti shoshonites mark the initiation of intra-arc rifting and they usually follow on from the eruption of arc tholeiites and calc-alkaline volcanic rocks. Initiation of the rifting is probably related to the northwestward subduction of young, hot oceanic lithosphere in the late Eocene–Oligocene (proto-Komandor Basin lithosphere) and late Miocene (less than 15 Ma-old oceanic lithosphere of the expanding Komandor Basin in the case of the north Kamchatka arc) (Bogdanov 1988; Stavsky *et al.* 1990). High-Ti shoshonites are temporally and spatially associated with slab-derived melts (adakites) erupted within the eastern volcanic zone of the northern Kamchatka arc in Late Miocene-Pliocene. These silica-rich melts are likely to react with a sub-arc mantle producing a hybrid source (Carroll & Wyllie 1989). Hybrid Al-augite series xenoliths of mantle origin have been discovered in high-Na, Nb-enriched basaltic lavas associated with high-Ti shoshonites in the north Kamchatka arc (Defant *et al.* 1993). The high Na concentrations of the slab-derived melts (Drummond & Defant 1990) can possibly account for the trenchward Na enrichment in shoshonite lavas observed in a transect across the northern Kamchatka arc. Partial melting of the hybrid source during further intra-arc rifting probably resulted in the generation of the high-Ti potassic magmas, which appear to mark the initiation of intra-arc extension analogous to the relationship between boninites and forearc rifting during incipient subduction (Pearce *et al.* 1992; Stern & Bloomer 1992).

Conclusions

(1) Late Eocene to Pliocene shoshonite lavas from the northern segment of the Kamchatka arc show west to east, across-arc variations in major and trace element chemistry suggesting their derivation from variable magma sources.

(2) Shoshonites from the western (rear-arc) volcanic zone exhibit lower HFSE concentrations, higher LREE/MREE and LREE/HREE ratios and lower Na_2O/K_2O ratios compared to the eastern (frontal arc) zone shoshonites, suggesting their derivation from a refractory source affected by a LILE and LREE-enriched component. Eastern zone shoshonites display higher HFSE contents and are enriched in Na along with only slight LREE-enrichment, and a source also identifiable by radiogenic isotope characteristics.

(3) North Kamchatka shoshonites were derived through partial melting of variably depleted mantle modified by addition of different slab components (hydrous fluid in the west and high-Na felsic melt in the east).

(4) Source hybridization took place prior to or during the initial stages of intra-arc rifting, which also triggered melting in the subduction-modified mantle wedge. High-Ti shoshonites may mark the initial stages of intra-arc extension in volcanic arcs.

This paper benefitted from discussions with A. Fedorchuk, A. Koloskov, M. Defant, M. Drummond and R. Maury. I thank J. Smellie for his comments and Editor's patience and D. Scholl and R. Stern for the discussions of regional tectonic setting and shoshonite petrogenesis during the Cambridge meeting. R. Taylor and two anonymous reviewers are gratefully acknowledged for the detailed comments which significantly improved the earlier version of the manuscript.

References

ALBAREDE, F. & TAMAGNAN, V. 1988. Modelling the recent geochemical evolution of the Piton de la Fournaise volcano, Reunion island, 1931–1986. *Journal of Petrology*, **29**, 997–1031.

ARCULUS, R.J., JOHNSON, R.W., CHAPPELL, B.W. *et al.* 1983. Ophiolite-contaminated andesites, trachybasalts and cognate inclusions of Mount Lamington, Papua New Guinea: anhydrite-, amphibole-bearing lavas and the 1951 cumuldome. *Journal of Volcanology and Geothermal Research*, **18**, 215-247.

BLOOMER, S.H., STERN, R.J., FISK, E. & GESCHWIND, C.H. 1989. Shoshonitic volcanism in the Northern Mariana arc. I. Mineralogic and major and trace element characteristics. *Journal of Geophysical Research*, **94**, 4469-4496.

BOGDANOV, N.A. 1988. Geology of the Komandorsky deep basin. *Journal of Physics of the Earth*, **36**, 65–71.

—— & KEPEZHINSKAS, P.K. In press. Lithospheric heterogeneity in the framing of the Komandorsky Basin, Bering Sea. *In*: BOURGEOIS, J. (ed.) *Tectonics of circum-Pacific margins*. VNU Publishers, Netherlands.

CARROLL, M.R. & WYLLIE, P.J. 1989. Experimental phase relations in the system tonalite-peridotite-H_2O at 15 kb; implications for assimilation and differentiation processes near the crust-mantle boundary. *Journal of Petrology*, **30**, 1351–1382.

DAVIDSON, J.P. 1985. Mechanisms of contamination in Lesser Antilles island arc magmas from radiogenic and oxygen isotope relationships. *Earth and Planetary Science Letters*, **72**, 163–174.

DEFANT, M.J., KEPEZHINSKAS, P.K. & DRUMMOND, M.S. 1993. Na metasomatism in the island arc mantle by slab melt-peridotite interaction. II. Origin of Nb-enriched arc basalts. *Journal of Petrology* (in press).

DOSTAL, J., ZENTILLI, M., CAELLAS, J.C. & CLARK, A.H. 1977. Geochemistry and origin of volcanic rocks of the Andes (26–28°S). *Contributions to Mineralogy and Petrology*, **63**, 113–128.

DRUMMOND, M.S. & DEFANT, M.J. 1990. A model for trondhjemite-tonalite-dacite genesis and crustal growth via slab melting: Archean to Modern comparisons. *Journal of Geophysical Research*, **95**, 21503–21521.

EDWARDS, C., MENZIES, M. & THIRLWALL, M. 1991. Evidence from Muriah, Indonesia, for the interplay of supra-subduction zone and intraplate processes in the genesis of potassic alkaline magmas. *Journal of Petrology*, **32**, 555–592.

ELLAM, R.M., MENZIES, M., HAWKESWORTH, C.J., LEEMAN, W.P., ROSI, M. & SERRI, G. 1988. The transition from calc-alkaline to potassic orogenic magmatism in the Aeolian Islands, Southern Italy. *Bulletin of Volcanology*, **50**, 386–398.

FEDORCHUK, A.V. & IZVEKOV, I.N. 1992. New data on the structure of the northern Sredinny Range, the Isthmus of Kamchatka. *Transactions of the Russian Academy of Science, Geological Series*, **2**, 147–151 (in Russian).

FIRSOV, L.V. 1987. Geochronology of magmatic rocks from the southwestern part of the Koryak upland (Olyutor Depression). *In*: NIKOLAEVA, I.V. (ed.) *Regional geochronology of Siberia and Soviet Far East*. Nauka Publishers, Novosibirsk, 7–32 (in Russian).

FODEN, J.D. 1986. The petrology of Tambora volcano, Indonesia: a model for the 1815 eruption. *Journal of Volcanology and Geothermal Research*, **27**, 1–41.

GILL, J.B. 1981. *Orogenic andesites and plate tectonics*. Springer-Verlag, New York, 390 pp.

—— & WHELAN, P. 1989. Early rifting of an oceanic island arc (Fiji) produced shoshonitic to tholeiitic basalts. *Journal of Geophysical Research*, **94**, 4561–4578.

HAWKESWORTH, C.J., PANKHURST, R.J., HAMILTON, P.J. & EVENSEN, N.M. 1977. A geochemical study of island-arc and back-arc tholeiites from the Scotia Sea. *Earth and Planetary Science Letters*, **36**, 253–262.

——, O'NIONS, R.K. & ARCULUS, R.J. 1979. Nd and Sr isotope geochemistry of island arc volcanics, Grenada, Lesser Antilles. *Earth and Planetary Science Letters*, **45**, 237–248.

HOCHSTAEDTER, A.G., KEPEZHINSKAS, P.K., DEFANT, M.J., DRUMMOND, M.S. & BELLON, H. In press. Tectonic significance of Neogene volcanism in Northern Kamchatka. *Journal of Geology*.

HOOPER, P.R. 1988. Crystal fractionation and recharge (RFC) in the American Bar flows of the Imnaha basalt, Columbia River Basalt Group. *Journal of Petrology*, **29**, 1097–1118.

JOLLY, W.T. 1971. Potassium-rich igneous rocks from Puerto Rico. *Geological Society of America Bulletin*, **82**, 399–408.

KELLER, J. 1974. Petrology of some volcanic rock series of the Aeolian Arc, Southern Tyrrhenian Sea: calc-alkaline and shoshonitic associations. *Contributions to Mineralogy and Petrology*, **46**, 29–47.

KEPEZHINSKAS, P.K. 1989. Origin of the hornblende andesites of northern Kamchatka. *International Geological Review*, **31**, 246–252.

—— 1990. *Cenozoic volcanic series in the framing of marginal seas*. Nauka Publishers, Moscow, 167pp (in Russian).

——, GULKO, N.I. & EFREMOVA, L.B. 1990. Geochemistry of Late Cenozoic high-K volcanites of the Isthmus of Kamchatka. *Geochemistry International*, **27**, 96–103.

——, KRAVCHENKO-BEREZHNOY, I.R. & GULKO, N.I. 1988. Cenozoic shoshonites from North Kamchatka: implications for the tectonic interpretation of island arc shoshonitic series. *In*: SOBOLEV, N.V. (ed.) *Mafic complexes at different stages of lithospheric evolution*. Nauka Publishers, Novosibirsk, 98–114 (in Russian).

——, BELLON, H., DEFANT, M.J., DRUMMOND, M.S., MAURY, R. & HOCHSTAEDTER, A.G. 1993. Temporal evolution of magmatism in western Bering Sea region: implications for the crustal growth in northwestern Pacific. *Geology* (in press).

KONTAK, D.J., CLARK, A.H., FARRAR, E., PEARCE, T.H., STRONG, D.F. & BAADSGAARD, H. 1986. Petrogenesis of a Neogene shoshonite suite, Cerro Moromoroni, Puno, Southeastern Peru. *Canadian Mineralogist*, **24**, 117–135.

KOVALENKO, D.V. 1990. Paleomagnetic investigation of arc-derived complexes from the Olutor zone and Karaginsky Island with tectonic implications. *Geotectonics*, **2**, 92–101 (in Russian).

KRAVCHENKO-BEREZHNOY, I.R., CHAMOV, N.P. & SHERBININA, E.A. 1990. MORB-like tholeiites in a late Eocene turbidite sequence on Karaginsky Island (the western Bering sea). *Ofioliti*, **15**, 231–250.

MACDONALD, R., UPTON, B.G.J., COLLERSON, K.D., HEARN, B.C. & JAMES, D. 1992. Potassic mafic lavas of the Bearpaw Mountains, Montana: mineralogy, chemistry, and origin. *Journal of Petrology*, **33**, 305–346.

MANETTI, P., PECCERILLO, A. & POLI, G. 1979. REE distribution in Upper Cretaceous calc-alkaline and shoshonitic volcanic rocks from Eastern Srednegorie (Bulgaria). *Chemical Geology*, **26**, 51–63.

MEEN, J.K. 1989. Formation of shoshonites from calc-alkaline basalt magmas: geochemical and experimental constraints from the type locality. *Contributions to Mineralogy and Petrology*, **97**, 333–351.

MORRISON, G.W. 1980. Characteristics and tectonic setting of the shoshonite rock association. *Lithos*, **13**, 97–108.

NAKAMURA, N. 1974. Determination of REE, Ba, Fe, Mg, Na and K in carbonaceous and ordinary chondrites. *Geochimica et Cosmochimica Acta*, **38**, 757–775.

NICHOLLS, J. & STOUT, M.Z. 1988. Picritic melts in Kilauea – evidence from the 1967–1968 Halemaumau and Hiiaka eruptions. *Journal of Petrology*, **29**, 1031–1057.

NOHDA, S. & WASSERBURG, G.J. 1981. Nd and Sr isotopic study of volcanic rocks from Japan. *Earth and Planetary Science Letters*, **52**, 264–276.

PEARCE, J.A. 1982. Trace element characteristics of lavas from destructive plate boundaries. *In*:

THORPE, R.S. (ed.) *Andesites*. John Wiley & Sons, Chichester, 525–548.

——, VAN DER LAAN, S.R., ARCULUS, R.J. *et al.* 1992. Boninite and harzburgite from ODP Leg 125 (Bonin-Mariana forearc): a case study of magma genesis during the initial stages of subduction. *Proceedings of the Ocean Drilling Program, Scientific Reports*, **125**, 623–659.

PECCERILLO, A. & TAYLOR, S.R. 1976. Geochemistry of Eocene calc-alkaline volcanic rocks from the Kastamonu area, northern Turkey. *Contributions to Mineralogy and Petrology*, **58**, 63–81.

PE-PIPER, G. 1980. Geochemistry of Miocene shoshonites, Lesbos, Greece. *Contributions to Mineralogy and Petrology*, **72**, 387–396.

RAPP, R.P. & WATSON, E.B. In press. Water-deficient partial melting of metabasalt at 8–32 kbar and continental growth. *Journal of Petrology*.

——, WATSON, E.B. & MILLER, C.F. 1991. Partial melting of amphibolite/eclogite and the origin of Archean trondhjemites and tonalites. *Precambrian Research*, **51**, 1–25.

SCHOLL, D.W., BUFFINGTON, E.C. & MARLOW, M.S. 1975. Plate tectonics and the structural evolution of the Aleutian-Bering Sea region. *In*: FORBES, R.B. (ed.) *Contributions to the geology of the Bering Sea Basin and adjacent regions*. Geological Society of America Special Papers, **151**, 1–32.

SHAPIRO, M.N., YERMAKOV, V.A., SHANTSER, A.E., SHULDINER, V.I., KHANCHUK, A.I. & VYSOTSKY, S.V. 1987. *Tectonic evolution of the Kamchatka arc*. Nauka Publishers, Moscow, 247 p. (in Russian).

SMELLIE, J.L. 1987. Geochemistry and tectonic setting of alkaline volcanic rocks in the Antarctic Peninsula. *Journal of Volcanology and Geothermal Research*, **32**, 269–285.

STAVSKY, A.P., CHEKHOVITCH, V.D., KONONOV, M.V. & ZONENSHAIN, L.P. 1990. Plate tectonics and palinspastic reconstructions of the Anadyr-Koryak region, northeast USSR. *Tectonics*, **9**, 81–101.

STERN, R.J., BLOOMER, S.H., LIN, P.N., ITO, E. & MORRIS, J. 1988. Shoshonitic magmas in nascent arcs: new evidence from submarine volcanoes in the northern Marianas. *Geology*, **12**, 426–430.

STERN, R.J. & BLOOMER, S.H. 1992. Subduction zone infancy: examples from the Eocene Izu-Bonin-Mariana and Jurassic California arcs. *Geological Society of America Bulletin*, **104**, 1621–1636.

SUN, S.S., NESBITT, R.W. & SHARASKIN, A.YA. 1979. Geochemical characteristics of mid-ocean ridge basalts. *Earth and Planetary Science Letters*, **44**, 119–138.

TAYLOR, R.N., MURTON, B.J. & NESBITT, R.W. 1992. Chemical transects across intra-oceanic arcs: implications for the tectonic setting of ophiolites. *In*: PARSON, L.M., MURTON, B.J. & BROWNING, P. (eds) *Ophiolites and their modern oceanic analogues*. Geological Society, London, Special Publications, **60**, 117–132.

VOLYNETS, O.N., ANTIPIN, V.S., PEREPELOV, A.B., CHUVASHOVA, L.A. & SMIRNOVA, E.V. 1986. Rare earth elements in Late Cenozoic high-K volcanic rocks of Kamchatka. *In*: TAUSON, L.V. (ed.) *Geochemistry of volcanic rocks from various tectonic settings*. Nauka Publishers, 149–165 (in Russian).

UFIMTSEV, G.F. 1975. Neotectonics of the continental part of the Soviet Far East. *Doklady Academii Nauk SSSR*, **221**, 932–934.

WHITE, W.M. & HOFMANN, A.W. 1982. Sr and Nd isotope geochemistry of oceanic basalts and mantle evolution. *Nature*, **296**, 821–825.

WHITFORD, D.J., WHITE, W.M. & JEZEK, P.A. 1981. Neodymium isotope composition of Quaternary island arc lavas from Indonesia. *Geochimica et Cosmochimica Acta*, **39**, 1287–1302

YOGODZINKSI, G.M., VOLYNETS, O.N., KOLOSKOV, A.V., SELIVERSTOV, N.I. & MATVEENKOV, V.V. 1993. Magnesian andesites and the subduction component in a strongly calc-alkaline series at Piip Volcano, far Western Aleutians. *Journal of Geophysical Research* in press.

The relationship between alkaline magmatism, lithospheric extension and slab window formation along continental destructive plate margins

M.J. HOLE[1], A.D. SAUNDERS[2], G. ROGERS[3] & M.A. SYKES[4]

[1] *Department of Geology and Petroleum Geology, University of Aberdeen, Meston Building, Aberdeen AB9 2UE, UK.*

[2] *Department of Geology, University of Leicester, University Road, Leicester LE1 7RH, UK*

[3] *Isotope Geology Unit, Scottish Universities Research and Reactor Centre, East Kilbride, Glasgow G72 0QU, UK.*

[4] *British Antarctic Survey, High Cross Madingley Road, Cambridge CB3 0ET and Department of Geology, University of Nottingham, UK.*

Abstract: Two distinct groups of Late Cenozoic (15–<0.1 Ma), OIB-like basalts are recognized along the Antarctic Peninsula. Small volumes of undersaturated basanites, tephrites, alkali and olivine basalts and 'within-plate' tholeiites, which post-date the cessation of subduction as a result of ridge crest-trench interactions by 40 to <10 Ma, are scattered along much of the southern part of the peninsula. At James Ross Island, in the north of the peninsula, alkali basalts, hawaiites and rare mugearites were erupted synchronously with subduction at the South Shetland Islands trench to the west. Both groups of basalts are remarkably similar in terms of their geochemical and isotopic characteristics, although they apparently owe their origin to two distinct combinations of tectonic processes. The south Antarctic Peninsula basalts are causally related to the cessation of subduction, the formation of slab windows and upwelling and decompressional melting of sub-slab asthenosphere. Correlated trace element-isotope systematics demonstrate that the slab window-related basalts exhibit little evidence for interaction with subduction-enriched mantle or continental lithosphere and must have been derived from sub-slab mantle that had not recently been affected by subduction; LILE/HFSE ratios (e.g. Th/Ta 1.0–2.5, Rb/Nb, 0.25–1.5, Ba/Nb 2.5–7.0) and Sr- and Nd-isotope ratios ($^{87}Sr/^{86}Sr$ 0.7027–0.7035, $^{143}Nd/^{144}Nd$ 0.51286–0.51296) are well within the range for ocean island basalts. The slab window-related basalts underwent rapid uprise from the mantle to the crust accompanied by limited fractional crystallization.

The syn-subduction alkalic basalts of James Ross Island were generated during slab roll-back, probably related to the formation of the extensive slab window to the south. The James Ross Island Volcanic Group (JRIVG) exhibits evidence of polybaric fractionation including ponding at the base of the crust. Again, there is little evidence for interaction of these asthenospheric magmas with subduction-enriched mantle or continental lithosphere, although in detail, there are recognizable differences between the JRIVG and the slab windowrelated basalts. Melting at James Ross Island was facilitated by slab roll-back and associated lateral and vertical asthenospheric migration into the locus of a pre-existing area of attenuated lithosphere ('thinspot'). This phase of extensional tectonism was probably originally initiated during the Late Jurassic, but other periods of extension may have taken place in latest Cretaceous times.

Neither of these suites of basalts were generated as a result of significant lithospheric extension and passive asthenospheric upwelling on a regional scale. In addition, there is no evidence for the existence of a mantle hotspot beneath the region. The commonly held assumption that alkaline volcanism along consuming plate margins results from periods of significant inter-arc extensional tectonism is, therefore, not necessarily valid. The unique tectono-magmatic regime resulting from the formation of slab windows is probably the only setting in which small degree melts of MORB source asthenosphere not associated with a plume are generated and preserved in the geological record.

From Smellie, J.L. (ed.), 1995, *Volcanism Associated with Extension at Consuming Plate Margins*, Geological Society Special Publication No. 81, 265–285.

The bulk of magma generation along continental destructive plate margins is initiated because of a lowering of the mantle peridotite solidus due to a flux of H_2O– CO_2-rich fluids derived from dehydration of the uppermost portion of the descending oceanic lithosphere (e.g. Gill 1981). Such calc-alkaline magmas carry the distinctive trace element and isotopic fingerprints of fluid involvement, such as high large ion lithophile element (LILE)/high field strength element (HFSE) ratios, and often relatively radiogenic Sr-isotope and unradiogenic Nd-isotope ratios compared with MORB. Interaction of mantle wedge-derived magmas with continental lithosphere of variable composition and age is also acknowledged as being an important factor in the evolution of continental subduction-related magmas (e.g. Pearce 1982; Ellam & Hawkesworth 1988; Rogers & Hawkesworth 1989; Hole *et al.* 1991*a*). However, a surprisingly common feature of both oceanic and continental destructive plate margins is the close spatial and temporal association between subduction-related calc-alkaline magmatism and 'within plate' alkaline basalts, which are almost indistinguishable from ocean island basalts (OIB). Examples of this association have been reported from a number of locations along the Pacific margin of the Americas and the Antarctic Peninsula (e.g. SW California, Johnson & O'Neil 1984; British Columbia, Bevier *et al.* 1979; Bevier 1983; Baja California, Rogers *et al.* 1985; Rogers & Saunders 1989; Storey *et al.* 1989; Ukinrek Maars, (Alaska), Keinle *et al.* 1980; Patagonia, Baker *et al.* 1981; Stern *et al.* 1990; Ramos & Kay 1992; Antarctic Peninsula, Smellie 1987; Hole 1988; 1990*a,b*; Hole *et al.* 1991*a,b,c*).

Generation of 'within plate' magmas requires the operation of quite different tectono-magmatic processes from calc-alkaline magmatism. Upwelling and decompressional melting of asthenosphere in areas of continental lithospheric attenuation is one way of generating within plate basalts (McKenzie & Bickle 1988) and this process apparently may follow-on from long periods of subduction-related magmatism (e.g. western USA). Additionally, if mantle hotspots are present beneath continental crust, alkaline magmas may be generated in much the same way as OIB, due to abnormally high asthenospheric temperatures. OIB-like continental alkalic magmas do, however, require some extension or transtension to allow their unimpaired uprise through the continental lithosphere. A manifestation of hotspot activity is lithospheric attenuation but Thompson & Gibson (1991) pointed out that in some cases (e.g. Parana basalts of Brazil and the Columbia River

basalts of the NW United States), thinning of the lithosphere appears to pre-date the initiation of plume activity by substantial periods of time; in this case magmatism requires a plume and a pre-existing 'thinspot' (Thompson & Gibson 1991). It is therefore no surprise that the recognition of closely associated 'within-plate' and calc-alkaline volcanic rocks in subduction complexes has often led to the assumption that such changes in the character of magmatism reflect periods of significant inter-arc extension and/or plume activity.

We will demonstrate that this is not necessarily the case, and that alkalic magmatism in this setting can owe its existence to changes in the geometry, convergence rate and overall configuration of the descending slab of oceanic lithosphere. A combination of periods of slab roll-back, the generation of slab windows and ridge crest-trench interactions may be the dominant processes in promoting alkaline magmatism.

Late Cenozoic alkaline magmatism at destructive plate margin

In detail, late Cenozoic OIB-like volcanism along destructive continental margins appears to occur in two distinct tectono-magmatic settings, either being closely associated with the cessation of subduction following ridge crest-trench collision (e.g. British Columbia, Baja California, SW California and Antarctic Peninsula) or occurring in a syn-subduction setting (e.g. James Ross Island (northern Antarctic Peninsula) and Patagonia). Thorkelson & Taylor (1989) and Hole *et al.* (1991*b*) noted that there was a spatial and temporal coincidence of the occurrence of the post-subduction alkalic volcanism with the generation of slab windows beneath continental margins. Slab windows are formed as a consequence of ridge crest-trench collisions, the locking of the trailing plate to the continental margin and continued subduction of the leading plate. Magmatism appears to be initiated by decompressional melting as a consequence of asthenospheric upwelling to fill the incipient void formed by the removal of oceanic lithosphere from beneath the continental margin. Significant syn-magmatic attenuation of the lithosphere in the hanging wall is therefore not a prerequisite to the formation of magmas in this setting, although the mechanical boundary layer (MBL) needs to be sufficiently thin to enable melting. Therefore, magmatism associated with slab windows is largely independent of the prevalent tectonic regime in the overlying continental lithosphere (Hole *et al.* 1991*b*).

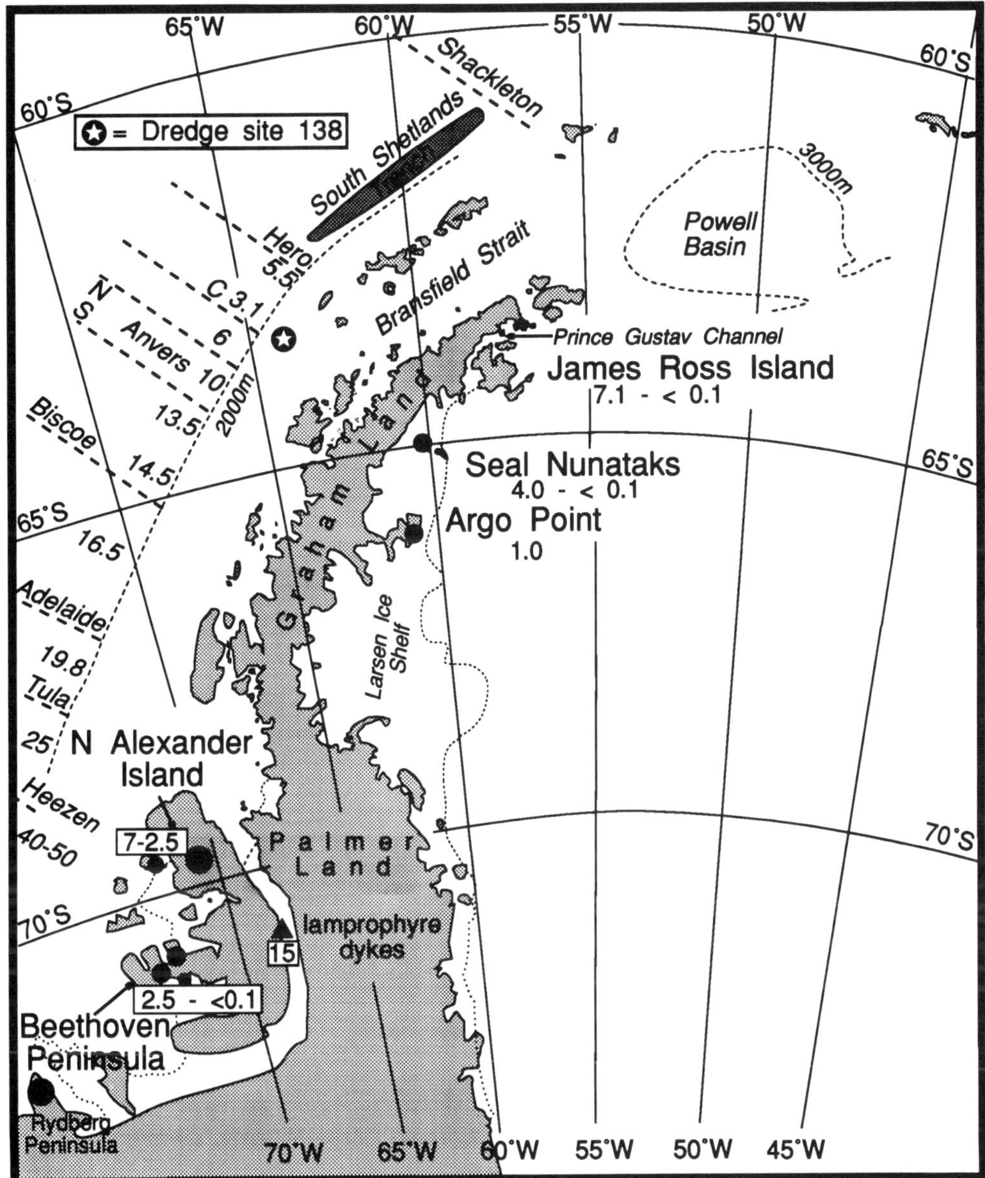

Fig. 1. Map of the Antarctic Peninsula showing the location and age range of the Late Cenozoic volcanic centres. The position of fracture zones, and ridge crest–trench collision times (in italics) for each segment of oceanic crust are also shown. Note that subduction is still active at the South Shetlands Islands trench. Data from Barker (1982), Hole (1988; 1990*a*), Larter & Barker (1991), Smellie *et al.* (1988) and Hole & Larter (1993).

However, in cases where oblique collision has occurred, transcurrent faulting may be generated, facilitating egress of magmas to the surface. Syn-subduction alkalic volcanism like that at James Ross Island and in Patagonia may not be directly related to slab window formation, and a variety of models have been put forward to explain its existence, including slab 'roll-back' and subduction-induced asthenospheric convection and upwelling (e.g. Stern *et al.* 1990). However, Ramos & Kay (1992) have suggested that slab windows generated as a consequence of the collision of the Chile Rise with the Chile trench may have been responsible for Patagonian alkaline magmatism, although subduction is still active along the adjacent portion of the Chile trench.

A significant feature of the late Cenozoic geology of the Antarctic Peninsula is that both syn-subduction and slab window-generated volcanism overlap in time and occur in different segments of the same arc (Fig. 1). The geophysical record of subduction processes and the timing and geometry of ridge crest trench interactions are well-known and tightly constrained (Barker 1982; Larter & Barker 1991) allowing a detailed examination of the relationship between alkalic magmatism and subduction history to be made. Here, we examine the tectono-magmatic controls on both post-subduction and syn-subduction alkalic volcanism along the Antarctic Peninsula are examined.

Slab window formation

Dickinson & Snyder (1979) showed that subduction of a triple junction will ultimately result in the generation of a slab-free region beneath a continental margin that is otherwise characterized by continued subduction of oceanic lithosphere. The geometry of slab windows is dependent on the convergence angle of the relevant ridge crest; oblique collision results in triangular slab windows whereas nearly orthogonal collision of a segmented length of ridge crest should result in the formation of a series of rectangular slab windows (Dickinson & Snyder 1979). Thorkelson & Taylor (1989) further developed this hypothesis and demonstrated that areas of the Pacific margin of the Americas had been underlain by slab windows during various periods throughout the Cenozoic; indeed, in addition to the Antarctic Peninsula, slab window models have been proposed to explain the generation of undersaturated alkalic and tholeiitic basalts in SW Calfornia, northern Baja California, British Columbia and Patagonia (Johnson & O'Neil 1984; Storey *et al.* 1989; Hole *et al.* 1991*a*; Ramos & Kay 1992). Baja California is a particularly important and well-constrained locus of slab window-related volcanism as it has been shown that the locus of generation of a slab window coincided spatially and temporally with the eruption of alkalic basalts at San Quintin, whereas in areas where subduction had stopped but ridge crests had been abandoned offshore, distinctive high-magnesium andesites (bajaites) were generated (Rogers *et al.* 1985; Saunders *et al.* 1987; Rogers & Saunders 1989; Storey *et al.* 1989). Therefore, alkaline magmatism in this setting appears to be specifically related to slab windows and not

simply due to changes in plate boundary forces following ridge crest-trench collision.

The tectonic evolution of the Antarctic Peninsula during the Late Cenozoic was such that an extremely large slab window system developed over the past 50 Ma (Hole 1990*a*; Hole *et al.* 1991*b*) (Fig. 2). Slab windows totalling more than 2000 km in length currently underlie the peninsula, and occurrences of undersaturated OIB-like basaltic rocks are scattered along the entire length of the peninsula.

Geometry and timing of slab window formation and associated volcanism along the Antarctic Peninsula

Two areas of slab window-related (i.e. post-subduction) alkalic magmatism are present along the Antarctic Peninsula. On Alexander Island, 15.0–<1.0 Ma spinel lherzolite- and kaersutite megacryst-bearing basanites, tephrites, alkali basalt and olivine basalts unconformably overlie a Mesozoic accretionary prism complex. At Seal Nunataks, 4.0–<1.0 Ma alkali basalts, olivine basalts and 'within-plate' tholeiites were erupted behind the Triassic-early Tertiary palaeo-arc at the site of a late Jurassic to Eocene back-arc basin (Larsen Basin; MacDonald *et al.* 1988; MacDonald & Butterworth 1990). Large (up to 25 cm diameter) spinel lherzolite xenoliths are common in the Seal Nunataks basalts.

Marine magnetic anomalies record the collision history along the west coast of the peninsula and the tectonic configuration of the destructive plate margin prior to the cessation of subduction is well defined (Barker 1982; Larter & Barker 1991). Subduction ceased after a series of northeastward-younging ridge crest-trench collisions along the NW coast of the peninsula. Prior to the cessation of subduction, the oceanic crust to the NW was divided into discrete segments by a series of NW–SE orientated transform fracture zones. Collision times decreased from *c.* 50 Ma in the SW at Alexander Island to *c.* 4 Ma west of Anvers Island, 150 km WNW of Seal Nunataks (Fig. 1). Prior to each collision event, calc-alkaline volcanism stopped (Barker 1982). For example, at the locus of a 25 Ma ridge crest-trench collision in northern Alexander Island, the youngest calc-alkaline volcanism occurred between 40 and 60 Ma ago (Burn 1981).

The destiny of the leading plate during such collision events is open to debate; it may simply sink into the asthenosphere possibly following its previous subduction trajectory, resulting in

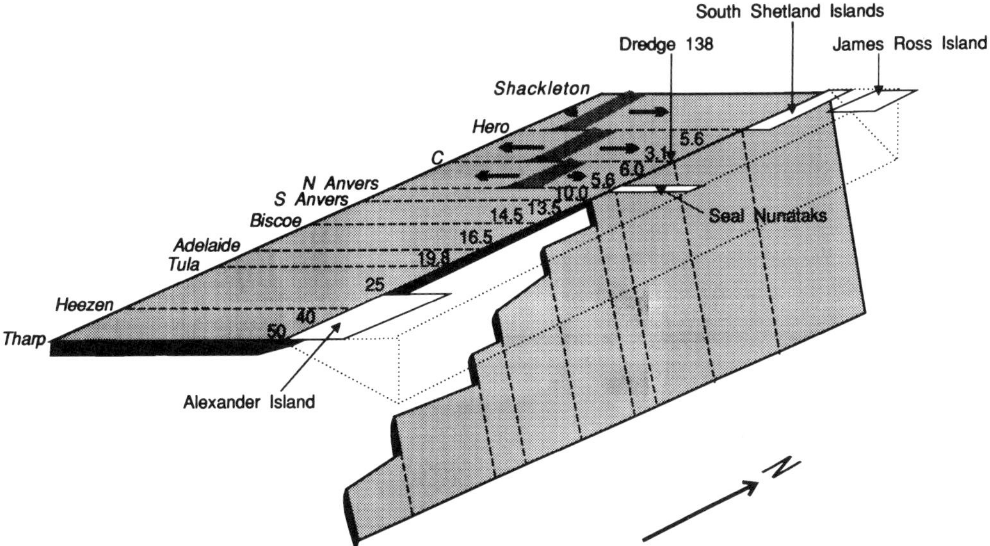

Fig. 2. Cartoon of the slab window system beneath the Antarctic Peninsula, with collision times shown for each window segment. Because of the complexity of the collision events at the northern end of the peninsula, these have been simplified to orthogonal collisions. A detailed discusion of the youngest collision events can be found in Larter & Barker (1991) and Hole & Larter (1993).

progressively greater separation from the previous spreading ridge, or it may simply roll back. In the Antarctic Peninsula, the collision of a segment of the ridge crest with the trench was most likely to have resulted in progressively greater separation of the trailing and leading plates along the line of greatest weakness, the spreading centre. The southerly leading plate segment remained coupled to the still-subducting northerly leading plate segment by way of a transform fracture zone, which would have represented a far more coherent, rigid couple between adjacent segments of oceanic lithosphere than a spreading centre (Fig. 2). The southern leading plate therefore continued to subduct after ridge crest-trench collision, resulting in the formation of a slab window. The sub-orthogonal nature of all but the earliest collision event resulted in the formation of a 'zig-zag' shaped window as predicted by Dickinson & Snyder (1979; Fig. 3).

Because collision times and subduction rates (equivalent to total spreading rate) are well known, it is possible to calculate the length of an individual slab window segment at a given time, assuming that the leading plate remained coupled to the next segment of leading plate to the north. Figure 3 illustrates the generation of slab windows in relation to the geographical distribution of post-subduction alkalic basalts. The assumptions that are used in the calculations

are that the angle of slab dip was 45°, and the rate of opening of an individual slab window segment was dependent on the spreading rate of the adjacent ridge crest to the north (see Barker 1982 and Larter & Barker 1991). It is also assumed that the subducted slab was planar and the slab dip did not vary over time. Reducing the slab dip to 30 increases the width scale from 1400 km to 1650 km. It is noticeable that the oldest volcanism, represented by basanitic camptonite dykes (15 Ma) in segment 1, and the volcanic centres at Seal Nuntaks are close to the landward trace of subducted transforms, whereas the remaining volcanic centres are in the centre of slab window segments.

The time between the passage of the edge of the slab window beneath the locus of volcanism and the initiation of volcanism at that locus can be estimated from Figure 3. For example, basanitic dykes formed at 15 Ma, which straddle the Heezen fracture zone, post-date the passage of the window margin at that location by approximately 7–10 Ma, depending on slab dip. Tephrites (c. 7 Ma old) in northern Alexander Island post-date window opening at that location by 7–9 Ma. Thus it is clear that magmatism did not occur immediately after the passage of the window margin beneath any one location. For the majority of the volcanic centres shown on Fig. 3, the edge of the slab window had passed the locus of the oldest volcanism associated with

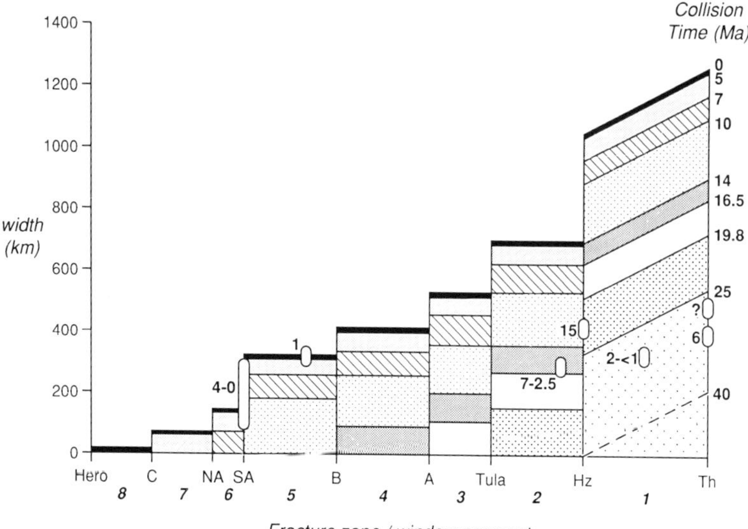

Fig. 3. Plan view of slab window growth with time, showing the geographical location of the slab window-related basalts (open symbols with K–Ar age adjacent). The *x*-axis represents the palaeo-trench and the different ornaments correspond to the amount of slab window formation associated with each spreading ridge, the solid lines separating the ornaments being 'isochrons' for the slab window as a whole. The assumptions that are used in the calculations are that the angle of slab dip was 45°, and the rate of opening of an individual slab window segment was dependent on the spreading rate of the adjacent ridge crest to the north (see Barker 1982 and Larter & Barker 1991). It is also assumed that the subducted slab was planar, and the slab dip did not vary over time. Reducing the slab dip to 30 increases the linear width scale from 0 to 1400 km to 0 to 1650 km.

window formation. However, volcanism appears to have continued for an extended period after slab window formation at the loci of the oldest collision episodes, such as that in southern Alexander Island where volcanism as young as 2.5 Ma occurred more than 30 Ma after slab window formation. The slab window-related basalts of San Quintin volcanic field, Baja California, were erupted 12.5 Ma after subduction stopped.

Syn-subduction alkalic magmatism at James Ross Island

Throughout the entire period of alkalic volcanism on James Ross Island, subduction was active to the NW, along a segment of the arc which is now represented by the South Shetlands Islands trench (Fig. 1). In this case, the JRIVG cannot be directly related to slab window formation in the same way as the rest of the alkalic volcanism along the Antarctic Peninsula.

The oldest volcanism on James Ross Island is recorded as clasts in sub-volcanic diamictites, dated at 7.1 Ma (Sykes 1988*a*). However, the majority of the volcanism occurred between 3.0 and 6.5 Ma with a migration of the magmatic focus north-eastward from James Ross Island towards Prince Gustav channel about 2 Ma ago. This migration was probably related to the opening of Bransfield Strait to the NW at around 4 Ma (Weaver *et al.* 1979; Sykes 1988*a*), at which time subduction rates at the South Shetlands Islands trench slowed dramatically from *c.* 20 mm a^{-1} to 2 mm a^{-1} at most (Barker 1982). Due to the migration of the trench to the NW, the pre-existing position of the JRIVG volcanism at 7.1–3.0 Ma was therefore superseded by a more trench-distal location at around 4 Ma. In terms of overall tectonic setting, the JRIVG of James Ross Island therefore bears strong similarities to that of the cratonic alkalic basalts of Patagonia (Baker *et al.* 1981; Stern *et al.* 1990; Ramos & Kay 1992). The JRIVG unconformably overlies more than 5 km of late Jurassic to Eocene sedimentary rocks deposited in the Larsen Basin, which was formed as a result of inter-arc lithospheric extension. In contrast to the small volumes of volcanic and volcaniclastic rocks erupted at Seal Nunataks and Alexander Island (Hole 1988; 1990*a*,*b*; Hole *et al.* 1991*a*,*b*),

Table 1. *Representatative analyses of James Ross Island Volcanic Group alkalic basalts*

Age	4.44 ± 0.25			5.4 ± 0.5		5.37 ± 0.35		4.36 ± 0.21	2.47 ± 0.38		5.15 ± 0.33	6.5 ± 0.6		
Rock type Sample No.	Hawaiite D.8710.4	Basalt D.8711.1	Basalt D.8725.1	Hawaiite D.8728.1	Ph-teph D.8728.2	Hawaiite D.8730.1	Basalt D.8749.1	Hawaiite D.8814.1	Basalt D.8818.1	Hawaiite D.8819.1	Hawaiite D.8822.1	Basalt D.8826.3	Picro-bas D.8827.1	Basanite D.8838.1
(wt%)														
SiO_2	49.61	46.74	47.79	49.68	50.99	48.84	47.78	49.52	48.86	49.39	48.78	47.43	42.36	48.99
TiO_2	2.06	1.91	1.70	2.00	3.11	1.70	1.86	2.79	1.84	1.86	1.93	1.88	0.71	2.06
Al_2O_3	16.29	14.51	14.84	16.44	14.90	15.84	14.91	14.94	16.63	16.36	16.68	15.65	10.99	17.10
Fe_2O_3*	10.07	11.55	12.90	9.82	11.26	12.59	12.83	11.64	10.38	10.78	11.34	11.70	14.86	9.68
MnO	0.14	0.15	0.16	0.14	0.17	0.17	0.17	0.17	0.14	0.14	0.14	0.16	0.18	0.13
MgO	6.53	9.91	9.65	5.12	2.66	7.24	8.45	3.04	7.06	6.52	6.26	8.22	18.25	4.66
CaO	7.07	8.65	8.82	7.77	6.11	7.37	8.98	7.36	8.12	8.03	8.93	8.99	4.70	7.82
Na_2O	4.37	3.08	3.01	4.42	5.31	3.78	3.34	4.81	3.99	4.27	3.91	3.56	3.49	4.56
K_2O	2.43	1.39	0.90	2.16	3.42	1.63	0.83	2.40	1.75	1.51	1.26	0.95	0.20	2.47
P_2O_5	0.80	0.44	0.35	0.67	0.17	0.60	0.34	0.78	0.54	0.59	0.44	0.35	0.10	0.76
LOI	0.90	2.50	0.00	1.30	0.54	0.13	0.10	2.34	0.64	0.53	0.00	1.01	3.77	1.82
Total	100.27	100.83	100.12	99.52	98.64	99.89	99.59	99.79	99.95	99.98	99.67	99.90	99.61	100.05
(ppm)														
Cr	152	316	308	104	3	147	257	8	176	169	133	248	430	67
Ni	110	222	198	37	9	128	123	9	97	78	58	111	536	25
Co	34.10	49.40				42.70	49.20		35.90	36.70	37.50		90.20	
Rb	24	13	14	35	51	12	10	31	16	13	17	12	6	24
Ba	301	158	129	274	374	132	140	304	205	248	161	149	76	246
Th	4.28	2.08				2.62	1.85	0.40	3.04	3.55	2.67			0.47
U	1.71	0.53				1.40			1.29	1.12	0.82			
Sr	1018	666	418	755	560	630	447	483	720	850	584	470	231	896
Zr	257	180	128	241	408	243	135	306	212	191	170	160	53	290
Hf	5.56	3.71				4.89	3.10		4.27	4.08	1.59		1.09	
Nb	53	29	21	44	79	32	23	59	38	42	26	24	5	56
Ta	3.36	1.73				2.08	1.35		2.33	2.40	1.59		0.30	
La	37.30	20.10		36.20	54.41	25.10	16.40	37.00	28.50	31.10	21.40	16.41	5.00	34.55
Ce	73.90	42.50		59.42	100.92	51.90	32.90	35.68	55.40	56.80	40.20	33.07	10.60	68.95
Nd	36.60	23.00		26.54	44.26	26.00	16.20	72.63	28.50	28.20	21.10	17.44	5.40	30.45
Sm	7.37	4.97		5.52	9.96	5.56	4.33	34.40	5.58	5.62	5.01	4.74	1.44	6.72
Eu	2.37	1.75		1.93	2.75	1.86	1.59	7.28	1.94	1.96	1.74	1.58	0.65	2.25
Gd				5.35	9.18			2.42				4.54		5.57
Dy				4.07	7.21			7.34				4.52		4.42
Ho				0.81	1.34			6.12				0.83		0.78
Er				2.22	3.75			1.26				2.24		2.17
Tb	0.93	0.85				0.93	0.77	3.56	0.91	0.74	0.85		0.34	
Yb	1.63	1.90		1.72	3.08	1.76	1.97		1.99	1.72	2.14	2.11	0.96	1.68
Lu	0.25	0.29		0.25	0.41	0.26	0.31	2.79	0.31	0.28	0.34	0.28	0.16	0.21
Y	23	22	24	24	39	22	24	37	26	21	24	24	11	21
Sc	17.70	24.10				18.70	26.60		22.50	20.60	26.70		17.00	
$^{87}Sr/^{86}Sr$	0.7033	0.7033	0.7034	0.7035	0.7034	0.7029	0.7034	0.7036	0.7034	0.7036	0.7033	0.7035	0.7033	0.7033
$^{143}Nd/^{144}Nd$	0.51289	0.512871	0.512836	0.512817		0.512907	0.512795	0.512781	0.512789	0.512812	0.512865	0.512795	0.512861	0.512863

Samples which have been analysed for Ta were analysed for the REE, Sc, Hf, Th, Co by INAA. Other REE data by ICP at Royal Holloway and Bedford New College, using the procedures of Walsh *et al.* (1981). K–Ar Ages are from Sykes (1988*a*). Sample location details can be found in Sykes (1988*a, b*). Errors on $^{87}Sr/^{86}Sr$ ratios ± 0.005% based on replicate analyses of NBS#987. Measured $^{143}Nd/^{144}Nd$ ratios normalized to a natural $^{146}Nd/^{144}Nd$ ratio of 0.7219. Errors on $^{143}Nd/^{144}Nd$ ratios based on replicate analyses of La Jolla; 0.511842 ± 19 ppm ($n = 37$). Errors on INAA data are less than 5% (2 c.v.) for all elements based on replicate analyses of Whin Sill dolerite.
* Total iron expressed as Fe_2O_3.

relatively large volumes of material were produced at James Ross Island; the estimated thickness of the JRIVG is 1500 m. There is clearly a temporal overlap of the volcanic activity at James Ross Island with that of the slab window-related volcanism along the rest of the Antarctic Peninsula. However, consideration of Fig. 1 & 3 shows that at the time of initiation of volcanism at James Ross Island (7.1 Ma), a slab of the subducting Phoenix plate occupied the subduction zone beneath it from the south Anvers to Shackleton fracture zones, whereas the majority of the Antarctic Peninsula was underlain by a slab window.

Geochemistry

The majority of elemental, isotopic and geochronological data for the post-subduction volcanic rocks have been presented elsewhere (Smellie 1987; Hole 1988; 1990*a,b*; Hole *et al.* 1991*b*). New data for James Ross Island are presented in Table 1. For the James Ross Island samples, major and trace elements were analysed by standard XRF techniques at the University of Nottingham, and the rare earth elements (REE), Th, Ta, Hf, Sc, Co and U by INAA at the Open University. Other REE data were analysed by ICP at Royal Holloway and Bedford New College, London. Sr- and Nd-isotope compositions for both James Ross Island and the rest of the Antarctic Peninsula were analysed at the NERC Isotope Geoscience Laboratories, Gray's Inn Road, London.

Basanites, tephrites, alkali and olivine basalts and 'within-plate' tholeiites are all represented in the slab window-related suite. Hy- and ne-normative alkali basalts predominate at Seal Nunataks, whereas strongly ne-normative compositions are a feature of Alexander Island (Fig. 4a). A striking feature of the majority of the slab window-related basalts is their relatively primitive nature: all fall in the range 4.8–11.7% MgO, with only a few samples having <6.0% MgO. Ni (54–357 ppm) and Cr (67–447 ppm) contents cover a broad range, but more than 90% of samples analysed have Ni >150 ppm and Cr >250 ppm. Fractional crystallisation seems largely to have been restricted to olivine and minor clinopyroxene (Hole 1988; 1990*a,b*); in general both CaO and Al_2O_3 exhibit progressive increases whilst MgO decreases with increasing $Fe_2O_{3t}/(Fe_2O_{3t}+MgO)$ (Fe_2O_{3t} = total iron as Fe_2O_3) ratio. The Seal Nunataks ne-normative samples plot close to the thermal divide (ol–di join) in an expanded normative tetrahedron (Fig. 4a) and there does not seem to be a continuum of data across the divide; a compositional gap exists between the undersaturated and saturated samples. A trend of

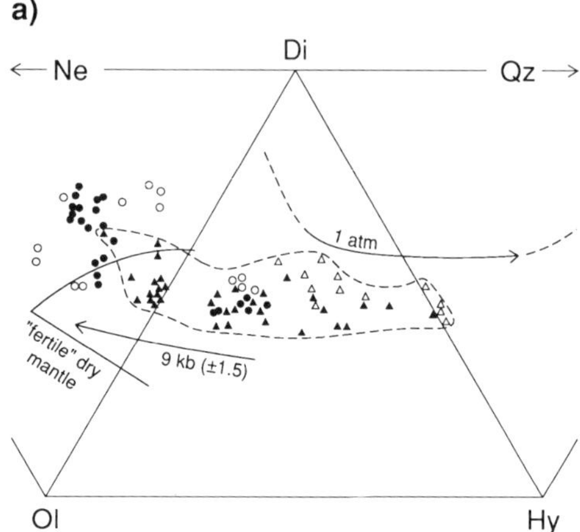

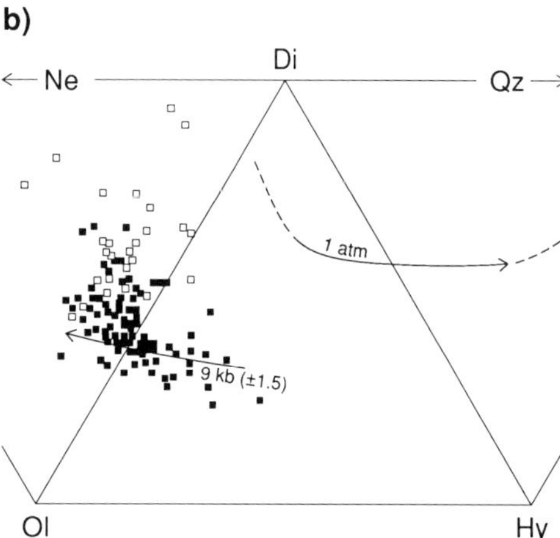

Fig. 4. Expanded normative tetrahedron of Thompson (1982) with data for (**a**) Alexander Island (open and filled circles) and Seal Nunataks (open and filled triangles) and (**b**) James Ross Island Volcanic Group (open and filled squares). In both cases open symbols are for samples with <7.0% MgO and filled symbols >7.0% MgO. Diagram constructed after Thompson (1982). Note that the 1 atm and 9 kbar fractionation pathways diverge with decreasing MgO.

decreasing MgO with increasing saturation is evident, indicating the probability of low pressure control on fractionation (Thompson 1982; Hole & Morrison 1992). Only a small number of the Alexander Island samples are hy-normative and these samples exhibit elevated Sr-isotope and large ion lithophile element (LILE)/high field strength element (HFSE) ratios compared with other Alexander Island samples, a likely

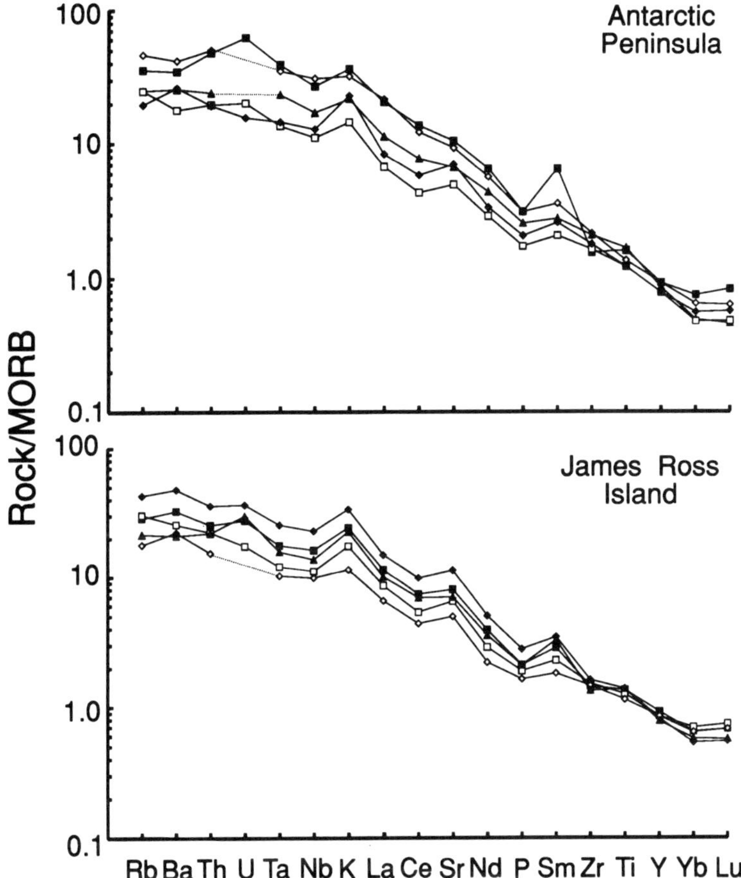

Fig. 5. MORB-normalized plots for (**a**) slab window-related basalts and (**b**) James Ross Island alkali basalts and hawaiites (see Table 1). Symbols for Fig. 5a: open diamond KG.3619.1, 5.5 Ma basanite, Rothschild Island; filled square KG.3719.17, 6.5 Ma tephrite, N Alexander Island; filled triangle <0.1 Ma alkali basalt, Pollux Nunatak, Seal Nunataks; open square R.3728.1, 4.0 Ma 'within plate' tholeiite, Evensen Nunatak, Seal Nunataks; filled diamond <1.0 Ma alkali basalt, Alexander Island. Data from Hole (1988, 1990*a*, 1991*a,b*); Symbols for Fig. 5b: filled square, D.8818.1; open square D.8822.1; filled diamond, D.8710.4; open diamond, D.8749.1; filled triangle, D.8730.1. Normalizing values from Sun & McDonough (1989).

of minor subduction enrichment and/or interaction with the continental lithosphere (Hole 1988). The JRIVG is dominated by alkali basalts and hawaiites and lacks the highly undersaturated basanitic volcanism which characterizes some of the slab window-related basalts from Alexander Island (Fig. 4b). The only exception to this are low MgO pegmatitic dolerites, which are local segregations from laccolithic intrusions formed by high-level, low-pressure crystallization. MgO in the JRIVG basalts varies from *c.* 11% to 5%. The majority of the most primitive JRIVG samples have more normative ol than all the Alexander Island samples; no compositional gap is evident at the ol–di join. There is a general trend towards increasing Si-undersaturation

with decreasing MgO content, the opposite to that of Seal Nunataks. Indeed, the JRIVG samples fall close to the projected pathway for high-pressure (9 ± 1.5 kb) fractionation of an initially Si-saturated basalt (Thompson 1982).

As the majority of both the slab window-related and syn-subduction basalts fall in the range 5–12% MgO, the effects of high-level fractional crystallization on inter-element ratios of HFSE and LILE can largely be ignored in the following discussions (Fitton & Dunlop 1985). All the alkalic basalts of the Antarctic Peninsula exhibit smooth profiles on MORB-normalized multi-element plots, with a steep pattern from Y to K, but a relatively flat pattern from K to Rb (Fig. 5). Such profiles characterize alkalic

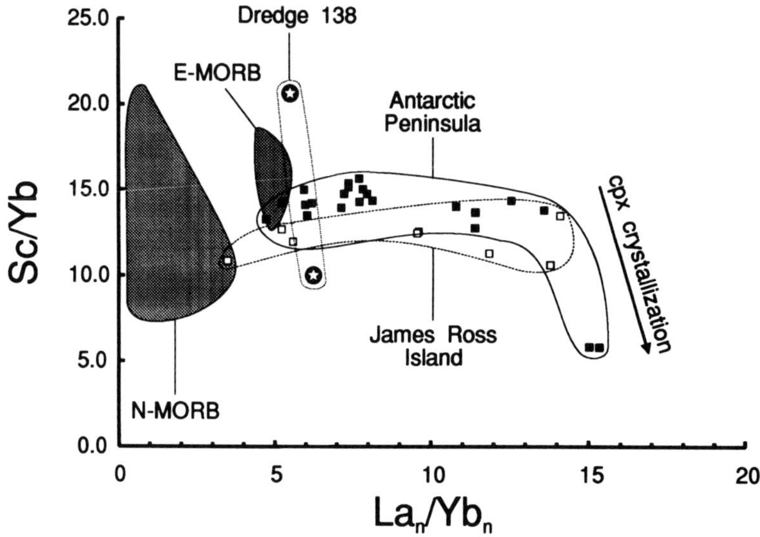

Fig. 6. Plot of Sc/Yb v. La_n/Yb_n for MORB, Antarctic Peninsula slab window-related basalts and James Ross Island Volcanic Group. The two Antarctic Peninsula basalts with Sc/Yb *c.* 5 have undergone clinopyroxene fractionation. Note that one sample from Dredge site 138 has Sc/Yb and La_n/Yb_n ratios similar to E-MORB. Data sources: MORB, Wood *et al.* (1978); LeRoex *et al.* (1983, 1985); Antarctic Peninsula slab window-related basalts, Hole (1988; 1990*a*); Dredge 138 samples Hole & Larter (1993).

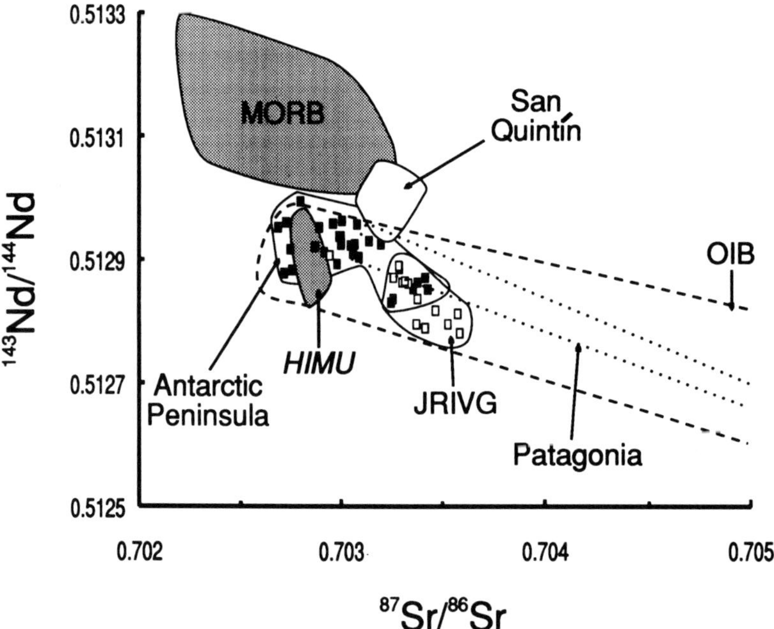

Fig. 7. Sr- and Nd-isotope variations in Antarctic Peninsula (filled squares) and JRIVG alkalic basalts (open squares). MORB data from Ito *et al.* (1987); San Quintin, Storey *et al.* (1989); Patagonia, Stern *et al.* (1990); OIB, Palacz & Saunders (1986) (Cook-Austral-Samoa chain) and Chaffey *et al.* 1989 (St Helena); Antarctic Peninsula slab window related basalts from Hole, (1988, 1990*a*). HIMU OIB (i.e. $^{206}Pb/^{204}Pb > 20.0$) are shown as a sub-set of the OIB data. Nd-isotope data are not available for the slab window-related basalts of British Columbia.

(Fig. 5). Such profiles characterize alkalic basalts from a number of continental as well as oceanic settings (e.g. Cameroon Line, Fitton & Dunlop 1985; western USA, Thompson *et al.* 1984; Cook-Austral-Samoa Islands, Palacz & Saunders 1986) and are generally considered to reflect low degrees of partial melting of a predominantly asthenospheric mantle source region. MORB-normalized abundances of the LILE exhibit significant absolute variations which are attributable to variable degrees of partial melting (Hole 1988, 1990a; Sykes 1988b), but all samples exhibit MORB-like LILE/HFSE ratios. Chondrite-normalized rare earth element (REE) plots are indicative of melting within the garnet stability field of the mantle as La_n/Yb_n ratios increase from tholeiites to basanites, with no coherent concomitant increase in Yb or Lu abundances. On a plot of La_n/Yb_n v. Sc/Yb (Fig. 6) the Antarctic Peninsula basalts define a sub-horizontal array with consistent Sc/Yb abundances regardless of La_n/Yb_n ratio, such that Sc/La ratios decrease rapidly with decreasing degrees of partial melting. Because Sc has a high distribution coefficient relative to both Yb and La in garnet ($k_D{}^{Sc} \approx 10$; $k_D{}^{Yb} \approx 4$; $k_D{}^{La} \approx 0.01$: see compilation in LeRoex *et al.* 1981). This is a predictable result of residual garnet during low degrees of partial melting. In contrast, MORBs, which are generally considered to have been derived by moderate degrees of partial melting of spinel lherzolite, exhibit variable Sc/Yb ratios and low La_n/Yb_n ratios, coupled with more uniform Sc/La ratios, as a result of the significantly lower k_Ds for Sc and Yb in spinel compared with garnet. These trace element ratios imply that the Antarctic Peninsula basalts equilibrated with mantle at a pressure of at least 3 GPa (*c.* 80 km).

The range of Sr- and Nd-isotope ratios is limited for all the alkalic basalts of the Antarctic Peninsula, whether post- or syn-subduction (Fig. 7). $^{87}Sr/^{86}Sr$ ratios vary from 0.70275 to 0.70357 and $^{143}Nd/^{144}Nd$ ratios from 0.51275 to 0.51300 (ϵ_{Nd} +2.8 to +6.9) with the James Ross Island samples tending to have the most enriched Sr- and Nd-isotope characteristics. In general, $^{143}Nd/^{144}Nd$ ratios tend to be lower at a given $^{87}Sr/^{86}Sr$ ratio compared with MORB, and so all of the Antarctic Peninsula samples must have been derived from a mantle reservoir which evolved with low time-integrated Rb/Sr but high time-integrated Sm/Nd ratios, broadly similar to the source of non-Dupal OIB (e.g. St Helena, Ascension Island, Mangaia, Tubai, Nunivak, Balleny: White & Hofmann 1982; Palacz & Saunders 1986; Hart 1988).

Because Sr- and Nd-isotope ratios are not fractionated during partial melting or melt extraction events, isotope ratios can be considered to reflect mantle source region heterogeneities or source mixing rather than being artifacts of degree of partial melting. In addition, incompatible trace element ratios which correlate with isotope ratios are similarly likely to reflect source region heterogeneities and mixing processes. Correlations between Th/Ta, La/Th, K/Rb, Rb/Nb and $^{87}Sr/^{86}Sr$ ratios are well defined for the slab window-related basalts (e.g. Fig. 8). La/Th (11–5) and K/Rb (1750–400) ratios decrease whereas Th/Ta (0.9–2.5), Rb/Nb (0.35–1.25) and Ba/Nb (3–7) ratios increase systematically with increasing $^{87}Sr/^{86}Sr$ ratio. It has been previously postulated that such trace element-isotope correlations are the effect of the mixing of predominantly asthenosphere-derived melts with subduction-enriched mantle or continental lithosphere (Hole 1988, 1990a; Hole *et al.* 1991b). However, this subduction enrichment has only affected a limited number of the samples, as is clear from the preservation of OIB-like isotopic characteristics and MORB-like LILE/HFSE ratios in the majority of the basalts (Hole 1988, 1990a; Hole *et al.* 1993). The syn-subduction alkalic basalts of James Ross Island do not fall on the well-defined trace element-isotope trends delineated by the slab window related samples, although the overall range in LILE/HFSE ratios are similar to that for the slab window-related basalts. For example, Th/Ta and La/Th ratios (1.2–1.7 and 8–10 respectively) for James Ross Island are within the range for the rest of the Antarctic Peninsula, but $^{87}Sr/^{86}Sr$ ratios are slightly higher (e.g. Fig.8). Rb/Nb v. $^{87}Sr/^{86}Sr$ ratio correlations for James Ross Island are broadly negative.

It is important to emphasize that geochemical variations both within and between the suites of slab window and syn-subduction basalts are extremely limited both in terms of isotopes and incompatible trace elements, and none of the slab window-related or syn-subduction alkali basalts exhibits the high LILE/HFSE ratios of the early Tertiary calc-alkaline basalts from the same region. In this respect, all the Antarctic Peninsula alkalic basalts differ markedly from Patagonian alkalic basalts which show an extended range of both isotopic (e.g. $^{87}Sr/^{86}Sr$ ratios 0.7033–0.7050) and LILE/HFSE ratios (e.g. Th/Ta 1.5–5.0; Ramos & Kay 1992; Stern *et al.* 1990). The variations which are discernable in the Antarctic Peninsula basalts can satisfactorily be explained by very minor 'subduction-enrichment' or interaction with the continental lithosphere (Hole 1988, 1990a,b; Sykes 1988b). The paucity of subduction or lithospheric 'fingerprints' in their geochemical composition is

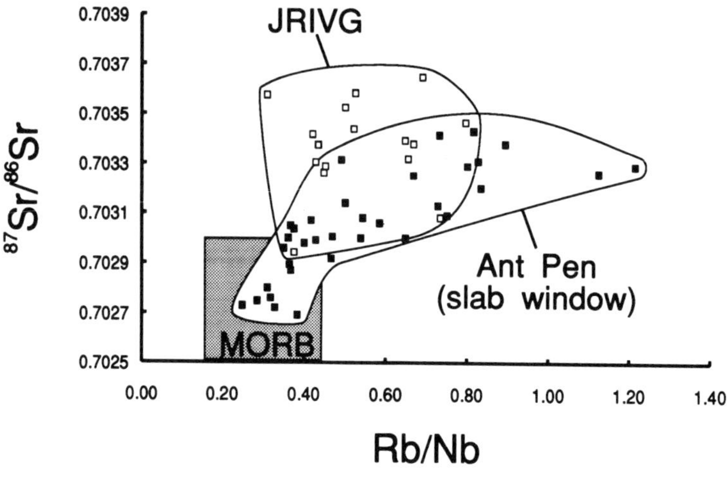

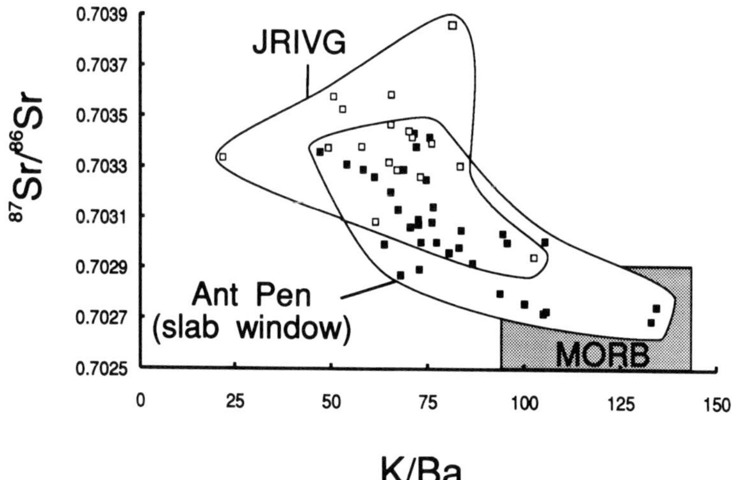

Fig. 8. (**a**) Rb/Nb and (**b**) K/Ba ratio v. Sr-isotope ratios for JRIVG and Antarctic Peninsula basalts. The well-defined trends for the slab window-related (Antarctic Peninsula) basalts is probably the result of very minor subduction enrichment or crustal interaction. Note that the JRIVG does not fall on these trends. Symbols as for Fig. 7

surprising because both slab window and syn-subduction basalts follow closely an extended period of arc magmatism. The implication here, is that the alkalic magmas were derived from asthenospheric sources that had not recently been affected by subduction. We will now consider the possible mechanisms for melt production in both settings.

Melt production

The generation of slab window-related basalts

It seems highly likely that the mantle wedge beneath the Antarctic Peninsula was thoroughly metasomatized by slab-derived fluids during the previous 200 Ma of subduction. Therefore, in order to produce undersaturated alkalic basalts with depleted isotopic signatures and consistently low LILE/HFSE ratios, their source must have been within asthenosphere which was originally either isolated below the subducted slab during subduction, or far removed from the subduction zone to the east. With portions of oceanic lithosphere in place beneath the continental margin such asthenosphere beneath the slab could not be tapped. However, by the generation of a slab window, the physical barrier between convecting asthenosphere, mantle wedge and continental lithosphere is removed, facilitating, at least in principle, the uprise of asthenospheric melts through to the continental lithosphere.

Significant slab roll-back may have induced lateral migration of asthenosphere from behind the arc, replacing subduction-modified mantle wedge material. The problem here lies in the mechanism of melt production. In many continental alkali basalt provinces (e.g Western USA), OIB-like magmas are considered to be produced by asthenospheric upwelling and decompressional melting, either because of elevated mantle temperatures resulting from local plume activity or significant lithospheric attenuation. Neither of these models is applicable to the slab window-related basalts.

A plume model is not likely because it would require the spontaneous and fortuitous initiation of plume activity following cessation of subduction (Hole *et al.* 1991*b*). It would appear to be highly coincidental for deep mantle plumes to be sited precisely at the location of cessation of subduction, particularly given that slab window-related basalts are so widely separated geographically, occurring from Antarctica to Canada. Extension of the lithosphere sufficient to cause decompression would need to exceed a β factor (β = original mechanical boundary layer thickness/post extension thickness) of about 2.5 (Latin & Waters 1991) and Late Cenozoic extension of this magnitude is not observed at any slab window locality. This precludes the generation of slab window-related basalts solely as a result of rift-associated passive asthenospheric upwelling on a regional scale. Additionally, significant lithospheric attenuation tends to result in the lithosphere becoming a potential magma source causing significant lithosphere-asthenosphere interaction (Fitton *et al.* 1988; Leat *et al.* 1989; Latin & Waters 1991). Such effects are not observed in the Antarctic Peninsula. However, this does not preclude limited, localized extension to allow egress of melts to the surface, but it is important to emphasize that limited extension would not be capable of melt generation unless the potential temperature of the mantle was higher than normal (e.g. McKenzie & Bickle 1988).

During the generation of a slab window, the subducting slab continues to be consumed, and is detached along the former ridge axis. This effectively causes the removal of a several km thickness of rigid lithosphere from beneath the continental margin; in this setting the subducting oceanic lithosphere can be considered to be part of the total lithospheric thickness. An effective β factor can therefore be calculated at the locus of

magmatism above a slab window. For example, consider the 25 Ma collision event off northern Alexander Island. Consideration of whether extension was instantaneous is important because it has been shown that for slow rates of extension, heat may have sufficient time to be dissipated by conduction, and melting would not result (Jarvis & McKenzie 1980). Thermal oceanic lithosphere thickness is generally considered to be approximately 9.4 times the square root of its age (Parker & Oldenberg 1973). At the locus of the 25 Ma collision, Barker (1982) showed that at *c.* 65 Ma, 11 Ma (*c.* 31 km thick) old oceanic lithosphere was present at a depth of 100 km in the subduction zone; at the time of ridge crest-trench collision 8 Ma old lithosphere (*c.* 27 km thick) was present at 100 km depth. By definition, at the spreading ridge at the time of collision the oceanic lithosphere was zero age and thickness. Thus, we can assume that the maximum vertical thickness of the oceanic lithosphere beneath the arc for the period 65 Ma to the present was *c.* 30 km. The apparent magmatic 'front' of the slab window-related magmatism at this location is 200 km from the trench. At the time of ridge crest-trench collision it is therefore reasonable to assume that the thickness of oceanic lithosphere beneath this point was *c.* 30 km. If the slab continued to subduct at the rate of spreading of the next ridge to the north (approximately 40 km Ma^{-1}), then 30 km of vertical lithospheric thickness would have been removed from beneath northern Alexander Island during the 7 Ma after collision. The limiting value for instantaneous stretching proposed by Jarvis & McKenzie (1980) is $60/\beta^2$ < *c.* 20 Ma. For a decrease in oceanic lithosphere thickness of *c.* 30 km in 7 Ma, and a continental lithosphere thickness of say, 50 km, the β factor would be *c.* 1.6 and $60/\beta^2 = 23$ Ma. Therefore, as long as the continental lithosphere was < *c.* 50 km thick, then the removal of 30 km of slab in *c.* 7 Ma would constitute instantaneous extension. These calculations necessarily require the assumption that the inter-relationship between thermal conductivity constant and β factor derived by Jarvis & McKenzie (1980) is applicable to this tectonic scenario.

It is likely that the total thickness of the lithosphere beneath Alexander Island is at present <50 km given that it is entirely formed of accreted material. Whilst the age of the accretionary complex is not well constrained, a lower Jurassic fauna has been identified associated with an accreted seamount (Thomson & Tranter 1986) giving a maximum age of early Jurassic (*c.* 190 Ma), although Storey & Nell (1988) suggested that the accretionary prism

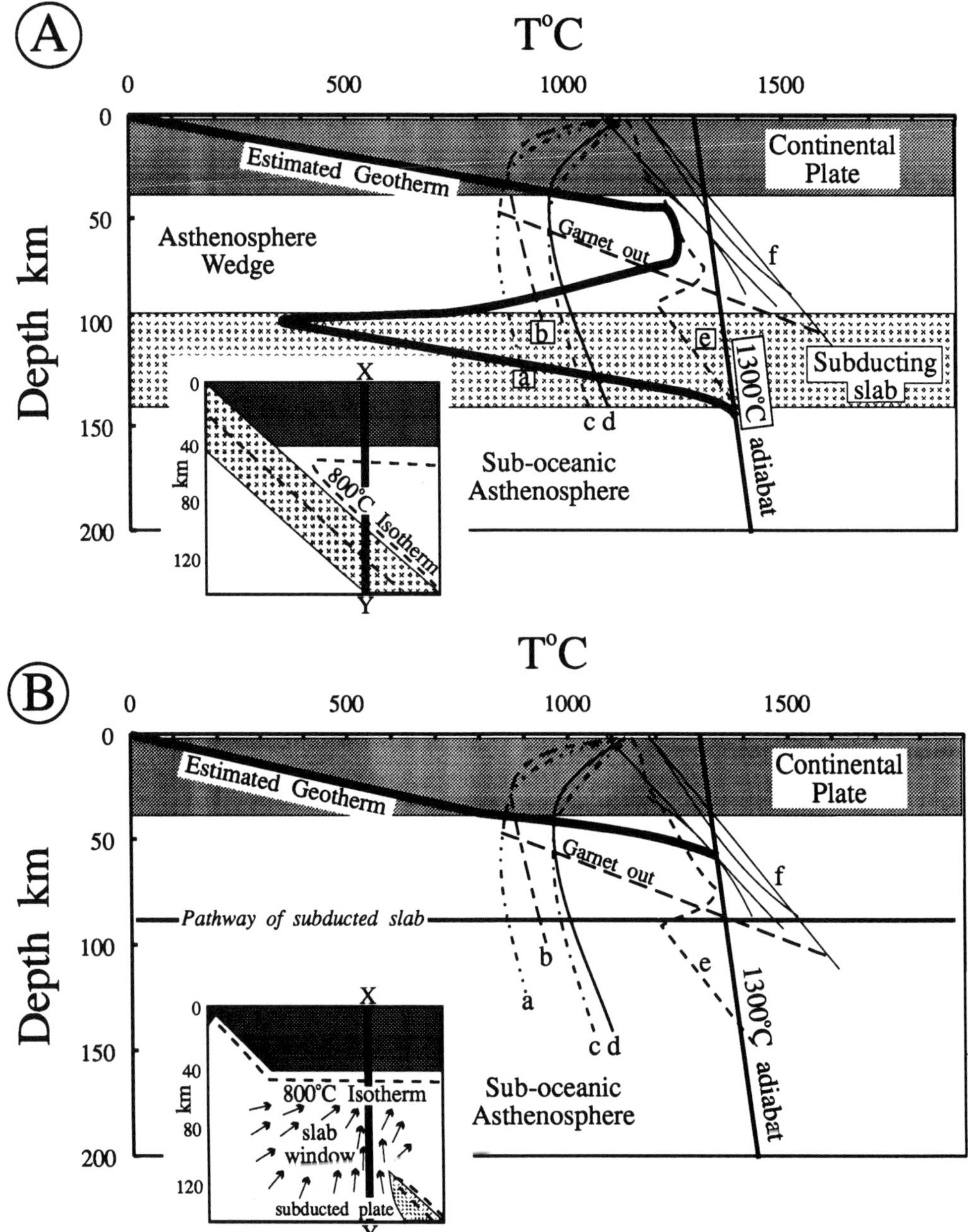

could be as young as mid-Cretaceous. Given these young ages, the continental lithosphere could be relatively thin in this area, although we acknowledge that at present this is difficult to constrain.

However, there are a number of problems with generating magma purely as a result of slab window formation. Firstly, if mantle material is coupled to the base of the subducting slab and is carried down with the slab, it will simply be compressed along the adiabat. In this case, unless it rises to a shallower level from that at which it started, or it was originally undergoing melting, it is difficult to see how it will intersect its solidus and produce melt. Indeed, Hole & Larter (1993) have shown that the only setting in which melt will be generated simply by slab window formation is in the trench proximal setting, where the total thickness of the continental lithosphere is less than that of the oceanic lithosphere (note also the higher Sc/Yb ratio of one of the dredge samples in Fig. 6). Additionally, for the more trench-distal settings, a problem arises in that under purely anhydrous and decarbonated conditions, mantle with a reasonable potential temperature of 1300°C cannot melt at pressures much greater than 1.5 GPa and certainly not at pressures corresponding to garnet peridotite (McKenzie & Bickle 1988; Fig. 9). Consequently, under these conditions it would not be possible to produce melts which were in equilibrium with garnet and which were depleted in the HREE, Y and Sc, like those of the Antarctic Peninsula. Figure 9 illustrates estimated thermal profiles of a convergent margin before and after the consumption of a half spreading axis. The estimated initial conditions are: a potential temperature within the convecting part of the mantle of 1300°C; a subduction angle of 45°; a 30 m thick slab; and a rate of subduction of 100 km Ma^{-1}. The subduction zone is shown 'face-on' so that temperature and pressure are projected onto one section of the zone. Sub-slab mantle, which was previously ascending and melting beneath the ridge axis will presumably convect into the void left by the subducting slab, and displace any remaining mantle wedge asthenosphere. However, because this convection is occurring at greater depths than the original depth of the spreading ridge and therefore the pressure is higher, melting will cease; because the temperature gradient in the asthenosphere is adiabatic, subduction of a spreading ridge will not result in a thermal anomaly in the asthenosphere. The peridotite solidus is depressed by the addition of volatiles, such that small-degree carbonated and hydrous melting of the sub-slab mantle may occur as it ascends into the void left by the subducted slab leading edge. Indeed, the occurrence of hydrous-phase megacrysts (kaersutite)

and scattered phlogopite-bearing xenoliths within the basanitic volcanic rocks of Alexander Island attests to the probable importance of volatile-bearing mantle. Whether volatiles were resident within the asthenosphere or were remnants from the subduction history of the area is open to question, although Bai & Kohlstedt (1992) pointed out that hydrogen has substantial solubility in mantle olivine and olivine may be a primary sink for volatiles in the upper mantle.

If the mantle was hydrous, melting could occur at high pressure within the garnet stability field of the mantle (e.g. McKenzie 1985, 1989). The small-degree melts will be strongly HREE, Y and Sc depleted and enriched in incompatible elements. As they percolate upward, the extent of melting will increase but this will drive the solidus temperature of the mantle to higher temperatures as the water and CO_2 are dissolved in the melt. Consequently, the extent of melting will be self-limiting, and all melt initially generated would presumably be Si-undersaturated and alkaline. We suggest that whilst the opening of a slab window may be insufficient alone to produce magma, it does provide the physical conditions to allow hydrous melting. Magmatism appears to have continued for a considerable period of time after ridge crest-trench collision along the Antarctic Peninsula, implying that upwelling and decompression melting are able to continue once initiated.

Tapping of small-degree melts from 'normal temperature', hydrous, carbonated asthenosphere may be unique to the slab window tectonic setting. At mid-ocean ridges they are unlikely to appear at the surface because they will be consumed and integrated into larger degree melts of spinel lherzolite beneath the ridge axis (e.g. Klein & Langmuir 1987). In areas of significant lithospheric attenuation (e.g. western USA), the tectono-thermal regime will

Fig. 9. Two diagrams showing possible P–T conditions in the mantle wedge during encroachment of a spreading axis. (**a**) 'Steady state' situation with 28 Ma – old ocean crust, subducting at an angle of 30° beneath a rigid plate with a 40 km-thick mechanical boundary layer. Subduction rate is 72 km/Ma. Plane of drawing is along line X–Y on inset. Viscous coupling between the slab and the overlying asthenosphere draws in hot mantle, but temperatures are insufficient to allow melting of dry peridotite within the plane of the diagram. (**b**) Development of a slab-window, with the sub-slab mantle ascending through the window (schematically shown on inset) and displacing the cool asthenosphere wedge. Note that the ascending mantle does not intersect the dry peridotite solidus unless the overlying mechanical boundary layer is less than about 50 km thick, and cannot intersect the dry solidus within the garnet-stable field unless the potential temperature is substantially higher than 1300°C. Intersection of the wet and CO_2-saturated solidi is, however, achieved, even within the garnet stability field. On both diagrams a range of solidus curves is drawn. a: water-saturated peridotite (Mysen & Boettcher 1975a,b); b: 'average' wet peridotite (Davies & Stevenson 1992); c: wet peridotite solidus (Thompson 1992); d: wet peridotite solidus (Wyllie 1982); e: carbonated peridotite solidus (Wyllie 1982). f: a selection of dry peridotite solidi (McKenzie & Bickle 1988; Davies & Stevenson 1992; Thompson 1992; Kinzler & Grove 1992).

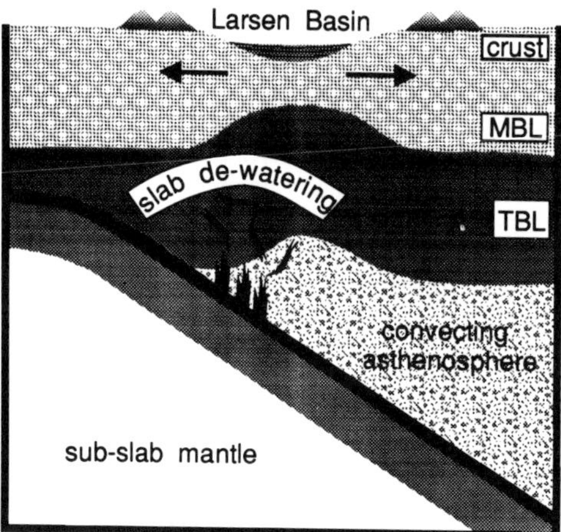

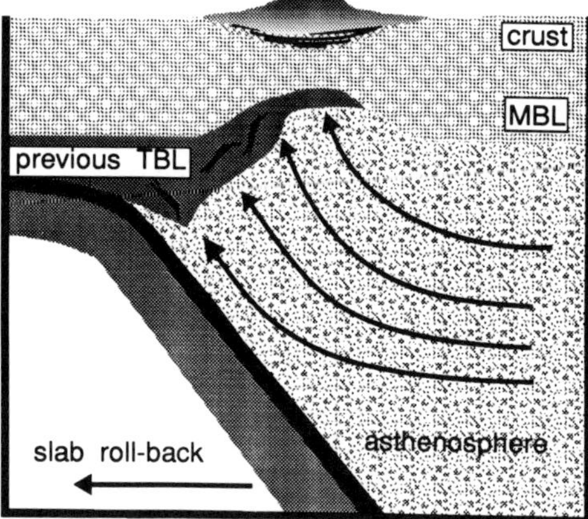

Fig. 10. Cartoons illustrating the evolution of the James Ross Island area between the Late Cretaceous and 7 Ma. (**a**) Extension and formation of the Larsen Basin resulted in lithospheric attenuation and the upwelling of the asthenosphere which eventually formed the thermal boundary layer (TBL). This mantle was subsequently enriched in slab dehydration products. (**b**) at *c.* 7 Ma significant slab roll-back occurred due to the formation of a major slab-window beneath the rest of the peninsula. 'Pristine' asthenosphere migrated from the east as a result and displaced the previous TBL, upwelling into the palaeo-rift or 'thinspot'. This upwelling resulted in melting and eruption of the JRIVG. There was potential for mixing with old TBL and lithosphere at this stage.

be sufficiently different to allow asthenosphere–lithosphere interaction, again causing integration of initial small degree melts with other magma sources. Essentially, the same melt-integration scenario applies to plume-related magmatism, where melting results from elevated mantle temperatures. The combination of low β factors and slab windows results in an unique tectonic, thermal and magmatic regime. The generation of melt requires neither significant lithospheric attenuation nor abnormal mantle temperatures to occur.

The control of lithospheric structure on syn-subduction alkalic magmatism

Post-subduction magmatism on Alexander Island and Seal Nunataks can be explained by the slab window model. However, the syn-subduction magmatism at James Ross Island cannot be related directly to slab window formation. Whereas there is some evidence for half-graben formation in the James Ross Island area during the late Tertiary, the degree and extent of lithospheric attenuation in this area is largely unknown. Important differences between the JRIVG and slab window-related basalts are: (1) the greater volume of basalts produced in the JRIVG than most other locations in the Antarctic Peninsula, (2) the more consistent chemical composition of the JRIVG, and its apparent high pressure fractionation history (Fig. 4b), (3) more radiogenic Sr- and unradiogenic Nd-isotopes; and (4), less well-defined trace element-isotope correlations than the slab window-related basalts.

The JRIVG unconformably overlies approximately 5 km of predominantly arc-derived Kimmeridgian–Eocene sedimentary rocks of the Larsen Basin, which was generated in the late Jurassic by dextral strike-slip, broadly correlating with the later stages of Gondwana break-up (MacDonald & Butterworth 1990; MacDonald *et al.* 1988; Storey & Nell 1988). There appears to be little evidence for magmatism associated with this extensional phase of arc history, although occurrences of middle to late Cretaceous (90–67 Ma ago) mildly alkalic acid and basic intrusive activity are recorded, and middle Jurassic high magnesian andesites may be an indication of the initial stages of arc extension (Hamer & Hyden 1984; Hole 1986; Hole *et al.* 1991a; Saunders *et al.* 1980; Alabaster & Storey 1990). During this stage of back-arc basin formation, extension of continental lithosphere would have occurred, allowing mantle upwelling and potentially, melt production. However, the existence of the subducted slab beneath the region at this time would have prevented the production of melt from asthenosphere unmodified by subduction processes and, only locally, magmas transitional between calc-alkaline and mildy alkaline compositions could have been produced giving rise to the mildly alkalic Cretaceous intrusions in this area.

Figure 10 is a cartoon illustrating the possible method of melt production at James Ross Island. The sub-continental lithospheric mantle forms part of the mechanical boundary layer (MBL) which is relatively cold and rigid and is unlikely to be replaced by asthenosphere unless it is mobilized by a plume. This is underlain by the thermal boundary layer (TBL), which is unstable and may be periodically replaced by asthenospheric mantle, although it is likely to have the same isotopic characteristics as the asthenosphere. Although the TBL will be undergoing very sluggish flow, it will be replenished and replaced by new asthenosphere within a time scale of <100 Ma (McKenzie & Bickle 1988). At about the time of the initiation of alkalic magmatism at James Ross Island ($c.$ 7.0 Ma), the majority of the Antarctic Peninsula was underlain by slab windows, the only area of active subduction being along the South Shetlands Islands trench. The potential for slab roll-back was thus extremely high. To accommodate slab roll-back, lateral replenishment of mantle wedge material with pristine (i.e. not subduction modified) asthenosphere from the east would have occurred in much the same way as suggested by Stern $et\ al.$ (1990) as a possible origin for the Patagonia plateau basalts. However, simple lateral replenishment of the asthenosphere may not result in melting because lateral migration would probably not cause intersection with the hydrous peridotite solidus. However, replacement of the TBL formed during the Cretaceous extensional phase with this pristine asthenosphere and its upwelling into the palaeo-rift system could have resulted in melt production. Indeed, Thompson & Gibson (1991) envisaged the utilization of pre-existing lithospheric thinspots by plumes in the generation of continental alkali basalts. In the case of James Ross Island, the authors envisage hydrous asthenosphere with a 'normal' potential temperature (1300°C) upwelling into a pre-existing thinspot, which would allow intersection of the hydrous peridotite solidus. Initiation of extensional tectonism at the locus of the Larsen Basin probably occurred during the Late Jurassic, but the record of sedimentation carries on into the Eocene implying that extension may have been active up to that time (MacDonald $et\ al.$ 1988; MacDonald & Butterworth 1990). During the final stages of ridge crest–trench collision, subduction rates decreased dramatically from $c.$ 76 mm a^{-1} at 9–6 Ma to 18 mm a^{-1} at

4.6–3.1 Ma and finally to 4 mm a^{-1} at present, enhancing the potential for slab roll-back, and once the opening of Bransfield Strait was initiated at $c.$ 4 Ma, the reactivation of half-grabens in Prince Gustav Channel (Fig. 1) may have resulted in the westward migration of magmatism within the younger members of the JRIVG (Sykes 1988a,b). The normative compositions of the JRIVG are suggestive of an origin by polybaric fractionation (Fig. 4b), with a maximum pressure of $c.$ 9 kbar and a minimum of 1 atm for differentiates associated with high-level laccolithic intrusions. In this case, the initial ponding of magma in a complex magmatic plumbing system at the base of the crust seems likely, and would be accommodated by upwelling into the palaeo-rift system of the Larsen Basin. This contrasts with the dominantly low pressure crystallization history of the slab window-related basalts which presumably underwent rapid uprise through the continental lithosphere. The uprise of the slab window-related basalts was tectonically controlled by both the geometry of the slab window system and pre-existing lithospheric discontinuities formed by strike-slip in the accretionary prism complex of Alexander Island (Storey & Nell 1989), and earlier tectonic segmentation of the arc (Hole 1986; Hole $et\ al.$ 1991a). Additionally, ponding of the JRIVG magmas at deep crustal levels may explain the paucity of mantle xenoliths compared with the common occurrence of spinel lherzolites in some of the slab window-related basalts.

The TBL formed during the late Jurassic to Cretaceous extensional phase beneath James Ross Island is likely to have been subduction modified such that during lateral replenishment, upwelling and polybaric fractionation there is the potential to tap both pristine asthenosphere and subduction modified TBL mantle, as well as MBL lithospheric mantle and upper crust. This is similar to the suggestion of Leat $et\ al.$ (1989) to explain some of the diversity in magma types formed during extension in parts of the Western USA. The reason that there is no clear evidence for major interaction of JRIVG magmas with 'old' lithosphere is simply because of the relatively young age (maximum age of mid-Palaeozoic; Milne & Millar 1989) and presumably geochemical 'immaturity' of the Antarctic Peninsula subcontinental lithosphere compared with areas such as the Western USA where Proterozoic crust and associated lithospheric mantle are present (Leat $et\ al.$ 1989).

However, the majority of the differences between the slab window-related basalts and the JRIVG can be reconciled with the generation of the latter in an area of older back-arc extension i.e. due to a thinspot. Greater volumes of magma could have been produced and the potential for mixing of a number of different sources gave rise to the geochemical contrasts between the two types of magmatism. Seal Nunataks also falls within the area of the Larsen Basin, but is more easily reconciled with the slab window model. There, a slab window was produced at *c.* 6–10 Ma and the apparent orientation of the Seal Nunataks volcanic centres along the landward trace of closely spaced ocean crust fracture zones, implies the operation of a 'slab edge' effect (cf. Natland 1980). Simple upwelling from beneath the slab, and not lateral migration of asthenosphere is likely to have dominated in this location, although the thinning of the crust beneath Seal Nunataks during the formation of the Larsen Basin may have allowed considerably shallower levels of upwelling than on Alexander Island. Integration of initial small degree melts with higher degree melts may therefore have been important in Seal Nunataks, but had little influence in Alexander Island. This is a possible explanation for the generally more saturated nature of the Seal Nunataks suite compared to Alexander Island, although the common occurrence of spinel lherzolites and the normative composition of the Seal Nunataks basalts suggests rapid uprise to high crustal levels and not ponding at the base of the crust as for the JRIVG.

ultimately resulted in decompression melting. Melting initially occurred in the garnet stability field of the mantle under hydrous, carbonated conditions, with the degree of melting being self-limiting as a result of progressive dissolution of CO_2 in the initial small melt-fractions, and a consequent increase in the solidus temperature of the mantle.

At James Ross Island, slab roll back facilitated lateral migration, upwelling and melting of asthenosphere, unmodified by subduction, into a thinspot formed between Late Jurassic and early Tertiary times. Slab roll-back was enhanced by the formation of a substantial slab window to the south, and by a rapid decrease in subduction rate at about 7 Ma.

In both cases, the generation of melt requires neither significant lithospheric attenuation nor high potential temperatures in the mantle. This combination of low β factors and slab windows results in a unique tectonic, thermal and magmatic regime. This is therefore potentially one of the few settings where low degree melts of the asthenosphere not associated with plumes are produced and preserved.

The authors thank J.G. Fitton, P.R. Kyle and P.F. Barker for thoughtful and constructive reviews. Rob Larter (British Antarctic Survey) and Alan Crane (Aberdeen) are thanked for informative discussion on many aspects of the tectonics of subduction zones. Barry Fulton drafted many versions of figs 3 & 4.

Conclusions

Alkalic magmatism along the Antarctic Peninsula occurs in two distinct tectonic settings. On James Ross Island, alkali basalts, hawaiites and mugearites were erupted synchronously with subduction at the South Shetland Islands trench to the west, whereas farther south along the Antarctic Peninsula, basanites, tephrites and alkali and olivine basalts were erupted following cessation of subduction by ridge crest-trench collision. Both the James Ross Island basalts and the other occurrences along the Antarctic Peninsula exhibit depleted Sr- and Nd-isotope characteristics and consistently low (MORB-like) LILE/Nb ratios, indicative of very limited interaction with enriched mantle sources, continental lithosphere or mantle material which previously constituted the mantle wedge during the subduction history of the region.

The generation of low volumes of alkalic basalts along the Antarctic Peninsula was causally related to slab window formation and upwelling of subslab asthenosphere into the incipient void left by the descending slab, which

References

ALABASTER, T. & STOREY, B.C. 1990. Antarctic Peninsula continental magnesian andesites: indicators of ridge-trench interaction during Gondwana break-up. *Journal of the Geological Society, London,* **147**, 595–598.

BAI, Q. & KOHLSTEDT, D.L. 1992. Substantial hydrogen solubility in olivine and implications for water storage in the mantle. *Nature,* **357**, 672–674.

BAKER, P.E., REA, W.J., SKARMETA, J., CAMINOS, R. & REX, D.C. 1981. Igneous history of the Andean cordillera and Patagonian Plateau around latitude 46°S. *Philosophical Transactions of the Royal Society, London,* **303**, 105–149

BARKER, P.F. 1982. The Cenozoic subduction history of the Pacific margin of the Antarctic Peninsula: ridge crest-trench interactions. *Journal of the Geological Society, London,* **139**, 787–801.

BEVIER, M.L. 1983. Implications of chemical and isotopic composition for petrogenesis of Chilcotin Group Basalts, British Columbia. *Journal of Petrology,* **24**, 207–226.

——, ARMSTRONG, R.L. & SOUTHERN, J.G. 1979. Miocene peralkaline volcanism in west central British Columbia: its temporal and plate tectonic setting. *Geology,* **7**, 289–392.

Burn, R.W. 1981. Early Tertiary calc-alkaline volcanism on Alexander Island. *British Antarctic Survey Bulletin*, **53**, 175–193.

Chaffey, D.J. Cliff, R.A. & Wilson, B.M. 1989. Characterization of the St Helena magma source. *In*: Saunders A.D. & Norry M.J. (eds) *Magmatism in the ocean basins*. Geological Society, London, Special Publications, **42**, 257–276.

Davies J.H. & Stevenson, D.J. 1992. Physical model of source region of subduction zone volcanics. *Journal of Geophysical Research*, **97**, 2037–2070.

Dickinson, W.R. & Snyder, W.S. 1979. Geometry of subducted slabs related to the San Andreas transform. *Journal of Geology*, **87**, 609–627.

Ellam, R.M. & Hawkwesworth, C.J. 1988. Elemental and isotopic variations in subduction-related basalts: evidence for a three-component mantle. *Contributions to Mineralogy and Petrology*, **98**, 72–80.

Fitton, J.G. & Dunlop, H.M. 1985. The Cameroon line, West Africa, and its bearing on the origin of oceanic and continental alkali basalt. *Earth and Planetary Science Letters*, **72**, 23–38.

——, James, D., Kempton, P.D., Ormerod, D.S. & Leeman, W.P. 1988. The role of lithospheric mantle in the generation of late Cenozoic magmas in the Western United States. *Journal of Petrology Special Lithosphere Issue*, 331–349.

Gill, J.B. 1981. *Orogenic andesites and plate tectonics*. Springer-Verlag, Berlin.

Hamer, R.D. & Hyden, G. 1984. The geochemistry and age of the Danger Islands pluton, Antarctic Peninsula. *British Antarctic Survey Bulletin*, **64**, 1–20.

Hart, S.R. 1988. Heterogeneous mantle domains: signatures, genesis and mixing chronologies. *Earth and Planetary Science Letters*, **90**, 273–296.

Hole, M.J. 1986. *Time controlled geochemistry of igneous rocks of the Antarctic Peninsula*. PhD thesis, University of London, UK.

—— 1988. Post-subduction alkaline volcanism along the Antarctic Peninsula. *Journal of the Geological Society, London*, **145**, 985–988.

—— 1990a. Geochemical evolution of Pliocene-Recent post-subduction alkalic basalts from Seal Nunataks, Antarctic Peninsula. *Journal of Volcanology and Geothermal Research*, **40**, 149–167.

—— 1990b. Chapters C3 & C4, Alexander Island. *In*: Lemasurier, W.E. & Thomson J.W. (eds) *Volcanoes of the Antarctic plate and southern oceans*. American Geophysical Union, Antarctic Research Series, **48**, 271–276.

—— & Larter, R.D. 1993. Trench proximal volcanism following ridge crest-trench collision along the Antarctic Peninsula. *Tectonics*, **12**, 897–910.

—— & Morrison, M.A. 1992. The differentiated dolerite boss, Cnoc Rhaonastil, Islay: a natural experiment in the low pressure differentiation of an alkali olivine-basalt magma. *Scottish Journal of Geology*, **28**, 55–69.

——, Kempton, P.D. & Millar, I.L. 1993. The isotopic and trace element composition of small-degree melts of the asthenosphere: evidence from the alkalic basalts of the Antarctic Peninsula. *Chemical Geology*, **109**, 51–68.

——, Pankhurst, R.J. & Saunders, A.D. 1991a. The geochemical evolution of the Antarctic Peninsula magmatic arc: the importance of mantle-crust interaction during granitoid genesis. *In*: Thomson, M.R.A., Crame, J.A. & Thomson J.W. (eds) *The geological evolution of Antarctica*. Cambridge University Press, Cambridge, England, 369–374.

——, Rogers, G., Saunders, A.D. & Storey, M. 1991b. The relationship between alkalic volcanism and slab window formation. *Geology*, **19**, 657–660.

——, Smellie, J.L. & Marriner, G.F. 1991c. Geochemistry and tectonic setting of Cenozoic alkaline basalts from Alexander Island, southwest Antarctic Peninsula. *In*: Thomson, M.R.A., Crame, J.A. & Thomson, J.W. (eds) *The geological evolution of Antarctica*. Cambridge University Press, Cambridge, England, 521–526.

Ito, E. White, W.M. & Gopel, C. 1987. The O, Sr, Nd and Pb isotope geochemistry of MORB. *Chemical Geology*, **62**, 157–176.

Jarvis, G.T. & McKenzie, D.P. 1980. Sedimentary basin formation with finite extension rates. *Earth and Planetary Science Letters*, **48**, 42–52.

Johnson, C.M. & O'Neil, J.R. 1984. Triple junction magmatism: a geochemical study of Neogene volcanic rocks in western California. *Earth and Planetary Science Letters*, **71**, 241–262

Keinle, J., Kyle, P.R., Self, S., Motkya, R.J. & Lorenz, V. 1980. Ukinrek Maars, Alaska, I. April 1977 eruption sequence, petrology and tectonic setting. *Journal of Volcanology and Geothermal Research*, **7**, 11–37.

Kinzler, R.J. & Grove, T.L. 1992. Primary magmas of mid-ocean ridge basalts. *Journal of Geophysical Research*, **97**, 6907–6926.

Klein, E.M. & Langmuir, C.H. 1987. Global correlations of ocean ridge basalt chemistry with axial depth and crustal thickness. *Journal of Geophysical Research*, **92**, 8089–8115.

Larter, R.D. & Barker, P.F. 1991. Effects of ridge-crest trench interaction on Antarctic-Pheonix spreading: forces on a young subducting plate. *Journal of Geophysical Research*, **96**, 19583–19607.

Latin, D. & Waters, F.G. 1991. Melt generation during rifting in the North Sea. *Nature*, **351**, 559–562.

Leat, P.T., Thompson, R.N., Dickin, A.P., Morrison, M.A. & Hendry, G.L. 1989. Quaternary volcanism in northwestern Colorado: implications for the roles of asthenosphere and lithosphere in the genesis of continental basalts. *Journal of Volcanology and Geothermal Research*, **37**, 291–310.

LeRoex, A.P., Erlank, A.J. & Needham, H.D. 1981. Geochemical and mineralogical evidence for the occurrence of at least three distinct magma types in the 'Famous' Region. *Contributions to Mineralogy and Petrology*, **77**, 24–37.

——, Dick, H.J.B., Ried, A.M., Frey, F.A., Erlank, A.J. & Hart, S.R. 1983. Geochemistry, mineralogy and petrogenesis of lavas erupted along the Southwest Indian ridge between the Bouvet triple junction and 11 degrees east. *Journal of Petrology*, **24**, 267–318.

——, ——, ——, ——, —— & ——. 1985. Petrology and geochemistry of basalts from the American-Antarctic Ridge, Southern Ocean: implications for the westward influence of the Bouvet mantle plume. *Contributions to Mineralogy and Petrology*, **90**, 367–380.

McKENZIE, D.P. 1985. The extraction of magma from the crust and mantle. *Earth and Planetary Science Letters*, **74**, 81–91.

—— 1989. Some remarks on the movement of small melt fractions in the mantle. *Earth and Planetary Science Letters*, **95**, 53–72.

—— & BICKLE, M.J. 1988. The volume and composition of melt generated by extension of the lithosphere. *Journal of Petrology*, **29**, 625–679.

MacDONALD, D.I.M. & BUTTERWORTH, P.J. 1990. The stratigraphy, setting and hydrocarbon potential of the Mesozoic sedimentary basins of the Antarctic Peninsula. *In*: ST. JOHN, B. (ed.) *Antarctica as an exploration frontier*. American Association of Petroleum Geology, Studies in Geology, **31**, 101–125.

——, BARKER, P.F., GARRETT, S.W. *et al.* 1988. A preliminary assessment of the hydrocarbon potential of the Larsen Basin, Antarctica. *Marine and Petroleum Geology*, **5**, 34–52.

MILNE, A.J. & MILLAR, I.L. 1989. The significance of mid-Palaeozoic basement in Graham Land, Antarctica. *Journal of the Geological Society, London*, **146**, 207–210.

MYSEN, B.O. & BOETTCHER, A.L. 1975*a*. Melting of hydrous mantle, I. Phase relations of a natural peridotite at high temperatures and high pressures with controlled activity of water, carbon dioxide and hydrogen. *Journal of Petrology*, **16**, 520–548.

—— 1975*b*. Melting of hydrous mantle, II. Geochemistry of crystals and liquids formed by anatexis of mantle peridotite at high pressures and high temperatures as a function of controlled activities of water, carbon dioxide and hydrogen. *Journal of Petrology*, **16**, 549–593.

NATLAND, J.H. 1980. The progression of volcanism in the Samoan linear volcanic chain. *American Journal of Science*, **280A**, 709–735.

PALACZ, Z.A. & SAUNDERS, A.D. 1986. Coupled trace element isotope enrichment in the Cook–Austral–Samoa islands, southwest Pacific. *Earth and Planetary Science Letters*, **79**, 270–280.

PARKER, R.L. & OLDENBERG, D.W. 1973. Thermal model of ocean ridges. *Nature Physical Science*, **242**, 137–139.

PEARCE, J.A. 1982. Trace element characteristics of lavas from destructive plate boundaries. *In*: THORPE, R.S. (ed.) *Andesites*. Wiley, New York, 525–547.

RAMOS, V.A. & KAY, S.M. 1992. Southern Patagonian plateau basalts and deformation: backarc testimony of ridge collisions. *Tectonophysics*, **205**, 261–282.

ROGERS, G. & HAWKESWORTH, C.J. 1989. A geochemical traverse across the Northern Chilean Andes: evidence for crust generation from the mantle wedge. *Earth and Planetary Science Letters*, **91**, 271–285.

—— & SAUNDERS, A.D. 1989. Magnesian andesites from Mexico, Chile and the Aleutian Islands: implications for magmatism associated with ridge-trench collisions. *In*: CRAWFORD, A.J. (ed.) *Boninites and related rocks*. Unwin Hyman, London, 417–445.

——, ——, TERRELL, D.J., VERMA, S. & MARRINER, G.F. 1985. Geochemistry of Holocene volcanic rocks associated with ridge subduction in Baja California, Mexico. *Nature*, **315**, 389–392.

SAUNDERS, A.D., TARNEY, J. & WEAVER, S.D. 1980. Tranverse geochemical variations across the Antarctic Peninsula: implication for the genesis of calc-alkaline magmas. *Earth and Planetary Science Letters*, **46**, 344–360.

——, ROGERS, G., MARRINER, G.F., TERRELL, D.J. & VERMA, S.P. 1987. Geochemistry of Cenozoic volcanic rocks, Baja California, Mexico: implications for the petrogenesis of post-subduction magmas. *Journal of Volcanology and Geothermal Research*, **32**, 223–245.

SMELLIE, J.L. 1987. Geochemistry and tectonic setting of alkaline volcanic rocks in the Antarctic Peninsula: a review. *Journal of Volcanology and Geothermal Research*, **32**, 269–285.

——, PANKHURST, R.J., HOLE, M.J. & THOMSON, J.W. 1988. Age, distribution and eruptive conditions of late Cenozoic alkaline volcanism in the Antarctic Peninsula and eastern Ellsworth Land: review. *British Antarctic Survey Bulletin*, **80**, 21–49.

STERN, C.R., FREY, F.A., FUTA, K., ZARTMAN, R.E., PENG, K. & KYSER, T.K. 1990. Trace-element Sr, Nd, Pb and O isotopic composition of Pliocene and Quaternary alkali basalts of the Patagonian Plateau lavas of southernmost South America. *Contributions to Mineralogy and Petrology*, **104**, 294–308.

STOREY, B.C. & NELL, P.A.R. 1988. Role of strike-slip faulting in the tectonic evolution of the Antarctic Peninsula. *Journal of the Geological Society, London*, **145**, 333–338.

STOREY, M., ROGERS, G., SAUNDERS, A.D. & TERRELL, D. 1989. San Quintin volcanic field, Baja California, Mexico: 'within-plate' magmatism following ridge subduction. *Terra Nova*, **1**, 195–202.

SUN, S-S. & McDONOUGH, W.F. 1989. Chemical and isotopic systematics of oceanic basalts: implications for mantle composition and processes. *In*: SAUNDERS, A.D. & NORRY, M.J. (eds) *Magmatism in the ocean basins*. Geological Society, London, Special Publications, **42**, 313–345.

SYKES, M.A. 1988*a*. New K-Ar age determinations on the James Ross Island Volcanic Group, Northeast Graham Land, Antarctica. *British Antarctic Survey Bulletin*, **80**, 51–56

—— 1988*b*. *The petrology and tectonic significance of the James Ross Island Volcanic Group, Antarctica*. PhD thesis, University of Nottingham, England.

THOMPSON, A.B. 1992. Water in the Earth's mantle. *Nature*, **358**, 295–302,

THOMPSON, R.N. 1982. Magmatism in the British Tertiary Volcanic Province. *Scottish Journal of Geology*, **18**, 49–107.

—— & GIBSON, S.A. 1991. Subcontinental mantle plumes, hotspots and pre-existing thinspots. *Journal of the Geological Society, London*, **148**, 973–977.

——, MORRISON, M.A., HENDRY, G.L. & PARRY, S.J. 1984. An assessment of the relative roles of crust and mantle in magmagenesis: an elemental approach. *Philosophical Transactions of the Royal Society, London*, **A310**, 549–590.

THOMSON, M.R.A. & TRANTER, T.H. 1986. Early Jurassic fossils from Alexander Island and their geological setting. *British Antarctic Survey Bulletin*, **70**, 23–40.

THORKELSON, D.J. & TAYLOR, R.P. 1989. Cordilleran slab windows. *Geology*, **17**, 833–836.

WALSH, J.N., BUCKLEY, F. & BARKER, J. 1981. The simultaneous determination of the rare-earth elements in rocks using inductively coupled plasma spectrometry. *Chemical Geology*, **33**, 141–153.

WEAVER, S.D., SAUNDERS, A.D., PANKHURST, R.J. & TARNEY, J. 1979. A geochemical study of magmatism associated with the initial stages of back-arc spreading: the Quaternary volcanics of Bransfield Strait, from South Shetland Islands. *Contributions to Mineralogy and Petrology*, **68**, 151–169.

WHITE, W.M. & HOFMANN, A.W. 1982. Sr and Nd isotope geochemistry of oceanic basalts and mantle evolution. *Nature*, **296**, 821–825.

WOOD, D.A., VARET, J., BOUGAULT, H. *et al.* 1978. Transition metal and trace element analyses of Leg 49 samples. *Initial Reports of the Deep Sea Drilling Project*, **49**, 897–902.

WYLLIE, P.J. 1982. Subduction products according to experimental predictions. *Geological Society of America Bulletin*, **93**, 468–476.

Index